Rarefied Gas Dynamics: Space-Related Studies

Edited by
E. P. Muntz
University of Southern California
Los Angeles, California

D. P. Weaver
Astronautics Laboratory (AFSC)
Edwards Air Force Base, California

D. H. Campbell
The University of Dayton Research Institute
Astronautics Laboratory (AFSC)
Edwards Air Force Base, California

Volume 116
PROGRESS IN ASTRONAUTICS AND AERONAUTICS

Martin Summerfield, Series Editor-in-Chief
Princeton Combustion Research Laboratories, Inc.
Monmouth Junction, New Jersey

Technical papers selected from the Sixteenth International Symposium on Rarefied Gas Dynamics, Pasadena, California, July 10-16, 1988, subsequently revised for this volume.

Published by the American Institute of Aeronautics and Astronautics, Inc., 370 L'Enfant, Promenade, SW, Washington, DC 20024-2518

American Institute of Aeronautics and Astronautics, Inc.
Washington, D.C.

Library of Congress Cataloging in Publication Data

International Symposium on Rarefied Gas Dynamics
(16th:1988:Pasadena, California)
Rarefied gas dynamics: space-related studies/edited by E.P. Muntz,
D.P. Weaver, D.H. Campbell.

p. cm. – (Progress in astronautics and aeronautics; v. 116)
"Technical papers selected from the Sixteenth International Symposium on Rarefied Gas Dynamics, Pasadena, California, July 10–16, 1988, Subsequently revised for this volume."
Includes index.
1. Rarefied gas dynamics – Congresses. 2. Space sciences – Congresses. I. Muntz, E. Phillip (Eric Phillip), 1934– .
II. Weaver, D.P. III. Campbell, D.H.
(David H.) IV. American Institute of Aeronautics and Astronautics
V. Title. VI. Series.
TL507.P75 vol. 116 89–15068
[QC168.86]
629.1 s – dc20
[629.132'.3]
ISBN 0-930403-53-3

Symposium Advisory Committee

J. J. Beenakker (Netherlands)
G. A. Bird (Australia)
V. Boffi (Italy)
C. L. Burndin (UK)
R. H. Cabannes (France)
R. Campargue (France)
C. Cercignani (Italy)
J. B. Fenn (USA)
S. S. Fisher (USA)
W. Fiszdon (Poland)
O. F. Hagena (FRG)
F. C. Hurlbut (USA)
M. Kogan (USSR)
G. Koppenwallner (FRG)
I. Kuscer (Yugoslavia)
E. P. Muntz (USA)
R. Narasimha (India)
H. Oguchi (Japan)
D. C. Pack (UK)
J. L. Potter (USA)
A. K. Rebrov (USSR)
Yu. A. Rijov (USSR)
B. Shizgal (Canada)
Y. Sone (Japan)
J. P. Toennies (FRG)
Y. Yoshizawa (Japan)

Technical Reviewers

K. Aoki
G. Arnold
V. Boffi
J. Brook
R. Caflisch
D. H. Campbell
C. Cercignani
H. K. Cheng
N. Corngold
J. Cross
R. Edwards
D. Erwin
W. Fiszdon
A. Frohn
O. Hagena
L. J. F. Hermans
W. L. Hermina
E. L. Knuth
G. Koppenwallner
K. Koura
J. Kunc
H. Legge
E. P. Muntz
K. Nanbu
D. Nelson
J. O uchi
J. L. Potter
A. Rebrov
Yu. A. Rijov
J. Scott
A. K. Sreekanth
B. Sturtevant
H. T. Yang

Local Organizing Committee

E. P. Muntz (Chairman)
D. P. Weaver
D. H. Campbell
R. Cattolica
H. K. Cheng
D. Erwin
J. Kunc
M. Orme
B. Sturtevant

Symposium Sponsors

Los Alamos National Laboratory
Strategic Defense Initiative Organization
University of Southern California
U. S. Army Research Office
U. S. Air Force Astronautics Laboratory

Table of Contents

Table of Contents for Companion Volume 117

Table of Contents for Companion Volume 118

Preface

The 16th International Symposium on Rarefied Gas Dynamics (RGD 16) was held July 10–16, 1988 at the Pasadena Convention Center, Pasadena, CA. As anticipated, the resurging interest in hypersonic flight, along with escalating space operations, has resulted in a marked increase in attention being given to phenomena associated with rarefied gas dynamics. One hundred and seventy-three registrants from thirteen countries attended. Spirited technical exchanges were generated in several areas. The Direct Simulation Monte Carlo technique and topics in discrete kinetic theory techniques were popular subjects. The inclusion of inelastic collisions and chemistry in the DSMC technique drew attention. The Boltzmann Monte Carlo technique was extended to more complex flows; an international users group in this area was formed as a result of the meeting.

Space-related research was discussed at RGD 16 to a greater extent than at previous meetings. As a result of Shuttle glow phenomena and oxygen-atom erosion of spacecraft materials, the subject of energetic collisions of gases with surfaces in low Earth orbit made a strong comeback.

There were 11 excellent invited lectures on a variety of subjects that added significantly to the exchange of ideas at the symposium.

The Symposium Proceedings have, for the first time, been divided into three volumes. A very high percentage of the papers presented at the symposium were submitted for publication (practically all in a timely manner). Because of the large number of high-quality papers, it was deemed appropriate to publish three volumes rather than the usual two. An additional attraction is that the size of each volume is a little more convenient. Papers were initially reviewed for technical content by the session chairmen at the meeting and further reviewed by the proceedings' editors and additional experts as necessary.

In the proceedings of RGD 16, there are 107 contributed papers and 11 extended invited papers dealing with the kinetic theory of gas flows and transport phenomena, external and internal rarefied gas flows, chemical and internal degree-of-freedom relaxation in gas flows, partially ionized plasmas, Monte Carlo simulations of rarefied flows, development of high-speed atmospheric simulators, surface-interaction phenomena, aerosols and clusters, condensation and evaporation phenomena, rocket-plume flows, and experimental techniques for rarefied gas dynamics.

The Rarefied Gas Dynamics Symposia have a well-deserved reputation for being hospitable events. RGD 16 continued this admirable tradition. It is with much appreciation that we acknowledge Jan Muntz's contributions to RGD history along with Jetty and Miller Fong and Noel Corngold.

Many people helped make the symposium a success. Gail Dwinell supervised organizational details for the symposium as well as the pre-symposium correspondence. Eric Muntz was responsible for the computerized registration. Kim Palos, whose services were provided by the University of Dayton Research Institute, assisted with registration and retyped several manuscripts. Jerome Maes also retyped a number of manuscripts for these volumes. Nancy Renick and Marilyn Litvak of the Travel Arrangers of Pasadena were tireless in their efforts to assist the delegates.

E. P. Muntz
D. P. Weaver
D. H. Campbell
May 1989

Chapter 1. Rarefied Atmospheres

Nonequilibrium Nature of Ion Distribution Functions in the High Latitude Auroral Ionosphere

B. Shizgal*
University of British Columbia, Vancouver, British Columbia, Canada
and
D. Hubert†
Observatoire de Meudon, Meudon, France

Abstract

The velocity distribution function of ions, under the influence of the geomagnetic field and crossed ionospheric electric field, is calculated from the Boltzmann equation for the r^{-4} ion-neutral (polarization) interaction. The distribution function departs significantly from Maxwellian for large electric fields and small ion-neutral collision rates. An important parameter that determines the extent of the departure from equilibrium is the ratio, ν/Ω, where ν is the ion-neutral collision rate and Ω is the ion gyration frequency. The distribution function is determined by its expansion in the Burnett functions, which are the eigenfunctions of the collision operator for the polarization interaction. A recursion relation valid for all ν/Ω is derived for the velocity moments of the Burnett functions. The ion velocity distribution function is then determined in terms of these velocity moments. For the small ν/Ω limit, the convergence of the distribution function in the Burnett basis set is compared with the convergence with the re-expansion of the distribution with a bi-Maxwellian weight function in the frame of reference moving with the ions. The basis functions in this case are the Hermite polynomials for the velocity parallel to the magnetic field direction and the Laguerre polynomials for the velocities perpendicular to the magnetic field direction. An alternate expansion based on the BGK solution of the Boltzmann equation as weight function is also considered.

Presented as an Invited Paper.

*Professor, Departments of Geophysics and Astronomy, and Chemistry.
†Research Scientist, Departement de Recherche Spatiale.

I. Introduction

At high latitudes, the partially ionized plasma of the terrestrial ionosphere can be subjected to large electric fields perpendicular to the geomagnetic field. These convection electric fields (50-180 mV/m) drive the ions through the neutral background gas at speeds approaching a few kilometers per second. It is now well established theoretically[1-9] and experimentally[10-12] that the velocity distribution functions of the ions depart significantly from Maxwellian, particularly at high altitudes (160-300 km). The extent of the departures from Maxwellian depend on the electromagnetic field strengths, the frequency of ion-neutral collisions, the temperature T of the neturals, and the ion- neutral mass ratio. The experimental evidence for the large departures from equilibrium is derived from incoherent scatter radar[10,11] and satellite measurements.[12] In particular, the temperature of the ions is greater than the temperature of the neutrals, and the temperature for the ion distribution perpendicular to the magnetic field direction is elevated over the temperature for the ion distribution parallel to the magnetic field direction.

For a spatially uniform system, the velocity dependence of the time-independent ion distribution function f(**v**) is given by the Boltzmann equation

$$(\mathbf{\Gamma} + \mathbf{v} \times \mathbf{\Omega}) \bullet \nabla f(\mathbf{v}) = J[f] \tag{1}$$

where $\mathbf{\Gamma} = e\mathbf{E}/m$, $\mathbf{\Omega} = e\mathbf{B}/mc$, e is the ion charge, m is the ion mass, and the distribution function of neutrals is assumed to be Maxwellian. The electric field **E** is perpendicular to the geomagnetic field **B**. The collision operator J describes the effects of ion-neutral collisions on the velocity distribution function. The ion density is assumed to be much less than the neutral density, and ion-ion collisions are neglected. An important parameter that determines the extent of the departure from equilibrium is the ratio ν/Ω, where ν is the collision frequency that is a measure of the rate of ion-neutral collisions. At high altitudes where the densities are small and $\nu/\Omega \to 0$, the ions drift in the $\mathbf{E} \times \mathbf{B}$ direction with drift velocity $D = Ec/B$, and the departures from Maxwellian can be large.

Schunk and Walker[1] considered a solution of the Boltzmann equation based on the expansion of the distribution function about a Maxwellian in the Burnett functions (products of Laguerre polynomials and spherical harmonics) in a frame of reference moving with the ions. Their calculations were for the $1/r^4$ interaction, which is a reasonable first-order approximation for the ion-neutral interaction based on an ion-induced dipole potential, often referred to as the pure polarization interaction. Only a very small number of terms in the expansion were retained, and their solution did not permit a consideration of large electric fields and small collision frequencies. In order to obtain qualitative information regarding the nature of the nonequilibrium ion distribution function, St. Maurice and Schunk[2] considered the BGK model of the collision operator introduced by Bahatnager, Gross and Krook.[13] However, the BGK model does not take into account the dependence on the ion-neutral mass ratio, and it overestimates the departures from Maxwellian. Nevertheless, the form of the solution of the Boltzmann equation for the BGK collision model is routinely employed in the interpretation of radar spectra and satellite data.[9–12]

The theoretical work to date, based on solutions of the Boltzmann equation, has often been restricted to the polarization interaction. For this interaction and the small ν/Ω limit, St. Maurice and Schunk[5] subsequently determined the fourth-order velocity moments of the ion distribution function in the frame of reference moving with the ions. Alternate polynomial methods were introduced by St. Maurice and Schunk[3,4] and by Hubert,[7–9] in which the expansion of the distribution function was based on anisotropic weight functions. These weight functions were chosen to approximate the non-Maxwellian distributions, with the expectation that the number of polynomials required to attain convergence would be small. In all studies to date, only solutions based on the fourth-order moments calculated by St. Maurice and Schunk[5] have been considered, and there is no knowledge of the rate of convergence for different weight functions. Also, the extension of these results to the study of other, more realistic ion-neutral interactions is not rigorous. Formulas for the fourth-order moments valid only for the polarization interaction were employed[5,7–9] for other interactions by inserting different values of the angle-averaged cross sections, as discussed in Sec. 2. Moreover, the re-expansion of the distribution function about different weight functions is possible only for this interaction for which the moments of lower order are not coupled to the higher-order moments. This is not a general property of realistic ion-neutral interactions.

Recently, Barakat et al.[14] have considered Monte-Carlo simulations for the calculation of the ion distribution function for the polarization interaction as well as for ion-neutral charge exchange collisions. The Monte-Carlo simulations were restricted to isotropic scattering (in the ion-neutral center of mass), and these workers did not use the actual differential scattering cross section. Also, the results of the Monte-Carlo simulations are seriously limited by the statistical fluctuations in the results and cannot be used to compare directly with incoherent radar spectra. A direct solution of the Boltzmann equation is preferred.

The purpose of the present paper is to consider a calculation of the ion distribution function for the r^{-4} interaction based on solutions of the Boltzmann equation. The results should provide the basis for the theoretical consideration of more realistic interactions and a comparison with observations. There is also an important overlap of the present study and the calculation of ion mobilities in the absence of a magnetic field. In the study of the mobility of ions in neutral gases, Lin et al.[15] employed a bi-Maxwellian weight function in the frame of reference moving with the ions. For general ion-neutral interactions, their calculations are iterative since the ion drift velocity is unknown. The objective of these calculations is the ionic mobility in contrast to the ionospheric applications for which the distribution function is desired. Skullerud[16] has recently commented on the behavior of polynomial solutions of the Boltzmann equation for ions in an external electric field in relation to the choice of weight function in speed. The present study, although for crossed electromagnetic fields, should prove useful in comparison with his work.

II. Calculations

For the interaction $V(r) = V_0(d/r)^4$, the differential elastic cross section is of the form[17,18]

$$\sigma(g,\theta) = d^2[\frac{2V_0}{\mu g^2}]^{1/2} I_4(\theta) \tag{2}$$

where the parameters d and V_0 measure the strength of the ion-neutral interaction and are related to the neutral polarizability. In Eq. (2),

$$I_4(\theta) = -\frac{1}{\sin(\theta)}\frac{\mathrm{d}\cot(2\phi)}{\mathrm{d}\phi}, \tag{3}$$

where θ is an implicit function of ϕ which is given by

$$\theta = \pi - 2\sqrt{\cos(2\phi)}K(\sin^2\phi) \tag{4}$$

where

$$K(\sin^2\phi) = \int_0^{\pi/2} [1 - \sin^2\phi \sin^2\alpha]^{-1/2} \mathrm{d}\alpha \tag{5}$$

The ion-neutral collision rate is proportional to $\nu = d^2(2V_0/\mu)^{1/2}$.

The coordinate system chosen has the magnetic and electric fields in the z and y directions, respectively as shown in Fig. 1a. With this choice and the use of the reduced velocity $\mathbf{v}' = (m/2kT)^{1/2}\mathbf{v}$, the Boltzmann equation is of the form,

$$D'\frac{\partial f}{\partial v'_y} - v'_x\frac{\partial f}{\partial v'_y} + v'_y\frac{\partial f}{\partial v'_x} = \frac{\nu}{\Omega}\hat{J}[f] \tag{6}$$

where $D' = D(m/2kT)^{1/2}$, and the collision operator $\hat{J} = J/\nu$. For the r^{-4} interaction, the drift velocity of the ions can be calculated exactly by taking the moments of the Boltzmann equation with v'_x and v'_y, and recognizing that these are eigenfunctions of the collision operator, that is,

$$\hat{J}[v'_x] = -\lambda_{0,1}v'_x \tag{7}$$

where $\lambda_{0,1} = \frac{M}{m+M} Q^{(1)}$, M is the neutral mass and

$$Q^{(\ell)} = 2\pi \int_0^\pi I_4(\theta)[1 - \cos^\ell \theta] \sin\theta d\theta \tag{8}$$

A general expression for the eigenvalues is given later [Eq. (17)]. From these two moments of the Boltzmann equation, we find that the drift velocity is given by

$$\langle v'_x \rangle = \frac{D'}{1 + ((\nu/\Omega)\lambda_{0,1})^2} \tag{9}$$

$$\langle v'_y \rangle = \frac{(\nu/\Omega)\lambda_{0,1} D'}{1 + ((\nu/\Omega)\lambda_{0,1})^2} \tag{10}$$

In the limit $\nu/\Omega \to 0$, the drift velocity in the y direction tends to zero whereas the drift velocity in the x direction tends to D', as shown in Fig. 1b. With decreasing ν/Ω, the drift velocity vector rotates in the (x, y) plane about the magnetic field direction.

It is anticipated that the ion distribution function will become anisotropic, with an increase in the electric field strength and a decrease in the collision frequency. There are many physical situations in gasdynamics for which the velocity distribution function is anticipated to become very anisotropic in velocity space.[19,20] If the anisotropy is large, the convergence of the expansion of the distribution function about an isotropic weight function, that is, with the Burnett functions, might be slow. However, since the collision operator is a scalar operator, it is diagonal in the spherical harmonic basis. If an anisotropic weight function is chosen, the representation of the collision operator in the new basis is more complicated, and a complete study of the convergence properties may become difficult.

Schunk and Walker[1] sought solutions to the Boltzmann equation in the coordinate system moving with the drift velocity of the ions. This can only be done rigorously for the r^{-4} interaction for which the

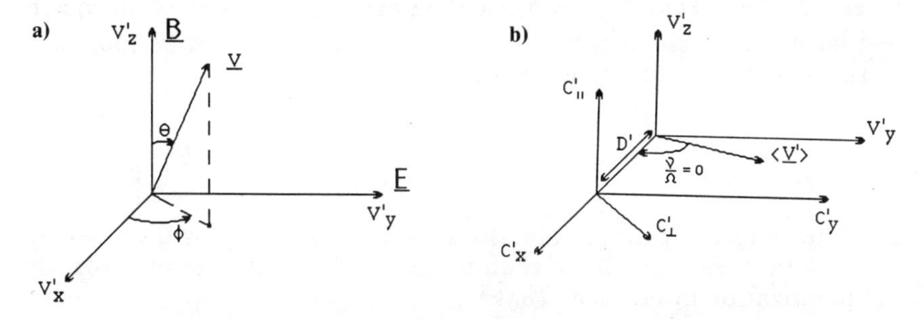

Fig. 1 Drift of ions in a crossed-electromagnetic field: (a) coordinate system in $\mathbf{v}'$ space; (b) displaced coordinate system about $\mathbf{E} \times \mathbf{B}$ drift point.

drift velocity is known exactly. In the study of ion mobilities in an external electric field, Lin et al.[15] chose to expand the ion distribution function about a zero-order anisotropic bi-Maxwellian drifting with the ions. But because realistic ion-neutral interactions are employed in this work, the calculations are iterative since the drift velocity is not known exactly. However, the treatment of the collision operator becomes more involved with this choice of anisotropic weight function. Skullerud[16] has suggested that the spherical harmonics can still be used as basis functions to describe the anisotropy but that one should choose a non-Gaussian weight function in the speed.

The present paper considers a study of the nature of the ion distribution function for the polarization interaction and the convergence of the expansion in Burnett functions, that is,

$$f(\mathbf{v}') = \frac{e^{-v'^2}}{\pi^{3/2}} \sum_{n=0}^{N} \sum_{\ell=0}^{L} \sum_{m=-\ell}^{\ell} \langle \psi_n^{\ell,m} \rangle^* \psi_n^{\ell,m}(\mathbf{v}') \tag{11}$$

where the Burnett functions are given by

$$\psi_n^{\ell,m}(\mathbf{v}') = N_{n,\ell} v'^{\ell} L_n^{\ell+1/2}(v'^2) Y_{\ell,m}(\theta, \phi) \tag{12}$$

where the quantities $N_{n,\ell}$ are given by

$$N_{n,\ell} = [\frac{2\pi^{3/2} n!}{\Gamma(n+\ell+3/2)}]^{1/2} \tag{13}$$

such that the Burnett functions satisfy the orthonormality condition,

$$\pi^{-3/2} \int \exp(-v'^2)(\psi_{n'}^{\ell',m'})^* \psi_n^{\ell,m} d\mathbf{v}' = \delta_{nn'}\delta_{\ell\ell'}\delta_{mm'} \tag{14}$$

where $L_n^{\ell+1/2}(v'^2)$ and $Y_{\ell,m}(\theta, \phi)$ are the Laguerre polynomials and spherical harmonics, resepectively. The quantities $\langle \psi_n^{\ell,m} \rangle$ are the moments of the distribution function, that is

$$\langle \psi_n^{\ell,m} \rangle = \int f(\mathbf{v}') \psi_n^{\ell,m} d\mathbf{v}' \tag{15}$$

The choice of the Burnett functions as basis functions is dictated by the fact that they are the eigenfunctions of the collision operator for the polarization interaction, that is,

$$\hat{J}[\psi_n^{\ell,m}] = -\lambda_{n,\ell} \psi_n^{\ell,m} \tag{16}$$

where the eigenvalues $\lambda_{n,\ell}$ are known[17,18] and given by

$$\lambda_{n,\ell} = 2\pi \int_0^\pi I_4(\theta)[(1 - 4M_iM_n \sin^2(\theta/2))^{n+\ell/2} \times P_\ell(\frac{1 - 2M_n \sin^2(\theta/2)}{\sqrt{1 - 4M_iM_n \sin^2(\theta/2)}}) - 1] \sin(\theta) d\theta. \quad (17)$$

where P_ℓ is the Legendre polynomial of order ℓ, $M_i = m/(m+M)$ and $M_n = M/(m+M)$.

With the substitution of Eq. (11) into the Boltzmann equation, Eq. (6), multiplication by $(\psi_n^{\ell,m})$ and integration over $\mathbf{v}'$, we get

$$D' \sum_{n'\ell'm'} \langle \psi_{n'}^{\ell',m'} \rangle \int \frac{e^{-v'^2}}{\pi^{3/2}} \psi_n^{\ell,m} \frac{\partial (\psi_{n'}^{\ell',m'})^*}{\partial v'_y} \mathrm{d}\mathbf{v}' - im\langle \psi_n^{\ell,m} \rangle = -\frac{\nu}{\Omega} \lambda_{n,\ell} \langle \psi_n^{\ell,m} \rangle \quad (18)$$

With the recurrence relations for the Laguerre polynomials and the addition theorem for the spherical harmonics, the integral term in Eq. (18) can be rewritten in terms of the lower-order moments. This calculation follows the work of Jancel and Kahan[21–23] and is not repeated here. The result is the recurrence relation for the moments given by,

$$\langle \psi_n^{\ell,m} \rangle = -\frac{m - i(\nu/\Omega)\lambda_{n,\ell}}{m^2 + ((\nu/\Omega)\lambda_{n,\ell})^2} D' \{ \sqrt{n} [C_{\ell,m} \langle \psi_{n-1}^{\ell+1,m+1} \rangle + E_{\ell,m} \langle \psi_{n-1}^{\ell+1,m-1} \rangle] + \sqrt{n+\ell+1/2} [D_{\ell,m} \langle \psi_n^{\ell-1,m+1} \rangle + F_{\ell,m} \langle \psi_n^{\ell-1,m-1} \rangle] \} \quad (19)$$

The quantities $C_{\ell,m} to F_{\ell,m}$ result from the coupling of spherical harmonics and are given by

$$C_{\ell,m} = \sqrt{\frac{(\ell+m+1)(\ell+m+2)}{(2\ell+1)(2\ell+3)}} \quad (20)$$

$$D_{\ell,m} = \sqrt{\frac{(\ell-m)(\ell-m-1)}{(2\ell-1)(2\ell+1)}} \quad (21)$$

$$E_{\ell,m} = \sqrt{\frac{(\ell-m+1)(\ell-m+2)}{(2\ell+1)(2\ell+3)}} \quad (22)$$

and

$$F_{\ell,m} = \sqrt{\frac{(\ell+m)(\ell+m-1)}{(2\ell-1)(2\ell+1)}} \tag{23}$$

The dependence on the ion-neutral mass ratio is contained in the eigenvalues as given by Eq. (17). The dependence on D' and the ratio ν/Ω is given explicitly in the recurrence relation, Eq. (19). The present approach differs from that employed by Schunk and Walker,[1] who considered all ν/Ω, and later by St. Maurice and Schunk[5] for the limit $\nu/\Omega \to 0$, in that the distribution function is determined by its expansion in basis functions in $\mathbf{v}'$ space rather than in the velocity space moving with the ions. In $\mathbf{c}'$ space ($c'_x = v'_x - \langle v'_x \rangle$, $c'_y = v'_y - \langle v'_y \rangle$, $c'_z = v'_z$), the collision operator cannot be treated as it was here in terms of the Burnett functions as eigenfunctions. Also, these earlier efforts[1,5] determined the Cartesian moments of the distribution function by taking the corresponding moments of the Boltzmann equation. The resulting equations for the Cartesian moments were coupled and an inversion of the moment equations was required. In the present approach, because the eigenfunctions of the collision operator are used as basis functions, the collision operator is diagonal, and its inversion is accomplished via the recurrence relation, Eq. (19). The previous calculations[1,5,9] were restricted to a very small number of basis functions or moments, and a convergence study of the previous solutions cannot be done. In the present work, the recurrence relation permits a detailed study of the convergence of the solution.

The unique feature of the pure polarization interaction is that the drift velocity is known exactly, and the lower-order moments are not coupled to the higher-order moments. This permits a re-expansion of the solution in an alternate basis set, as specified by some alternate weight function. These generalized polynomial techniques were employed by St. Maurice and Schunk[4,5] and Hubert[9] and rely on this unique property of the polarization interaction. In addition to the study based on the Burnett functions, we include here a reexamination of these alternate generalized polynomial approximations. The basic approach is to make the transformation from the moments $\langle \psi_n^{\ell,m} \rangle$ to the equivalent set of Cartesian moments in $\mathbf{v}'$ space and then to $\mathbf{c}'$ space, which is based on a knowledge of the drift velocity. The previous applications by St. Maurice and Schunk[5] and Hubert[9] were for the $\nu/\Omega \to 0$ limit, for which the distribution function is symmetric about the c'_z axis and depends only on the two velocity variables $c'_{\parallel} (= c'_z)$ and $c'_{\perp}$. St. Maurice and Schunk[5] considered the expansion of the distribution function about a bi-Maxwellian, with the temperatures $T_{\parallel}$ and $T_{\perp}$ as determined by the second-order moments $\langle c'^2_{\parallel} \rangle$ and $\langle c'^2_{\perp} \rangle$. With the bi-Maxwellian as weight function, the basis functions are the Hermite polynomials $H_m(c''_{\parallel})$ and the Laguerre polynomials $L_n^0(c''^2_{\perp})$. As mentioned earlier, the calculations did not go beyond the velocity moments of order four so that, with the ion temperatures determined by the bi-Maxwellian weight function, their solution of the Boltzmann equation for the polarization interaction in the $\nu/\Omega \to 0$ limit is given by

$$f(c''_{\parallel}, c''_{\perp}) = \pi^{-3/2} \exp(-c''^2_{\parallel} - c''^2_{\perp})[a_{00}H_0L^0_0 + a_{02}H_0L^0_2$$
$$+a_{21}H_2L^0_1 + a_{40}H_4L^0_0] \quad (24)$$

where $c''_{\parallel} = c'_{\parallel}\sqrt{T/T_{\parallel}}$, $c''_{\perp} = c'_{\perp}\sqrt{T/T_{\perp}}$, and the terms in a_{01} and a_{20} are zero, owing to the definition of the temperatures $T_{\parallel}$ and $T_{\perp}$ by the weight function. Hubert[9] considered expressing the solution of the Boltzmann equation with the BGK collision model as the weight function w($c^*_{\perp}$), and the corresponding solution to fourth-order in the velocity moments is

$$f(c''_{\parallel}, c^*_{\perp}) = \pi^{-3/2} \exp(-c''^2_{\parallel}) w(c^*_{\perp})[b_{00}H_0M_0 + b_{21}H_2M_1$$
$$+b_{40}H_4M_0] \quad (25)$$

where $c^*_{\perp} = c''_{\perp}\sqrt{T_{\perp}/\hat{T}}$ and $\hat{T}$ is an adjustable temperature parameter. The weight function $w(c^*_{\perp})$ is the solution of the Boltzmann equation with the BGK collision term and is given by,[3,4,9]

$$w(c^*_{\perp}) = \exp(-(\alpha D')^2 - c^{*2}_{\perp}) I_0(2\alpha D' c^*_{\perp}) \quad (26)$$

The polynomials $M_n(c^*_{\perp})$ are orthonormal with this weight function, that is,

$$\int_0^{\infty} w(c^*_{\perp}) M_n(c^*_{\perp}) M_{n'}(c^*_{\perp}) \mathrm{d}c^*_{\perp} = \delta_{nn'} \quad (27)$$

The Laguerre and Hermite polynomials are also normalized to unity. The term in b_{02} is zero, owing to the choice of α, such that the fourth-order moment is determined by the weight function.[9] Also, the temperature parameter $\hat{T}$ in the weight function is given by $\hat{T} = T_{\perp}/[1+(\alpha D')^2]$.

In the present work, we study the convergence of both orthogonal expansions for the $\nu/\Omega \to 0$ limit, that is,

$$f(c''_{\parallel}, c''_{\perp}) = \pi^{-3/2} \exp(-c''^2_{\parallel} - c''^2_{\perp}) \sum_{m=0}^{M} \sum_{n=0}^{N} a_{mn} H_m(c''_{\parallel}) L^0_n(c''^2_{\perp}) \quad (28)$$

and

$$f(c''_{\parallel}, c^*_{\perp}) = \pi^{-3/2} \exp(-c''^2_{\parallel}) w(c^{*2}_{\perp}) \sum_{m=0}^{M} \sum_{n=0}^{N} b_{mn} H_m(c''_{\parallel}) M_n(c^{*2}_{\perp}) \quad (29)$$

Table 1 $Q^{(\ell)}$ for the Polarization Interaction

	$Q^{(\ell)}/Q^{(1)a}$			
ℓ	This work	Model A[(b)]	Model B[(b)]	Model C[(b)]
2	1.0342	0.80	0.95	1.05
3	1.3866	1.2	1.3	1.45
4	1.4151	1.1	1.3	1.45
5	1.6345			
6	1.6593			
7	1.8206			
8	1.8428			

[a] $Q^{(1)} = 3.7443\nu$, [b] St. Maurice and Schunk, (Ref. 5)

The expansion coefficents are determined with the explicit definitions for the polynomials, that is,

$$H_m(c''_{\parallel}) = \sum_{k=0}^{m} A_{mk} c''^{k}_{\parallel} \tag{30}$$

$$L^0_m(c''^2_{\perp}) = \sum_{\ell=0}^{m} B_{m\ell} c''^{2\ell}_{\perp} \tag{31}$$

and

$$M_m(c^{*2}_{\perp}) = \sum_{\ell=0}^{m} C_{m\ell} c^{*2\ell}_{\perp} \tag{32}$$

With the orthonormality of the polynomials, we have, for example,

$$a_{mn} = \frac{\sqrt{\pi}}{2} \sum_{k=0}^{m} \sum_{\ell=0}^{n} A_{mk} B_{n\ell} \langle c'^{k}_{\parallel} c'^{2\ell}_{\perp} \rangle (T^{k+2\ell}/T^{k}_{\parallel} T^{2\ell}_{\perp})^{1/2} \tag{33}$$

where the temperature factor in Eq. (33) converts from the use of $T_{\parallel}$ and $T_{\perp}$ to T in the definition of reduced speed. The moments $\langle c'^{k}_{\parallel} c'^{2\ell}_{\perp} \rangle$ are calculated from the moments in $\mathbf{v}$ space as given by

$$\langle c'^{k}_{\parallel} c'^{2\ell}_{\perp} \rangle = \sum_{i=0}^{\ell} \sum_{j=0}^{2i} C^{\ell}_{i} C^{2i}_{j} (-D')^{2i-j} \langle v'^{j}_{x} v'^{2(\ell-i)}_{y} v'^{k}_{z} \rangle \tag{34}$$

where C^{j}_{i} are the binomial coefficients. The Cartesian moments $\langle v'^{j}_{x} v'^{2(\ell-i)}_{y} v'^{k}_{z} \rangle$, that occur in Eq. (34) were determined from the spher-

Table 2 Convergence of the distribution function Burnett function basis set[a]

N/L	2	4	6	8
		$\hat{c}_\perp = \hat{c}_\parallel$=0.5		
2	0.8431(-1)	0.8543(-1)	0.8541(-1)	
3	0.8537(-1)	0.8634(-1)	0.8632(-1)	
4	0.8495(-1)	0.8595(-1)	0.8593(-1)	
5	0.8505(-1)	0.8605(-1)	0.8602(-1)	
6	0.8504(-1)	0.8603(-1)	0.8601(-1)	
		$\hat{c}_\perp = \hat{c}_\parallel$=1.5		
2	0.1640(-2)	0.1604(-2)	0.1585(-1)	0.1585(-2)
3	0.1612(-2)	0.1576(-2)	0.1556(-2)	0.1556(-2)
4	0.1602(-2)	0.1566(-2)	0.1546(-2)	0.1546(-2)
5	0.1605(-2)	0.1568(-2)	0.1548(-2)	0.1548(-2)
6	0.1605(-2)	0.1568(-2)	0.1548(-2)	0.1548(-2)
		$\hat{c}_\perp = \hat{c}_\parallel$=2.0		
2	0.4308(-4)	0.4015(-4)	0.3874(-4)	0.3873(-4)
3	0.4899(-4)	0.4608(-4)	0.4470(-4)	0.4469(-4)
4	0.4929(-4)	0.4502(-4)	0.4501(-4)	0.4502(-4)
5	0.4906(-4)	0.4617(-4)	0.4479(-4)	0.4478(-4)
6	0.4905(-4)	0.4616(-4)	0.4478(-4)	0.4477(-4)

[a] $m/M = 1, \nu/\Omega = 0, D' = 0.5$

ical moments $\langle\psi_n^{\ell,m}\rangle$ by considering the expansion of these moments in the Burnett functions and using the orthogonality relation, Eq. (14), to determine the coefficients that give a particular Cartesian moment in terms of the spherical moments. The integration over ϕ that occurs can be done analytically, whereas the integrations over θ and v' were performed numerically with Gaussian quadrature points. Although the calculation, is in principle, straightforward, the matrix relating the Cartesian moments to the spherical moments is six-dimensional and, for the higher order-moments, the matrix becomes very large. The calculations of the moments $\langle c_\parallel'^k c_\perp'^{2\ell}\rangle$ were carried out by storing the particular coefficients for a given moment in a file and were read as required. Because in the $\nu/\Omega \to 0$ limit, the distribution function is symmetric in $c_\parallel'$, only even k occur. Following the procedure adopted by St. Maurice and Schunk[5] and Hubert[9], we consider the moments of a particular order p = k+2ℓ, and the convergence of the solution is considered vs the order p rather than in terms the number of basis functions M and N.

III. Results and Discussion

The ion distribution function was calculated vs the system variables, $D', \nu/\Omega$ and the mass ratio m/M with the recurrence relation for

Table 3 Convergence of the Distribution Function
Burnett function basis set[a]

N/L	2	4	6	8	10
			$\hat{c}_\perp = \hat{c}_\parallel$=0.5		
2	0.431(-1)	0.471(-1)	0.470(-1)	0.469(-1)	0.469(-1)
3	0.704(-1)	0.734(-1)	0.733(-1)	0.733(-1)	0.733(-1)
4	0.570(-1)	0.590(-1)	0.590(-1)	0.589(-1)	0.589(-1)
5	0.597(-1)	0.634(-1)	0.633(-1)	0.633(-1)	0.633(-1)
6	0.608(-1)	0.634(-1)	0.633(-1)	0.633(-1)	0.633(-1)
7	0.594(-1)	0.625(-1)	0.624(-1)	0.624(-1)	0.624(-1)
8	0.602(-1)	0.631(-1)	0.630(-1)	0.630(-1)	0.630(-1)
			$\hat{c}_\perp = \hat{c}_\parallel$=1.5		
2	0.110(-2)	0.110(-2)	0.103(-2)	0.103(-2)	0.103(-2)
3	0.134(-2)	0.134(-2)	0.129(-2)	0.128(-2)	0.129(-2)
4	0.120(-2)	0.145(-2)	0.113(-2)	0.112(-2)	0.112(-2)
5	0.114(-2)	0.113(-2)	0.117(-2)	0.106(-2)	0.107(-2)
6	0.118(-2)	0.118(-2)	0.112(-2)	0.112(-2)	0.112(-2)
7	0.118(-2)	0.118(-2)	0.112(-2)	0.111(-2)	0.111(-2)
8	0.118(-2)	0.117(-2)	0.111(-2)	0.110(-2)	0.111(-2)
			$\hat{c}_\perp = \hat{c}_\parallel$=2.0		
2	0.110(-4)	0.933(-5)	0.623(-5)	0.600(-5)	0.606(-5)
3	0.243(-4)	0.222(-4)	0.186(-4)	0.184(-6)	0.185(-4)
4	0.343(-4)	0.325(-4)	0.292(-4)	0.290(-4)	0.290(-4)
5	0.361(-4)	0.343(-4)	0.312(-4)	0.310(-4)	0.311(-4)
6	0.316(-4)	0.324(-4)	0.392(-4)	0.290(-4)	0.294(-2)
7	0.337(-4)	0.318(-4)	0.286(-4)	0.284(-4)	0.284(-4)
8	0.340(-4)	0.322(-4)	0.290(-4)	0.288(-4)	0.289(-4)

[a] $m/M = 1, \nu/\Omega = 0, D' = 0.8$.

the moments, Eq. (19), and the expansion, Eq. (11). The eigenvalues that occur in the recurrence relation were calculated with a Gauss-Legendre quadrature for the θ integral in Eq. (17). The exact differential scattering cross section for the polarization interaction, given by Eqs. (2-5) was retained in these calculations. Instead of showing the eigenvalues that vary with mass ratio, we show in Table 1 the quantities $Q^{(\ell)}$ defined by Eq. (8) in comparison with the values given by St. Maurice and Schunk,[5] chosen to model realistic systems. Table 1 shows the ratios of these angle-averaged cross sections relative to $Q^{(1)}$. It is important to mention that the validity of the use of values of $Q^{(\ell)}$ that depart from the rigorous values for the polarization interaction in the analytic expressions for the fourth-order velocity moments (see the Appendix of Ref. 5) has never been investigated.

The main objective of the present paper is the examination of the convergence of the ion distribution in the different representations as discussed in Sec. 2. None of the other researchers[1,5,9] was able to comment on the validity of the solutions [Eqs. (24) and (25)] owing to the very limited basis sets used. The convergence of the ion distribution

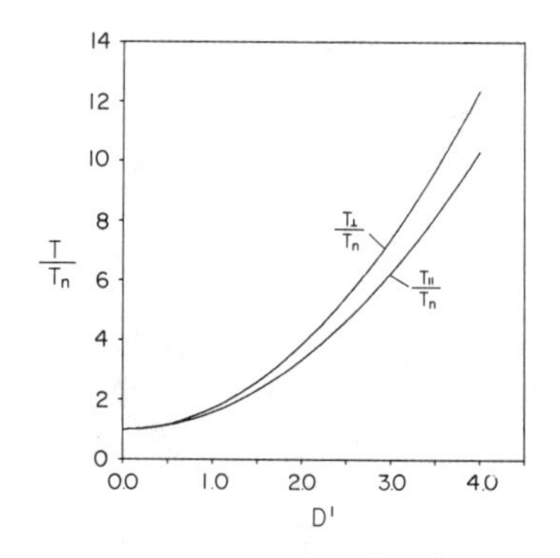

Fig. 2 Ion temperatures perpendicular and parallel to the magnetic field direction vs D': $m/M = 1, \nu/\Omega$=0.

function given by Eq. (11) is shown in Tables 2 and 3 for several velocities $\hat{\mathbf{c}} = \mathbf{c}'\sqrt{T/T_i}$, where $T_i = (2T_{\parallel} + T_{\perp})/3$ is the ion temperature. For the smaller electric field in Table 2, the convergence is rather rapid, even for the "tail" of the distribution at a reduced velocity of 2. As anticipated, the convergence is slower for the larger electric field in Table 3. Even for this case, the ion distribution function is converged to 2-3 figures for the maximum size of the basis functions shown in the table. Although, the convergence appears to slow down quickly with increasing D', the distribution function for this interaction is not strongly anisotropic, at least in terms of the temperatures for the distributions parallel and perpendicular to the magnetic field shown in Fig. 2. These temperatures are determined from the second-order velocity moments.[5]

The ion distribution function is three-dimensional, and it is difficult to show all details vs the three velocity variables. It is useful here to demonstrate the symmetry of the distribution function for $\nu/\Omega \to 0$. Figures 3 and 4 show contours of the distribution function vs ν/Ω. Figure 3 shows the distribution function vs $\hat{c}_x$ and $\hat{c}_y$ with $\hat{c}_z = 0$, and the symmetry about the $\hat{c}_z$-axis in the $\nu/\Omega \to 0$ limit is clear. Figure 4 shows the distribution function vs $\hat{c}_x$ and $\hat{c}_z$ with $\hat{c}_y = 0$ and the symmetry of the distribution in the small collision frequency limit is also shown.

Previous workers[4–6,9] have referred to the expansion of the distribution function in a basis set as Grad's moment method,[25] a terminology that we feel is not appropriate for these calculations. The Boltzmann equation under consideration here applies to a spatially uniform system. Grad's moment method is a technique for finding a closure relation for the system of differential equations in space and time, which is not of concern here. The order of the solutions employed to show the convergence in Tables 2 and 3 is extremely high, and to say that these are solutions of whatever order in the sense of Grad's moment method is misleading.

The convergence becomes rather slow for values of $D' > 1$ as would be expected because the expansion in Burnett functions yields a representation of the distribution function as a power series in D'. This aspect has been known for some time in connection with the calculation

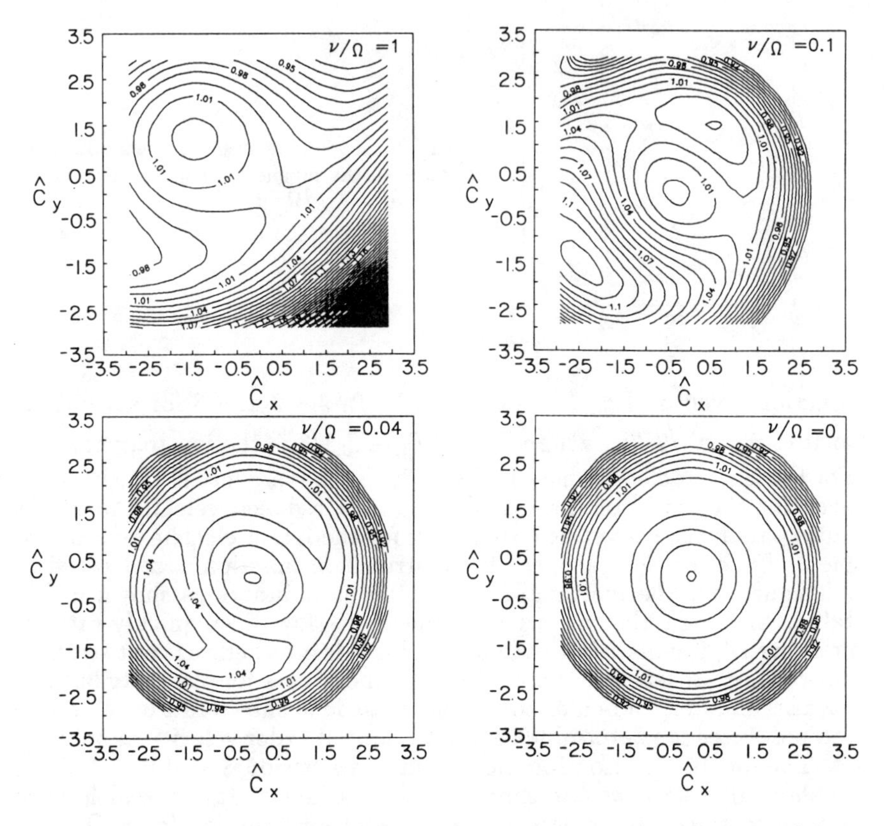

Fig. 3 Contours of the distribution function f/f_0 in the $(\hat{c}_x, \hat{c}_y)$ plane. f_0 is the Maxwellian; m/M=1; $\hat{c}_z$=0; D'=0.6.

of ion mobilities in gases[24]. For this reason, it is preferable to consider the expansion of the distribution in the frame of reference moving with the ions, that is, with Eqs. (28) and (29). It should be emphasized that this procedure is straightforward for the polarization interaction for which the lower-order moments are not coupled to the higher-order moments, and the drift velocity is known exactly. This is not the case for realistic interactions.

Tables 4-7 summarize the convergence of the expansion about a bi-Maxwellian and the BGK solution as weight function, for $D' = 1$ to $D' = 4$, with $\hat{c} = \hat{c}_{||} = \hat{c}_{\perp}$. For the BGK weight function, the value of α is chosen so that the fourth-order moment is given exactly by the weight function and the term in b_{02} is zero. Both weight functions give comparable rates of convergence and the same converged results. The convergence is much faster than that obtained with the Burnett functions. The fourth-order solutions ($p = 4$), Eqs. (24) and (25), can be in error by factors of 0.74 to 2.74, depending on the values of D' and $\hat{c}$. This can be seen by comparing the converged result with the p = 4 result.

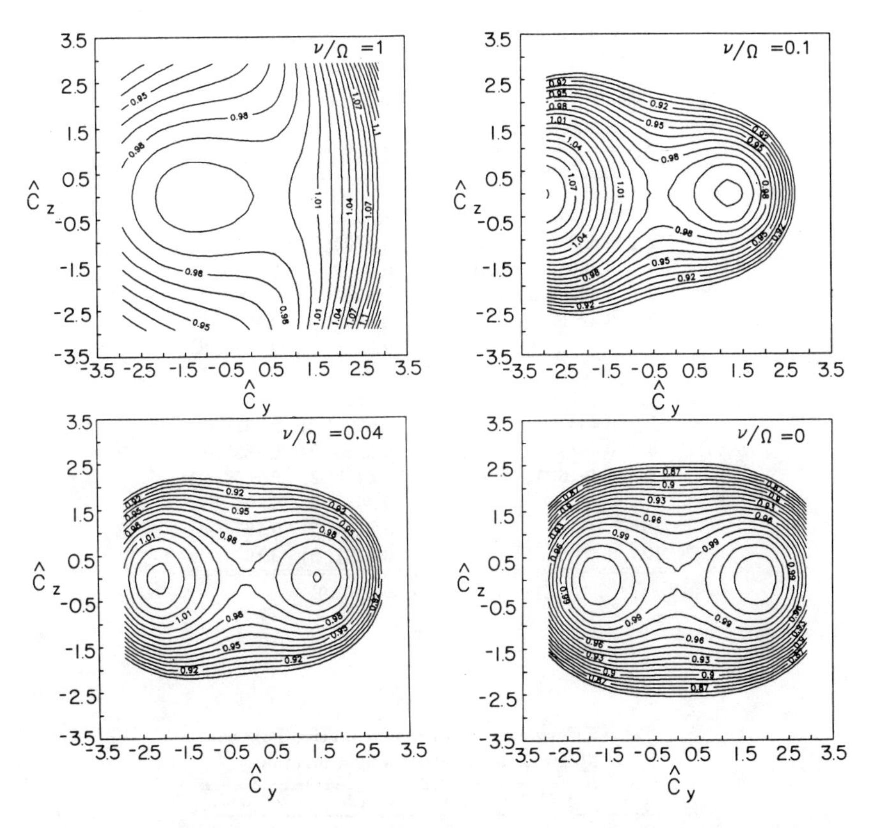

Fig. 4 Contours of the distribution function f/f_0 in the $(\hat{c}_z, \hat{c}_y)$ plane. f_0 is the Maxwellian; m/M=1; $\hat{c}_x$=0; D'=0.6.

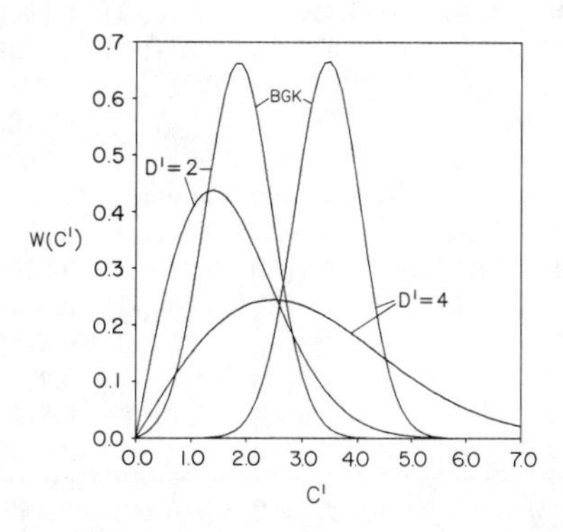

Fig. 5 Comparison of the BGK weight function with the Maxwellian.

Table 4 Convergence of the Distribution Function bi-Maxwellian and BGK weight functions[a]

p/ĉ	0.5	1.0	1.5	2.0
	bi-Maxwellian weight function			
2	0.5031(-1)	0.1099(-1)	0.8703(-3)	0.2500(-4)
4	0.4936(-1)	0.1099(-1)	0.8609(-3)	0.1889(-4)
6	0.4917(-1)	0.1124(-1)	0.8504(-3)	0.1953(-4)
8	0.4916(-1)	0.1124(-1)	0.8495(-3)	0.1976(-4)
10			0.8498(-3)	0.1973(-4)
	BGK weight function			
2	0.4936(-1)	0.1127(-1)	0.8921(-3)	0.2325(-4)
4	0.4918(-1)	0.1129(-1)	0.8548(-3)	0.1830(-4)
6	0.4916(-1)	0.1126(-1)	0.8472(-3)	0.1959(-4)
8			0.8483(-3)	0.1983(-4)
10			0.8494(-3)	0.1978(-4)
12			0.8497(-3)	0.1974(-4)

[a] $m/M = 1, \nu/\Omega = 0, D' = 1, \alpha = 0.5719$.

Table 5 Convergence of the Distribution Function bi-Maxwellian and BGK weight functions[a]

p/ĉ	0.5	1.0	1.5	2.0
	bi-Maxwellian weight function			
2	0.154(-1)	0.328(-2)	0.251(-3)	0.685(-5)
4	0.144(-1)	0.349(-2)	0.237(-3)	0.792(-5)
6	0.141(-1)	0.357(-2)	0.218(-3)	0.222(-5)
8	0.141(-1)	0.357(-2)	0.215(-3)	0.296(-5)
10		0.356(-2)	0.217(-3)	0.278(-5)
12			0.217(-3)	0.272(-5)
14				0.278(-5)
	BGK weight function			
2	0.143(-1)	0.366(-2)	0.261(-3)	0.466(-5)
4	0.141(-1)	0.367(-2)	0.216(-3)	0.636(-5)
6	0.141(-1)	0.361(-2)	0.206(-3)	0.315(-5)
8		0.357(-2)	0.213(-3)	0.315(-5)
10		0.355(-2)	0.217(-3)	0.279(-5)
12		0.356(-2)	0.218(-3)	0.271(-5)
14				0.274(-5)

[a] $m/M = 1, \nu/\Omega = 0, D' = 2, \alpha = 0.4426$.

Table 6 Convergence of the Distribution Function bi-Maxwellian and BGK weight functions[a]

$p/\hat{c}$	0.5	1.0	1.5	1.8
	bi-Maxwellian weight function			
2	0.581(-2)	0.123(-2)	0.924(-4)	0.119(-5)
4	0.534(-2)	0.134(-2)	0.846(-4)	0.450(-5)
6	0.516(-2)	0.139(-2)	0.724(-4)	0.372(-5)
8	0.513(-2)	0.139(-2)	0.710(-4)	0.517(-5)
10	0.512(-2)	0.138(-2)	0.725(-4)	0.534(-5)
12			0.729(-4)	0.507(-5)
14			0.723(-4)	0.504(-5)
16				0.515(-5)
	BGK weight function			
2	0.527(-2)	0.145(-2)	0.944(-4)	0.871(-5)
4	0.516(-2)	0.146(-2)	0.686(-4)	0.190(-5)
6	0.514(-2)	0.141(-2)	0.649(-4)	0.500(-5)
8	0.513(-2)	0.138(-2)	0.708(-4)	0.593(-5)
10		0.137(-2)	0.731(-4)	0.540(-5)
12			0.728(-4)	0.515(-5)
14			0.723(-4)	0.512(-5)
16			0.729(-4)	0.511(-5)

[a] $m/M = 1, \nu/\Omega = 0, D' = 3, \alpha = 0.3426$.

Table 7 Convergence of the Distribution Function bi-Maxwellian and BGK weight functions

$p/\hat{c}$	0.5	1.0	1.5	1.8
	bi-Maxwellian weight function			
2	0.270(-2)	0.568(-3)	0.424(-4)	0.543(-5)
4	0.245(-2)	0.625(-3)	0.381(-4)	0.151(-5)
6	0.235(-2)	0.652(-3)	0.312(-4)	0.111(-5)
8	0.233(-2)	0.652(-3)	0.304(-4)	0.199(-5)
10	0.233(-2)	0.646(-3)	0.314(-4)	0.210(-5)
12		0.645(-3)	0.317(-4)	0.192(-5)
14		0.647(-3)	0.312(-4)	0.189(-5)
16				0.198(-5)
18				0.203(-5)
	BGK weight function			
2	0.241(-2)	0.690(-3)	0.426(-4)	0.264(-5)
4	0.235(-2)	0.693(-3)	0.285(-4)	0.207(-5)
6	0.234(-2)	0.666(-3)	0.271(-4)	0.204(-5)
8	0.233(-2)	0.645(-3)	0.307(-4)	0.243(-5)
10	0.232(-2)	0.641(-3)	0.318(-4)	0.208(-5)
12		0.645(-3)	0.314(-4)	0.199(-5)
14		0.649(-3)	0.311(-4)	0.198(-5)
16		0.652(-3)	0.310(-4)	0.196(-5)
18		0.653(-3)	0.310(-4)	0.196(-5)

[a] $m/M = 1, \nu/\Omega = 0, D' = 4, \alpha = 0.2743$.

It is useful to note that the choice of $\alpha < 1$ makes the BGK weight function approach the Maxwellian. Figure 5 compares the BGK weight function($\alpha = 1$) with a Maxwellian for two values of D'. The reduced speed here is defined in terms of the neutral temperature so that the "heating" of the distribution function with increasing D' is clearly shown. With $\alpha < 1$, the BGK weight function broadens and approaches the Maxwellian. Calculations with larger values of α gave poorer convergence than that shown in Tables 4-7.

The present paper has presented the first detailed study of the convergence of the ion distribution function for the polarization interaction. This is an important endeavor toward a consideration of the use of realistic elastic and charge exchange collision cross sections for collisions between different ion-neutral pairs. We have demonstrated that accurate ion distribution functions can be calculated directly from the Boltzmann equation. Such distribution functions are required in order to calculate incoherent radar spectra and to compare with observations. The results of recent Monte Carlo simulations[14] cannot be easily compared with the observations because of the statistical fluctuations in the output data. Work is in progress to consider the calculation of the incoherent radar spectra[10,11] with the solutions presented here. A methodology based on the present work for the polarization interaction is being developed for a similar study with realistic collision cross sections.

Acknowledgment

This research is supported by a grant from the Natural Sciences and Engineering Research Council of Canada.

References

[1] Schunk, R. W. and Walker, J. C. G.,"Ion Velocity Distributions in the Auroral Ionosphere," Planetary and Space Science, Vol. 20, 1972, pp. 2175-2191.

[2] St. Maurice, J. P. and Schunk, R. W., "Auroral Ion Velocity Distributions Using a Relaxation Model," Planetary and Space Science, Vol. 21, 1973, pp. 1115-1130.

[3] St. Maurice, J. P. and Schunk, R. W., "Behaviour of Ion Velocity Distributions for a Simple Collision Model," Planetary and Space Science, Vol. 22, 1974, pp. 1-18.

[4] St. Maurice, J. P. and Schunk, R. W., "Use of Generalized Orthogonal Polynomial Solutions of Boltzmann's Equation in Certain Aeronomy Problems: Auroral Ion Velocity Distributions," Journal of Geophysical Research, Vol. 81, 1976, pp. 2145-2154.

[5] St. Maurice, J. P. and Schunk, R. W., "Auroral Ion Velocity Distributions for a Polarization Collision Model," Planetary and Space Science, Vol. 25, 1977, pp. 243-260.

[6] St. Maurice J. P. and Schunk, R. W., "Ion Velocity Distributions in the High-Latitude Ionosphere," Reviews of Geophysics and Space Physics, Vol. 17, 1979, pp. 99-134.

[7] Hubert, D., "Convergence and Approximation of Auroral Ion Velocity Distribution Function," Journal of Geophysical Research, Vol. 87, 1982, pp. 8255-8262.

[8] Hubert, D., "Auroral Ion Velocity Distribution Function: The Boltzmann Model Revisited," Planetary and Space Science, Vol. 11, 1982, pp. 1137-1146.

[9] Hubert, D., "Auroral Ion Velocity Distribution Function: Generalized Polynomial Solution of Boltzmann's Equation," Planetary and Space Science, Vol. 31, 1983, pp. 119-127.

[10] Lockwood, M., Bromage, B. J. I., Horne, R. B., St. Maurice, J. P., Willis, D. M. and Cowley, S. W. H., "Non-Maxwellian Ion Velocity Distributions Using Eiscat," Geophysical Research Letters, Vol. 14, 1987, pp. 111-114.

[11] Perraut, S., Brekke, A., Baron, M. and Hubert, D., "EISCAT Measurements of Ion Temperatures which Indicate non-Isotropic ion Velocity Distributions," Journal of atmospheric and terrestrial Physics, Vol. 46, 1984, pp. 531-543.

[12] St. Maurice, J. P., Hanson, W. B. and Walker, J. C. G., "Retarding Potential Analyzer Measurements of the Effect of Ion-Neutral Collisions on the Ion Velocity Distribution in the Auroral Ionosphere," Journal of Geophysical Research Vol. 81, 1976, pp. 5438-5446.

[13] Bahatnager, P. L., Gross, E. P. and Krook, M., "A Model for Collision Processes in Gases. I. Small Amplitude Processes in Charged and Neutral one Component Systems," Physical Review, Vol. 94, 1954, pp. 511-525.

[14] Barakat, A. R., Schunk, R. W. and St. Maurice, J. P., "Monte Carlo Calculations of the O^+ Velocity Distribution in the Auroral Ionosphere," Journal of Geophysical Research, Vol. 88, 1983, pp. 3237-3241.

[15] Lin, S. L., Viehland, L. A. and Mason, E. A., "Three-Temperature Theory of Gaseous Ion Transport," Chemical Physics, Vol. 37, 1979, pp. 411-424.

[16] Skullerud, H. R., "On the Calculation of Ion Swarm Properties by Velocity Moment Methods," Journal of Physics B: Atomic and Molecular Physics, Vol. 17, 1984, pp. 913-929.

[17] Lindenfeld, M. J. and Shizgal, B., "Matrix Elements of the Boltzmann Collision Operator for Gas Mixtures," Chemical Physics, Vol. 41, 1979, pp. 81-95.

[18] Ford, G. W., "Matrix Elements of the Lineraized Collision Operator," Physics of Fluids, Vol. 11, 1968, pp. 515-521.

[19] Weinert, U. and Shizgal, B., "Half-Range Expansions for the Solution of the Boltzmann Equation; Application to the Escape of Atmospheres," Rarefied Gas Dynamics, Vol. 15, 1986, pp. 214- 225.

[20] Shizgal, B., Wienert, U. and Lemaire, J., "Collisional Kinetic Theory of the Escape of Light Ions from the Polar Wind," Rarefied Gas Dynamics, Vol. 15, 1986, pp. 374-383.

[21] Jancel, R. and Kahan, T., "Etude Theorique de la Distribution Electronique dans un Plasma Lorentzien Heterogene et Anisotrope," Journal de Physique et le Radium, Vol. 20, 1959, pp. 35-42.

[22] Jancel, R. and Kahan, T., "Mecanique Statistique d'un Plasma Lorentzien Inhomogene et Anisotrope: Etude de la Distribution Electronique," Journal de Physique et le Radium, Vol. 20, 1959, pp. 804-811.

[23] Jancel, R. and Kahan, T., Electrodynamics of Plasmas, Wiley, New York, 1966.

[24] McDaniel, E. W. and Mason, E. A., The Mobility and Diffusion of Ions in Gases, Wiley, New York, 1973.

[25] Grad, H., "On the Kinetic Theory of Rarefied Gases", Communications in Pure and Applied Mathematics, Vol. 2, 1949, pp. 231-407.

VEGA Spacecraft Aerodynamics in the Gas-Dust Rarefied Atmosphere of Halley's Comet

Yu. A. Rijov,* S. B. Svirschevsky,† and K. N. Kuzovkin‡
Moscow Aviation Institute, Moscow, USSR

Abstract

The aerodynamics of the VEGA spacecraft moving with the speed of about 80 km/s in the rarefied atmosphere of Halley's Comet has been investigated. Under consideration is the interaction of particles in the comet's gas and dust environment and of solar electromagnetic radiation with the vehicle's surfaces.

Introduction

Halley's Comet is a comparatively active periodic comet with the opposite motion around the Sun. Under the action of solar radiation at a distance less than 3-4 a.u. from the sun the comet begins to change its appearance. The comet's small nucleus is surrounded by a visible coma, which is a dust-gas layer appearing from the evaporation of the volatiles in the nucleus, such as ice and stony meteoritic materials, and by other processes. A typical visible coma diameter is 10^5 km. Subjected to solar radiation the dust particles carried by the gas flow are thrown away from their radial trajectories back to the comet's tail, in the direction opposite to the sun. The diameter of the invisible gaseous coma is nearly 10^7 km; its chemical composition is mainly hydrogen atoms.

The relative speed of the VEGA spacecraft in its rendezvous with Halley's Comet was extremely high, ~ 80 km/s. With such speeds, the encounter of gas molecules with the spacecraft surface results in sputtering. The interaction of the comet's dust with the vehicle creates powerful shock waves with pressure amplitudes of tens of millions of atmospheres and temperatures of hundreds of thousands of degrees. For example, a dust particle with the mass $m_* = 10^{-3}$ g is capable of penetrating an aluminium screen of $h \geq 1$ cm thickness. At a hypervelocity impact the spacecraft's response impulse is 5-10 times greater than the quantity of motion of a colliding particle. This may lead to the spatial disorientation of the spacecraft.

A direct experiment within the framework of the VEGA project enables us to establish the degree of particle action on the vehicle and its dynamics while nearing the comet. The FOTON measuring device was installed on board the VEGA spacecraft supplying experimental data about the behavior of the dust shell of the comet and parameters of the high-speed impact.[6,7]

Halley's Comet Gas-dust Atmosphere

On March 6 and 9, 1986 the automatic interplanetary stations VEGA-1 (V-1) and VEGA-2 (V-2) passed through the head of Halley's Comet. The main characteristics of the VEGA spacecraft flight are given in Fig. 1. The V-1 and V-2 entered the comet envelope at sun-nucleus-spacecraft angles of 111.2 and 113.4 deg, respectively.

Figure 1 shows the Halley's comet gas-dust envelope structure. The points of the plot correspond to experimental data received by scientific instruments on VEGA, GIOTTO, SUISEI and SAKIGAKE vehicles on their entering and leaving the gas-dust mantle.[2,3,7]

A cometocentric dependence is constructed in Fig. 2 on concentrations of neutral and ionized gases in the vicinity of Halley's comet, where 1: H_2O (V-1; Ref. 3, pp. 165-174); 1a, b: H_2O (V-2; Ref. 3, pp. 165-174); 2: H_2O after model (Ref. 4); 3: heavy ions (entry Ref. 8, pp. 183-188); c: V-1; d: V-2; 4.1: H_2O (GIOTTO Ref. 2, pp. 326-329); 4.2: CO_2 (GIOTTO; Ref. 2, pp. 326-329); 5: ions (V-1, V-2; Ref. 6, pp. 914-920); 6: ions (GIOTTO; Ref. 8, pp. 199-202); 7: ions (SUISEI; Ref. 8, pp. 71-80); 8: ions (GIOTTO; Ref. 8, pp. 71-80); 9: photodestruction scale length for H_2O (GIOTTO; Ref. 2, pp. 326-329]); 10: photodestruction scale length for H_2O after model (Ref. 4); 11: shock front; 12: cometopause; 13: boundary of free molecular scattering gas; 14: cold cometary plasma boundary; 15: ionopause, Π: dust envelope.

Registration of Dust Particles in the Coma of the Comet by FOTON

Registration of dust particles by FOTON was accomplished by an optical method involving particle penetration of a nickel screen of $h = 0.1$ mm thickness and of area 137 cm^2. Measurement was made of the flux of solar radiation, which fell through the holes onto a photodiode. The flash signal intensity has also been measured (its duration and amplitude), appearing at the moment of screen penetration, as well as the impulse transmitted to the screen. The most dangerous for the apparatus were the larger particles, which are suitable for registration by FOTON, but exceeded the bounds of capacity of the rest of the onboard instrumentation. A detailed description of the device is given in Refs. 1 and 7.

The lowest moment from the dust particle distribution function in radius a for various distances R from the nucleus $f_*(a,R)$ determines the particle concentration of given size (mass):

$$n_*(a_j,a_{j+1}) = \int_{a_j}^{a_{j+1}} f_* \, da = A_*(R)\,(u-1)^{-1}\,(a_j^{\,1-u} - a_{j+1}^{\,1-u}) \quad (1)$$

$$f_*(a,R) = A_*(R)\, a^{-u} \quad (2)$$

where $A_*(R)$ is the definition of ratio, whose numerical value is derived from a condition of function normalization $f_*(a,R)$ for the assumed dust production rate of the comet $\dot{M}_* = 4.153 \times 10^6$ g/s;[4,7] u = 4 is the exponent for Halley's comet. It follows from Eq. (2) that the particle concentration for a given size at a distance R from the nucleus can be written as:

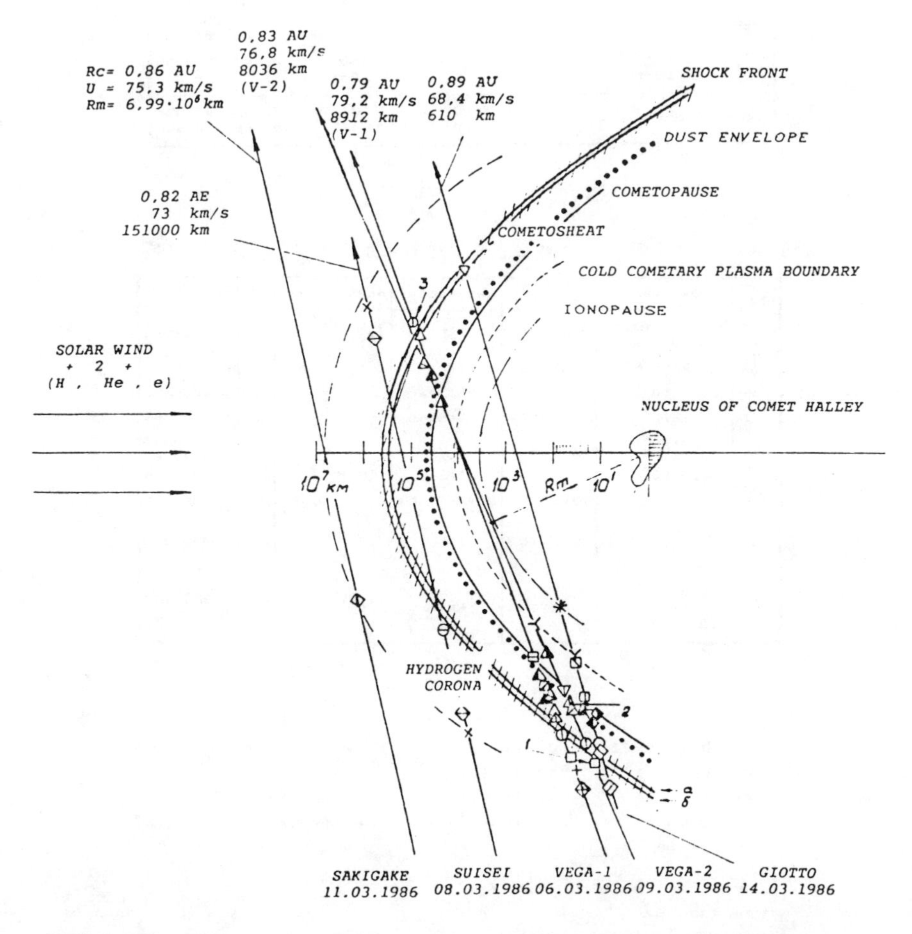

Fig. 1 The gas-dust atmosphere structure of Halley's Comet. The symbols are shown in tables.

Fig. 1 continued.

NEUTRAL GAS

Sym-bol	R, km	Spacecraft	Experiment	References
	$3\cdot10^6$	V-1, V-2	RFC	6, pp. 895-899
	$2,3\cdot10^5$	V-1	PLASMAG	3, pp. 149-154
	$6\cdot10^4$	V-1	ING	2, pp. 273-274
	$1,5\cdot10^5$	GIOTTO	NMS	2, pp. 326-329
	$(3,86\cdot10^4)$*	GIOTTO	NMS	2, pp. 326-329

*the fotodestruction scale lenght for H_2O

COMETARY BOW SHOCK

Sym-bol	R, km	Spacecraft	Experiment	References
	$4,5\cdot10^5$	SUISEI	ESP	2, pp. 299-303
	$1,1\cdot10^6$, entry $5,5\cdot10^5$, exit	V-1, V-2 V-1	PLASMAG, MISCHA	3, pp. 155-174 2, pp. 292-293 2, pp. 285-288
	$1,15\cdot10^6$ $1,13\cdot10^6$	GIOTTO	MAG JPA	2, pp. 352-355 2, pp. 344-347
Shock front apex:	a - $4,5\cdot10^5$ km (GIOTTO) b - $2,7\cdot10^5$ km (VEGA)			2, pp. 349-352 6, pp. 900-906

COMET IONS

Sym-bol	R, km	Space-craft	Experiment	Measurement	References
	$3,6\cdot10^6$, entry $1,2\cdot10^6$, exit	SUISEI	ESP	H_2O^+, OH^+, O^+ fly by	3, pp. 55-64
	$7\cdot10^6$	SAKIGAKE	IMF	H_2O^+, O^+	3, pp. 55-64
	$5\cdot10^6$	V-1, V-2	PLASMAG		2, pp. 282-285
	$1\cdot10^7$	V-1	TÜNDE-M		2, pp. 285-288
	$7,8\cdot10^6$ $7,5\cdot10^6$ $7,5\cdot10^6$	GIOTTO	JPA EPA IMS	H^+ H^+	2, pp. 344-347 2, pp. 347-349 2, pp. 330-334
	$1,05\cdot10^6$	GIOTTO	RPA2-PICCA	H_2O^+, CO^+, CO_2^+	3, pp. 221-225
	$3\cdot10^5$	GIOTTO	IMS	$H_2O^+(OH^+, O^+)$ C^+, CO^+, S^+	2, pp. 330-334
	$5,5\cdot10^5$ $4\cdot10^5$ $3,5\cdot10^5$	GIOTTO	NMS	O^+ $^{12}C^+$ OH^+	2, pp. 326-329

1. SW (H^+); $R=(2\div3)\cdot10^6$ km

V-1 (entry)
V= 510 km/s
n= 12 sm^{-3}
T= $1,2\cdot10^5$ K

V-2 (entry)
V= 620 km/s
n= 11 sm^{-3}
T= $3\cdot10^5$ K

2. SW (H^+); $R=2,8\cdot10^5$ km; V-2

V= 230 km/s
T= $2\cdot10^5$ km/s

3. SW (H^+); $R=4,5\cdot10^5$ km

V-1 (exit)
V= 380 km/s
n= 17 sm^{-3}
T= $4\cdot10^4$ K

Fig. 1 continued.

Symbol	R, km	Spacecraft	Experiment	References
⩡	$1,6\cdot10^5$	V-2	PLASMAG	3, pp. 165-174
▽	$1,35\cdot10^5$, entry $2,63\cdot10^5$ $1,5\cdot10^5$, entry $1,4\cdot10^5$, entry	GIOTTO	MAG MAG IMS RPA	2, pp. 352-355 2, pp. 352-355 2, pp. 330-334 3, pp. 221-225
Cometopause apex: $(4\div5)\cdot10^4$ km (GIOTTO)				2, pp. 349-352

COLD COMETARY PLASMA BOUNDARY

Symbol	R, km	Spacecraft	Experiment	References
⤙	$1,5\cdot10^4$	V-2	PLASMAG	3, pp. 165-174
Y	$3\cdot10^4$	GIOTTO	RPA-PICCA	3, pp. 221-225
⧅	$3,9\cdot10^4$	GIOTTO	NMS	2, pp. 326-329

I O N O P A U S E

Symbol	R, km	Spacecraft	Experiment	References
✳	$4,6\cdot10^3$ $4,7\cdot10^3$ $< 6\cdot10^3$	GIOTTO	IMS NMS RPA	2, pp. 330-334 2, pp. 326-329 3, pp. 221-225

D U S T

Symbol	R, km	Spacecraft	Experiment	References
△	$5\cdot10^5$, entry	V-1	PUMA	6, pp. 840-859
◬	$2,6\cdot10^5$ $3,2\cdot10^5$, entry $1\cdot10^5$, exit	V-1 V-2 V-2	SP-1	6, pp. 867-883
▲	$2,8\cdot10^5$, entry	V-1	SP-2	2, pp. 276-278
◭	$6,4\cdot10^5$, entry $2,6\cdot10^5$, entry	V-1 V-2	DUSMA	2, pp. 278-280
◭	$1,28\cdot10^5$, entry $7\cdot10^4$, exit	V-1	FOTON	6, pp. 860-866
◮	$4,6\cdot10^4$, entry $3\cdot10^4$, exit	V-2	FOTON	6, pp. 860-866
◬	$2,9\cdot10^5$	GIOTTO	DIDSY	2, pp. 338-341

Fig. 1 concluded.

$$n_*(a_j,a_{j+1},R) = n_o(R_m/R)^2 \qquad (3)$$

where n_o is the particle concentration of a given size at the shortest distance from the nucleus R_m when flying via the comet's coma ($t_o=R_m/U$); U is speed of motion of the vehicle [$R^2 = R^2{}_m+L^2 = R^2{}_m+(tU)^2$]; and t and L are the running time of the flyby and the distance, respectively. The number of particle of the given size per second, coming onto the surface F_i (cm^2), the normal to which coincides with the vector of the particle, and is calculated at time t in the following way

$$N_*(t) = n_oF_iU[1+ (t+(t/t_o)^2]^{-1} \qquad (4)$$

The number of particles encountered within the area F_i at time t is equal to

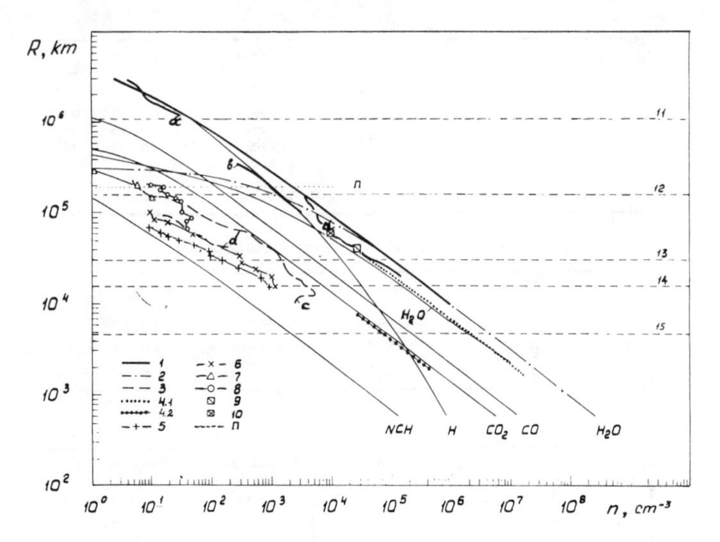

Fig. 2 Profile comparison of concentration of neutral and ionized gases in Halley's Comet.

$$N(t) = \int_{-\infty}^{t} N_*(t)\,dt = (\pi/2)n_o t_o F_i U[1+(2/\pi)\arctan(Ut/R_m)] \tag{5}$$

The total number of collisions of particles of a given size, which fall on the area F_i during the flyby at the distance R_m from the nucleus of the comet, is calculated by

$$N = \int_{-\infty}^{+\infty}\int_{a_j}^{a_{j+1}} F_i U f_* da\, d(L/U) = = 0.448\pi\,(u-1)^{-1}(a_j^{1-u} - a_{j+1}^{1-u})R_m^{-1} F_i \tag{6}$$

The results of calculations of the preceding formulas [Eq. (1-6)] for $R_m = 10^4$ km are given in Ref. 7. Calibration of the device is described also in Ref. 7.

The minimum dust particle mass registered by FOTON made up 6×10^{-10} g. A group of particles is discussed below, which is determined by all three characteristics of their interaction with the screen: energy burst, impulse, and hole size. On V-1 there were registered $N_o = 275$ of such particles. The minimum detectable area of a hole in the screen, equaling 2.2×10^{-4} cm^2, was pierced by particles with $m_* = 5\times10^{-9}$ g. Figure 3 shows the number of dust particles entering the screen of the device as a function of time (the summation was extended over a time interval of 30 s). The flux of particles experiences considerable fluctuations. However, the full number of particles N(t) registered at t (t = 0 corresponds to a point of closest rendezvous of the vehicle with the nucleus) is well approximated by

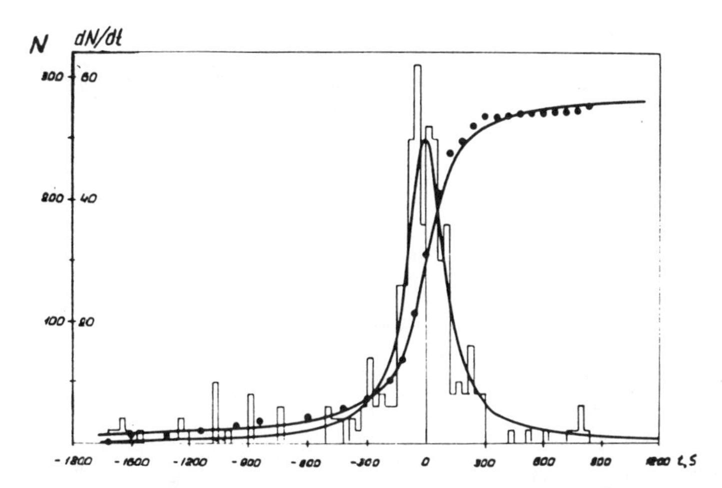

Fig. 3 Measurement of dust particle flow in time registered by FOTON for V-1.

the following formula

$$N(t) = (N_o/2)\,[1+(2/\pi)\arctan(U_\infty t/R_m)] \tag{7}$$

where U_∞ = 79.27 km/s for V-1, 76.8 km/s for V-2; R_m is the minimum distance from the nucleus of the comet (8912 km for V-1, 8036 km for V-2). Comparison of Eq. (7) with the experimental data, depicted by dots, shows that the average concentration of particles lessens with distance from the nucleus of the comet according to the law R^{-2}. Differentiating Eq. (7) in time, it is possible to obtain the mean flow of dust particles onto the screen. Comparing that value with the measured value (see Fig. 3), it is seen that at the trajectory of the VEGA spacecraft in the vicinity of point t = 0 there appeared a region with a density of particles significantly different from the mean. This variation is connected with intensive local ejections of dust off the surface of the nucleus, i.e. jets, which were registered by other instruments as well.[2,3]

During the flyby of V-2 both the general quantity of dust and the fluctuations of the particle flow density were less than those measured during the V-1 flyby.[6,7]

From the measurements conducted it is possible to calculate the mass distribution of dust particles. With regard to Ref. 7 the number of particles in mass interval $m_1 < m_* < m_2$ displayed by the device can be presented as

$$N(m_1,m_2) = C(m_1^{-x} - m_2^{-x}) \tag{8}$$

Direct processing of the experimental data gives a value x = 0.58, which is noticeably different from x = 1 derived from astronomical observations. The reason for this divergence can be understood in view of the fact that

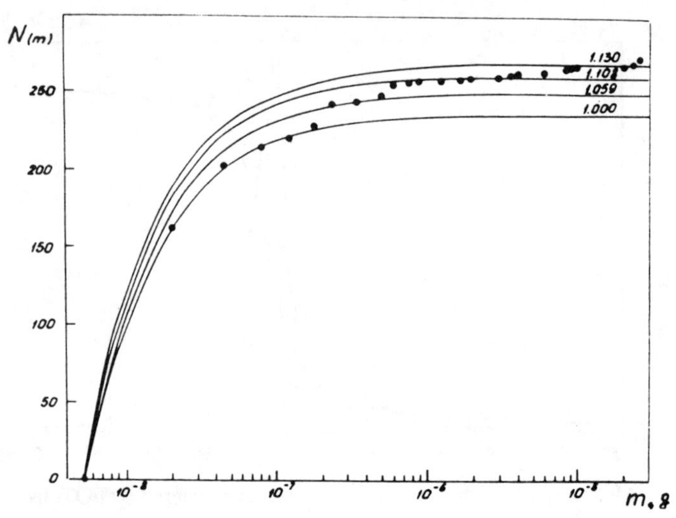

Fig. 4 Distribution of comet dust particles in masses for V-1.

the time span between the dust particle ejection and its registration is comparable in the VEGA experiment with the characteristic period of the comet's activity change. The comet's activity falls with time after passing through perihelion. Since the heavy particles move slower and the detection of them is delayed compared to the lighter particles, the distribution observed in the experiment must show an excess of heavy particles, which causes a lessening of the value of effective exponent x. To find out the particle distribution emitted by the nucleus, it is best to use data related to the group of lightest particles, for which the delay effect is minimum. For mass interval $m_* < 10^{-7}$ g we can obtain values of the exponent x = 0.86 for V-1 and x = 0.8 for V-2. The corresponding distributions are shown in Fig. 4. The depicted curves correspond to different values of the constant C, the data for particles of greater mass being plotted in the curves with greater values of C. Note that, according to GIOTTO data[2] in the vicinity of the comet nucleus, x = 0.83, whereas at a distance of 2200 km x = 0.66.

Different distribution functions of dust particles for masses in the interval 10^{-9} g < m_* < 10^{-5} g according to V-1 data can be represented as

$$dN = A m_*^{-1.86}\, dm_* \tag{9}$$

where $A = 1.5\times10^{-9}$ cm^{-2} g$^{0.86}$. The rate of dust production with regard to Eq. (9) can be calculated as follows:

$$M_* = 4R_m \int_{m_1}^{m_2} m_* U(m_*) dN = 1.5\times10^{7}\,(m_1^{-0.027} - m_2^{-0.027}) \text{ g/s} \tag{10}$$

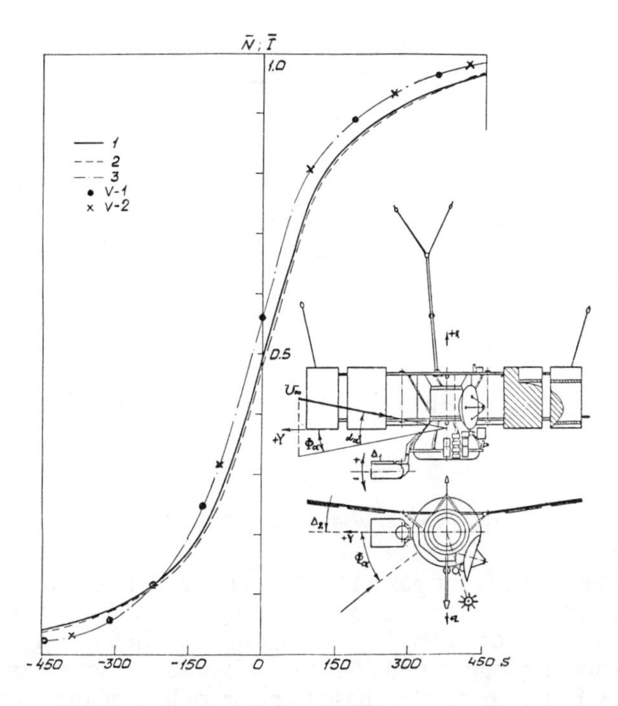

Fig. 5 Influence of dust particle flow on $\bar{I}$ and $\bar{N}$ in time.

where $U(m_*) = 10^3\rho_*^{-1/3}m_*^{-1/6}$ cm/s is the dust speed of motion and ρ_* = 1 g/cm is the dust particle density. For each mass interval, the distribution (9) is valid, $\dot{M}_*$ = 5 t/s (in the comet's model $\dot{M}_*$ was equal to 4.153 t/s). Since $\dot{M}_*$ is weakly dependent of the limits of integration and at $m_* < 10^{-9}$ g the number of particles slowly increases with the decrease of mass[2] (e.g., according to data of SP -2 x ~ 0.5-0.6), the total mass of dust, originating from the nucleus of the comet per second, does not exceed 10 according to V-1 data and 5 according to V-2 data. The value for the production of dust and gas by the nucleus of the comet was different from the model representation[1] and is estimated to be within $\dot{M}_* / \dot{M}_g$ = 0.1 - 0.25.

The large contribution of dust mass from the relatively low number of heavy particles causes a high degree of risk to the spacecraft that can produce damage or change the position of the spacecraft (e.g., GIOTTO[2]). The unprotected panels of solar cells on the VEGA spacecraft lost 60 to 80% of their power as a result of bombardment by dust particles.[6] Figure 5 depicts the drop in the normalized current $\bar{I}$ in time (line 3) due to this bombardment . For comparison, the figure illustrates the nature of change of the normalized particle number $\bar{N}$, which have been registered by FOTON (line 1) and in model calculations within the framework of the fountain model (with u = 4, ρ_* = 0.5 (line 2)).

The Vehicle's Aerodynamic Characteristics in Cometary Atmosphere

The dynamics of gas and dust flows is one of the urgent, quickly developing topics in the mechanics of rarefied media.[10] Strict description of heterogeneous systems is based on a kinetic approach. Considering a rarefied gas-dust medium as a mixture of gas molecules of one kind and monodisperse solid particles, the kinetic equations in dimensionless form can be written as

$$D\bar{f} = (1/Kn_{gg})J_{gg}(\bar{f},\bar{f}) + wJ_{gd}(\bar{f},\bar{f}_*) \quad (11a)$$

$$D\bar{f}_* = (1/Kn_{dd})(1/(\theta b))J_{dd}(\bar{f}_*,\bar{f}_*) + (1/Kn_{gg})(1/(\theta b))J_{dg}(\bar{f}_*,\bar{f}) \quad (11b)$$

$$Kn_{ii} = l_{ii}/L \ (i = g,d);\ b = n_d/n_g \quad (11c)$$

$$\theta = v_d/v_g;\ w = (v_{gd}/v_1)Kn_*^{-1} F \quad (11d)$$

Here, g, d are indices for gas and solid particles, respectively; $J_{gg}(\bar{f},\bar{f})$, $J_{dd}(\bar{f}_*,\bar{f}_*)$ are local Boltzmann collision integrals and $J_{gd}(\bar{f},\bar{f}_*)$, $J_{dg}(\bar{f}_*,\bar{f})$ are the integrals of interphase collisions.[10] In Eq.. (11) $f(v_g,r,t)$, $f_*(v_d,r,t,a)$ are distribution functions for molecules and solid particles; $Kn_* = l_{ii}/a$ is a Knudsen number based on the radius of the particle; $F = (4/3)\pi a^3 n_d$ is the volume fraction of solid particles; n_g, n_d are the concentration of gas and dust particles; and v_{gd} is the velocity of interphase interactions. In the case of strong rarefaction of the medium the distribution f_* changes little due to collisions and, hence, at Kn_{gg}, Kn_{dd}, $Kn_* >> 1$ a free molecular regime of the gas-disperse medium is realized. The solution of equations of the type in Eqs. (11) is similar to the problem of solving a system of kinetic equations for a polyatomic gas at different values of the similitude diagnostic parameters.

The establishment of a type of distribution function is important, in particular, for disperse particles. Because of the complexity of the problems addressed in the analysis given earlier, special importance is given to experiments to obtain distribution functions. For a particular class of problems we can refer to the type of distribution function f_*, found in the FOTON experiment.

In the process of rendezvous the vehicle will move in the rarefied gas-dust atmosphere of the comet. The regime of its free molecular motion is determined from the condition

$$K = Kn_\infty\{[(G-1)/G]^{1/2}S_\infty\}^{-1}T_w/T_\infty)^{1/2} \geq 10 \quad (12)$$

where $Kn_\infty = l_\infty/L_p$ is the Knudsen number, equaling the ratio of mean free path of gas particles $l_\infty = (2^{1/2}\pi d^2 n_\infty)^{-1}$ to a specific aerodynamic dimension of the vehicle $L_p = 10$ m; $d = 4.0 \times 10^{-8}$ cm the effective diameter of H_2O molecules; $n_\infty = 10^6$ cm^{-3} at the distance R = 8036 km (V-2) from the nucleus of the comet; $S_\infty = U_\infty(2kT_\infty/m)^{-1/2} = 131.37$ the relative velocity;

$G = C_p/C_v = 1.33$; $T_w = 363°K$ the temperature of the probe surface on the sunlit side; and $T_\infty = 370°$ K the temperature of the flow. For the considered parameters of motion K = 2130.4.

Condition (12) is satisfied along the whole route of the flyby of the comet's coma for $R \approx 10^4$ km; hence, collisions of particles in the vicinity of the vehicle can be neglected.

Calculations of the aerodynamic characteristics for VEGA in the gaseous dust atmosphere of the comet have been conducted by the method of numerical integration along the vehicle's surface of the corresponding components of forces acting upon the element of the vehicle's surface, for free molecular conditions of flow with the shadow effects and interference included.

On calculation of aerodynamic characteristics of the spacecraft VEGA there have been considered different schemes for interaction of cometary particles with its surfaces. The process of sputtering was included as it was caused by bombardment of the spacecraft surfaces by particles of high energy (the energy of H_2O molecules falling on the surface made up 552.47 eV for V-1 and 519.5 eV for V-2). Data obtained by scientific instruments on the VEGA's proved the existence of sputtering process on their surfaces [Grard et al., Ref. 3, pp. 149-154; Klimov et al. Ref. 2, pp. 292-293).

A boundary distribution function for particles drifting off the surface was determined using Nochilla's model. Reconstruction of the parameters of this function was accomplished by an algorithm[11] based on available experimental indices of sputtering and scattering for various pairs of atom-metal, by means of which were determined $S_{nr}{}^R$, $S_{nr}{}^Y$, the normal components of speed ratio for reflected (index R) and sputtering (index Y) particles and angles of emission $\theta_r{}^R$, $\theta_r{}^Y$ (see Fig. 6).

Figure 7 gives (for angle of incidence of ions $\theta_i = 0$) the generalized dependences for $S_{nr}{}^R$ and $S_{nr}{}^Y$, which depend on the reduced energy of particles e for different combination of ions and atoms in the target.

$$e = E_i a_1 M_2/(e^2 Z_1 Z_2 (M_1+M_2)) \tag{13a}$$

$$a_1 = 0.885 a_o (Z_1{}^{2/3}+Z_2{}^{2/3})^{-1/2} \tag{13b}$$

where the value $Z_1Z_2e^2/a_1$ is a specific energy of interaction of two nuclei with charges Z_1 and Z_2 at a distance of the order of the atomic radius;[12,14] M_1 and M_2 are the corresponding masses of the ion and target, expressed in amu; $a_o = 0.529$ Å. Dependencies cover a wide spectrum of values of the reduced energy e.

Coefficients of forces and moments are expressed in the form of[5]

$$C_F^v = 2F^v(m^v n_\infty^v U_\infty^2 A)^{-1} = (0.5\rho_\infty^v U_\infty^2 A)^{-1} \sum_{\psi=1} \int_{A_{K,\psi}} \zeta(r)\,{}^1P_n^v \bar{n} + {}^1P_t^v \bar{t})\,dA_k \tag{14a}$$

$$m_F{}^v = 2M^v(m^v n_\infty{}^v U_\infty{}^2 AL)^{-1} \tag{14b}$$

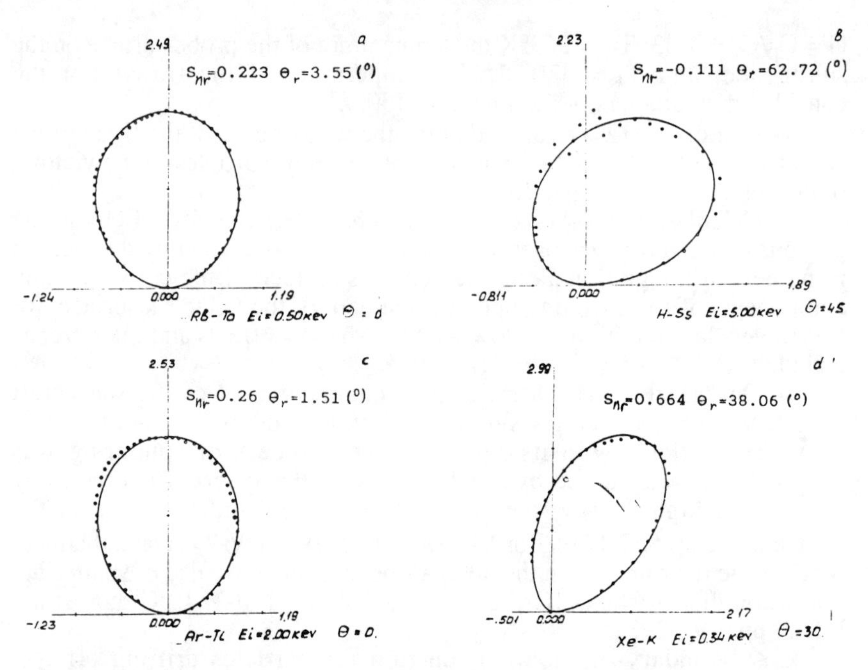

Fig. 6 Angular dependence of scattered (a,b) and sputtered fluxes for ions incident on different surfaces: a) Na-W (Ref. 16): S^R_{nr} = 0.354, θ^R_r = 92.17 deg; b) Rb-Ta (Ref. 13): S^R_{nr} = 0.223, θ^R_r = 93.55 deg; c) Ar-Ti (Ref. 17): S^Y_{nr} = 0.26 θ^Y_r = 91.51 deg; d) Xe-K (Ref. 15): S^Y_{nr} = 0.664, θ^Y_r = 51.94 deg.

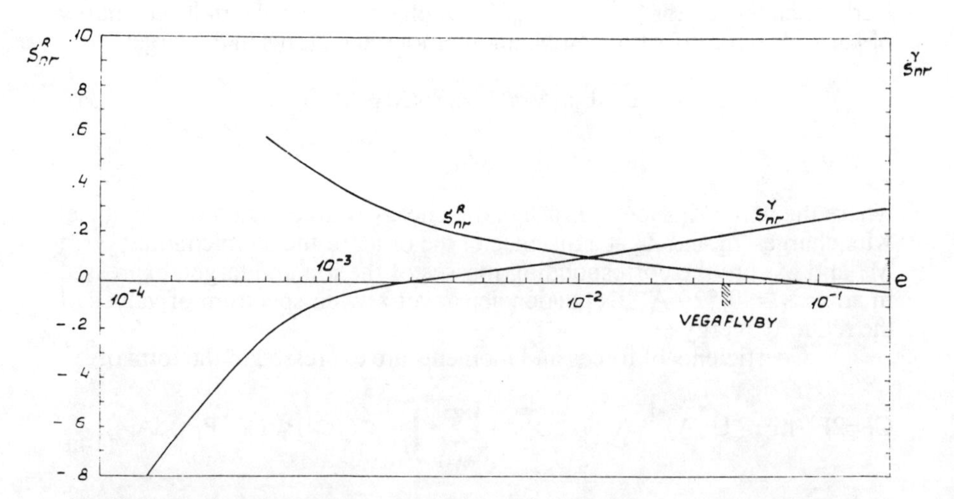

Fig. 7 Generalized dependences of normal speed relations for scattered and sputtered particles on the reduced energy.

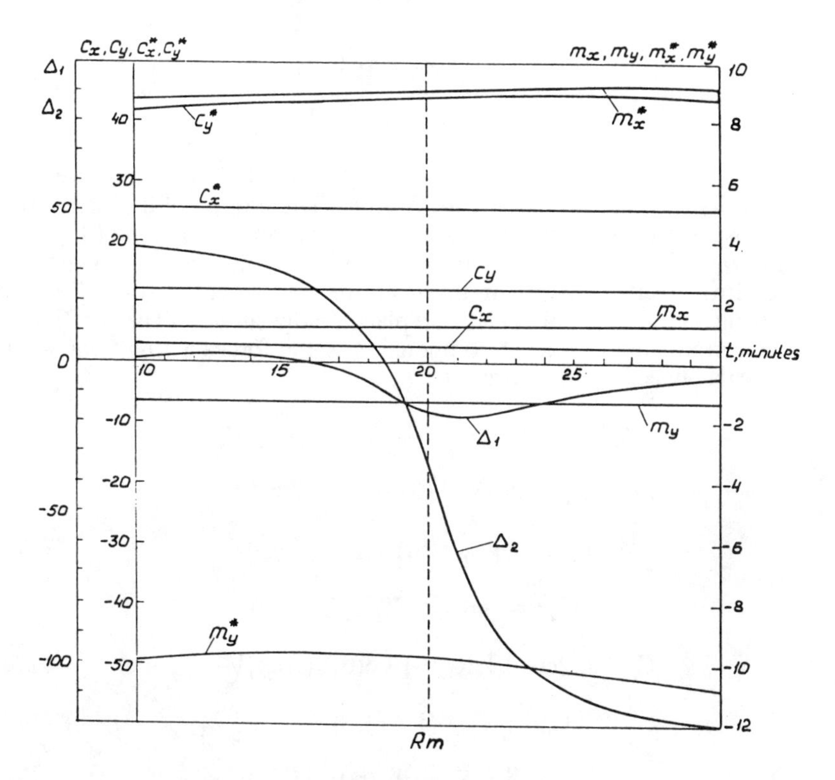

Fig. 8 Aerodynamic coefficients for forces and moments for VEGA at different position of the platform ASP during passage through the comet's coma.

where v = $*$ corresponds to a dust and v = g is for a gas, $\zeta(r)$ is the shielding coefficient, A_k is the element of the surface, ${}^1P_n^v$ and ${}^1P_t^v$ are the normal and tangent components of forces upon a unit area, A and L are the specific area and size, respectively. Moments of forces were determined with respect to the center of mass of the vehicle.

In calculating the interaction of energy of solar radiation with the vehicle's surfaces, the coefficients of normal and tangent components of forces acting upon a unit area were determined in the following way

$$P_n{}^c = 2\cos\theta_c[(\cos\theta_c + 2/3) + C_1(1\text{-}A_c)(\cos\theta_c - 2/3)] \tag{15a}$$

$$P_t{}^c = 2\cos\theta_c \sin\theta_c[1\text{-}C_1(1\text{-}B_1)] \tag{15b}$$

where $C_1 = 1\text{-}a_c$ is the reflection factor; a_c is the absorption factor; A_c and B_c are factors characterizing the fraction of radiation re-emitted diffusely; and θ_c is the angle between the normal to the surface element and direction to the sun. As distinguished from Eq. (14) the coefficients of components of forces and moments in this case were calculated by the formulas

$$C_F^c = 2F^c(R_c/R_o)^2 c(S_o A)^{-1} \tag{16a}$$

$$m_F^c = 2M(R_c/R_o)^2 c(S_o AL)^{-1} \tag{16b}$$

where R_c is the distance between the vehicle and the sun, c is the velocity of light, R_o = 1AU, and S_o = 1400 W/m^2. The motion of a spacecraft in an atmosphere of water molecules results in shock dissociation of H_2O and formation of two atoms of hydrogen H and monatomic oxygen O. Depending on the energy of these atoms the processes of reflection, sputtering, and implantation can take place on the surface. The coefficients of the normal and tangent of components of the forces acting upon a unit area are calculated by the formulas

$$P_n^R = 2\cos^2\theta_i + 1_1 C_{nr}^R(H) + l_2 C_{nr}^R(O) \tag{17a}$$

$$P_t^R = 2\sin\theta_i\cos\theta_i - 1_1 C_{tr}^R(H) - 1_2 Ctr^R(O) \tag{17b}$$

$$P_n^Y = 1_2 C_{nr}^Y(Si) + 1_2 C_{nr}^Y(O) \tag{17c}$$

$$P_t^Y = -1_2 C_{tr}^Y(Si) - 1_2 C_{tr}^Y(O) \tag{17d}$$

$$C_{nr}^R = 2\cos\theta_i J_2 R_E^{1/2}[R_N(\theta_i)/(J_1 J_e)]^{1/2} \tag{17e}$$

$$C_{tr}^R = C_{nr}^R(J_1/J_2)(S_{tr}^R) \tag{17f}$$

$$S_{tr}^R = S_{nr}^R(\theta_i)tg(\theta_r^R)' \tag{17g}$$

$$C_{nr}^Y(Si) = 2\cos\theta_i J_2 Y_E^{1/2}(Si)[Y_N(Si)/(J_1 J_e)]^{1/2} \tag{17h}$$

$$C_{nr}^Y(O) = 2\cos\theta_i J_2 Y_E^{1/2}(O)[Y_N(O)/(J_1 J_e)]^{1/2} \tag{17i}$$

$$C_{tr}^Y = C_{nr}^Y(J_1/J_2)\,(S_{tr}^Y) \tag{17j}$$

$$S_{tr}^Y = S_{nr}^Y(\theta_i)tg(\theta_r^Y)' \tag{17k}$$

Since the interaction of oxygen with the vehicle's surface (mainly SiO_2) results in sputtering of Si and O,[14] this process in Eq. (17) is taken into account by the corresponding components. Here l_1 = 0.111, l_2 = 0.888 are coefficients accounting for the partial composition of hydrogen and oxygen; R_N, R_E are the coefficients of reflection of particles and energy, being calculated by formulas depending on e and θ_i;[12,14] and Y_N, Y_E are the coefficients of particle sputtering and energy reflection in sputtering.[14] The action of dust particles onto the vehicle result in complex physical and chemical processes taking place at the surface.[4] The kinetic energy of the dust particles is many times greater than the energy of the bond of atoms in the crystal lattice. At impact of the particle upon the surface there appears a powerful shock wave, causing its local heating, and sublimation of the surface is taking place in the wave of relaxation. This

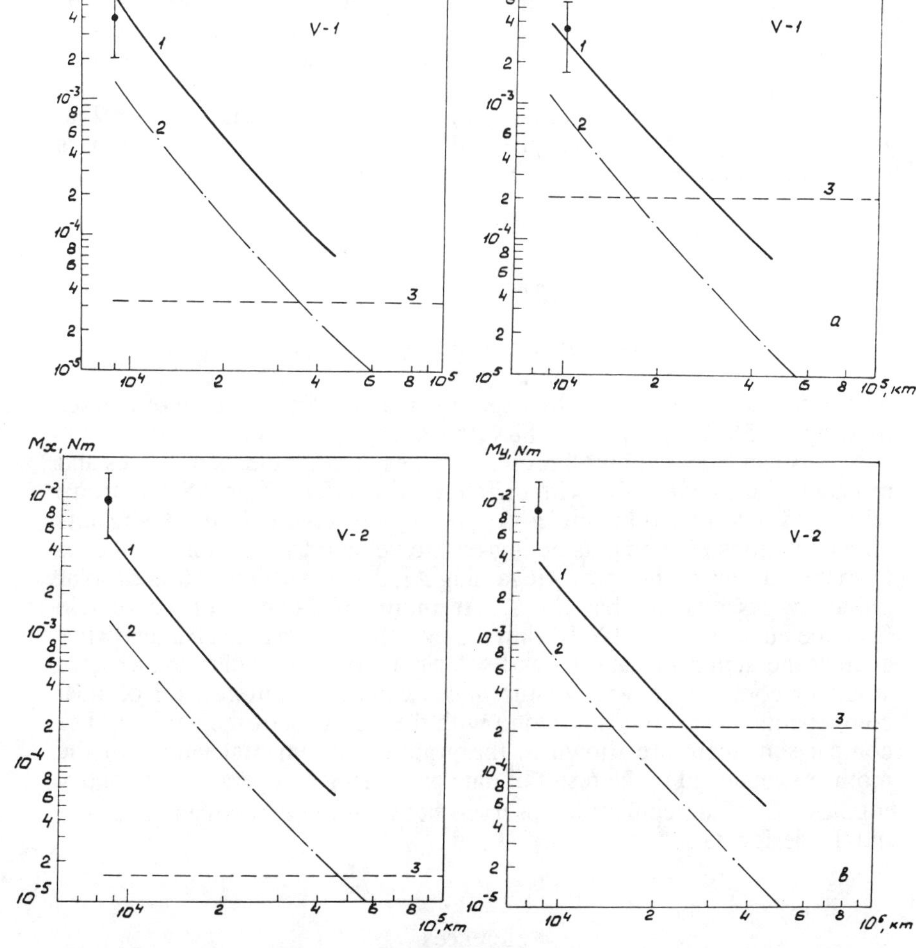

Fig. 9 Moments of forces acting on VEGA. Dependance on distance to the nucleus of the comet: 1,dust; 2,gas; 3,solar radiation.

process is similar to a burst of evaporation from a crater and produces further scattering of ionized products.

The components of coefficients of forces acting on an element of the surface, oriented with respect to the incoming dust flow and the normal at an angle θ_i, are (assuming the substance evaporated to scatter in the mean along the normal towards the surface at the point of impact):

$$P_{N*} = 2\cos^2\theta_i + 2\cos\theta_i r \tag{18a}$$

$$P_{t*} = 2\sin\theta_i \cos\theta_i \tag{18b}$$

where $r = m_k V_k/(m_* U)$ is the impulse factor, the value of which was specified based on the results of the FOTON experiment (Anisimov's estimation); and m_* and U are the mass and mean velocity of the struck particle, respectively. If the impact occurs without piercing with crater formation, then r = 9.6 at h > 24a. If a complete piercing is realized with ejection of a struck particle onto the opposite side of the construction (i.e., a complete piercing of the solar battery), then $r = 0.273(h/D)^{3/2}$ at h< 24a. Assuming the particles to be spherical with $\rho_* = 1$ g/cm, the mean mass flow density was determined by the formula[4] (u = 4):

$$\rho^*_\infty = m_* n_* = \int_{a_{min}}^{a_{max}} (4/3)\pi a^3 \rho_* A_*(R) a^{-u}\, da$$

$$= (4/3)\,\pi\rho_*(b/R^2)[1na_{max} - lna_{min}] \qquad (19)$$

where, for example, for V-1 by experimental data (in the range of masses $6X10^{-17}$ - $2.5X10^{-5}$ g) $a_{max} = 2.88X10^{-2}$ cm, $a_{min} = 3.85X10^{-6}$ cm.

As an example Fig. 8 presents values of coefficients of forces and moments during the flyby with different orientation of the ASP platform with TVS determined by angles Δ_1 and Δ_2.[1,2] Figure 9 presents results on calculations of moments acting on the vehicle for different distances from the nucleus of the comet (regarding Δ_1, Δ_2) against the action of solar radiation, gaseous and dust flows. At minimum distances of the vehicle from the nucleus R = 8-9 X 10^3 km the overall aerodynamic characteristics against the action of gaseous and solar components of disturbances are small in comparison with disturbances caused by interaction of dust components of the cometary cloud with the elements of the vehicle. For comparison, there are shown in the graphs experimental values of the moments, obtained in the result of data processing on operation of control engines. Good agreement is observed between the experimental results and the design data.

References

[1]Sagdeev, R. Z., (editor), Venus-Halley mission, "Experiment Description and Scientific Objectives of the International Project VEGA (1984-1986)," Louis-Jean, GAP, France ,1984.

[2]"Encounters with Comet Halley. The First Results," Nature, London, Vol. 321, No. 6067, 1986, pp. 259-366.

[3]"Comets Halley and Giacobini-Zinner," Advances in Space Research, Vol. 5, No. 12, 1985.

[4]Rijov, Yu. A., Bass, V. P., Karyagin, V. P., Kovtunenko, V. M., Kuzovkin, K. N., and Svirschevsky, S. B., "Aerodynamic Problems of Space Probes in Comet Atmosphere," Rarefied Gas Dynamics, Plenum, New York, 1985, Vol. 1, p. 503-515.

[5]Bass, V. P. et al., "Aerodynamic Characteristics of a Vehicle Designed for Studying Halley's Comet," Applied Problems of Flying Vehicles Aerodynamics. Naukova Dumka, Kiev, 1984, pp. 11-15 (in Russian).

[6]Space Investigations, Vol. 25, No. 6, pp. 820-957 (n Russian).

[7]Anisimov, S. I., et al., Investigation of the Dust Atmosphere of Comet Halley FOTON Experiments on the Interplanetary stations VEGA," The Moscow Aviation Institute, Moscow, 1987 (preprint, in Russian).

[8]"Exploration of Halley's Comet," ESA SP-250, Vol. 1, 1986.

[9]Divine, N., et al., "The Comet Halley Dust and Gas Environment," Space Science Reviews, Vol. 43, No. 1/2, 1986, pp. 1-104.

[10]Bogdanov, A. V., et al., "Kinetic Theory of Gas Mixture with Solid Particles, II," Preprint 989, FTI USSR Academy of Science, Leningrad, 1985, (in Russian).

[11]Atamanenko, A. V., "Reconstruction of Parameters Distribution Function of Reflected Molecules by Measurements of Forces and Indices of Scattering in Molecular Flow," Scientific Notes TsAGI, Vol. 28, No. 3, 1987, pp. 69-78 (in Russian).

[12]Martinenko, Yu. V., "Interaction of Plasma with Surfaces. Summary of Science and Technology," Vol. 3. VINITI. Moscow, 1982, pp. 119-175 (in Russian).

[13]Arifov, U. A., Interaction of Atomic Particles with Surfaces of Solids, Nauka, Moscow, 1968 (in Russian).

[14]Sputtering by Particle Bombardment, Vol. I, Vol II, edited by R. Rehrish, Springer-Verlag, Berlin, 1981, 1983.

[15]Stein, R. P. and Hurlbut, F. C., "Angular Distribution of Sputtered Potassium Atoms," Physical Review, Vol. 123, No. 3, 1961, pp. 790-796.

[16]Gruitch, D. D., et al., "The Study of Angular Distribution of Secondary Ions with W Depending on Angular Incidence and Energy Falling Ions," Interaction of Gas with Surfaces of Solids, ITPM USSR SO Academy of Sciience, Novosibirsk, USSR, 1971, pp. 63-68 (in Russian).

[17]Matsuda, Y., et al., "Detailed Measurements of Differential Sputtering Yields by Ion-Beam Bombardment Using Laser Fluorescence Spectroscopy," Journal of Nuclear Materials, Vol. 145/147, No. 2, 1987, pp. 421-424.

Oscillations of a Tethered Satellite of Small Mass Due to Aerodynamic Drag

E. M. Shakhov*
USSR Academy of Sciences, Moscow, USSR

Abstract

This paper deals with the plane transverse oscillations due to aerodynamic drag of a tethered satellite of small mass ($\sim$ 1 kg) at low orbit. Observed from the orbiter, the oscillations may give useful information. The period of the oscillations is a single-valued function of the aerodynamic drag at the altitude of flight. In an atmosphere of nonuniform density the oscillations are asymmetric. Both the difference of half-periods and the amplitude difference in deviations up and down from a position of relative equilibrium are defined by the density gradient. The orbiter is assumed to move in a circular orbit. The tether is treated as a long flexible nonelastic and noninertial thread. Some possible ways of using the satellite-pendulum as a facility for experimental rarefied gasdynamics and aeronomy are discussed. Necessary modifications of the mathematical model are proposed.

Introduction

One of the promising new projects for detailed studies related to flight in low-density atmosphere is the program called the Shuttle Continuous Open Wind Tunnel.[1] The project makes use of a satellite attached to the orbiter by a long tether with the satellite being more or less massive ($\sim$ 500 kg) and having necessary instrumentation for scientific measurements. The tethered system is treated as a solid body in stable motion. Oscillations that may occur in the

*Computing Center.

system are considered as undesirable phenomena because they may result in destruction of the system.[2] In this paper the tethered satellite is treated as a passive object that may give useful information when observed from the orbiter.

Imagine first that the satellite has become free at an initial time. Being free, the satellite looses speed (and hence altitude) due to aerodynamic drag with the acceleration being inversely proportional to its mass. That means that the light satellite may have visible speed moving away from the orbiter. The characteristic mass of the satellite considered later will be $m \approx 1$ kg. The aerodynamic drag $D = (1/2)c_D \rho V^2 S$. The parameters corresponding to the orbit will be denoted by the subscript 0 so that the value of the drag at the orbit is D_0, which is equal to $(1/2)c_D \rho_0 V_0^2 S$. The velocity of the orbiter is $V_0 \approx 8000$ m/s. At the altitude of flight $H = 230$ km the density $\rho_0 \approx 10^{-10}$ kg/m^3 and the flow regime is free molecular. For a satellite of spherical form, for example, $c_D \approx 2.6$. Assuming for the characteristic area $S = 1$ m^2 we obtain the following value for the drag: $D_0 \approx 10$ gm/s^2. Thus, due to aerodynamic drag the satellite with mass $m = 1$ kg has an acceleration $a = 0.01$ m/s^2. Hence, after 100 s it goes away from the orbiter with a velocity of 1 m/s and after traveling 5 km away the satellite has a relative velocity of 10 m/s. It is evident that the effect of aerodynamic drag is stronger for altitudes of flight less than 230 km. The observation of the free motion of the released light satellite may be useful. The measurements of the trajectory and velocity may give the product $c_D \rho_0$. For the drag coefficient c_D, the density of the atmosphere ρ_0 can be determined, and for the given density ρ_0 the drag coefficient c_D for various surface materials can be defined. As is well known, free spherical balloons are fruitfully used in meteorology.

The first stage of motion of the tethered satellite depends on the regime of deployment of the system. The tethered satellite released at initial time together with noninertial flexible tether of length L moves as if it were free until the distance from the orbiter is equal to the length of the tether. After the deployment, oscillations may occur in the system along the tether. We assume that such oscillations are effectively damped. Then, due to aerodynamic forces the tethered satellite is set in a regime of steady transverse oscillations. The study of such oscillations is the object of the paper.

Preliminary studies have indicated that the satellite during its free motion may lose altitude rapidly; hence, it may have a significant angular deviation. This means that the oscillations may be nonlinear. When the tether is long enough, the nonuniformity of the atmosphere should be taken into account. In the paper, the regime of steady transverse oscillations (up and down in the orbital plane) of the satellite-pendulum in an atmosphere of exponential density is investigated. The tether is treated as a flexible, nonelastic, and noninertial thread. The orbiter is assumed to move in a circular orbit. Both the period and half-periods in displacement up and down from the position of the relative equilibrium are determined. The corresponding amplitudes are defined in terms of the density and density gradient with the velocity of the orbiter being taken into account. If these values are measured from the orbiter, they may give useful information on stream-surface interaction, density, and density gradient of the atmosphere.

Governing Equations

Consider the relative motion of the satellite-pendulum with mass m tethered by a noninertial, nonelastic flexible thread of length L to an orbiter with mass being much larger than m. Both the orbiter and the tethered satellite are treated as mass points. Assume that the orbiter moves in a circular orbit with radius R_0 and center at point O (which coincides with the center of gravity) with constant peripheral velocity V_0. Introduce a polar system of coordinates with OP as polar axis. Let the position of the orbiter and the satellite be defined by the points K and M, respectively. In the coordinate system adopted, the coordinates of the orbiter are (R_0, Θ_k) and coordinates of the satellite (R, Θ) (Fig. 1). In addition, we introduce a polar system of coordinates relative to the orbiter with the polar axis KO being rotated with the angular rate $\omega_0 = V_0/R_0$. Location of the point M in the rotating coordinate system is defined by the distance KM and the angle α. The tether is assumed to be stretched so that KM is a segment of straight line and $KM = L$.

For sake of simplicity we assume that wind is absent so that the oscillations may take place in the plane of the orbit. Moreover, the aerodynamic drag is assumed to be directed along the tangent to the circle with the radius R and the center at point O. The gravity is directed to the center and has the modulus $G = mg_0(R_0/R)^2$; here g_0 is the acceleration of gravity at the orbit R_0.

The system of governing equations consists of the equation of the oscillations in the rotating frame of reference and the expression for the tether tension. They can be written as follows (Fig. 1):

$$\ddot{\alpha} - \frac{D}{mL}\sin\beta + \frac{g_0}{L}\left(\frac{R_0^2}{R^2}\cos\beta - \sin\alpha\right) = 0 \tag{1}$$

$$N = mL(\omega_0 + \dot{\alpha})^2 + D\cos\beta + mg_0\left(\frac{R_0^2}{R^2}\sin\beta - \cos\alpha\right) \tag{2}$$

where D is the aerodynamic drag, N is the tether tension, and β is the angle between the gravity vector and the tangent to the circle of the satellite relative to the orbiter.

For the tethered satellite-pendulum (SP) there exists a position of equilibrium in the vicinity of the value $\alpha = 1/2\pi$ so that it is located at the orbit R_0 behind the orbiter. In this case, the drag generates the angular momentum returning SP to equilibrium. The last term in the equation of motion [Eq. (2)] characterizes the resulting effect of the competition between centrifugal force and gravity. Note that this term is not equal to zero even for the case of homogeneous gravity.

Let us use the following geometric relations:

$$\varphi_K = \Theta_K - \Theta, \qquad \beta = \gamma - \varphi, \qquad \gamma = 1/2\pi - \alpha$$

$$\sin\varphi = \frac{L}{R}\sin\alpha \qquad \Delta R = R - R_0 = -L\sin\gamma$$

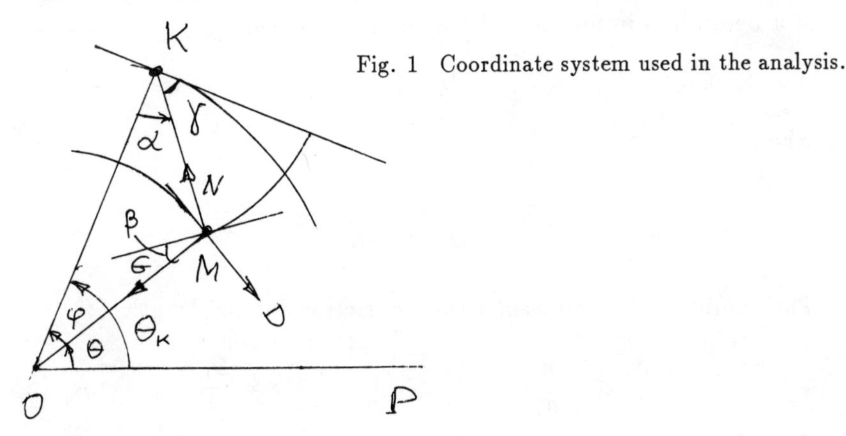

Fig. 1 Coordinate system used in the analysis.

Transform the system (1) into another form by using γ instead of α and using the condition that the tether length L is much smaller than the radius of the orbit $R_0 : L << R_0$. Keeping the main terms of expansion in the L/R_0 series, we obtain

$$\ddot{\gamma} + \left(\frac{D}{mL} - 3\omega_0^2 \cos\gamma\right) = 0 \tag{3}$$

$$\frac{N}{mL} = (\omega_0 - \dot{\gamma})^2 + 2\omega_0^2 \sin^2\gamma - \omega_0^2 \cos^2\gamma + \frac{D}{mL} \cos\gamma \tag{4}$$

The set of Eqs. (3) and (4) will be closed when an expression for aerodynamic drag D is given. Here, we assume an empirical atmospheric density exponential law. Since the drag is proportional to density, the following expression can be written:

$$D = D_0 e^{\delta \sin\gamma} \tag{5}$$

Here D_0 is the aerodynamic drag at the orbit R_0. The numerical value of the parameter δ depends on the length of the tether L and weakly depends on the altitude of flight H, with $\delta = 1$ for $L = 10$ km.

The equations of oscillation [Eqs. (1), (2), (3), and (4)] include the effect of circulation around the center. When the frequency ω of oscillations forced by aerodynamic drag is much larger than the circulation frequency ω_0, then the second term in brackets is small and can be omitted. Under this condition, Eqs. (3) and (4) can be simplified and transformed into the form of the equation of motion of a pendulum in an inertial frame of reference:

$$\ddot{\gamma} + \frac{D}{mL} \sin\gamma = 0 \tag{6}$$

when

$$\omega^2 = \frac{D_0}{mL} >> \frac{g_0}{R_0} = \omega_0^2 \tag{7}$$

The condition (7) represents the restriction for the length L:

$$L << \frac{a}{g_0} R_0, \qquad a = \frac{D_0}{m} \tag{8}$$

At the altitude of flight $H = 230$ km, the ratio $a/g_0 = 10^{-3}$. The value of a/g_0 increases up to 10^{-2} when the altitude H decreases to 180 km.

Linear and Weakly Nonlinear Oscillations

Let us consider first the oscillations of small amplitude. The principal distinction of the Eqs. (3) and (4) from the classical equation of a mathematical pendulum is that the factor D is an asymmetric function of γ. But this property of asymmetry will be retained even for the case of weakly nonlinear oscillations of small amplitude. Using the expansion of the functions in Eqs. (3) and (4) in the power series of γ and keeping powers up to γ^2, we obtain from Eqs. (3) and (4) the simplified equation of the oscillations:

$$\ddot{\gamma} + (\omega^2 - 3\omega_0^2)\gamma + \delta\omega^2\gamma^2 = 0 \tag{9}$$

In pointing out the case of linear oscillations, assume that $\delta = 0$ and we obtain harmonic oscillations with frequency Ω and period T_0:

$$\Omega^2 = \omega^2 - 3\omega_0^2, \qquad T_0 = 2\pi\Omega^{-1} \tag{10}$$

When the tether is short enough, the term with ω_0 will be negligible. For example, for $a = 0.01$ m/s^2 and $L = 100$ m, we obtain the value $T_0 \sim 10$ min. Remembering that the period of circulation around the Earth at altitude $H = 200$-300 km is about 90 min, we are able to evaluate the conditions under which the frequency ω_0 must be taken into account.

If the period of oscillations T_0 is measured, it enables us to find aerodynamic drag by the formula

$$D_0 = mL(T_0/2\pi)^2 \tag{11}$$

When the drag coefficient c_D is known, we can determine the density ρ_0 at the altitude of flight. But if ρ_0, is known we can use this relation to find c_D for various materials of the SP or various forms of the SP and so on.

Thus, even linear regimes of oscillations of the SP with a short tether may give useful information. But the experiment may be more informative if the tether is long enough and the oscillations are nonlinear.

To demonstrate the principal features of nonlinearity and non-symmetry, we consider first the simplified Eq. (9), which includes asymmetry of the oscillations. We shall solve the problem for Eq. (9) with the initial data

$$\gamma = \gamma_+, \qquad \dot{\gamma} = 0 \text{ when } t = 0 \tag{12}$$

Transform Eqs. (9) and (12) introducing new variables from the formulas

$$\gamma = \gamma_+ Y(t_1), \qquad t_1 = \Omega t \tag{13}$$

For the new unknown function Y, we obtain the problem

$$\ddot{Y} + Y + \varepsilon Y^2 = 0, \quad Y(0) = 1, \quad \dot{Y}(0) = 0 \quad \varepsilon = \delta\gamma_+\omega^2/\Omega^2 \tag{14}$$

Keeping in mind that $\varepsilon << 1$, we can solve the problem (14) using the method of singular perturbations. The solution is sought in the form of an ε-power series:

$$Y(t_1, \varepsilon) = Y_0(\tau) + \varepsilon Y_1(\tau) + \ldots \tag{15}$$

$$t_1 = \tau + \varepsilon f_1(\tau) + \ldots \tag{16}$$

As the zeroth approximation, we have

$$Y_0(\tau) = \cos\tau \tag{17}$$

For the first approximation we obtain

$$f_1 = 0 \qquad Y_1'' + Y_1 = -\cos^2\tau \tag{18}$$

where the double prime denotes the derivative with respect to τ. The solution of Eq. (18) with zero initial data is

$$Y_1 = -(1/2) + (1/3)\cos\tau + (1/6)\cos 2\tau$$

The solution in the first approximation has the same period T_0 as the zeroth approximation Y_0. However, the asymmetry appears even in this solution, because

$$Y_1(0) = 0, \qquad Y_1(\pi) = -2/3$$

In the second approximation

$$f_2 = (5/12)\tau$$

$$Y_2 = -(1/3) + (29/144)\cos\tau + (1/9)\cos 2\tau + (1/48)\cos 3\tau$$

Thus, the period of oscillations differs from T_0 starting with the second approximation, but the amplitude difference for displacement upward and downward from the position of relative equilibrium appears even in the first approximation. In the second approximation the amplitude difference can be written in the form

$$\Delta Y = Y(0) + Y(\pi) = -(2/3)\varepsilon - ((2/3)\varepsilon)^2 \tag{19}$$

Let γ be the maximum deviation of the SP in its motion up and $\gamma_+\Delta$ the modulus of the amplitude difference. Since $\gamma_- < 0$, we have $\gamma_+\Delta = -(\gamma_- + \gamma_+)$, and then from Eq. (19) we obtain

$$\gamma_+\Delta = (2/3)\varepsilon\gamma_+(1 + (2/3)\varepsilon) \tag{20}$$

As will be shown later, Eq. (20) is valid for the amplitudes γ_+, which are essentially larger than expected from the analysis given earlier.

It is worth noting that Eq. (20) could be obtained more easily, namely, by use of the integral of energy. This method will be used for general nonlinear cases.

Nonlinear Oscillations

Substituting Eq. (5) for aerodynamic drag into Eqs. (3) and (4), we obtain a completely nonlinear equation for oscillations in the form

$$\ddot{\gamma} + (\omega^2 e^{\delta \sin\gamma} - 3\omega_0^2 \cos\gamma)\sin\gamma = 0 \tag{21}$$

$$\omega^2 = D_0/(mL), \qquad \omega_0^2 = g_0/R_0 \tag{22}$$

The integral of energy can be obtained from this equation. Assuming that $\gamma = \gamma_+$ and $\dot{\gamma} = 0$ at initial time $t = 0$. Multiplication of Eqs. (20) and (21) by $\dot{\gamma}$ and integration over t yields the integral

$$\dot{\gamma}^2 + \int_{\gamma_+}^{\gamma} F \, \mathrm{d}\gamma = 0 \tag{23}$$

$$F(\gamma) = (\omega^2 e^{\delta \sin \gamma} - 3\omega_0^2 \cos \gamma) \sin \gamma \tag{24}$$

The integral corresponding to the second term in integrand F can be easily evaluated. But it is not necessary to isolate it because we shall use a numerical approach.

When γ attains the maximum value γ_- in deviation of the SP upward then $\dot{\gamma} = 0$. Accordingly, the integral in Eq. (23) becomes equal to zero. It is this condition from which the amplitude γ_- is defined as a function of the variable γ_- and the parameters δ and $p = \omega_0^2/\omega^2$.

Using the fact that $\delta << 1$, we can represent the integral in Eq. (23) as an expansion series in δ_- powers. Restricting by linear approximation in δ, we obtain instead of Eq. (24) the following equation

$$\begin{aligned} &- \cos \gamma_- + \cos \gamma_+ + (1/2)\delta(\gamma_- - \gamma_+) - (1/4)\delta(\sin 2\gamma_- - \sin 2\gamma_+) \\ &+(3/2)p(\cos^2 \gamma_- - \cos^2 \gamma_+) = 0 \end{aligned} \tag{25}$$

Recalling that $\gamma_- = -\gamma_+(1 + \Delta)$, substitute γ_- into Eq. (25); then linearize with respect to Δ. The result is

$$\gamma_+\Delta = \frac{\gamma_+ - \sin \gamma + \cos \gamma_+}{F(\gamma_+)}\delta \tag{26}$$

When $\gamma_+ << 1$, formula (26) transforms into a linear version of formula (20).

The process of evaluation of successive approximations of Δ as a δ -power series expansion can be continued, of course. But the evaluations are tedious and are omitted here.

In the general case, Eqs. (23) and (24) were solved numerically. The solution depends on two parameters: δ and p. The expression of integrand F shows that there exists a restriction on the domain of the parameters for the oscillations near $\gamma = 0$ to be stable. For the case of a uniform atmosphere when $\delta = 0$, the expression in brackets is always positive when

$$\omega^2 > 3\omega_0^2, \qquad \text{i.e.,} \quad p < 1/3 \tag{27}$$

However, under conditions of nonuniform atmospheres, i.e., when $\delta = 0$, the drag D is an increasing function of γ. That means that the expression in parentheses [Eq. (25)] may vanish, even if the condition in Eq. (27) is satisfied. Preliminary studies show that for $\delta < 0.5$ the inequality (27) must be replaced by a little stronger one, namely,

$$\omega^2 > 4\omega_0^2, \qquad \text{i.e.,} \quad p < 1/4 \tag{28}$$

Under this condition the position $\gamma = 0$ will be the position of stable relative equilibrium.

For numerical integration the differential Eq. (24) has been transformed into the form

$$t = \int_{\gamma_+}^{\gamma} \frac{d\gamma}{\sqrt{2I}}, \qquad I = -\int_{\gamma_+}^{\gamma} F d\gamma \tag{29}$$

The integral I was calculated by Simpson's rule. The first integral in Eq. (29) was evaluated by use of the same method with nonuniform steps in γ except for the first and last intervals.

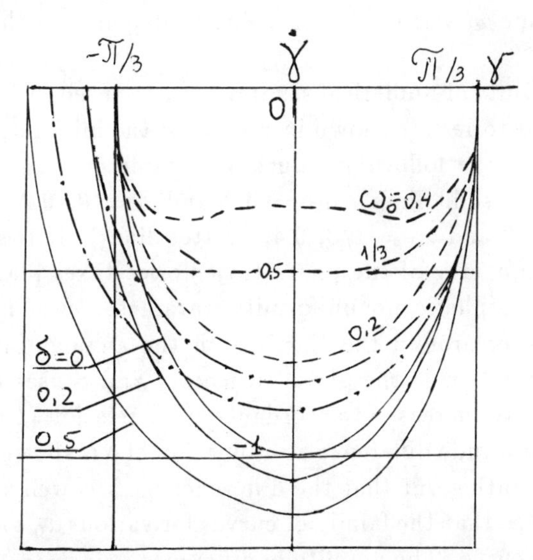

Fig. 2 Phase trajectories for initial deviations $\gamma_+ = \pi/3$ with $\omega_0 = 0$ and $\delta = 0, 0.2$, and 0.5 (dashed lines); $\delta = 0$ and $\omega_0 = 1/3$ and 0.4 (dotted lines).

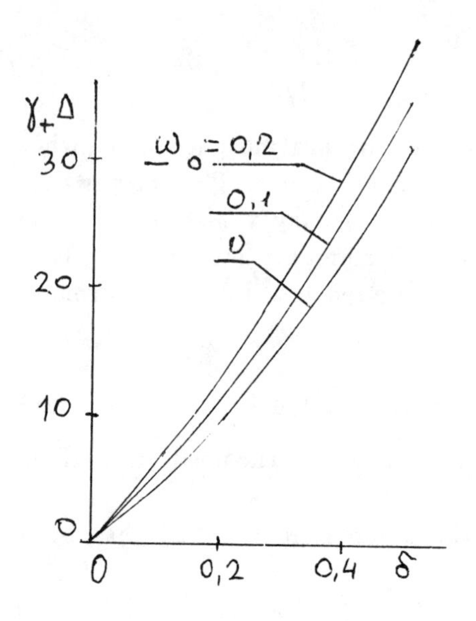

Fig. 3 Amplitude difference as a function of δ.

When $\gamma \to \gamma_+$, the integral I was replaced by the expression $I = -F(\gamma_+)(\gamma-\gamma_+)$, which was substituted into Eq. (29) and integrated so that $t = 2\sqrt{\gamma_+ - \gamma}/\sqrt{2F(\gamma_+)}$ as $t \to 0$. The same method was used for the limit $\gamma \to \gamma_-$.

In the calculations we assumed $\omega = 1$, which means that dimensionless time ωt was used. The solution depends on the parameters δ and ω_0.

The results of numerical studies are presented in Figs. 2-4. The phase trajectories are shown in Fig. 2 for the initial deviation $\gamma_+ = \pi/3$ and for the following values of parameters: $\omega_0 = 0$ and $\delta = 0,\ 0.2,\ 0.5$, (solid lines); $\omega_0 = 0.2$ and $\delta = 0,\ 0.2,\ 0.5$, (dashed lines); $\delta = 0$ and $\omega_0 = 1/3,\ 0.4$, (dotted lines). It is seen that the visible asymmetry of the phase trajectories takes place depending on the atmospheric nonuniformity parameter δ. The circulation of the orbiter around the Earth with the angular rate ω_0 results in transformation of the phase trajectory and causes the period of oscillations to increase in accordance with relation (10).

The amplitude difference as a function of δ is shown in Fig. 3. It is worth pointing out that the difference $\gamma_+\Delta$ is well approximated by Eq. (9) so that the family of curves for various ω_0 can be reduced to a single curve. The amplitude difference being measured enables us to define the exponent of nonuniformity of the atmosphere density.

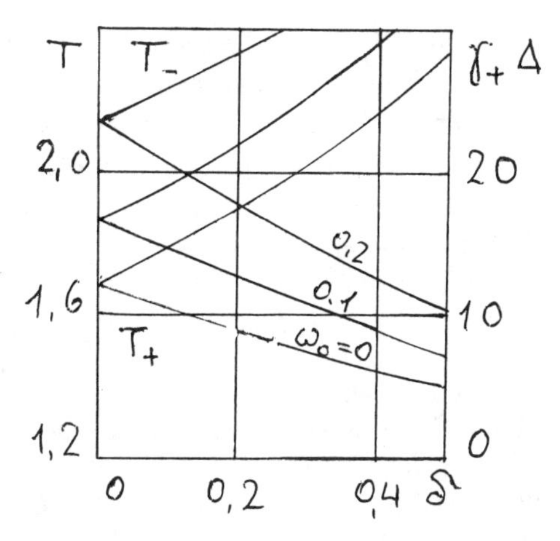

Fig. 4 Dependence of the half-periods, T_+ and T_- on non-uniformity parameter δ for various ω_0.

Another effect of the atmospheric nonuniformity where the oscillations of the SP may take place is that the lower part of the path is passed by the pendulum faster than the upper one. In Fig. 4 the curves of the half-periods T_+ and T_- in deviations downward and upward vs the nonuniformity parameter δ are presented. Again, using these dependencies (or simply the difference $\Delta T = T_- - T_+$, which differs slightly from the linear relation), we can determine the nonuniformity exponent when the difference of half-periods ΔT is measured. In the motions previously considered the tether tension N did not vanish.

Conclusion

The purpose of this paper was to demonstrate the principal possibility for utilization of a light SP tethered to an orbiter at low orbit as a very simple facility in experimental rarefied gasdynamics and aeronomy. We have considered only the simplest oscillations in the plane of the orbit. Actually, the situation is even more complicated for the case of plane oscillations because the atmosphere density at the altitude of flight depends on time. In general, of course, the oscillations should be nonplanar. One of the main reasons for the SP motion being nonplanar is the crosswind due to the rotating atmosphere. Observing the oscillation components out of the plane of orbit may illuminate the structure of winds at high altitudes.

Mathematical models of the SP should be improved, of course. Both the inertia and finite diameters are of importance for long tethers. The nonstationary processes of deployment as well as transient motions with a loose tether are of primary interest. In addition, an electrically charged SP may interact with the electromagnetic field of the Earth and its atmosphere.

References

[1]Bevilacqua, F. and Chiarelli, C., "Tethered Space System: a New Facility for Experimental Rarefied Gas Dynamics," *Proceedings of the XVth International Symposium on Rarefied Gas Dynamics*, edited by V. Boffi and C. Cercignani, Teubner, Stuttgart, FRG, 1986, Vol. 1, pp. 558-573.

[2]Carlomagno, G. M., et al., "Low Density Aerothermodynamics Studies Performed by Means of the Tethered Satellite System," *Proceedings of the XVth International Symposium on Rarefied Gas Dynamics*, edited by V. Boffi and C. Cercignani, Teubner, Stuttgart, FRG, 1986, Vol. 1, pp. 600-609.

Chapter 2. Plasmas

Semiclassical Approach to Atomic and Molecular Interactions

Joseph A. Kunc*
University of Southern California, Los Angeles, California

Abstract

A general approach, combining quantum and classical mechanics, is used to determine the electron position and velocity distributions in atoms and atomic ions (positive and negative). The Hartree-Fock electronic wave functions and the classical central field approximation are used for evaluation of the dynamic properties of the localized electrons. The distributions, which are of fundamental importance in applications of the binary encounter approximation to description of atomic and ionic collisions, are obtained in the form of simple analytical expressions. The quantum-classical distributions of this work are compared with several other distributions in Ne, Ar, and Al atoms in the ground states.

Introduction

Modeling of atomic and molecular phenomena in nonequilibrium gases and plasmas consists of a formulation of a microscopic physical model and a solution of the model with a set of simplifying assumptions. The formulation of the model includes a description of atomic and molecular collisions and a statistical formalism capable of superposing efficiently the large number of collisions in the gas. In principle, the collisions can be described by various approximations of the quantum-mechanical scattering theory; this however,

---

Presented as an Invited Paper.

***Professor, Departments of Aerospace Engineering and Physics.**

requires complex and time-consuming calculations, even with modern computers. As a result, it is often very difficult, or impossible, to obtain an acceptable solution, especially when collisions involving heavy atoms and complex molecules are considered. Therefore, it is often necessary to apply approximate methods of the semiclassical scattering theory, especially in large-scale modeling of gases and plasmas when the description of the collisions is only a small part of the modeling. These approximations are expected to have an acceptable accuracy and high computational efficiency. One such approach, the subject of this work, is the *binary encounter approximation* (b.e.a.) originating from the semiclassical scattering theory.[1]

A system of two, A and B, interacting particles (atoms, ions, or molecules) can consist of a substantial number of electrons and nuclei, called hereafter "components" of the particles. In simplified models, the components may be considered as groups of species, each group containing several electrons and a nucleus (e.g., atoms as components of a molecule). Before collision, the system is defined by a set C of n parameters (variables and constants): a set C_A, describing the internal motion of the components in particle A; a set C_B, describing the internal motion of the components in particle B; and a set C_{AB}, describing the initial relative motion of particles A and B. During the collision, set C is transformed into a set C' ($C \rightarrow C'$), where the prime denotes the state of the colliding system after the collision. In general, the set C includes all the dynamical and geometrical parameters characterizing the system, such as the particle masses, charges, velocities, and impact parameter.

The probability that the collision system will change from state C to state C' can be determined if the relation between sets C and C' is known. This relation can be given by a function F such that

$$C' = F(w, D, \Theta, C_4, \ldots, C_n) \tag{1}$$

where, for reasons of convenience, we separate (Fig.1) the three parameters, $C_1 \equiv w$ (the relative velocity of particles A and B), $C_2 \equiv D$ (the impact parameter of the collision), and $C_3 \equiv \Theta$ (the azimuthal orientation of the "shot line" of the incident particle before the collision), from the rest of the parameters of set C.

In principle, the relation (1) can be obtained from the Schrödinger equation for the collision system or from the motion equations of classical dynamics, if the potentials for the interactions between the components of the A–B system are known. These potentials are often approximated by central-force potentials with the components represented as structureless centers of scattering: the Coulomb potential (for electron-electron and electron-nuclei interaction), the two- or three-parameter potentials such as the generalized Lennard-Jones and Morse potentials (for interactions involving atoms and molecules), or more general series expansions emphasizing contributions of different terms of multipole interaction at different interparticle distances.

A *particular* cross section for a collision of particles A and B is defined[1] as a measure of the probability of a *definite change* in C during the collision, that is, the probability of a definite change in the state of the colliding system. This change is characterized by a change of one or several parameters (energy ΔE, momentum $\Delta\vec{p}$, etc.) of set C (this subset of parameters is denoted symbolically by ξ). Since set C can be large, the number of possible cross sections with respect to ξ, for the collision of particles A and B, can also be large.

Taking the preceding into account, we can define the cross section with respect to ξ as

$$\sigma_\xi(w,\xi,\ldots,C_n) = \int_0^{2\pi}\int_0^{\infty} \delta[\xi - F_\xi(w,D,\Theta,\xi,\ldots,C_n)]D\mathrm{d}D\mathrm{d}\Theta \tag{2}$$

Thus, the cross section σ_ξ is the small shadowed area (Fig.1), selected from the entire range of D and Θ, where the integral (2) is nonzero. Adding all the cross sections that are nonzero in some practically useful range of ξ, we obtain the so-called differential cross section with respect to ξ,

$$Q_\xi(w,\ldots,C_n) = \int_{\xi_1}^{\xi_2} \sigma_\xi(w,\xi,\ldots,C_n)\mathrm{d}\xi \tag{3}$$

which is represented by the large shadowed area in Fig.1. For example, the cross section for collision of A and B in which the particles exchange more energy than $\Delta E = U_B$ (U_B being the ionization potential of the particle B) is

$$Q_{\Delta E}(w,\ldots,C_n) = \int_{U_B}^{\infty} \sigma_{\Delta E}(w,\xi\equiv\Delta E,\ldots,C_n)\mathrm{d}\Delta E \tag{4}$$

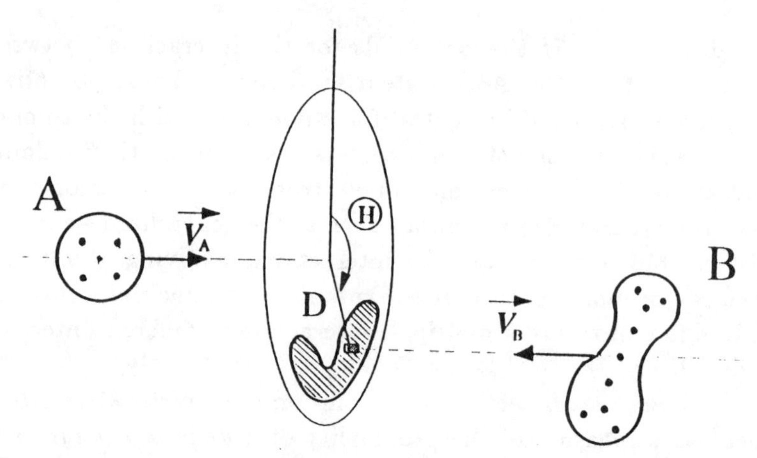

Fig.1 Geometry of collision between two, A and B, multicomponent particles: $\vec{v}_A$ and $\vec{v}_B$ are the velocities of the particles, D is the impact parameter of the collision, and Θ is the azimuthal orientation of the shot line before the collision.

which corresponds to a collision of A and B in which the particle B may be ionized.

In practice, we need average cross sections that are functions of a few parameters only. These average cross sections are obtained by integrating the cross sections (3) over some variables of set C. With these variables denoted by subscripts $j+1, \ldots, n$, the average cross section with respect to ξ is

$$\langle Q_\xi(w, \cdot\cdot, C_j)\rangle = \int \cdot\cdot \int Q_\xi(w, C_{j+1}, \cdot\cdot, C_n) f_{j+1}(C_{j+1}) \cdot\cdot f_n(C_n) \mathrm{d}C_{j+1} \cdot\cdot \mathrm{d}C_n \quad (5)$$

where $f_{j+1}(C_{j+1}), \cdot\cdot, f_n(C_n)$ are distributions of variables $C_{j+1}, \cdot\cdot, C_n$.

Determination of the relations (1) - (5) is a difficult (often impossible) and time-consuming task. The complexity of the problem can be reduced substantially by using the binary encounter approximation. In this approximation, the collision of the two particles A and B is treated as a statistical superposition (using the concept of the cross section) of all the pairwise interactions (binary collisions) between the components of the colliding particles (one component belonging to particle A and the other belonging to particle B). The cross section $Q_{\xi,kl}$ for the pairwise interaction between any pair of such components, with respect to ξ, can be calculated from the general relation (2) by replacing particles A and B by the two k and l components. If the components are treated as structureless centers of scattering, interacting through central-force potentials, then the cross sections $Q_{\xi,kl}$ for interaction between the two centers can be described quite accurately by both classical and quantum mechanics.[1-2]

Further simplification of the binary encounter approximation can be made by using the semiclassical approach[1] to description of the dynamical properties of the interacting components. Such an approach uses classical mechanics for description of the motion of the colliding particles and quantum-mechanical requirements for quantization of rotational, vibrational, and electronic energies. In addition, it is often justified to consider only a small number of the most important binary interactions between the components. A good example of such a situation is the collision between a low- or medium-energy electron and an atom (molecule) when it is quite sufficient to consider only the binary interactions between the incident electron and the electrons of the outer atomic (molecular) shell. At low and medium impact energies, the outer-shell electrons are the major contributors to the electron-impact atomic (molecular) transitions.

The averaging (5) is often done over the variables $\vec{v}_k(v_{k_x}, v_{k_y}, v_{k_z})$ and $\vec{v}_l(v_{l_x}, v_{l_y}, v_{l_z})$, the velocity vectors of the kth component of the incident particle A and the lth component of the target particle B, respectively. Then, the average cross section with respect to ξ, for collision of particles A and B, is

$$\langle Q_\xi(w, \cdot\cdot, C_j)\rangle = \sum_{k=1}^{N_A}\sum_{l=1}^{N_B}\int\cdot\cdot\int Q_{\xi,kl}(w, \cdot\cdot, C_j, \vec{v}_k, \vec{v}_l) f_A(\vec{v}_k) f_B(\vec{v}_l) d\vec{v}_k d\vec{v}_l \quad (6)$$

where N_A and N_B are the numbers of the components and $f_A(\vec{v}_k)$ and $f_B(\vec{v}_l)$ are velocity distributions of the components in particles A and B, respectively.

Quite often the target medium is weakly polarized. In such a case, the spatial distributions $p_{A,B}(\theta_{k,l}, \vartheta_{k,l})$ of the electron velocity vectors $\vec{v}_{k,l}$ in atoms and atomic ions can be assumed to be isotropic. Consequently, the distributions $p_{A,B}(\theta_{k,l}, \vartheta_{k,l})$ and $f_{A,B}(v_{k,l}, \theta_{k,l}, \vartheta_{k,l})$ (in the spherical coordinate systems, located at the centers of particles A and B, with the angles $\theta_{k,l}$ measured from the vector $\vec{w}$)

are

$$p_{A,B}(\theta_{k,l},\vartheta_{k,l})\mathrm{d}\theta_{k,l}\mathrm{d}\vartheta_{k,l} = \sin\theta_{k,l}\mathrm{d}\theta_{k,l}\mathrm{d}\vartheta_{k,l}/4\pi \tag{7}$$

and

$$f_{A,B}(\vec{v}_{k,l})\mathrm{d}\vec{v}_{k,l} = p_{A,B}(\theta_{k,l},\vartheta_{k,l})g_{A,B}(v_{k,l})\mathrm{d}v_{k,l}\mathrm{d}\theta_{k,l}\mathrm{d}\vartheta_{k,l} \tag{8}$$

where $g_{A,B}(v_{k,l})$ are absolute velocity distributions of the electrons in particles A and B. Thus, the averaging over the electron distributions $f_{A,B}(\vec{v}_{k,l})$ can be reduced to averaging over the distributions $g_{A,B}(v_{k,l})$.

In the important case of collisions of structureless charges (electrons, positrons, and completely stripped ions) with atoms and molecules, the statistical superposition includes all the binary collisions between the incident charge (particle A) and all the electrons of the target particle (particle B). Then the average cross section with respect to ξ can be given as

$$\langle Q_\xi(w,\cdot\cdot,C_j)\rangle = \sum_{l=1}^{N_B}\int\cdot\cdot\int Q_{\xi,1l}(w,\ldots,C_j,\vec{v}_l)f_B(\vec{v}_l)\mathrm{d}\vec{v}_l \tag{9}$$

The main advantages of the classical and semiclassical distributions of the electron velocity are that they have simple, sometimes analytical, forms and that they represent the average dynamic properties of the atomic electrons with accuracy acceptable in a broad range of applications.

A few remarks should be added about the binding energies of the atomic electrons participating in the pairwise collisions. The energy of each atomic electron can be determined from the quantum-mechanical orbital theory discussed in the next section. However, since the Koopman criterion is valid in atoms and atomic ions, we can assume that, for an ith atomic orbital,

$$W_i = U_i, \tag{10}$$

where $-W_i$ is the total energy of an electron belonging to the ith orbital and U_i is the ionization potential for the orbital.

The situation is more complicated when an entire atomic shell is considered. Then, the average energy of each of the shell electrons can be taken as equal to $-W_o$, where[3]

$$W_o = N_o^{-1} \sum_{i=1}^{N_o} U_i \tag{11}$$

with N_o representing the number of electrons in the shell.

In order to determine the dynamical properties of the atomic electrons, we combine the quantum-mechanical theory of atomic orbitals with the laws of classical dynamics. When doing this we assume that each electron is moving in the field of a time-averaged, spherically-symmetric Coulomb force resulting from the presence of the nucleus screened by the other atomic electrons. Consequently, we evaluate semiclassical velocity distributions $g(v)$ for electrons in outer (L and M) shells of several ground-state atoms (Ne, Ar, and Al). The distributions for some other species, including positive and negative ions, can be found in Ref. 4. (It should be noted that the approach discussed here can be applied in a straightforward way to excited atoms and atomic ions). The cross sections σ_ξ and $Q_{\xi,1l}$ for various interactions are discussed in Refs. 1 – 3 and 5 – 7.

Quantum-Classical Distributions for Outer (L and M) Shells

A. Atomic Wave Functions

The electronic wave functions for the ground-state species discussed here are calculated by using the Hartree-Fock (HF) approach.[8] The approach gives the one-electron wave function φ_i for an ith atomic orbital (AO) as a linear combination of Slater-type functions (STF s)

$$\varphi_i(\vec{r}) = R_i(r)Y_i(\theta, \vartheta) = \sum_p C_{i,p}\chi_{i,p}(\vec{r}) \tag{12}$$

where the sum is over the basis set of STF's for the orbital, $R_i(r)$ is the radial part of the wave function, and $Y_i(\theta, \vartheta)$ are the normalized spherical harmonics in complex form. Here, $\vec{r}$ is the position of the orbital electron in the spherical coordinate system (r, θ, ϑ) centered on the nucleus, $C_{i,p}$ are the coefficients of the expansion, and $\chi_{i,p}$ are the STF s,

$$\chi_{i,p}(\vec{r}) = (2\zeta_{i,p}/a_o)^{n_i+1/2}((2n_i)!)^{-1/2} r^{n_i-1} \exp(-\zeta_{i,p} r/a_o) Y_i(\theta, \vartheta) \tag{13}$$

where n_i is the principal quantum number, a_o is the Bohr radius, and $\zeta_{i,p}$ are the exponents of the basis set.

B. Radial Distributions

The radial distribution $\varphi_i^2(r)$ of an electron belonging to the ith orbital can be given as

$$\varphi_i^2(r)\mathrm{d}r = \int_0^{2\pi}\int_0^{\pi}\varphi_i^2(r,\theta,\vartheta)r^2\sin\theta\mathrm{d}\theta\mathrm{d}\vartheta\mathrm{d}r = R_i^2(r)r^2\mathrm{d}r \qquad (14)$$

(This distribution and all the other distributions in this work are normalized to 1).

The radial distribution $\psi^2(r)$ of the electron in the atom is

$$\psi^2(r)\mathrm{d}r = N_t^{-1}\sum_i N_i\varphi_i^2(r)\mathrm{d}r \qquad (15)$$

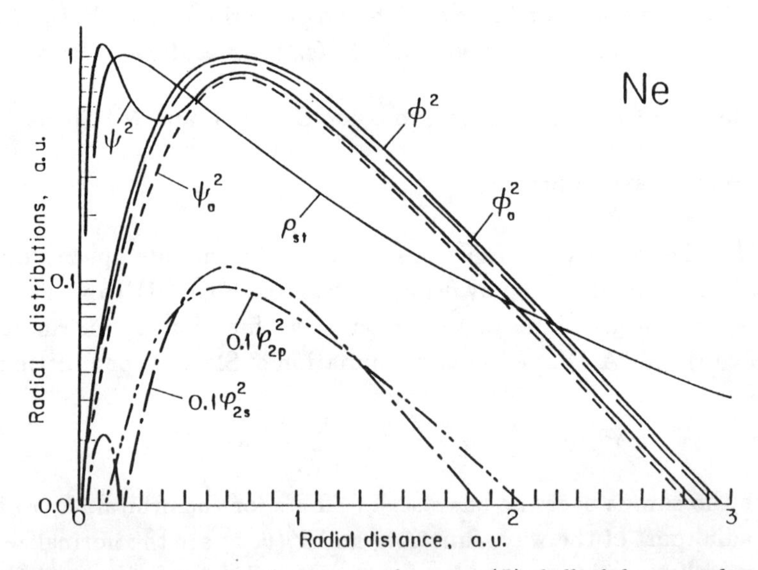

Fig.2 Electron radial distributions in the outer (L) shell of the ground-state neon atom: $\psi^2(r)$ is the atomic distribution obtained from the Hartree-Fock approach, $\psi_a^2(r)$ the part of the atomic distribution $\psi^2(r)$ corresponding to the atomic L shell, $\rho_{st}(r)$ the atomic distribution obtained from the Thomas-Fermi-Dirac statistical theory; $\phi^2(r)$ is the L-shell distribution obtained from the RHF approach, and $\phi_a^2(r)$ the L-shell distribution given by Eq. (16); $\varphi_{2s}(r)$ and $\varphi_{2p}(r)$ are the electronic wave functions for the $2s$ and $2p$ orbitals, respectively.

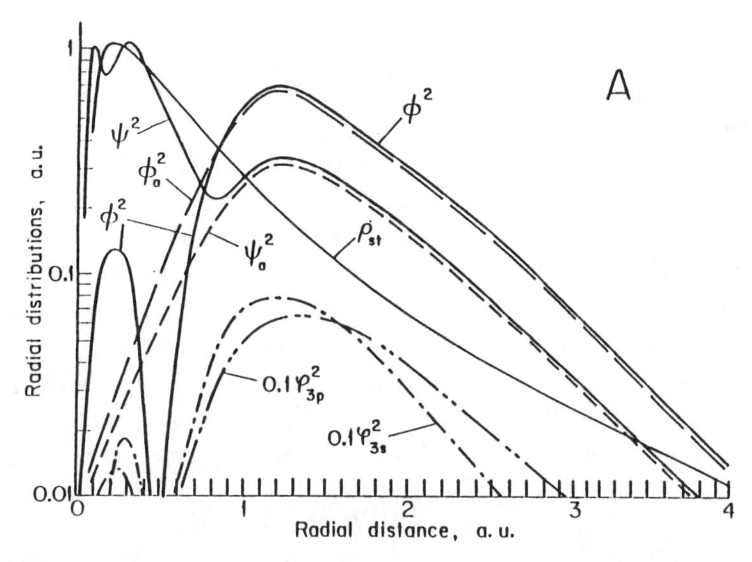

Fig.3 Electron radial distributions in the outer (M) shell of the ground-state argon atom: $\varphi_{3s}(r)$ and $\varphi_{3p}(r)$ are the electronic wave functions for the $3s$ and $3p$ orbitals, respectively. The meaning of the other symbols is the same as in Fig. 2.

where N_i is the number of electrons occupying the ith orbital, N_t is the total number of electrons in the atom, and the sum is over all the atomic orbitals. The radial parts of the electronic wave functions considered in this work are calculated by using the sets of the coefficients $C_{i,p}$ and the exponents $\zeta_{i,p}$ obtained by Clementi and Roetti.[8] The results of numerical calculations for the distributions $\psi^2(r)$ for some atoms are shown in Figs. 2 – 4.

The quantum-classical radial distributions $\phi^2(r)$ of the outer-shell electrons are also shown in Figs. 2 – 4. The distributions were obtained from the relation (15), with the sum on the right-hand side taken over the outer-shell orbitals only. In species considered in this work, these are either $2s$ and $2p$ (L-shell) or $3s$ and $3p$ (M-shell) orbitals. Another way of obtaining the distributions of the outer-shell electrons is extrapolating to zero the parts of the distributions $\psi^2(r)$ that correspond to the outer shells. The resulting distributions $\psi_a^2(r)$ can then be used to determine the approximate radial distributions $\phi_a^2(r)$ of the outer-shell electrons by renormalization

$$\phi_a^2(r)\mathrm{d}r = \frac{\sum_{i=1}^{N_t} N_i}{\sum_{j=1}^{N_o} N_{o,j}} \psi_a^2(r)\mathrm{d}r \qquad (16)$$

where N_i and $N_{o,j}$ are numbers of electrons in the ith and the jth atomic orbitals of the outer-shell, respectively, and N_o is the number of electrons in the outer shell. We can see from Figs. 2 – 4 that such an extrapolation procedure is very accurate at large and medium radial distances. At small distances, it is also quite accurate when the outer-shell electrons belong to $2s$ and $2p$ orbitals, but it becomes less accurate when the outer-shell electrons belong to $3s$ and $3p$, or higher, orbitals. In general, the extrapolation is less accurate in the case of the outer shells with multimodal wave functions $\varphi_i(r)$.

We can compare the radial distributions of this work with those obtained from the Thomas-Fermi-Dirac (TFD) statistical theory,[9–10] which gives the electron radial distribution $\rho_{st}(r)$ as

$$\rho_{st}(r)\mathrm{d}r = Zb^2(N_t\mu^2)^{-1}r\Psi^2(r)\mathrm{d}r \tag{17}$$

with $b = 1.864$ and $\mu = 0.885a_oZ^{-1/3}$ (Z is the charge number of the nucleus). The TFD screening function $\Psi(r)$ is[9,11]

$$\Psi(r) = (1 + cr)^{-2} + \kappa\eta_o(r) \tag{18}$$

where the first term on the right-hand side represents the TFD screening function for the parental atom, the second term represents

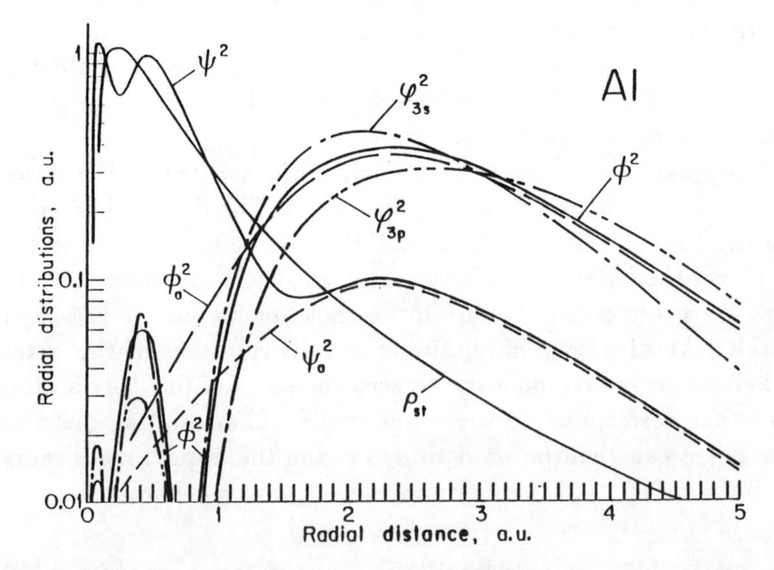

Fig.4 Electron radial distributions in the outer (M) shell of the ground-state aluminum atom. The meaning of the symbols is the same as in Fig. 3.

the ionic correction, $c = 0.761/\mu$, and

$$\kappa = 0.083(Z_q/Z)^3 \tag{19}$$

with Z_q being the surplus (or deficit) of electrons in the particle. The function $\eta_o(r)$ in Eq. (18) is the ionic function of the TFD theory. Eq. (17) should be used with a caution because the TFD theory overestimates the distribution $\rho_{st}(r)$ at larger radial distances. Comparison of the results of TFD theory and the results of HF calculations for various atoms and ions shows that the discrepancy can be significant (especially for negative ions), even for closed-shell electron configurations, when the TFD model is expected to give the best results. The results of numerical calculations for the distributions $\psi^2(r)$ [Eq. (15)] and $\rho_{st}(r)$ [Eq. (17)] are shown in Figs. 2 – 4.

C. Analytical Approximations for the Radial Distributions

The radial part of an ith (nl) orbital can be approximated by a single STF

$$R_{nl}(r) = C_{nl} r^{n-1} \exp(-\zeta_{nl} r/a_o) \tag{20}$$

where the normalization constant C_{nl} is

$$C_{nl} = (2\zeta_{nl}/a_o)^{n+1/2}((2n)!)^{-1/2} \tag{21}$$

Representation of the electronic radial wave functions $R_{nl}(r)$ by single STF is quite accurate in the case of single-mode $2p$ wave functions. In the case of multimodal wave functions, such representation is crude, especially at small radial distances. The situation can be improved by using the ζ_{nl} exponents that would lead to the best possible approximation of the distributions. Such exponents, called "single-zeta" exponents, have been determined by an optimalization process for many atomic species.[8] If the single-zeta exponents are not available, then the original Slater relation can be used:

$$\zeta_{nl} = Z'_{nl}/n \tag{22}$$

where Z'_{nl} is the charge number of the atomic "core" (Subsection D).

Using Eq. (15), with the sum taken over all the outer-shell orbitals, and the electronic radial wave functions (20), we obtain an analytical expression for the radial distribution $\phi_o'^2(r)$ of the electrons belonging to L or M outer shells[4]

$$\phi_o'^2(r)\mathrm{d}r = \frac{r^{2n}G_{nl}(r)}{N_{ns}+N_{np}}\mathrm{d}r \tag{23}$$

where

$$G_{nl}(r) = N_{ns}C_{ns}^2\exp(-2\zeta_{ns}r/a_o) + N_{np}C_{np}^2\exp(-2\zeta_{np}r/a_o) \tag{24}$$

and N_{ns} and N_{np} are numbers of electrons in the ns and np orbitals, respectively.

Values of the single-zeta exponents for ns and np orbitals considered here are very close to each other, so that we can use an average exponent ζ_o, one for the entire shell, such that

$$\zeta_o = (\zeta_{ns} + \zeta_{np})/2 \tag{25}$$

Substituting Eq. (25) into Eq. (23), we obtain an analytical expression for the radial distribution of the outer-shell electrons:

$$\phi_o^2(r)\mathrm{d}r = C_o d_o r^{2n}\exp(-2\zeta_o r/a_o)\mathrm{d}r \tag{26}$$

where

$$d_o = (2\zeta_o/a_o)^{2n+1}/(2n)! \tag{27}$$

and where the renormalization constant C_o has been introduced to balance the small inaccuracy resulting from approximations (23) and (25). The constant has an analytical form,

$$C_o = \left(\int_0^{r_o}\phi_o^2(r)dr\right)^{-1} = [d_o(\alpha_1-\alpha_2)]^{-1} \tag{28}$$

where

$$\alpha_1 = \exp\left(-\frac{2\zeta_o r_o}{a_o}\right)\sum_{k=0}^{2n}(-1)^k\frac{(2n)!r_o^{2n-k}}{(2n-k)!(-2\zeta_o/a_o)^{k+1}} \tag{29}$$

and

$$\alpha_2 = (2n)!(a_o/2\zeta_o)^{2n+1} \tag{30}$$

with r_o (the classical radius of the shell), determined from the requirement for the electron velocity to be a real and positive quantity, as

$$r_o = \frac{Z_o'e^2}{W_o} \tag{31}$$

Numerical calculations show that approximation (26) gives results (not shown in Figs. 2–4) in good agreement with the corresponding values of $\phi_a^2(r)$. This results from the fact that using single-zeta STF s for multimodal wave functions in evaluation of Eq. (26) plays the same role as the extrapolation leading to the distributions $\phi_o^2(r)$; it produces an averaging effect that describes pretty accurately the average properties of the electron radial distributions at small radial distances.

D. Velocity Distributions

The velocity distribution $g_i(v)$ for the electrons of the *ith* orbital is calculated by applying classical dynamics to descriptions of the motion of the electrons in the atomic species. We assume that each such electron is moving in a spherically symmetric Coulomb field of an atomic core containing a nucleus and all the other electrons screening the interaction of the electron with the nucleus. As a result, the effective charge of the core can be given as $Z_i'e$, with the charge number of the core

$$Z_i' = Z - \zeta_i \tag{32}$$

where ζ_i is the screening constant for the electrons of the ith orbital. The screening constants are calculated using the "Slater rules", which can be summarized as follows (the AO's are divided into the following groups: $(1s), (2s, 2p), (3s, 3p), (3d),...$): 1) All the electrons in groups outside the one being considered do not affect the screening constant, 2) Each other electron in the group considered increases ζ_i by 0.35, except that 0.30 is used in the $1s$ group, 3) For an s or p orbital, each electron with n one less than the group considered increases ζ_i by 0.85, and each electron with n two or more less than the group considered increases ζ_i by 1.00. Using these rules, we find that the screening constants for the $2s$ and $2p$ orbitals of the outer shells have the same values. A similar conclusion applies to the screening constants for the $3s$ and $3p$ electrons of the outer shells. Therefore, the screening effect for the entire outer (L or M) shell can be represented by one value (ζ_o) of the screening constant (then the charge number of the core is $Z_o' = Z - \zeta_o$).

The total energy $-W_i$ of an electron belonging to the ith orbital and moving in the Coulomb field of the atomic core is given in classical dynamics by

$$\frac{mv^2}{2} - \frac{Z_i'e^2}{r} = -W_i \tag{33}$$

where r, as before, is the radial distance from the nucleus and m is the electron mass. Consequently, we obtain

$$\left|\frac{dr}{dv}\right| = mvZ_i'e^2(\frac{1}{2}mv^2 + W_i)^{-2} \tag{34}$$

In the central field approximation, the probability of finding an electron belonging to the ith orbital with velocity between v and $v+dv$ (v is the electron velocity at r) is equal to the probability of finding the electron at a radial distance between r and $r+dr$. This leads to

$$g_i(v) = \varphi_i^2(r(v))\left|\frac{dr}{dv}\right| \tag{35}$$

where $g_i(v)$ is the electron velocity distribution and

$$r(v) = Z_i'e^2(\frac{1}{2}mv^2 + W_i)^{-1} \tag{36}$$

(The radius r_i of the ith orbit is given by relation (31) when $Z_o' \equiv Z_i'$ and $W_o \equiv W_i$).

Because the screening of the outershell electrons can be described by one screening constant ξ_o, the quantum-classical velocity distribution $g(v)$ for the outershell electrons is

$$g(v) = \phi_a^2(r)\left|\frac{dr}{dv}\right| \tag{37}$$

where $\phi_a^2(r)$ is the outershell electron radial distribution discussed in the previous sections. Examples of the distribution (37) are shown in Figs. 5 – 7.

E. Analytical Approximations for the Velocity Distributions

Combining Eqs. (37) and (26) leads to an analytical expression for the velocity distribution of the outer-shell electrons,

$$g_o(v)dv = C_o\frac{2b^{2n+1}v}{(2n)!(v^2+v_o^2)^{2(n+1)}}\exp\left(-\frac{b}{v^2+v_o^2}\right)dv \tag{38}$$

where

$$b = \frac{4\zeta_o Z'_o e^2}{m a_o} \tag{39}$$

with

$$v_o = \left(\frac{2W_o}{m}\right)^{1/2} \tag{40}$$

The distributions (38) are shown in Figs. 5 – 7.

δ Distributions

The simplest electron distributions of classical atomic physics are the δ distributions:

Radial distribution:

$$\rho_\delta(r)\mathrm{d}r = \delta(r - r_o)\mathrm{d}r \tag{41}$$

where

$$r_o = \frac{Z'_o e^2}{2W_o} \tag{42}$$

Velocity distribution:

$$g_\delta(v)\mathrm{d}v = \delta(v - v_o)\mathrm{d}v \tag{43}$$

where W_o and v_o are given by Eqs. (11) and (40), respectively. These distributions are identical with the distributions for the Bohr model of hydrogenic species if W_o is equal to the ionization potential of the species. The distributions (43) are shown in Figs. 5 – 7.

Other common distributions of an electron moving in a Coulomb field of charge $Z'e$ are those originating from the microcanonical distribution of classical statistical mechanics[12]

$$f(P, W) = c(4\pi)^{-2}\delta(H + W) \tag{44}$$

where

$$H = \frac{P^2}{2m} - \frac{Z'e^2}{r} \tag{45}$$

is the classical Hamiltonian, c is the normalization factor, and $-W$ and P are the total energy and total momentum of the atomic electron, respectively. Integrating the distribution (44) over the coordinates of the classical phase space[13], we obtain the radial and velocity microcanonical distributions of an electron in the multielectron shell,

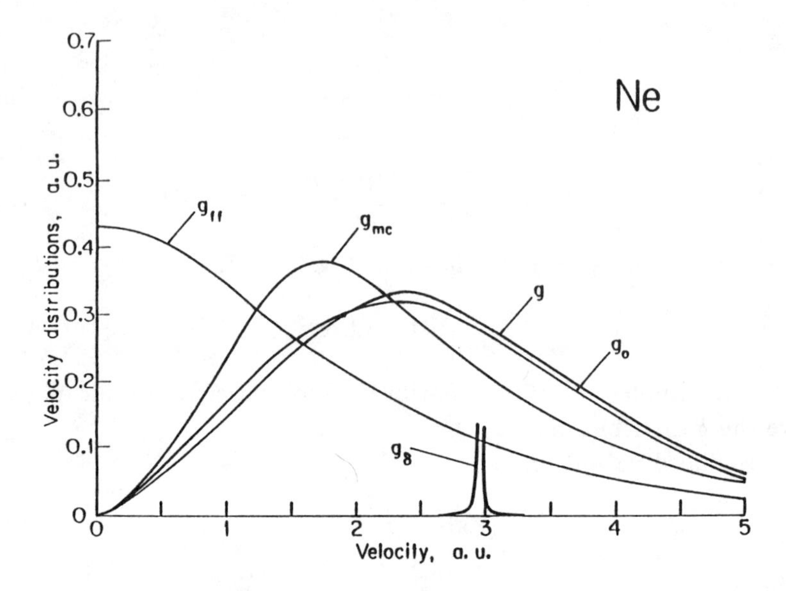

Fig.5 Velocity distributions of the L-shell electrons in the ground-state neon atom: $g(v)$ is the quantum-classical distribution given by Eq. (37), and $g_o(v)$ is the analytical quantum-classical distribution given by Eq. (38); $g_{mc}(v)$, $g_{ff}(v)$, and $g_\delta(v)$ are microcanonical, free-fall, and δ distributions, respectively.

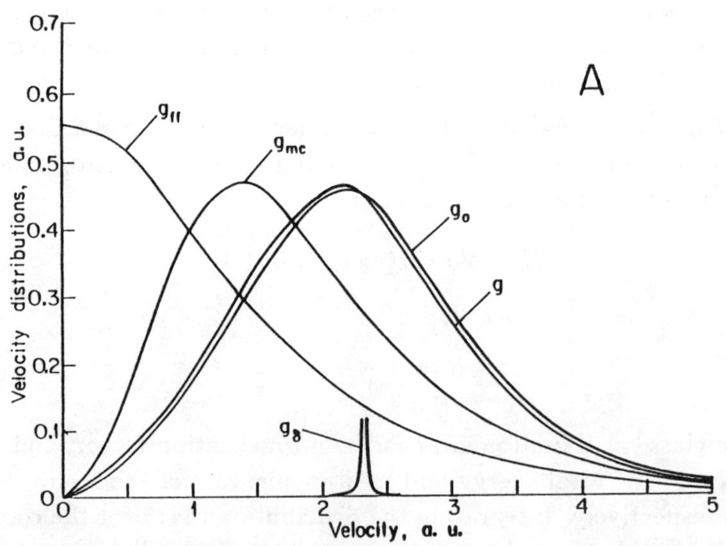

Fig.6 Velocity distributions of the M-shell electrons in the ground-state argon atom. The meaning of the symbols is the same as in Fig. 5.

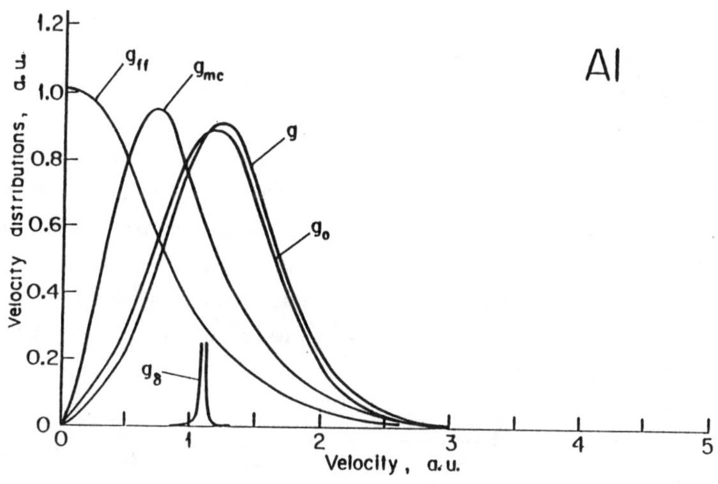

Fig.7 Velocity distributions of the *M*-shell electrons in the ground-state aluminum atom. The meaning of the symbols is the same as in Fig. 5.

assuming that $W = W_o$ [Eq. (11)]. Then, the microcanonical radial distribution has the form

$$\rho_{mc}(r)\mathrm{d}r = \frac{16}{\pi r_m^3} r^{3/2}(r_m - r)^{1/2}\mathrm{d}r \qquad 0 \le r \le r_m, \tag{46}$$

whereas the microcanonical velocity distribution is

$$g_{mc}(v)\mathrm{d}v = \frac{32 v_o^5 v^2}{\pi(v^2 + v_o^2)^4}\mathrm{d}v \tag{47}$$

where r_m, equal to r_o [Eq. (31)], is the radius of the electron classical orbit. The distributions (47) are shown in Figs. 5 – 7.

It may be added that the distribution (47) has the same functional form as the quantum-mechanical distribution for a hydrogen atom[2]

$$\frac{4\pi}{n^2}\sum_{l,m} |\psi_{nlm}(\vec{p})|^2 v^2 \mathrm{d}v = \frac{32 w_o^5 v^2}{\pi(v^2 + w_o^2)^4}\mathrm{d}v \tag{48}$$

where ψ_{nlm} is the electron momentum wave function for the state, with the principal quantum number n and the angular momentum quantum numbers l and m and where

$$w_o = \frac{2\pi e^2}{nh} \tag{49}$$

with h the Planck constant.

Free-Fall Distributions

Another classical distribution that has been quite successful in the physics of atomic collisions is the free-fall distribution[14] resulting from a zero angular momentum solution of the Kepler problem in a Coulomb field when the magnetic moment of the electron is taken into account. This distribution was used recently in a modified form by Gryzinski and Kunc[3] to investigate a large class of inelastic collisions involving electrons of inner and outer shells in number of atoms. The modified free-fall radial distribution for the outer-shell electrons can be written as

$$\rho_{ff}(r)\mathrm{d}r = \frac{2}{\pi r_s}\left(\frac{r}{r_s - r}\right)^{1/2} \mathrm{d}r, \qquad 0 \le r \le r_s \tag{50}$$

where

$$r_s = \frac{Z_s e^2}{W_o} \tag{51}$$

with the effective charge number Z_s estimated as[3]

$$Z_s = \begin{cases} Z - 1.25 & \text{outer-shells} \\ Z - 0.25 & \text{inner-shells} \end{cases} \tag{52}$$

The modified free-fall velocity distribution of the outer-shell electrons is

$$g_{ff}(v)\mathrm{d}v = \frac{4v_s^3}{\pi(v^2 + v_s^2)^2}\mathrm{d}v \tag{53}$$

where $v_s = v_o$. The distributions (53) are shown in Figs. 5 – 7.

Summary and Discussion

The results shown in Figs. 5 – 7 indicate that the quantum-classical approach presented here leads to velocity distributions that seem to be good approximations of the average dynamical properties of the electrons in L and M shells of atoms and atomic ions. In higher shells $(O, N, ...)$, the number of modes in the electronic wave functions increases. Then, the extrapolation of the radial distributions at small radial distances and the representation of the orbitals by the single-zeta STF [in evaluation of the analytical expressions (26) and (38)] is usually less accurate (in general, the approach of this work is more accurate for atoms with a small number of shells). In such a case, the outer-shell distributions of the present work should be treated as first-order approximations.

The analytical distributions (38), shown in Figs. 5 – 7, are in very good agreement with the corresponding distributions obtained from numerical calculations. These distributions are of special importance and convenience for b.e.a. descriptions of various elastic and inelastic collisions involving atoms and atomic ions.

The calculations of this work have been done for the electron distributions in ground-state atoms and ions. However, the present approach can also be applied, in a straightforward way, to excited species. In particular, the expression (38) can be used directly to determine the electron velocity distributions in the outer and inner shells of excited atoms and ions.[15] The present approach can also be used for descriptions of some inelastic processes involving molecules, assuming the δ distribution for electron velocities in outer molecular shells.[16] This results from the fact that the cross sections for some of these processes (e.g., electron-impact ionization and slowing down) are rather insensitive to the shape of the electron velocity distribution in the molecular outer shell (in contrast, for example, to the cross section for electron capture, which is sensitive to the velocity distribution at high velocities of relative motion).

Some remarks should be made about comparison of modified free-fall distributions with quantum-classical and microcanonical distributions. Quantum-classical and microcanonical distributions are statistical averages over the entire set of the shell electronic orbits; thus they represent the average dynamical properties of each electron in the shell. The free-fall distribution is characterized by the average (over all the electrons in the shell) energy of the shell electron, but the electron orbit is uniquely defined as a zero angular momentum solution of the Kepler problem. This is why the free-fall distribution has nonzero value at zero electron velocity.

It should be added that the semiclassical binary encounter approximation is very suitable for Monte Carlo simulations of atomic processes because of its high computational efficiency and because Monte Carlo sampling of the parameters characterizing interactions requires well-defined collisions (in the b.e.a., all the components are *localized* species). In addition, the b.e.a. allows us to consider the interaction of the incident particle A with electrons of *individual* orbitals of the target particle B. This is particularly useful for development of new methods for plasma diagnostics and for investigation of high-energy scattering when contribution of the inner shells can be significant. Another convenience of the semiclassical b.e.a. results from the fact that the calculation procedure for atomic and molecular differential cross sections can be defined in a comprehensive and straightforward way, even for multiply differential cross sections.

Acknowledgments

This work was supported by the National Aeronautics and Space Administration, Grant NAGW - 1061, and the Air Force Office for Scientific Research, Grants 88 - 0119 and 88-0146.

References

[1]Gryzinski, M., "Two particle Collisions. I. General Relations for Collisions in the Laboratory System," Physical Review, Vol. 138, 1965, p. 305.

[2]Massey, H.S.W. and Burhop, E.S.W., Electronic and Ionic Impact Phenomena, Oxford University Press, London, 1969.

[3]Gryzinski, M. and Kunc, J.A., "Collisional Ionization and the Atomic Model," Journal of Physics B, Vol. 19, 1986, p. 2479.

[4]Kunc, J.A., "Quantum-Classical Electron Distributions in Atoms and Atomic Ions," Journal of Physics B, Vol. 21, 1988, p. 3619.

[5]Rapp, D. and Kassal, T., "The Theory of Vibrational Energy Transfer Between Simple Molecules in Nonreactive Collisions," Chemical Reviews, Vol. 69, 1969, p. 61.

[6]Bates, D.R. and Kingston, A.E., "Use of Classical Mechanics in the Treatment of Collisions Between Massive Systems," Advances in Atomic and Molecular Physics, Vol. 6, 1970, p. 269.

[7]Vriens, L., "Binary-Encounter Collision Theory," Case Studies in Atomic Collision Physics, edited by E.W. McDaniel and M.R. Mc Dowell, North-Holland, Amsterdam, 1970.

[8]Clementi, E. and Roetti, C., "Roothaan-Hartree-Fock Atomic Wave Functions," Atomic Data and Nuclear Data Tables, Vol. 14, 1974, p. 177.

[9]Gombas, P., "Die Statistische Theorie Des Atoms," Handbuch der Physik, edited by S. Flugge, Springer-Verlag, Berlin, 1956.

[10]Abrahamson, A.A., "Statistical Electron-Density Distributions and Thomas-Fermi-Dirac Screening Functions for Positive Ions with Degree of Ionization One through Four," Physical Review A, Vol. 185, 1969, p. 44.

[11]Umeda, K. and Kobayashi, S., "Systematization of the Approximate Solutions of the Thomas-Fermi Equation," Journal of Physical Society of Japan, Vol. 10, 1955, p. 749.

[12]Landau, L.D. and Lifshitz, E.M., Statistical Physics, Pergamon, Oxford, 1978.

[13]Bates, D.R. and Mapleton, R.A., "On a Classical Distribution Used in Electron Capture," Proceedings of the Physical Society, London, Vol. 85, 1965, p. 605.

[14]Gryzinski, M, "Concept of Free-Fall Multi-Electron Atomic Model," Physics Letters, Vol. 44A, 1973, p. 131.

[15]Kunc, J.A., "Electron Ionization Cross Sections of Excited Atoms and Ions," Journal of Physics B, Vol. 13, 1980, p. 587.

[16]Erwin, D.A. and Kunc, J.A., "Transport of Low- and Medium-Energy Electron and Ion Beams in Seawater and Its Vapors," Physical Review A, Vol. 38, 1988, p. 4135.

Monte Carlo Simulation of Electron Swarm in a Strong Magnetic Field

Katsuhisa Koura*
National Aerospace Laboratory, Chofu, Tokyo, Japan

Abstract

The electron swarm in rare gases (He and Ar) under the influence of the strong magnetic (and electric) field is studied using the null-collision test-particle Monte Carlo method. Comparisons with the Monte Carlo results confirm that the conventional two-term expressions for the swarm parameters derived from the Boltzmann equation in the absence of electronic excitations and ionizations are sufficiently accurate even for the strong magnetic field. Bernstein's experimental values of the reduced magnetic field strength for which the ratio of the perpendicular to transverse drift velocities is equal to 10 in the crossed electric and strong magnetic fields are larger than the Monte Carlo values, but the discrepancy is within 20% for He.

Introduction

The transport phenomena of the electron swarm in gases under the influence of the strong magnetic field (SMF) is of interest because electrons are confined in a spatially narrow range and the electron density gradient is large, which may raise some questions about the validity of the conventional two-term expressions[1] for the swarm parameters derived from the Boltzmann equation on the assumption that the electron velocity distribution function is almost isotropic and the density gradient is small and about the validity of the usual expression for the electron-molecule collision frequency.[2] The electron swarm in the SMF, however, may seldom be investigated; Bernstein[3] measured the ratio of the perpendicular ($\mathbf{E}\times\mathbf{B}$) to transverse ($-\mathbf{E}$)

*Chief of Rarefied Gas Dynamics Laboratory.

drift velocities in H_2, D_2, and He with the crossed electric (E) and strong magnetic (B) fields and obtained as a function of E/B the reduced magnetic field strength for which the ratio of drift velocities is equal to 10.

In this paper, the electron swarm in rare gases under the influence of the strong magnetic (and electric) field is studied using the null-collision test-particle Monte Carlo method.[4] The electron density is assumed to be so small in heat-bath atoms that electron-electron collisions are negligible. The electron-atom collision frequency is taken as the usual expression. The validity of the two-term expressions for the swarm parameters in the SMF is investigated for low-energy electrons in He and Ar, where electronic excitations and ionizations are negligible. Comparisons with Bernstein's experimental values of the reduced magnetic field strength are made for He, where excitations and ionizations may occur at a large value of E/B because the mean electron energy increases with increasing E/B.

Monte Carlo Method

The test-particle Monte Carlo method with the null-collision concept[5] is generally described in Ref. 4. Here, the null-collision test-particle method is extended to the ionization process. A large number of simulation (primary and ejected) electrons are followed through collisions with heat-bath (ground state) atoms during the time step Δt from time t to $t+\Delta t$:

1) All of the primary electrons are followed individually in the subsequent procedures 2-5, which are repeated until the summation of the collision time interval Δt_c given by Eq. (1) exceeds Δt in procedure 3.

2) A time interval Δt_c between successive collisions of the electron with heat-bath atoms is assigned by the probability density function

$$p(\Delta t_c) = \nu_c \exp(-\nu_c \Delta t_c) \quad (1)$$

where ν_c is the constant (between successive collisions) collision frequency given by

$$\nu_c = nS_{max} \quad (2)$$

where n is the atomic number density assumed to be constant and S_{max} is a constant (between successive collisions) defined by

$$S_{max} = \max[g\Sigma_i \sigma_i(g)], \quad g \leq g_{max} \quad (3)$$

where $g=|\mathbf{v}-\mathbf{w}|$ is the electron speed relative to the colliding atom with the velocity $\mathbf{w}$, $\mathbf{v}$ is the electron velocity, g_{max} is the effective maximum value of g between successive collisions, and $\sigma_i(g)$ is the total cross section for the collision event i (elastic, excitation, and ionization). The g_{max} is estimated as

$$g_{max} = v_{max} + w_{max} \tag{4}$$

where v_{max} is the maximum electron speed between successive collisions evaluated as $v_{max}=|\mathbf{v}|+v_E$, in which $v_E[\leqq(eE/m)\Delta t]$ is the maximum speed accelerated by the electric field during the remaining time ($\leqq\Delta t$), e and m being the electron charge and mass, respectively, and w_{max} is the effective maximum speed of heat-bath atoms evaluated as $5(2kT/M)^{1/2}$, in which k is the Boltzmann constant, T is the constant heat-bath temperature, and M is the atomic mass.

The null-collision cross section $\sigma_{null}(g)$ for which no real collision occurs is defined by

$$S_{max} = g[\Sigma_i\sigma_i(g) + \sigma_{null}(g)] \tag{5}$$

3) When the summation of Δt_c is less than Δt ($\Sigma\Delta t_c\leqq\Delta t$), the electron velocity $\mathbf{v}$ and position $\mathbf{r}$ are changed into the new velocity $\mathbf{v}'$ and position $\mathbf{r}'$, respectively, corresponding to the motion during the time interval $\Delta t_m=\Delta t_c$. The electron motion is governed by the equation of motion

$$m\, d\mathbf{v}/dt = q(\mathbf{E} + \mathbf{v}\times\mathbf{B}) \tag{6a}$$

$$d\mathbf{r}/dt = \mathbf{v} \tag{6b}$$

where q=-e. If both $\mathbf{E}$ and $\mathbf{B}$ are uniform in space and constant in time, then the analytical expressions for $\mathbf{v}'$ and $\mathbf{r}'$ are obtained:

$$\mathbf{v}' = \mathbf{v}'_{//} + \mathbf{v}'_{+} \tag{7a}$$

$$\mathbf{v}'_{//} = \mathbf{a}_{//}\Delta t_m + \mathbf{v}_{//} \tag{7b}$$

$$\mathbf{v}'_{+} = (\mathbf{v}_{+} - \mathbf{u}_{+})\cos(\omega\Delta t_m) + (\mathbf{a}_{+}/\omega)\sin(\omega\Delta t_m) + \mathbf{u}_{+} \tag{7c}$$

$$\mathbf{r}' = \mathbf{r}'_{//} + \mathbf{r}'_{+} \tag{8a}$$

$$\mathbf{r}'_{//} = (\mathbf{a}_{//}/2)(\Delta t_m)^2 + \mathbf{v}_{//}\Delta t_m + \mathbf{r}_{//} \tag{8b}$$

$$\mathbf{r}'_{+} = (1/\omega)\{(\mathbf{v}_{+} - \mathbf{u}_{+})\sin(\omega\Delta t_m) + (\mathbf{a}_{+}/\omega)[1 - \cos(\omega\Delta t_m)]\}$$
$$+ \mathbf{u}_{+}\Delta t_m + \mathbf{r}_{+} \tag{8c}$$

where the subscripts // and + denote the vector components parallel and perpendicular to $\mathbf{B}$, respectively, $\mathbf{a}_{//}=q\mathbf{E}_{//}/m$ and $\mathbf{a}_{+}=(q/m)(\mathbf{E}_{+}+\mathbf{v}_{+}\times\mathbf{B})$ are the accelerations, $\mathbf{u}_{+}=\mathbf{E}_{+}\times\mathbf{B}/B^2$ is the electric drift velocity, and $\omega=qB/m$ is the cyclotron frequency.

When the summation of Δt_c exceeds Δt ($\Sigma\Delta t_c>\Delta t$), $\mathbf{v}$ and $\mathbf{r}$ of the electron are changed into the new values $\mathbf{v}'$ and $\mathbf{r}'$, respectively, corresponding to the motion during the remaining time $\Delta t_m=\Delta t_c-(\Sigma\Delta t_c-\Delta t)$ and the simulation procedure for the electron ends.

4) The velocity $\mathbf{w}$ of an atomic collision partner is chosen from the atomic velocity distribution function assumed to be the Maxwellian distribution at the heat-bath temperature T.

5) The collision event i, which includes the null collision, is assigned by the probability

$$p_i = g\sigma_i(g)/S_{max} \tag{9}$$

If the event i is the null collision, then no (real) collision occurs. If the event i is the elastic or excitation collision, then the electron velocity after collision is calculated from the momentum and energy conservation by assigning a scattering solid angle from the differential cross section. If the event i is the ionization, then the velocities of ejected electrons are calculated by assigning their energies and scattering solid angles from the differential ionization cross section. The velocity after collision of the primary electron is then calculated from the energy conservation by assigning the scattering solid angle, where the kinetic energies of the colliding atom and resultant ion are neglected because of the low atomic (heat-bath) energy and large ion mass and the excited states of ion are also neglected. The time t' when the ejected electrons are generated is memorized for the following procedure of ejected electrons.

6) Each of the ejected electrons, which are regarded as a primary electron this time, is followed in the same procedures 2-5 until the summation of Δt_c exceeds the remaining time $\Delta t'=\Delta t-(t'-t)$ in procedure 3, where $\Delta t'$ is substituted for Δt.

7) When all of the successively generated electrons have been followed in procedures 2-6, the simulation procedure during Δt ends.

Cross Sections

It is assumed that all of the scatterings of electrons by atoms are isotropic. The elastic cross section for Ar is taken as the data of the momentum-transfer cross section $\sigma_m(\varepsilon)$ measured by Milloy et al.[6] in the range of electron energy $0 \leqq \varepsilon \leqq 4$ eV and recommended by Hayashi ($5 \leqq \varepsilon \leqq 10^4$ eV).[7] The elastic cross section for He is taken as the data of $\sigma_m(\varepsilon)$ measured by Crompton et al. ($0 \leqq \varepsilon \leqq 3$ eV),[8] Milloy and Crompton ($4 \leqq \varepsilon \leqq 12$ eV),[9] and Register et al. ($20 \leqq \varepsilon \leqq 200$ eV),[10] and recommended by Hayashi ($250 \leqq \varepsilon \leqq 10^4$ eV).[7] Because excitations and ionizations are taken into account for He, it is assumed that multiple ionizations and excited ion states are negligible and the singlet or triplet excitations are collectively treated. The set of ionization and singlet or triplet excitation cross sections is taken as the data evaluated by de Heer and Jansen ($\varepsilon \leqq 4$ KeV)[11] together with the threshold energies for the ionization (I=24.59 eV) and singlet (20.61 eV) or triplet (19.82 eV) excitation. The differential ionization cross section (ejected electron spectrum) $\sigma(\varepsilon_p, \varepsilon_s)$, from which the ejected (secondary) electron energy ε_s is chosen for the given incident (primary) electron energy ε_p, is taken as the model form given by Opal et al.,[12]

$$\sigma(\varepsilon_p, \varepsilon_s) = C(\varepsilon_p)/[1 + (\varepsilon_s/E_s)^{\alpha}], \quad \varepsilon_s \leqq (\varepsilon_p - I)/2 \tag{10}$$

with the exponent α=2.1, where $C(\varepsilon_p)$ is the normalization constant and the shape parameter E_s is taken to be 15.8 eV for He. It is assumed for simplicity that the secondary electron is generated isotropically.

Results and Discussion

In the Monte Carlo calculation, the cross sections are approximated by the linear interpolation between the data points vs ε and extrapolation to $\varepsilon \to \infty$. The ε is taken as the relative electron energy $\varepsilon = (\mu/2)g^2 [\cong (m/2)v^2]$, where $\mu = mM/(m+M)$ is the reduced mass. The velocity $\mathbf{v}$, position vector $\mathbf{r}$=(x, y, z), and time t are normalized by the (heat-bath) electron speed $v_0 = (2kT/m)^{1/2}$, free path $\lambda_0 = (n\sigma_0)^{-1}$, and collision time $t_0 = \lambda_0/v_0$, respectively, where σ_0 is the unit of cross section taken as 1 Å^2. The number of simulation electrons is taken as $N = 10^4$.

Strong Magnetic Field in He and Ar

In order to investigate the validity of the two-term expression for the diffusion coefficient in the SMF, the electron swarm in the He or Ar heat bath under the influence of the constant magnetic field $\mathbf{B}$ (directed along the x axis) is simulated. The initial (t=0) electron positions and velocity distribution function are taken as the origin of coordinates x=y=z=0 and the Maxwellian distribution at the heat-bath temperature T=300°K, respectively. The gas pressure p is taken as 100 Torr. Because the electron energy is low, neither excitations nor ionizations occur.

The two-term expressions for the parallel ($D_{//}$) and perpendicular (D_{+}) diffusion coefficients are given by[1]

$$D_{//} = \langle v^2/3\nu_m \rangle \tag{11}$$

$$D_{+} = \langle \nu_m v^2/3(\nu_m{}^2 + \omega^2) \rangle \tag{12}$$

where $\nu_m = nv\sigma_m(v)$ is the usual expression for the momentum-transfer collision frequency and $\langle\phi\rangle$ denotes the average value of ϕ over the overall electron velocity distribution function, which remains the initial Maxwellian distribution because the electric field is absent; the mean electron energy $\langle\varepsilon\rangle$ also remains the initial electron energy (3/2)kT.

The parallel $[\langle(x-\langle x\rangle)^2\rangle/\lambda_0{}^2]$ and perpendicular $[\langle(y-\langle y\rangle)^2+(z-\langle z\rangle)^2\rangle/\lambda_0{}^2]$ mean-square displacements for $0 \leqq B \leqq 1$ megagauss (MG) are, respectively, compared with the

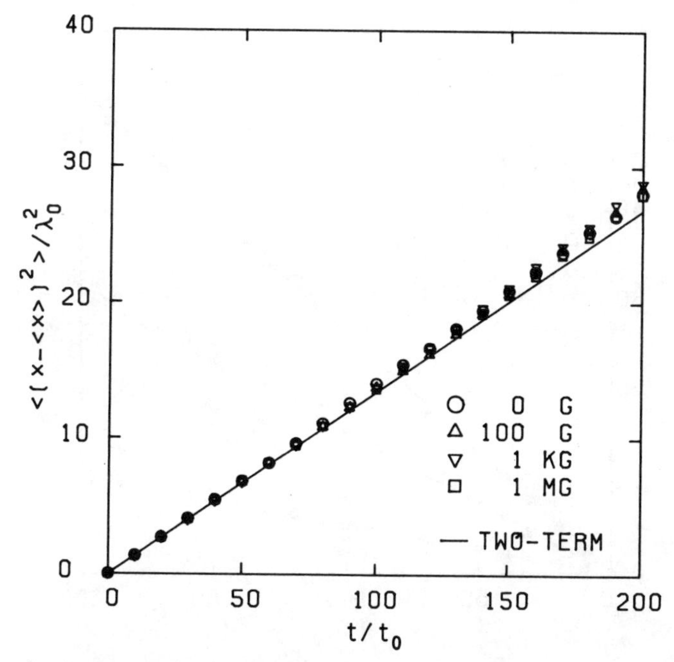

Fig. 1 Comparison of parallel mean-square displacement in He for $0 \leqq B \leqq 1$ MG between the Monte Carlo and two-term results.

two-term results in Figs. 1 and 2 for He and in Figs. 3 and 4 for Ar, where the parallel and perpendicular two-term results are indicated by the solid lines with the gradients corresponding to the parallel ($2D_{//}$) and perpendicular ($4D_{+}$) diffusion coefficients, respectively. The discrepancy between the Monte Carlo and two-term results observed particularly for the parallel displacement at $t/t_0>100$ and 50 for He and Ar, respectively, is considered to be statistical errors, because the parallel displacement is, in principle, independent of B. The gradient of the perpendicular displacement decreases with increasing B, because the perpendicular displacement is more strongly restricted by the stronger magnetic field ($D_{+}\propto B^{-2}$ for $\omega^2>>\nu_m{}^2$). The Monte Carlo results are in agreement with the two-term solution within statistical errors and, therefore, confirm that the two-term expressions are sufficiently accurate even for the SMF [$\omega^2>>\nu_m{}^2$; B>1 kilogauss (kG)].

Crossed Electric and Strong Magnetic Fields in He

Compared with the experimental results of Bernstein,[3] the electron swarm in the He heat bath at T=293°K and p=1

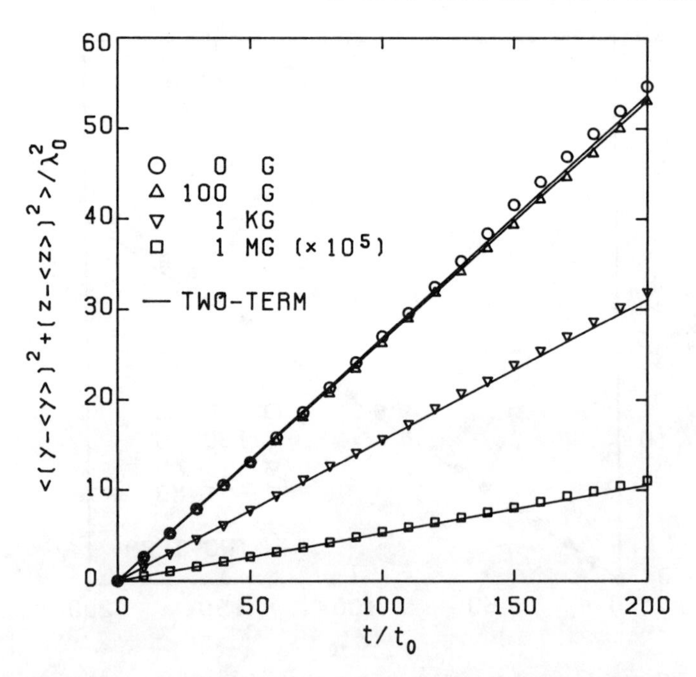

Fig. 2 Comparison of perpendicular mean-square displacement in He for 0 ≤ B ≤ 1 MG between the Monte Carlo and two-term results.

Torr under the influence of the crossed constant electric (directed along the z axis) and magnetic (directed along the x axis) fields in the range of strength ratio $0.5\times10^6 \leqq E/B \leqq 10\times10^6$ cm/s is simulated. The initial (t=0) electron positions and velocity distribution function are, respectively, taken as the origin of coordinates x=y=z=0 and the isotropic δ function distribution with the electron speed corresponding to nearly steady-state mean electron energy (estimated from the two-term value).

In the SMF limit ($\omega^2 >> \nu_m{}^2$), the two-term expressions for the perpendicular (v_+) and transverse (v_T) drift velocities in the absence of electronic excitations and ionizations are given by[1,3]

$$v_+ = E/B \tag{13}$$

$$v_T = (E/B)\langle\nu_m\varepsilon\rangle/(\langle\varepsilon\rangle\omega) \tag{14}$$

where $\langle\phi\rangle$ denotes the average value of ϕ over the overall electron velocity distribution function, which is the Maxwellian distribution at the electron temperature $T_e = 2\langle\varepsilon\rangle/3k$, and $\langle\varepsilon\rangle$ is the mean electron energy given by $\langle\varepsilon\rangle = (M/2)(E/B)^2 + 3kT/2$.

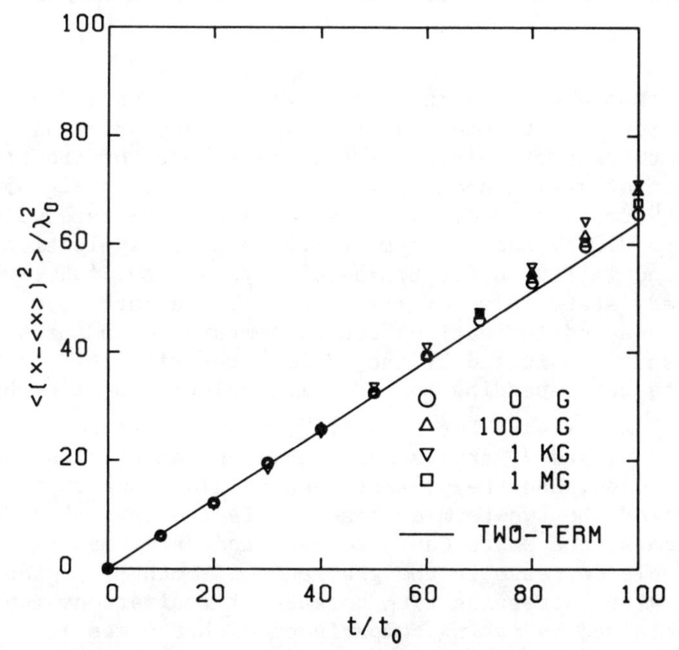

Fig. 3 Comparison of parallel mean-square displacement in Ar for $0 \leqq B \leqq 1$ MG between the Monte Carlo and two-term results.

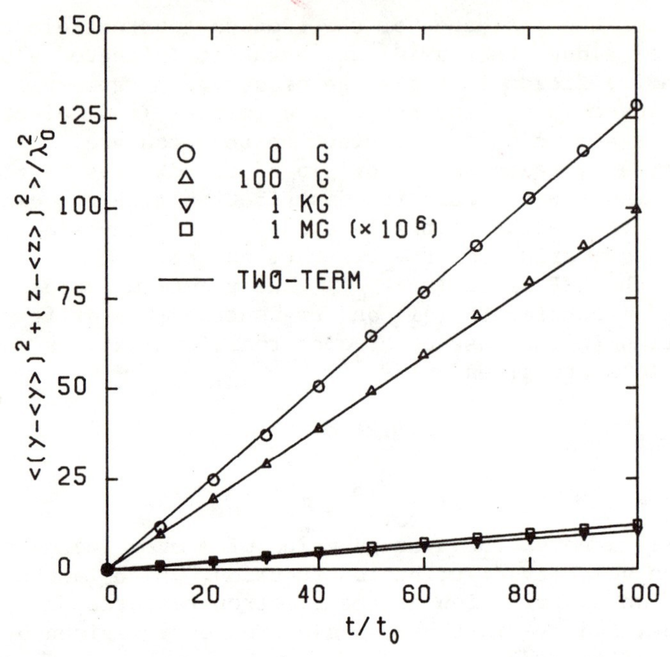

Fig. 4 Comparison of perpendicular mean-square displacement in Ar for 0 ≦ B ≦ 1 MG between the Monte Carlo and two-term results.

The Monte Carlo values of v_+ and v_T are obtained from the gradients of the mean drift displacements $\langle y\rangle$ and $-\langle z\rangle$ vs time t, respectively. The typical results of the time evolution of $\langle y\rangle/\lambda_0$ and $-\langle z\rangle/\lambda_0$ are presented in Fig. 5 for $E/B=1\times10^6$ cm/s and B=1.2 kG, where the steady-state value of $\langle\varepsilon\rangle$ is 2.0 eV and neither excitations nor ionizations occur, and in Fig. 6 for $E/B=5\times10^6$ cm/s and B=1.4 kG, where the steady-state value of $\langle\varepsilon\rangle$ is 8.4 eV and both excitations and ionizations occur, compared with the two-term results indicated by the solid lines with the gradients corresponding to the drift velocities given by Eqs. (13) and (14). For $E/B=1\times10^6$ cm/s, the Monte Carlo values of v_+ and v_T are in reasonable agreement with the two-term ones, and the present results also support the accuracy of the two-term expressions in the SMF. For E/B $=5\times10^6$ cm/s, the Monte Carlo results indicate the appreciable decrease in the gradients of both $\langle y\rangle/\lambda_0$ and $-\langle z\rangle/\lambda_0$ with increasing t/t_0 because of ionizations (which is ascertained by making comparisons with the results obtained by neglecting excitations or ionizations) and deviate significantly from the two-term results.

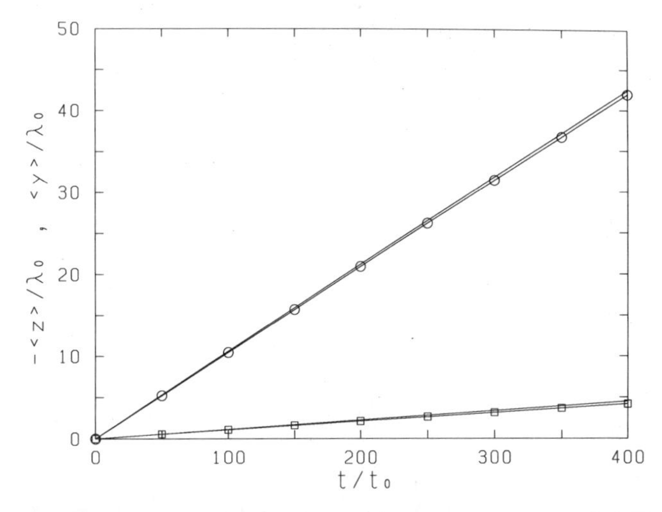

Fig. 5 Comparison of perpendicular and transverse mean drift displacements in He for E/B = 1 × 10^6 cm/s and B = 1.2 kG; circles, Monte Carlo ($\langle y\rangle/\lambda_0$); squares, Monte Carlo ($-\langle z\rangle/\lambda_0$); solid lines, two-term.

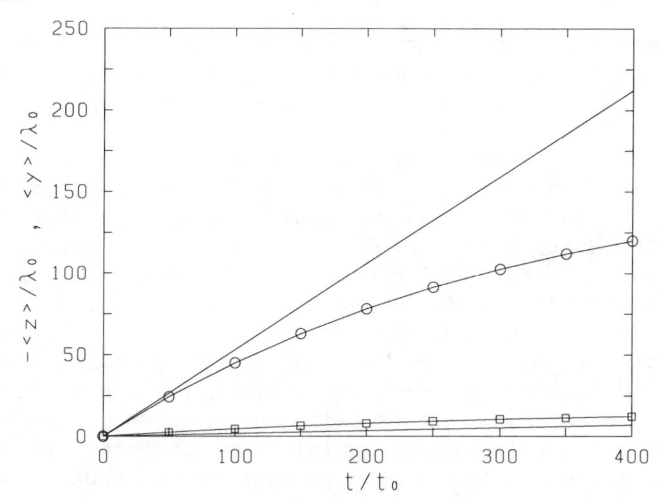

Fig. 6 Comparison of perpendicular and transverse mean drift displacements in He for E/B = 5 × 10^6 cm/s and B = 1.4 kG; circles, Monte Carlo ($\langle y\rangle/\lambda_0$); squares, Monte Carlo ($-\langle z\rangle/\lambda_0$); solid lines, two-term.

The Monte Carlo values of the magnetic field strength B for $v_+/v_T=10$ are compared with Bernstein's experimental ones in Fig. 7, where the uncertainty in the Monte Carlo values is represented by the width of circle. The Monte Carlo values are obtained after a sufficiently long time ($t/t_0 \gtrsim 300$), when the influences of the initial state are

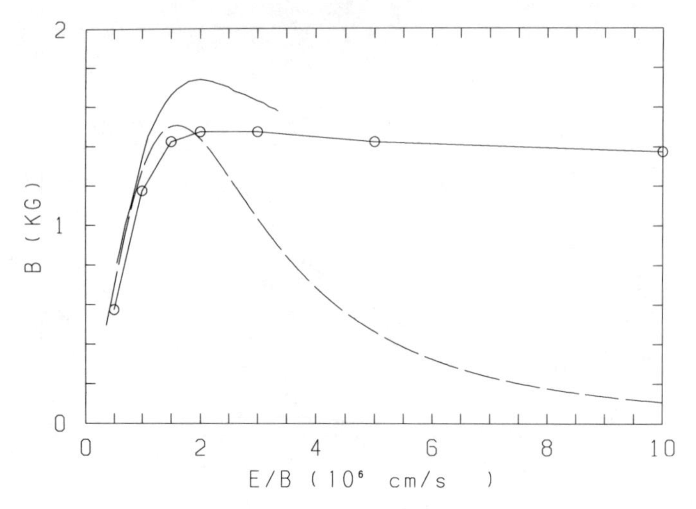

Fig. 7 Comparison of magnetic field strength for v_+/v_T = 10 in He; circles, Monte Carlo; solid line, experiment (Ref. 3); dashed line, two-term.

considered to disappear. The Monte Carlo values are smaller than the experimental ones, but the discrepancy is within 20%. Although the real cause of the discrepancy is not found out, it should be pointed out that the experimental value of v_T is somewhat larger than the true value because of the transverse diffusion and suffers from molecular impurities as was indicated by Bernstein, and that the Monte Carlo results are obtained by neglecting the electron-electron collisions and employing the usual expression for the collision frequency. The two-term results obtained from Eqs. (13) and (14) are also shown for comparison. At the low value of $E/B<2\times10^6$ cm/s, where $<\varepsilon><$ 6 eV and both excitations and ionizations are negligible, the Monte Carlo, experimental, and two-term values are in reasonable agreement and rapidly increase with the increase in E/B corresponding to the increasing function ν_m of the low ε. At $E/B\gtrsim3\times10^6$ cm/s, where $<\varepsilon>>7$ eV and both excitations and ionizations occur, the Monte Carlo value depends very weakly on E/B, but the two-term value remarkably decreases with the increase in E/B, corresponding to the decreasing function ν_m of the high ε, and is much smaller than the Monte Carlo value.

Concluding Remarks

The null-collision test-particle Monte Carlo method is shown to be applicable to the simulation of electrons undergoing rather complicated molecular processes such as electronic excitations and ionizations under the influence of electric and magnetic fields. Comparisons with the Monte Carlo results confirm that the conventional two-term expressions for the electron swarm parameters in rare gases derived from the Boltzmann equation in the absence of excitations and ionizations are sufficiently accurate even for the strong magnetic field. Bernstein's experimental values of the reduced magnetic field strength for which the ratio of perpendicular to transverse drift velocities is equal to 10 in the crossed electric and strong magnetic fields are larger than the Monte Carlo values, but the discrepancy is within 20% for He. In this study, the usual expression for the electron-atom collision frequency is employed, but the validity of the usual expression in the strong magnetic field should be verified.

References

[1] Huxley, L. G. H. and Crompton, R. W., The Diffusion and Drift of Electrons in Gases, Wiley, New York, 1974.

[2] Imazu, S., "Collision Frequency of Charged Particles in a Weakly Ionized Gas in a Strong Magnetic Field," Physical Review A, Vol. 23, May 1981, pp. 2644-2649; "Collision Frequencies between Charged and Neutral Particles in a Magnetic Field," Journal of Applied Physics, Vol. 57, March 1985, pp. 1602-1608.

[3] Bernstein, M. J., "Electron Drift and Diffusion Measurements in H_2 and D_2 with Crossed Electric and Strong Magnetic Fields," Physical Review, Vol. 127, July 1962, pp. 335-341.

[4] Koura, K., "Monte Carlo Simulation of Electron Thermalization in Gases. VIII. Thermalization Distance and Microwave Conductivity in Rare Gases," Journal of Chemical Physics, Vol. 87, Dec. 1987, pp. 6481-6487.

[5] Skullerud, H. R., "The Stochastic Computer Simulation of Ion Motion in a Gas Subjected to a Constant Electric Field," Journal of Physics D, Vol. 1, Nov. 1968, pp. 1567-1568.

[6] Milloy, H. B., Crompton, R. W., Rees, J. A., and Robertson, A. G., "The Momentum Transfer Cross Section for Electrons in Argon in the Energy Range 0-4 eV," Australian Journal of Physics, Vol. 30, Feb. 1977, pp. 61-72.

[7] Hayashi, M., "Recommended Values of Transport Cross Sections for Elastic Collision and Total Collision Cross Section for Electrons in Atomic and Molecular Gases," Institute of Plasma Physics, Nagoya University, Nagoya, Rept. IPPJ-AM-19, Nov. 1981.

[8] Crompton, R. W., Elford, M. T., and Robertson, A. G., "The Momentum Transfer Cross Section for Electrons in Helium Derived

from Drift Velocities at 77°K," Australian Journal of Physics, Vol. 23, Oct. 1970, pp. 667-681.

[9]Milloy, H. B. and Crompton, R. W., "Momentum-Transfer Cross Section for Electron-Helium Collisions in the Range 4-12 eV," Physical Review A, Vol. 15, May 1977, pp. 1847-1850.

[10]Register, D. F., Trajmar, S., and Srivastava, S. K., "Absolute Elastic Differential Electron Scattering Cross Sections for He: A Proposed Calibration Standard from 5 to 200 eV," Physical Review A, Vol. 21, April 1980, pp. 1134-1151.

[11]de Heer, F. J. and Jansen, R. H. J., "Total Cross Sections for Electron Scattering by He," Journal of Physics B, Vol. 10, No. 18 1977, pp. 3741-3758.

[12]Opal, C. B., Peterson, W. K., and Beaty, E. C., "Measurements of Secondary-Electron Spectra Produced by Electron Impact Ionization of a Number of Simple Gases," Journal of Chemical Physics, Vol. 55, Oct. 1971, pp. 4100-4106.

Collisional Transport in Magnetoplasmas in the Presence of Differential Rotation

Massimo Tessarotto*
Università degli Studi di Trieste, Trieste, Italy
and
Peter J. Catto†
Lodestar Research Corporation, Boulder, Colorado

Abstract

The aim of this paper is to propose a generalization of the treatment of collisional transport theory in a strongly toroidally rotating magnetoplasma recently developed by Catto et al. ("Ion transport in toroidally rotating Tokomak discharges", Physics of Fluids, Vol.30, 1987, pp.2784-), in order to allow the theoretical description of the decay of relative rotation in an axisymmetric plasma. By investigating the macroscopic transport equations describing the angular momentum and the parallel current density balance, the problem of determination of the divergenceless inductive electric self-field ($\vec{E}^{rot}$) is addressed. It is found that, although $\vec{E}^{rot}$ in a low-beta plasma does not influence transport directly, being of higher order in the Larmor radius, its determination is actually relevant for closing the transport equations. In particular it is pointed out that $\vec{E}^{rot}$ can always be chosen in such a way to produce a stationary current density vector. However, since such a constraint is not authomatically fulfilled by the kinetic distribution function, it is found that a generalization of the strong rotation drift-kinetic equation becomes necessary in order to describe first-order velocity perturbations of the equilibrium distribution function.

---

*Associate Professor in Mathematical Physics, Dipartimento di Scienze Matematiche.

†Research Plasma Physicist, Member of Professional Staff.

Introduction

It is well known that the current kinetic transport theories are unable to account for electron transport in weakly collisional magnetically confined plasmas, such as in Tokamaks, which results in unexpectedly large fluxes even in apparently quiescent plasmas,in contrast to theoretical predictions based on kinetic theory[1-3].

In reference to this problem, we want to point out the possible influence on enhanced electron transport produced by purely collisional effects on a quiescent magnetoplasma, related to collisional relaxation and diffusion of angular momentum in weakly collisional systems.

We recall that, in customary transport theory[1-3] it is assumed that collisions produce a rapid relaxation in the plasma differential (toroidal) rotation and consequently that rotation speed is small.

Recent theoretical work of Hinton and Wong[4],and Catto et al.[5], have systematically extended the theory to strong toroidal rotation speeds (i.e. comparable to the thermal speed of some particle species), showing that the inclusion of such effects in the model may yield more acccurate predictions as far as the ion transport is concerned.

In the present paper, instead, by an extension of the model developed in Ref.5, we address the problem of relaxation of plasma relative toroidal rotation (i.e. of the relative rotation speeds of the various particle species present).

We find that a correct description of this relaxation process requires a generalization of the drift-kinetic equation in order to include first order velocity perturbations with respect to a Larmor radius expansion. Inspection of appropriate moment equations shows that in a transport regime first-order velocity perturbations are uniquely determined in terms of the relevant thermodynamics forces and that they decay on a time scale comparable to that of customary collisional transport.

Moment Equations and Constraints

In order to discuss transport it is convenient to investigate, first of all, some general consequences stemming from appropriate moment equations of the Fokker-Planck kinetic equation in the presence of spatial symmetry (i.e., an axisymmetric torus) and taking into account a transport ordering appropriate to describe a magnetically confined plasma in the presence of strong drifts, i.e. in the presence of toroidal rotation speeds comparable to the ion thermal velocity (see e.g.,Refs.4 and 5).

It sufficies for this purpose to take into account the moment equations corresponding to $X_s=1, M_s R\vec{v}\cdot\vec{\theta}$ where R is the distance from the principal axis of the torus and $\vec{\theta}$ the toroidal unit vector. By denoting, therefore, $N_s=\int d^3v f_s$ the number density, and $P_s=M_s\Omega_s R^2 N_s= R\int d^3v M_s\vec{v}\cdot\vec{\theta} f_s$ the toroidal angular momentum, with $f_s=f_s(\vec{r},\vec{v},t)$ the kinetic distribution function, we obtain the moment equations of continuity and toroidal angular momentum balance, i.e. respectively:

$$<\frac{\partial}{\partial t}N_s> + (V')^{-1}(\frac{\partial}{\partial\psi}V'\Gamma_s)=0 \qquad (1)$$

$$<\frac{\partial}{\partial t}(P_s)> +(V')^{-1}(\frac{\partial}{\partial\psi}V'L_s)= S_s \qquad (2)$$

with $\Gamma_s=<\int d^3v\vec{v}\cdot\nabla\psi f_s>$ the particle flux and

$$L_s=<\int d^3v M_s R\vec{\theta}\cdot\vec{v}\vec{v}\cdot\nabla\psi f_s>$$

the angular momentum flux, S_s denoting the "source" term:

$$S_s=<\int d^3v M_s R\vec{v}\cdot\vec{\theta}C_s> - \frac{Z_s e}{c}[\ \Gamma_s - c<N_s R\vec{\theta}\cdot\vec{E}>\ ] \qquad (3)$$

and $<X>$ the surface average

$$<X>=(V')^{-1}\oint d\chi X(\vec{B}\cdot\nabla\chi)^{-1}$$

with

$$V'=\oint d\chi(\vec{B}\cdot\nabla\chi)^{-1}$$

Eq. (2) yields then immediately the transport equations for the flux averaged total toroidal angular momentum i.e., $<P>=<\Sigma_s P_s>$,and the toroidal current density

$$J_T= \Sigma_s Z_z e\Omega_s R N_s$$

In fact summing Eq. (2) over species and using Ampere's law delivers

$$<\frac{\partial}{\partial t}\{P+ R^2B_p^2\Sigma_s\Omega_s/4\pi c^2\}> + (V')^{-1}\frac{\partial}{\partial\psi}(V'\{\Sigma_s L_s + - <\nabla\psi.\vec{E}\vec{E}.\vec{\theta}R>/4\pi\}>)=0 \tag{4}$$

whereas multiplying Eq. (2) by $Z_z e/M_s$ and summing over species yields

$$<\frac{\partial}{\partial t}J_T R> + (V')^{-1}\frac{\partial}{\partial\psi}(V'\Sigma_s Z_s eL_s/M_s) =\Sigma_s Z_s eS_s/M_s \tag{5}$$

Using the Faraday equation, neglecting the displacement current and invoking again Ampere's law yields, moreover

$$\frac{\partial}{\partial t}\vec{J} =- \frac{c^2}{4\pi}\nabla x(\nabla x\vec{E}^{rot}) \tag{6}$$

where $\vec{E}^{rot}$ is the divergenceless part of the electric self-field. Notice that electric field $\vec{E}$ is assumed of the form

$$\vec{E}= E_T\vec{\theta} + \vec{E}^{rot}-\nabla\Phi \tag{7}$$

with E_T an externally produced inductive electric field and Φ the electrostatic potential. Equations (4) and (5) can be simplified recalling that fluxes Γ_s and L_s, to leading order with respect to a Larmor radius expansion, read[5]

$$\Gamma_s= c<N_s R\vec{\theta}.\vec{E}>+ \frac{c}{Z_s e}<\int d^3vM_s R\vec{v}.\vec{\theta}C_s\{f-f_M\}> + O(\epsilon_s^3) \tag{8}$$

$$L_s= cM_s<N_s R^3\vec{\theta}.\vec{E}>+ \frac{c}{Z_s e}<\int d^3v+(M_s R\vec{v}.\vec{\theta})^2C_s\{f-f_M\}> + O(\epsilon_s^3) \tag{9}$$

Here ϵ=max$\{\epsilon_s$,s=1,n$\}$, $\epsilon_s=r_s/L<<1$, r_s is the Larmor radius, n are the particle species and to leading order $N_s \cong N_s+O(\epsilon_s)$.

Invoking plasma quasineutrality through second order, i.e.:

$$\Sigma_s Z_s e N_s = 0 + O(\epsilon^3) \tag{10}$$

and recalling that it is possible to select the kinetic distribution function f_s, so that

$$\frac{\partial}{\partial t}\tilde{f}_s = 0 \;; \qquad \frac{\partial}{\partial t}\bar{f}_s \sim O(\epsilon_s^3) \tag{11}$$

(where $\tilde{f}_s = f_s - \bar{f}_s$ and $\bar{f}_s$ is the gyrophase average of f_s), it follows from Equations (4) and (5)

$$P \sim O(\epsilon^0) \qquad \langle \frac{\partial}{\partial t} P \rangle \sim O(\epsilon^3)$$
$$J_T \sim O(\epsilon) \tag{12}$$

whereas

$$S_s \sim (V')^{-1} \frac{\partial}{\partial \psi}(V' L_s) \sim O(\epsilon_s^3) \tag{13}$$

Equation (12) implies, in particular, that in order evaluate S_s it is necessary to determine the kinetic distribution function $f_s(\vec{r},\vec{v},t)$ up to second order in $O(\epsilon_s)$ (i.e., approximating $f_s \cong f_{o,s} + \epsilon_s f_{1,s} + \epsilon_s^2 f_{2,s}$). On the contrary, L_s (see Eq. (9)) and J_T can be safely determined neglecting second order corrections.

In fact, expanding the angular velocity Ω_s in power series with respect to ϵ, setting

$$\Omega_s = \Omega_{o,s} + \epsilon_s \Omega_{1,s} + O(\epsilon^2) \tag{14}$$

it can be proven[5] that to lowest order Ω_s must be species independent, i.e. $\Omega_{o,s} = \omega_o$. Thus considering the case of a two-species plasma, J_T results in

$$J_T = [R \Delta v_{ei} Z_i e N_i](1 + O(\epsilon)) \tag{15}$$

where Δv_{ie} is the relative toroidal velocity defined as

$$\Delta v_{ie} = R(\Omega_{1,i} - \Omega_{1,e}) \tag{16}$$

Equation (5) can be viewed as a constraint f_s to be fulfilled by selecting appropriately the parallel component of the electric self-field, i.e. $\vec{b}.\vec{E}^{rot}$ with $\vec{b}=\vec{B}/B$, a condition that evidently determines non uniquely $\vec{E}^{rot}$. The remaining components of $\vec{E}^{rot}$ are then given by Eq.(6).

In fact, expanding $\vec{E}^{rot}$ in power series of ϵ ($\vec{E}^{rot}= =\vec{E}_0^{rot}+\vec{E}_1^{rot}+ \epsilon\vec{E}_2^{rot} + +\dots$), results in $\vec{E}_0^{rot}=0$, since, by assumption, $\frac{\partial}{\partial t}\bar{f}_s\sim O(\epsilon_s^3)$; $\vec{E}_1^{rot}$ and $\vec{b}\cdot\vec{E}_2^{rot}$ are determined by $\frac{\partial}{\partial t}J_{\parallel}\Big|_3$, and the remaining components of $\vec{E}_2^{rot}$ are given by $\frac{\partial}{\partial t}J_{\parallel}\Big|_4$, etc.

Notice that $\vec{E}_1^{rot}$ results are therefore determined by $\vec{E}_2^{rot}$, and in turn $\vec{E}_2^{rot}$ by $\vec{E}_3^{rot}$, etc.- In particular, it can be proven that $\vec{b}\cdot\vec{E}_2^{rot}$ can always be chosen in such a way that

$$\frac{\partial}{\partial t}\vec{J}=0 \tag{17}$$

and

$$\vec{E}_1^{rot}=0 \tag{18}$$

a requirement consistent with the assumption of a low-β plasma. Denoting

$$J_{\parallel}= \Sigma_s Z_s e\int d^3v\vec{v}.\vec{b}f_s=\Sigma_s Z_s e\int d^3v\vec{v}.\vec{b}\bar{f}_s$$

Eqs. (11) and (17) imply:

$$\frac{\partial}{\partial t}J_{\parallel}= \Sigma_s Z_s e\int d^3v\vec{v}\frac{\partial}{\partial t}\bar{f}_s = \Sigma_s Z_s e\int d^3v\vec{v}\frac{\partial}{\partial t}\bar{f}_s=0 \tag{19}$$

and therefore also:

$$\frac{\partial}{\partial t}J_T=0 \tag{20}$$

whereas the right hand side of Eq. (7) and Eq. (22) yield two additional constraint equations for J_T and Γ_s. This conclusion represents a generalization to the case of an axisymmetric plasma in the presence of strong rotation of an analogous result obtained in the case of a non-rotating plasma[6].

In particular, we find that the first constraint delivers generally an integral expression for the electron particle flux Γ_e, valid through third order in $O(\epsilon)$, i.e.:

$$\Gamma_e= c\langle N_e\vec{R}\vec{\theta}.\vec{E}\rangle+ \frac{c}{e}\langle\int d^3vM_e\vec{R}\vec{v}.\theta C_e\{f-f_M\}\rangle + \\ + \frac{c}{e}(V')^{-1}\frac{\partial}{\partial\psi}(V'\Sigma_s Z_s eL_s/M_s) + O(\epsilon_e^4) \tag{21}$$

On the other hand, in view of Eq. (15), J_T results a linear function of the thermodynamic forces to leading order in ϵ, and moreover $\frac{\partial}{\partial t}J_T\cong\frac{\partial}{\partial t}J_{T,1}(1 + O(\epsilon))$. Thus, Eq. (20) yields a constraint among the thermodynamic forces to be fulfilled in addition to the relevant transport equations. In customary kinetic theories[1-5], however, such a constraint cannot generally be fulfilled since the thermodynamic forces

$$A_{0,s}=\frac{Z_se}{T_s}(E+\nabla\Phi).\vec{\theta};\quad A_{1,s}=\frac{M_sc}{Z_se}[\frac{\partial}{\partial\psi}\ln N_s - \frac{3\partial}{2\partial\psi}\ln T_s]; \\ A_{2,s}=\frac{M_sc}{Z_seT_s}\frac{\partial}{\partial\psi}\ln T_s;\quad A_{3,s}=\frac{M_s^2c}{2Z_seT_s}\frac{\partial}{\partial\psi}\ln\omega_0; \tag{22}$$

are uniquely determined by the transport equations.

In the sequel we intend to propose a possible solution to this problem, pointing out, for this purpose, an extension of the kinetic approach earlier developed by Catto et al.[5],in order to permit the description of first-order velocity perturbations. In fact, perturbations of this type are expected to contribute to the current density J_T, thus allowing, in principle, the fulfilment of the previous constraint on $\partial J_T/\partial t$.

Derivation of a Drift-Kinetic Equation for a Strongly Rotating Plasma

In an axisymmetric magnetic confinement configuration it may be shown (see Ref.5) that to lowest order in ϵ the kinetic distribution function for each particle species s must be a toroidally rotating Maxwellian of the form

$$f_{M,s}=N_{o,s}(\psi,t)(M_s/2\pi T_s)^{3/2}\exp[-H_s/T_s] \tag{23}$$

where

$$H_s= E -\omega_s P_s + \frac{Z_s e}{c}(\omega_s\psi - \int_{\psi_0}^{\psi} d\psi\omega_s) \tag{24}$$

is the energy in a frame rotating with angular velocity ω_s,

$$E=+M_s v^2+Z_s e$$

is the energy in the laboratory frame and

$$P_s=\frac{Z_s e}{c}\psi +M_s R\vec{v}\cdot\vec{\theta}$$

is the conserved canonical momentum (conjugate to θ), moreover, $\vec{V}_s=\omega_s R\vec{\theta}$ is a perturbed toroidal velocity (i.e., the mass flow velocity carried by $f_{M,s}$), with R denoting the distance from the principal axis. The ω_s result constant on a given magnetic surface and to lowest order in ϵ also species independent; i.e. expanding ω_s in a formal power series of ϵ

$$\Omega_s=\sum_{i=0}^{\infty}\epsilon^i\Omega_{i,s}$$

results in

$$\Omega_{o,s}=\omega_o=c\frac{\partial\langle\Phi\rangle}{\partial\psi}$$

In order to derive a drift-kinetic equation accurate through order $O(\epsilon)$, we adopt here an ansatz analogous to that introduced in Ref. 5. Thus we define the following transformation for the kinetic distribution function

$f_s(\vec{r},\vec{v},t)$:

$$f_s(\vec{r},\vec{v},t) = f_{*,s}(\psi_{*,s},H_{*,s},t) + h_s(\vec{r},H_s,\mu,\varphi,t) \qquad (25)$$

where $f_{*,s}$ is the auxiliary distribution function

$$f_{*,s} = N_{*,s}(M_s/2\pi T_{*,s})^{3/2}\exp[-H_{*,s}/T_{*,s}] \qquad (26)$$

with the notations

$$N_{*,s}=N_{0,s}(\psi_{*,s},t)$$
$$T_{*,s}=T_s(\psi_{*,s},t) \qquad (27)$$
$$\psi_{*,s}=cP_s/Z_s e$$

and moreover

$$H_{*,s}=E-(Z_s e/c)\int_{\psi_0}^{\psi_{*,s}} d\psi\omega_s. \qquad (28)$$

In Eq. (25) $\mu=M_s w/2B$, is the magnetic moment per unit mass, with $\vec{w}=|\vec{b}x(\vec{v}-\vec{V})|$, φ the gyrophase and $\vec{V}=\omega_0 R\vec{\theta}$. We remark that the function $f_{*,s}$ differs from $f_{M,s}$ by terms of order $O(\epsilon)$ at most, while through first order h_s is gyrophase independent.

Hence assuming

$$|\Omega_{c,s}^{-1}\frac{\partial}{\partial t}\ln f_s|\sim O(\epsilon^3)$$

(with $\Omega_{c,s}$ the Larmor frequency), as appropriate to transport theory, the Fokker-Planck kinetic equation delivers to first order in ϵ, upon taking its gyrophase-average

$$u\vec{b}.\nabla h_s = \langle C_s\{f-f_M\}\rangle_\varphi + \langle\dot{H}_{*,s}\rangle_\varphi f_{M,s}/T_s \qquad (29)$$

Neglecting higher-order contributions one obtains

$$<\dot{H}_{*,s}>_{\varphi} \cong Z_s e u \vec{b}.(\vec{E}+\nabla\Phi) \tag{30}$$

where the gyrophase-average of the collision operator delivers, by linearization

$$<C_s\{f-f_M\}>_{\varphi} \cong C_s\{f_M|\bar{g}\} \tag{31}$$

Here we have introduced the definition

$$g_s = f_{*,s} - f_{M,s} + h_s \tag{32}$$

so that $f_s = f_{M,s} + g_s$ results; furthermore we have denoted $\bar{g}_s$ as

$$\bar{g}_s = h_s + \bar{g}_s^{(D)} \tag{33}$$

with

$$\bar{g}_s^{(D)} = <f_{*,s}>_{\varphi} - f_{M,s} \cong \frac{M_s c}{Z_s e}[(\omega_0 R^2 + uR\vec{\theta}.\vec{b})\frac{\partial}{\partial\psi} f_{M,s} + \frac{Z_s e}{2cT_s} f_{M,s}\frac{\partial\omega_0}{\partial\psi}[(\omega_0 R^2 + uR\vec{\theta}.\vec{b})^2 + \frac{1}{M_s B}\mu R^2 B_p^2]] + \frac{1}{T_s} f_{M,s}[M_s(\omega_0 R^2 + uR\vec{\theta}.\vec{b})\omega_{1,s} - \frac{Z_s e}{c}\int_{\psi_0}^{\psi} d\psi\ \omega_{1,s}] \tag{34}$$

Eq. (28) finally yields

$$u\vec{b}.\nabla h_s - C_s\{f_M|h+\bar{g}^{(D)}\} = Z_s e[u\vec{b}.(\vec{E}+\nabla\Phi)]f_{M,s}/T_s \tag{35}$$

which is a generalization of Eq. (29) of Catto et al.[5], to include first order velocity perturbations for each species. We notice that new source terms appear in Eq. (27), related to a possible explicit dependence from the first order velocity perturbation $\omega_{1,s}$.

Thanks to the invariance of the Landau collision operator with respect to a Galilei transformation, from Eq. (35)

it results that h_s may then depend only on $\Delta v_{sk} = R(\omega_{1,s} - + \omega_{1,k})$ (for k=1,N). Recalling the definitions (22) for the generalized forces A_{js} (for j=0,3) and introducing as a new effective generalized force

$$A_{4s} = \frac{MR\omega_{1,s}}{T_s} \tag{36}$$

we obtain

$$\bar{g}_s = f_{M,s} \sum_{j=0}^{3} \sum_{k=1,n} A_{jk} h_{sk}^{(j)} \tag{37}$$

Eq.(35) then yields:

$$u\vec{b}.\nabla h_{sk}^{(j)} - K_s h_k^{(j)} = \beta_{sk}^{(j)} \tag{38}$$

where the operator K_s operating on the arbitrary function X is defined according to

$$K_s X = (f_{M,s})^{-1} C_s(f_M|X) \tag{39}$$

and for j=4 the source term $\beta_{sk}^{(4)}$ reads in particular

$$\beta_{sk}^{(4)} = u\vec{b}.\nabla(\omega_0 R^2 + uR\vec{\theta}.\vec{b})\delta_{sk} \tag{40}$$

Equations (38) and (39) represent a generalization of Eq. (43) of Ref.5 to include first-order velocity perturbations, in the framework of a general strong rotation theory allowing rotation speeds comparable to the ion thermal speed.

Conclusions

We stress that the solution of the kinetic equation (35) (or (38)) now contains the yet undetermined generalized force Δv_{ei}.

Such a freedom can be used to require $\partial J_T/\partial t=0$, consistently with an assumption of a low-- plasma (implying

$|\vec{E}^{rot}|/|\vec{E}_T|<<1$). In such a case Eq. (20) yields a first-order partial differential equation for $\partial\Delta v_{ei}/\partial t$, which, supplemented with appropriate initial conditions, uniquely determines Δv_{ei}. Since the assumption $\Delta v_{ei}\neq 0$ implies that a net contribution is added to the toroidal current density J_T, this means that such an initial condition, related to $\vec{E}^{rot}$, as previously pointed out, may significantly influence transport. Because of the higher mobility of the electron species, it is reasonable to expect that the present theory may predict enhanced electron fluxes.

On the other hand, we remark that other choices of $\vec{E}^{rot}$ are permitted in the present theory, as may be of interest for the investigation of finite-β plasmas.

It follows that the present theory appears to generalize previous kinetic theories in two respects, i.e. allowing both the presence of first-order velocity perturbations as well as the effect of a rotational electric self-field $\vec{E}^{rot}$. In view of such features, the theory seems susceptible of possible applications to investigate a wide class of problems in collisional transport theory, related both to space and laboratory plasmas.

Acknowledgments

The present work has been performed in the framework of the Gruppo Nazionale per la Fisica Matematica, of the Italian Consiglio Nazionale delle Ricerche, and the MPI 40% National Research Program Modelli e problemi non lineari in Fisica Matematica of the Italian Ministero della Pubblica Istruzione.

References

[1]Rosenbluth M.N.,Hazeltine R.D. and Hinton F.L., "Plasma transport in toroidal confinement systems", Physics of Fluids,Vol.15,1972,pp.116-.

[2]Bernstein I.B.,"Transport in axisymmetric systems", Physics of Fluids,Vol.17,1974,pp.547.-.

[3]Tessarotto M.,"Variational theory of collisional transport for toroidal axisymmetric plasmas in the weakly collisional regime", Nuovo Cimento, Vol.75B, 1983, pp.19-.

[4]Hinton F.L. and Wong K.S.,"Neoclassical transport in rotating axisymmetric plasmas", Physics of Fluids, Vol. 28, 1985, pp.3082-.

[5]Catto P.J., Bernstein I.B. and Tessarotto M.,"Ion transport in toroidally rotating Tokomak plasma", Physics of Fluids, Vol.30, 1987, pp.2784-.

[6]Nocentini A.,"Neoclassical diffusion of finite--Tokomak plasma", Journal of Plasma Physics, Vol.7, 1972, pp.427-.

Electron Oscillations, Landau, and Collisional Damping in a Partially Ionized Plasma

V. G. Molinari* and M. Suminit†
Università di Bologna, Bologna, Italy
and
B. D. Ganapol‡
University of Arizona, Tucson, Arizona

Abstract

The aim of this paper is to analyze electron plasma oscillations and the associated damping effects (both Landau and collisional) when binary collisions are included. Starting from the Boltzmann-Vlasov equation to describe the dynamics of a partially ionized plasma and using a suitable dimensionless formulation, we have pursued a rigorous analysis of the location of the zeros of the dispersion relation, showing explicitly the physical connection between the collision frequency and Landau damping. Finally, the expression of the Fourier transform for the self-consistent electric field has been explicitly shown in the time domain through an original numerical inversion technique for Laplace transforms.

Introduction

The Landau damping (LD) phenomenon was one of the most interesting and studied phenomena in the early times of the plasma physics.[1,2]

The aim of these studies was to show the ability of a mathematical analysis to explain some unsuspected physical properties and also to determine the limits of the description of the two standard approaches to the plasma modeling, i.e., kinetic equations and macroscopic theory. Then, the LD problem (at least in his original formulation) became a standard topic

---

* Full Professor, Laboratorio di Ingegneria Nucleare di Montecuccolino.

† Associate Professor, Laboratorio di Ingegneria Nucleare di Montecuccolino.

‡ Full Professor, Department of Nuclear and Energy Engineering.

of well-assessed plasma physics textbooks [3,4] and moved only in the context of some experimental studies validations[5] or very peculiar theoretical approaches.[6,7]

Our effort in reviving the LD problem is mainly to examine a new way of analysis, together with an application of a new numerical Laplace transforms inversion technique, offering an original point of view also useful for teaching purposes.

We study this phenomenon in this paper (which follows a series of previously published works of our group [8,9]) in a general situation to stress the influence of binary collisions on LD. Starting from an integral equation for the distribution function, we carry out a complete analysis of the resulting dispersion relation, showing explicitly the time behavior of the selfconsistent electric field.

Theoretical Background

Let us consider a partially ionized plasma where we can assume that ions form a fixed homogeneous positive background and neutral atom-electron collisions or nuclear elastic scattering cannot be disregarded. The starting point of our analysis is the integral operational form of the Boltzmann-Vlasov equation [8] for the electron distribution:

$$f(\mathbf{r},\mathbf{v},t) = f_0(\mathbf{r},\mathbf{v})\ exp[-\nu t] + \int_{\Re_3} d\mathbf{v}' \Big\{ \int_0^t d\tau\ exp[-\nu\tau]$$

$$K[\mathbf{v}',\mathbf{v}(\tau)]\ exp\{-[\mathbf{x}(\tau)\cdot\nabla + \tau\mathcal{D}]\} \Big\} f(\mathbf{r},\mathbf{v}',\tau) \qquad (1)$$

where ν is the collision density, $f_0(\mathbf{r},\mathbf{v})$ is the initial electron distribution, $K[\mathbf{v}',\mathbf{v}(t)]$ is the collision kernel, $D = \partial/\partial t$, and $\mathbf{x}(t)$ and $\mathbf{v}(t)$ result from the integration of the motion equations.

We can now adopt the BGK kernel as a suitable approximation for K in the present situation of close binary collisions, i.e.,

$$K[\mathbf{v}',\mathbf{v}] = \nu\left(\frac{\beta}{\pi}\right)^{3/2} exp[-\beta v^2] \qquad (2)$$

where $\beta = m/kT$. Moreover, we seek the normal modes of plasma evolution; therefore, we can forget the initial conditions and write down an asymptotic form for Eq. (1),[8] namely,

$$f(\mathbf{r},\mathbf{v},t) = \nu \int_0^\infty d\tau\ exp[-\nu\tau]\ (\frac{\beta}{\pi})^{3/2}\ exp\{-[\mathbf{x}\cdot\nabla + \tau\mathcal{D}]\}\ n(\mathbf{r},\tau) \qquad (3)$$

$$n(\mathbf{r},t) = \int f\, d\mathbf{v} \quad (electron\ density) \tag{4}$$

The subsequent steps consist of some approximations introduced in the equations of motion:

$$\mathbf{x}(\tau) \simeq \mathbf{r} - \mathbf{v}\tau \tag{5}$$

$$\mathbf{v}(\tau) \simeq \mathbf{v} + \int_0^\tau d\tau' \frac{e}{m}\mathbf{E}(\mathbf{r} - \mathbf{v}\tau', t - \tau') = \mathbf{v} - \mathbf{v}^*(\tau) \tag{6}$$

and in the space dependence of the self-consistent electric field (long wavelength approximation),

$$\mathbf{E}(\mathbf{r},t) = \mathbf{E}_0(t) + \epsilon\mathbf{E}_1(\mathbf{r},t) \tag{7}$$

where ϵ is a small quantity. As fully explained in Ref.[8], this entails the following consequences:

$$exp[-\tau\mathbf{v}\cdot\nabla] \simeq 1 - \epsilon\tau\mathbf{v}\cdot\nabla \tag{8}$$

and, under the hypothesis of small electric field (small-amplitude oscillations), we find

$$exp[-\beta v^2(\tau)] \simeq exp[-\beta v^2](1 + 2\beta\mathbf{v}\cdot\mathbf{v}^*(\tau)) \tag{9}$$

This leads to a simplified expression for $f(\mathbf{r},\mathbf{v},t)$:

$$f(\mathbf{r},\mathbf{v},t) \simeq n\nu\left(\frac{\beta}{\pi}\right)^{3/2} exp^{-\beta v^2}\int_0^\infty d\tau\, exp[-\nu\tau](1 + 2\beta\mathbf{v}\cdot\mathbf{v}^*(\tau)) \tag{10}$$

Equation (10), jointly with the Ampere equation

$$D\mathbf{E} = -\frac{1}{\epsilon}\mathbf{J} = \frac{e}{\epsilon_0}\int_{\Re_3}\mathbf{v} f\, d\mathbf{v} \tag{11}$$

leads to an equation for the self-consistent electric field of the form

$$\begin{aligned} DE = -2\omega_p^2\nu\beta\left(\frac{\beta}{\pi}\right)^{1/2}\int_{-\infty}^{+\infty} du\, u^2\, exp[-\beta u^2]\int_0^\infty d\tau\, exp[-v\tau] \\ \int_0^\tau E(x - u\tau', t - \tau')\, d\tau' \end{aligned} \tag{12}$$

where u and x are the components of $\mathbf{v}$ and $\mathbf{x}$ parallel to $\mathbf{E}$.

Now the problem can be easily handled by means of Fourier and Laplace transforms with respect to space and time, respectively:

$$\tilde{E}_k(s) = \frac{E_k(t=0)}{s + 2\omega_p^2\beta\left(\frac{\beta}{\pi}\right)^{1/2} \cdot \int_{-\infty}^{+\infty} \frac{u^2\, exp[-\beta u^2]}{\nu+s+iku}\, du} \tag{13}$$

where s is the complex argument of the Laplace transforms and k is the real variable of the Fourier transforms.

Analysis of the Dispersion Relation

To give the solution in the time domain, the knowledge of the poles of $\tilde{E}_k(s)$ is necessary to evaluate the inversion integral:

$$E_k(t) = \frac{1}{2\pi i} \int_{s_0 - i\infty}^{s_0 + i\infty} \frac{E_k(t=0)}{\Delta(k,s)} e^{st}\, ds \tag{14}$$

i.e., we need to solve the equation:

$$\Delta(k,s) = s + 2\omega_p^2\beta\left(\frac{\beta}{\pi}\right)^{1/2} \int_{-\infty}^{+\infty} \frac{u^2\, exp[-\beta u^2]}{\nu + s + iku} du \tag{15}$$

the so-called dispersion relation. We can simplify the analysis introducing some dimensionless quantities through the group $\beta^{1/2}/k$, which measures the time employed by a particle at thermal velocity to travel the wavelength distance. This choice gives the dimensionless parameters

$$\hat{s} = s\frac{\beta^{1/2}}{k}, \quad \hat{\omega}_p = \omega_p\frac{\beta^{1/2}}{k}, \quad \hat{\nu} = \nu\frac{\beta^{1/2}}{k} \tag{16}$$

and completely scale out k (the Fourier transform variable) from the dispersion relation [from now on, we will use the variables as defined in Eq. (16)]:

$$\Delta(s) = s + \omega_p^2 \int_{-\infty}^{+\infty} \frac{u^2\, e^{-u^2}}{\nu + s + iu} du = 0 \tag{17}$$

where the integration variable has been obviously redefined. With the integration performed, we obtain[9]

$$s + 2\omega_p^2(s+\nu) - 2\sqrt{\pi}\omega_p^2(s+\nu)^2\, exp[(s+\nu)^2]\,\{\, erfc(s+\nu) + \delta_1(s+\nu)$$

$$+ 2\mathcal{N}[-Re(s+\nu)]\} = 0 \tag{18}$$

where

$$\delta_1(x) = \begin{cases} 1 \Leftrightarrow x = 0 \\ 0 \;\; elsewhere \end{cases}, \quad \mathcal{N}(x) = \begin{cases} 1 \Leftrightarrow x > 0 \\ 0 \;\; elsewhere \end{cases}$$

A first investigation of Eq. (18) shows that we always have at least one real zero in $(-\nu, 0)$ entailing a purely exponentially decaying mode for the electric field, which results only because of collisions.

An approximate evaluation of the most important real zero can be obtained in some special situations (see Appendix A), namely,

$$\nu \ll \omega_p \Rightarrow \;\; s_1 = -\nu \frac{2\omega_p^2}{1 + 2\omega_p^2} \tag{19}$$

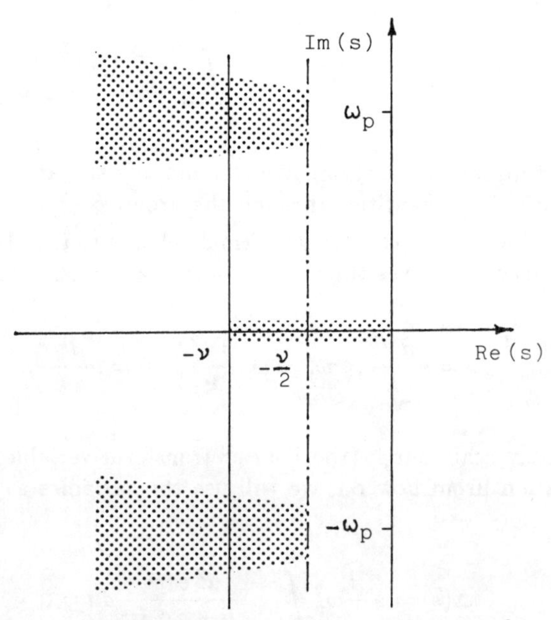

Fig. 1 Qualitative picture showing the distribution of the zeros of Δ on the complex plane.

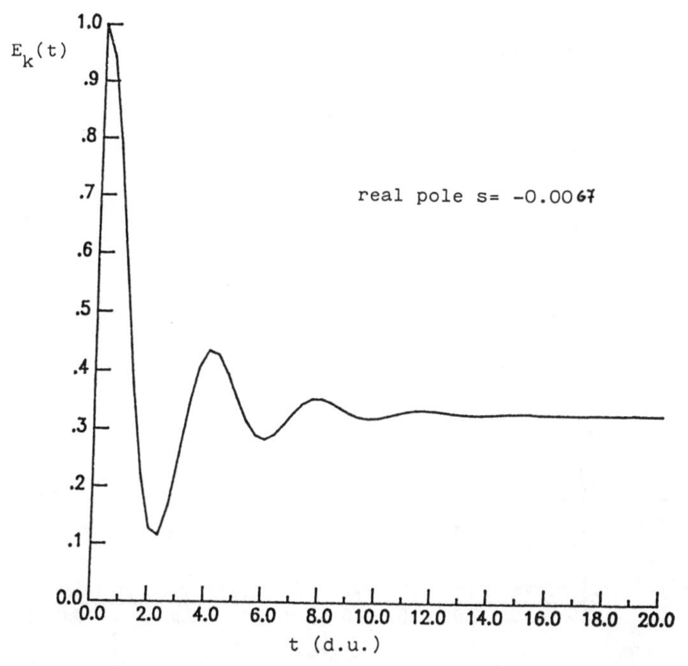

Fig. 2 Electric field evolution, $\hat{\omega}_p = 1$, $\hat{\nu} = 0.01$, $4\gamma_L = 0.582$.

$$\nu \gg \omega_p \Rightarrow s_1 = -\frac{\omega_p^2}{\nu} \tag{20}$$

The localization of the complex zeros of Δ shows some intriguing features. The collisions introduces an upper bound in $-\nu/2$ for the complex zeros (see Appendix B) and, if $\nu \gg \omega_p$, inhibit any oscillatory mode with real-time constant $> -\nu$ (see fig. 1). Introducing appropriate approximations for the erfc in Eq. (17), it is also possible to give an evaluation of the complex zeros connected to the oscillatory modes of the self-consistent electric field. Such an analysis (see Appendix B and Ref.[9] for the details) shows that the typical LD effects can be easily identified in the expression for the complex zeros of Δ lying to the left of $-\nu$ on the complex plane, as is physically obvious. What is less evident is that we have found situations that have as a first zero of Δ (that connected with the asymptotic time behavior of the electric field), not a real one (that heavily connected to the collisions, as could be reasonable), but a couple of complex conjugate ones, giving rise to a persistent oscillatory mode.

Numerical Results

In this section we try to give some comments on the numerical analysis of the zeros of Eq. (18) and on the results obtained from a direct numerical inversion technique for the Laplace transforms.

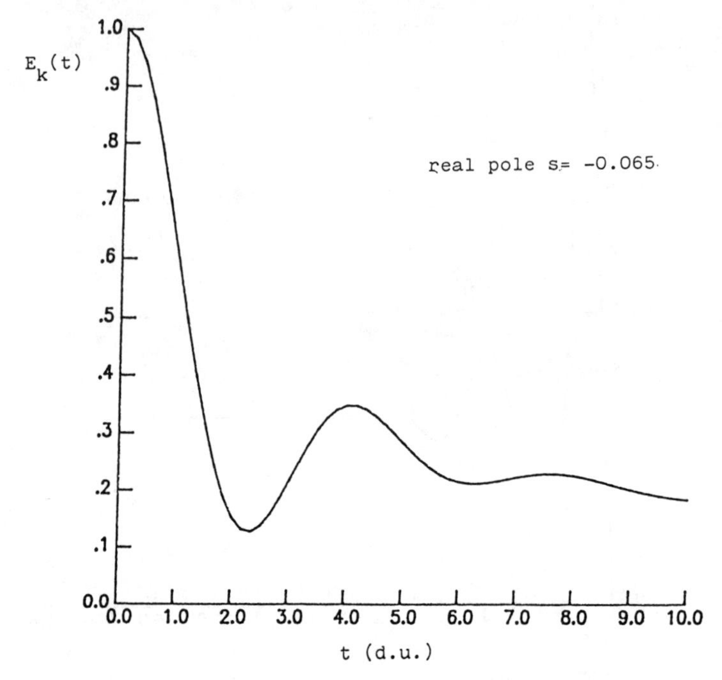

Fig. 3 $\hat{\omega}_p = 1$, $\hat{\nu} = 0.1$, $4\hat{\gamma}_L = 0.582$.

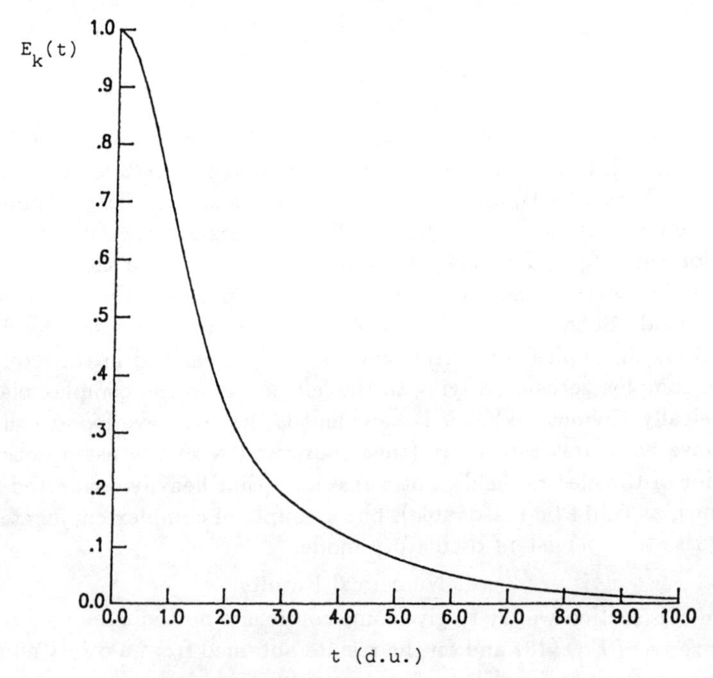

Fig. 4 $\hat{\omega}_p = 1$, $\hat{\nu} = 1$, $4\hat{\gamma}_L = 0.582$.

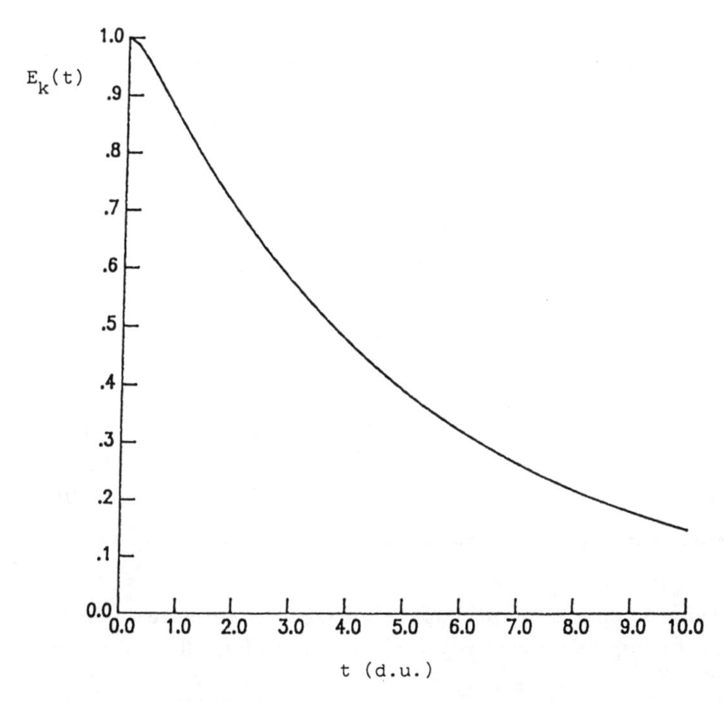

Fig. 5 $\hat{\omega}_p = 1$, $\hat{\nu} = 5$, $4\hat{\gamma}_L = 0.582$.

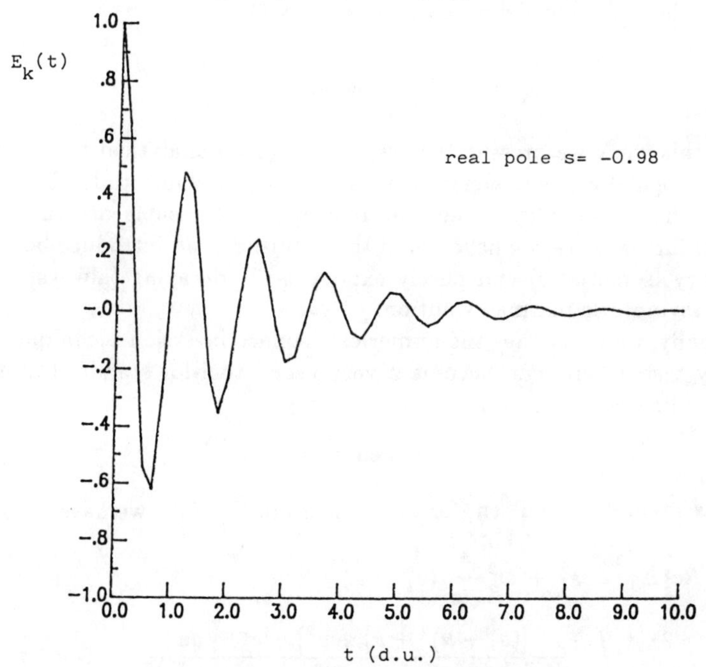

Fig. 6 $\hat{\omega}_p = 5$, $\hat{\nu} = 1$, $4\hat{\gamma}_L = 1.4E - 7$.

Table 1 Time constants (dimensionless units)

ν \ ω_p	0.1	1.0	2.0	5.0
0.01	-.0002	-.0067	-0.0089	-0.0098
		$-0.009 \pm 1.58i$	$-.009 \pm 2.24i$	$-0.005 \pm 5.15i$
0.1	-.017	-0.065	-0.089	-0.098
		$-0.099 \pm 1.58i$	$-0.089 \pm 2.24i$	$-0.053 \pm 5.15i$
1.0	-0.048	-0.46	-0.86	-0.98
	$-3.3 \pm 1.9i$	$-0.99 \pm 1.5i$	$-0.75 \pm 2.3i$	$-0.53 \pm 5.04i$
2.0	-0.038	-0.43	-1.63	-1.96
			$-1.45 \pm 1.98i$	$-1.06 \pm 5.04i$
5.0	-.002	-0.2	-0.9	-4.88
				$-2.66 \pm 4.42i$
10.0	-.001	-0.01	-0.41	-4.01

Table 1 reports the principal time constants [the real one and the first couple of complex zeros of Eq. (18) in $(-\nu, 0)$] for different values of ω_p and ν.

Figures 2 - 6 show the time behavior of $E_k(t)/E_k(t=0)$ produced through the technique described in Appendix C.

Conclusions

In this work we present the results of a deep analytical investigation of a paradigmatic phenomenon in plasma physycs. Our study shows that the LD effects can play a relevant role even if collisions are taken into account; furthermore, we have found that collisions can introduce both new oscillatory (sometimes) and purely exponentially decaying (always) modes for the asymptotic plasma evolution.

Finally, we think that the numerical Laplace inversion technique satisfactorily tested here, can become a very useful tool for standard dynamic problems in plasma physics.

Appendix A

Let $s = s_1 + is_2$. Then, for $s_1 > -\nu$, from Eq. (17) we have

$$Re\{\Delta\} \equiv s_1 + \omega_p^2 \frac{4}{\sqrt{\pi}} (s_1 + \nu) \int_0^\infty \frac{[(s_1+\nu)^2 + s_2^2 + u^2]\, u^2 e^{-u^2}\, du}{[(s_1+\nu)^2 - s_2^2 + u^2]^2 + 4s_2^2(s_1+\nu)^2} = 0 \tag{A.1}$$

$$Im\{\Delta\} \equiv s_2\Big\{1 + \omega_p^2 \frac{4}{\sqrt{\pi}} \int_0^\infty \frac{[u^2 + s_2^2 - (s_1+\nu)^2]\, u^2\, e^{-u^2}\, du}{[(s_1+\nu)^2 - s_2^2 + u^2]^2 + 4s_2^2(s_1+\nu)^2}\Big\} = 0 \tag{A.2}$$

From (A.2) it follows immediately that $s_2 = 0$ is a solution that has as a counterpart the zeros of

$$\Delta(s_1, 0) = s_1 + \omega_p^2 \frac{4}{\sqrt{\pi}} \int_0^\infty \frac{u^2\, e^{-u^2}\, du}{(s_1+\nu)^2 + u^2} \tag{A.3}$$

An analysis of expression (A.3) shows that

$$\left.\frac{\partial \Delta}{\partial s_1}\right|_{\substack{s_1<-\nu \\ s_1>0}} > 0, \quad \lim_{s_1 \to -\nu+} \Delta < 0, \quad \Delta\Big|_{s_1=0} > 0$$

implying the existence of at least one real zero in $(-\nu, 0]$. To obtain an approximate evaluation of the fundamental real time constant, we start from

$$\Delta(s_1, 0) = s_1 + 2\omega_p^2\,(s_1+\nu)\{1 - \sqrt{\pi}(s_1+\nu)e^{(s_1+\nu)^2}[1 - erf(s_1+\nu)]\} = 0$$

$$s_1 \in (-n, 0] \tag{A.4}$$

Then, if $\nu \ll \omega_p$,

$$\Delta(s_1, 0) \simeq s_1 + 2\omega_p^2(s_1+\nu) = 0 \;\Rightarrow\; s_1 = -\frac{2\nu\,\omega_p^2}{1 + 2\omega_p^2} \tag{A.5}$$

or viceversa, if $\omega_p \ll \nu$ and $\nu > 1$, using for the erf function an asymptotic expansion, [10] we obtain

$$\Delta(s_1, 0) \simeq s_1 + \frac{\omega_p^2}{\nu} \;\Rightarrow\; s_1 = -\frac{\omega_p^2}{\nu} \tag{A.6}$$

Appendix B

Defining the two functions (≥ 0),

$$G(s_1, s_2) = \frac{\omega_p^2}{\sqrt{\pi}} \int_0^\infty \frac{u^4\, e^{-u^2}\, du}{[(s_1+\nu)^2 - s_2^2 + u^2]^2 + 4s_2^2(s_1+\nu)^2} \tag{B.1}$$

$$H(s_1, s_2) = \frac{\omega_p^2}{\sqrt{\pi}} \int_0^\infty \frac{u^2\, e^{-u^2}\, du}{[(s_1+\nu)^2 - s_2^2 + u^2]^2 + 4s_2^2(s_1+\nu)^2} \qquad (B.2)$$

Equation (A.1) and (A.2) can be written as

$$\begin{cases} s_2\, G + (s_1+\nu)^2 G = 1 + H \\ s_1(1 + (s_1+\nu)^2 G + s_2^2 G + H) = -\nu((s_1+\nu)^2 G + s_2^2 G + H) \end{cases} \qquad (B.3)$$

By eliminating the H dependence from the first equation, we obtain the relation

$$s_2 = \pm \left[\frac{\nu/2 - (s_1+\nu)^2 G}{(s_1+\nu)^2} \right]^{1/2} \qquad (B.4)$$

which indicates that complex solutions of system (B.3) can disappear. Moreover, substituting G into the second equation gives

$$s_1 = -\frac{\nu}{2}\left(2 - \frac{1}{1+H}\right) \qquad (B.5)$$

from which the presence of an upper bound in $-\nu/2$ for the real zero can be seen. Some approximate solutions can also be obtained from equation (17), provided that the erfc is written as[10]

$$\lim_{x \to \infty} erfc(x) \approx \frac{e^{-x^2}}{\sqrt{\pi}\, x}\left(1 - \frac{1}{2\,x^2} + \frac{3}{4\,x^4} \cdots \right) \qquad (B.6)$$

When this asymptotic expression is valid (in the dimensionless formulation this implies ω_p at least $\geq$ 2), the analysis carried out in Ref. [8] can be retained. We recall here the main results of that discussion.

For $s_1 > -\nu$ equation (18) reduces to

$$s + \frac{\omega_p^2}{(s+\nu)} - \frac{3\omega_p^2}{2(s+\nu)^3} = 0 \qquad (B.7)$$

yielding the following approximation as a representation of the zeros:

$$s \simeq -\frac{\nu}{2} \pm i\left[\omega_p^2 + \frac{3}{2}\right]^{1/2} \qquad (B.8)$$

Similarly, for $s_1 < -\nu$, we have

$$\Delta \simeq s + \frac{\omega_p^2}{(s+\nu)} - \frac{3\omega_p^2}{2(s+\nu)^3} - 4\sqrt{\pi}\,\omega_p^2\,(s+\nu)\,e^{(s+\nu)^2} = 0 \qquad (B.9)$$

which produces the following results:

$$s \simeq -\frac{\nu}{2} - 2\gamma_L \pm i\left[\omega_p^2 + \frac{3}{2}\right]^{1/2} \qquad (B.10)$$

provided that $\nu < 4\gamma_L$, where γ_L is the dimensionless expression of the Landau damping decrement, i.e.,

$$\gamma_L = \sqrt{\pi}\,\omega_p^2\, exp\left[-\left(\omega_p^2 + \frac{3}{2}\right)\right] \qquad (B.11)$$

Appendix C

In this Appendix we will give a brief outline of an original technique for the numerical inversion of the Laplace transforms, namely, for the computation of the inversion integral

$$f(t) = \frac{1}{2\pi i}\int_{s_0-i\infty}^{s_0+i\infty} \tilde{f}(s)\, e^{st}\, ds \qquad (C.1)$$

where s_0 is the real part of the Bromwich contour.

As is easily seen, we can write Equation (C.1) as

$$\begin{aligned} f(t) &= \frac{2e^{s_0 t}}{\pi t}\int_0^{\infty} du\; Re\{\tilde{f}(s_0 + iu/t)\} cos\, u = \\ &= \frac{2}{\pi t}e^{s_0 t}\left\{\int_0^{\pi/2} du\; Re\{\tilde{f}(s_0 + iu/t)\} cos\, u \right. \\ &\left. + \sum_{k=1}^{\infty}(-1)^k \int_{-\pi/2}^{\pi/2} du\; Re\{\tilde{f}(s_0 + i(u+k\pi)/t)\} cos\, u\right\} \end{aligned} \qquad (C.2)$$

We now need a procedure for numerical integration (we used a Romberg quadrature routine) and an acceleration technique for the series (the Euler transformation for alternating series, if rapidly convergent) coupled with an iteration scheme for the abscissa s_0 , i.e.,

$$\cdots,\ s_{j-1} \rightarrow (s_{j-1} + s_0)/2\ , j = 1, 2, \ldots \qquad (C.3)$$

where s_0 represents the contour closest to the first singularity of $\tilde{f}$.

References

[1] L.D.Landau, Journal of Physics, 10, 25-32 (1946).

[2] P.L.Bhatnagar, E.P.Gross, M.Krook, Physical Review, 94, 511-525 (1954).

[3] N.A.Krall, A.W.Trivelpiece, Principles of Plasma Physics, Mc Graw Hill, New York, 1973.

[4] D.C.Montgomery, Theory of Unmagnetized Plasma, Gordon & Breach, New York, 1971.

[5] R.J.Armstrong, W.J.Weber, J.Trulsen, Physics Letters, 74A, 319-322 (1979).

[6] J.R.Cussenot, M.Fabry, M.Richard, Plasma Physics, 15, 1-5 (1973).

[7] H.Hebenstreit, K.Suchy, Journal of Plasma Physics, 35, 151-164 (1986).

[8] V.G.Molinari, P.Peerani, Nuovo Cimento, 5, 527-540 (1985).

[9] V.G.Molinari, P.Peerani, "Dispersion Characteristics and Landau Damping in a Plasma with an External Field," Proceedings of the Wing Conference on Invariant Imbedding and Integral Equations, Santa Fe, 1988, in press.

[10] M.Abramowitz, I.Stegun, Handbook of Mathematical Tables, Dover, New York, 1972.

Bifurcating Families of Periodic Traveling Waves in Rarefied Plasmas

James Paul Holloway* and J. J. Dorning†
University of Virginia, Charlottesville, Virginia

Abstract

A necessary condition for the existence of undamped spatially periodic traveling waves near spatially uniform equilibria is derived. When the necessary condition is satisfied, a bifurcation analysis of Bernstein-Greene-Kruskal modes is used to construct branches of such waves with wavelength equal to the fundamental wavelength. Expansions of the bifurcating potential for these waves are derived and the reconstruction of the distribution functions is discussed.

1. Introduction

The longitudinal motion of a rarefied multispecies plasma along a straight magnetic field is described by the one-dimensional Vlasov-Maxwell equations

$$\frac{\partial f_\alpha}{\partial t} + u\frac{\partial f_\alpha}{\partial x} + \frac{q_\alpha}{m_\alpha}E\frac{\partial f_\alpha}{\partial u} = 0 \tag{1}$$

$$\frac{\partial E}{\partial x} = 4\pi \sum_\alpha q_\alpha \int f_\alpha \, du \tag{2}$$

$$\frac{\partial E}{\partial t} + 4\pi \sum_\alpha q_\alpha \int u f_\alpha \, du = 0 \tag{3}$$

that couple the distribution functions $f_\alpha(x, u, t)$ to the magnetic field aligned electric field $E(x, t)$. The subscript α denotes the particle species, and particles of species α have charge q_α and mass m_α. Integrals are over all velocity, except when otherwise noted. These equations result when the three-dimensional Vlasov-Maxwell equations are integrated over the transverse particle velocities in the case where the distribution functions and magnetic field aligned component of the electric field depend only on the field aligned coordinate x and the transverse electric field has zero divergence.

---

*Research Assistant Professor of Engineering Physics and of Applied Mathematics.

†Whitney Stone Professor of Nuclear Engineering and Professor of Engineering Physics; also Center for Advanced Studies.

This system of equations possesses an infinite family of spatially uniform equilibria at which the electric field parallel to the magnetic field is zero. These equilibria are characterized by distribution functions $F_\alpha(u)$ that satisfy the zero charge and zero current conditions, $\sum_\alpha q_\alpha \int F_\alpha(u)\,du = 0$ and $\sum_\alpha q_\alpha \int uF_\alpha(u)\,du = 0$. In this paper we present results on spatially periodic traveling wave solutions of these equations close to members of this infinite family of spatially uniform equilibria.

Because of the Galilean invariance of Eqs. (1–3), a traveling wave solution can be transformed into an equilibrium in the wave frame; thus, for a solution of the form $f_\alpha(x - Vt, u - V)$, $E(x - Vt)$, the functions f_α and E must satisfy the stationary equations

$$u\frac{\partial f_\alpha}{\partial x} + \frac{q_\alpha}{m_\alpha}E\frac{\partial f_\alpha}{\partial u} = 0 \tag{4}$$

$$\frac{dE}{dx} = 4\pi \sum_\alpha q_\alpha \int f_\alpha\,du \tag{5}$$

$$0 = 4\pi \sum_\alpha q_\alpha \int uf_\alpha\,du \tag{6}$$

Therefore, the study of spatially periodic traveling waves, with phase velocity V near the Vlasov equilibrium $F_\alpha(u)$, is reduced to the study of spatially periodic equilibria near the shifted spatially uniform equilibrium $F_\alpha(u + V)$.

2. A Necessary Condition

We now wish to answer the question: When can there be spatially periodic traveling waves $f_\alpha(x - Vt, u - V)$, $E(x - Vt)$, such that the perturbations $f_\alpha(x - Vt, u - V) - F_\alpha(u)$ and $E(x - Vt)$ can be made as small as desired? In other words, when are there plasma waves that are arbitrarily small perturbations of the equilibrium F_α? In this section we shall describe the derivation of a condition that must be satisfied for such waves to exist.

To formulate the necessary condition, we first derive a second order differential equation for the electric field. Suppose that f_α and E are C^2 (twice continuously differentiable) functions that satisfy the stationary Vlasov equation, Eq. (4), and that there is a function $B \in L^1(\mathbf{R})$ such that $|\partial f_\alpha/\partial x| < B(u)$ for all x. Then, for $u \neq 0$

$$\frac{\partial f_\alpha}{\partial x}(x, u) = -E(x)\frac{q_\alpha}{m_\alpha}\frac{1}{u}\frac{\partial f_\alpha}{\partial u}(x, u) \tag{7}$$

Differentiating Eq. (5) with respect to x and using the L^1 bound on $\partial f_\alpha/\partial x$ we find

$$\frac{d^2 E}{dx^2} + 4\pi \sum_\alpha \frac{q_\alpha^2}{m_\alpha}\int \frac{1}{u}\frac{\partial f_\alpha}{\partial u}E\,du = 0 \tag{8}$$

When $E \not\equiv 0$, Eq. (4) can be used to show that $\partial f_\alpha/\partial u|_{u=0} = 0$ for all x; therefore $\lim_{u\to 0}(\partial f_\alpha/\partial u)/u = \partial^2 f_\alpha/\partial u^2|_{u=0}$, and the integral is well defined and non-singular.

Now define $g_\alpha(x,u) = f_\alpha(x,u) - F_\alpha(u+V)$ as the deviation of the distribution function from the Vlasov equilibrium as measured in the wave frame. While the functions $F_\alpha(u+V)$ and $g_\alpha(x,u)$ do not, in general, satisfy $F'_\alpha(V) = 0$ and $\partial g_\alpha/\partial u|_{u=0} = 0$, they do satisfy $F'_\alpha(V) + \partial g_\alpha/\partial u|_{u=0} = \partial f_\alpha/\partial u|_{u=0} = 0$. It follows from this last fact and the Lebesgue dominated convergence theorem that

$$\int \frac{1}{u}\frac{\partial f_\alpha}{\partial u} E\,du = \left[\mathrm{P}\!\!\int \frac{F'_\alpha(u+V)}{u}\,du + \mathrm{P}\!\!\int \frac{1}{u}\frac{\partial g_\alpha}{\partial u}(x,u)\,du \right] E(x) \tag{9}$$

where $\mathrm{P}\!\int$ denotes the principal value integral. So, defining

$$\kappa^2 = 4\pi \sum_\alpha \frac{q_\alpha^2}{m_\alpha} \mathrm{P}\!\!\int \frac{F'_\alpha(u+V)}{u}\,du \tag{10}$$

and

$$\gamma(x) = 4\pi \sum_\alpha \frac{q_\alpha^2}{m_\alpha} \mathrm{P}\!\!\int \frac{1}{u}\frac{\partial g_\alpha}{\partial u}(x,u)\,du \tag{11}$$

Eq. (8) can be written as

$$\frac{d^2E}{dx^2} + \kappa^2 E + \gamma E = 0 \tag{12}$$

In order for a periodic traveling wave to exist near F_α, Eq. (12) must have a periodic solution for *some* small γ; conversely, if Eq. (12) does not have any (non-trivial) periodic solution for *any* small γ, then there can be no traveling wave with phase velocity V near F_α.

Theorem 1. *Let $F_\alpha : \mathbf{R} \to \mathbf{R}$ denote a C^2 Vlasov equilibrium, let $V \in \mathbf{R}$ be given, and suppose that $\kappa^2 < 0$. If $f_\alpha : [0, 2\pi/k] \times \mathbf{R} \to \mathbf{R}$ and $E : [0, 2\pi/k] \to \mathbf{R}$ are C^2 functions, periodic over $[0, 2\pi/k]$, such that*

1) *$\hat{f}_\alpha(x,u,t) = f_\alpha(x - Vt, u - V)$ and $\hat{E}(x,t) = E(x - Vt)$ satisfy the one-dimensional Vlasov-Maxwell equations, and*
2) *there exists a $B \in L^1(\mathbf{R})$ such that $\left|\frac{\partial f_\alpha}{\partial x}(x,u)\right| < B(u)$ for all $x \in [0, 2\pi/k]$,*

then

$$0 < |k|^2 \frac{\kappa^4\sqrt{-\kappa^2}}{4|k|\sqrt{-\kappa^2} - \pi\kappa^2} \le 2\pi \sup_x |\gamma(x)|^2$$

The key to the proof of this theorem[1] is the demonstration that, with $\kappa^2 < 0$, Eq. (12) does not have any $2\pi/k$ periodic solutions if $\gamma(x)$ is sufficiently small; this demonstration is based on a standard result in the perturbation theory of linear operators.[2] The lower bound for $\sup_x |\gamma(x)|^2$ is then derived by estimating the norm of the inverse of the operator $d^2/dx^2 + \kappa^2$. While in this paper we have justified Eq. (12) and stated Theorem 1 for the case of C^2 distribution functions, the equation and theorem actually hold when the distribution functions are only C^1 and have velocity derivatives that satisfy appropriate Hölder conditions near zero velocity.[1]

Theorem 1 tells us that a non-trivial periodic traveling wave cannot exist near the spatially uniform equilibrium F_α unless $\kappa^2 \geq 0$.

3. Construction of Families of Waves

Theorem 1 indicates that for a branch of traveling wave solutions to pass through F_α it must be that $\kappa^2 \geq 0$. In this section we shall hypothesize that $k^2 = \kappa^2 > 0$ for a given equilibrium F_α and phase velocity V. Based on this hypothesis we shall describe the construction of bifurcating families of traveling waves with wavelength $2\pi/k$, which in fact do go to F_α in the small amplitude limit. Thus, we shall show by explicit construction that whenever $\kappa^2 > 0$ there do exist families of traveling wave solutions that pass through (i.e., bifurcate from) the equilibrium F_α in the small amplitude limit.

Let F_α denote any spatially uniform Vlasov equilibrium, and for each α and any V define two functions $G_\alpha^\pm$ by

$$G_\alpha^\pm(\eta) = F_\alpha(\pm\sqrt{2\eta} + V) \tag{13}$$

Then

$$F_\alpha(u+V) = \begin{cases} G_\alpha^+\left(\dfrac{u^2}{2}\right), & u \geq 0 \\ G_\alpha^-\left(\dfrac{u^2}{2}\right), & u \leq 0 \end{cases} \tag{14}$$

is a representation of F_α in the wave frame. The functions $G_\alpha^\pm$ can easily be singular, even when F_α is smooth, but the function $R_\alpha(\eta) = G_\alpha^+(\eta) + G_\alpha^-(\eta)$ is smooth; in fact, when F_α is C^{2r}, R_α is C^r, although R_α is defined only for positive values of its argument.

Based on this smoothness, it is easy to see that

$$\begin{aligned} \mathrm{P}\int \frac{F_\alpha'(u+V)}{u}\,du &= \lim_{\epsilon\to 0}\int_\epsilon^\infty \frac{F_\alpha'(u+V) - F_\alpha'(-u+V)}{u}\,du \\ &= \lim_{\epsilon\to 0}\int_\epsilon^\infty \frac{dG_\alpha^+}{d\eta}\left(\frac{u^2}{2}\right) + \frac{dG_\alpha^-}{d\eta}\left(\frac{u^2}{2}\right) du \\ &= \int_0^\infty \frac{dR_\alpha}{d\eta}\left(\frac{u^2}{2}\right) du \end{aligned}$$

and so from the definition of κ^2, Eq. (10)

$$\kappa^2 = 4\pi\sum_\alpha \frac{q_\alpha^2}{m_\alpha}\int_0^\infty \frac{dR_\alpha}{d\eta}\left(\frac{u^2}{2}\right) du \tag{15}$$

Hence, κ^2 is very naturally related to R_α.

To show that plasma waves exist under the hypothesis $k^2 = \kappa^2$, it is not necessary to find the most general wave-form; a special case will suffice. The construction here will be based on the simple BGK representation[3] in which the distribution function (in the wave frame) is of the form

$$f_\alpha(x,u) = \begin{cases} g_\alpha^+\left(\frac{u^2}{2} + \frac{q_\alpha}{m_\alpha}\phi(x)\right), & u \geq 0 \\ g_\alpha^-\left(\frac{u^2}{2} + \frac{q_\alpha}{m_\alpha}\phi(x)\right), & u \leq 0 \end{cases} \tag{16}$$

where the functions $g_\alpha^\pm$ satisfy

$$g_\alpha^+(\eta) = g_\alpha^-(\eta) \quad \forall\, \eta \in \left[\Phi_\alpha^{min}, \Phi_\alpha^{max}\right] \tag{17}$$

and ϕ is a C^2 potential such that $E = -d\phi/dx$, $\Phi_\alpha^{min} = \inf_x\{(q_\alpha/m_\alpha)\phi(x)\}$ and $\Phi_\alpha^{max} = \sup_x\{(q_\alpha/m_\alpha)\phi(x)\}$. This representation is sufficiently general that all spatially uniform equilibria can be studied, as is evident from Eqs. (13) and (14). Every function f_α that admits such a representation with the functions $g_\alpha^\pm$ satisfying $dg_\alpha^\pm/d\eta = o\left(\sqrt{2\eta - 2\Phi_\alpha^{min}}\right)^{-1}$ as $\eta \longrightarrow \Phi_\alpha^{min}$ will automatically satisfy the Vlasov equation. The matching condition, Eq. (17), accounts for those particles whose direction of motion is reversed by the field (e.g., trapped particles), thereby relating the distribution function for positive velocities to the distribution function for negative velocities. This representation is physically suited to finding periodic distribution functions whose wavelength is identical to the fundamental wavelength $2\pi/\kappa$. In general, a different representation of this form could be used in each of several different, neighboring, potential wells, thereby producing several different trapped particle populations and, by appropriate matching, giving the system a longer wavelength. This more general possibility will not be considered further here.

Using Eq. (16) in Eq. (5) yields

$$\frac{d^2\phi}{dx^2} + 4\pi\sum_\alpha q_\alpha \int_0^\infty r_\alpha\left(\frac{u^2}{2} + \frac{q_\alpha}{m_\alpha}\phi(x)\right) du \tag{18}$$

where $r_\alpha(\eta) = g_\alpha^+(\eta) + g_\alpha^-(\eta)$. Similarly, using Eq. (16) in Eq. (6) and making a change of integration variable gives

$$0 = 4\pi\sum_\alpha q_\alpha \int_{(q_\alpha/m_\alpha)\phi(x)}^\infty \left[g_\alpha^+(\eta) - g_\alpha^-(\eta)\right] d\eta \tag{19}$$

But, according to Eq. (17), $g_\alpha^+(\eta) - g_\alpha^-(\eta) = 0$ for $\Phi_\alpha^{min} \leq \eta \leq \Phi_\alpha^{max}$; therefore, this zero current condition can be written as

$$0 = 4\pi\sum_\alpha q_\alpha \int_{\Phi_\alpha^{max}}^\infty \left[g_\alpha^+(\eta) - g_\alpha^-(\eta)\right] d\eta \tag{20}$$

It is possible to develop a general, but abstract, generic bifurcation theory for Eq. (18),[1] but here we wish to consider a more explicit construction. We shall introduce families of the functions r_α, parameterized in such a way that $\phi = 0$ gives a trivial solution of Eq. (18) for all values of the parameter μ and so that $r_\alpha = R_\alpha$ at $\mu = 0$, and determine the non-trivial solutions that bifurcate from $\phi = 0$, $\mu = 0$.

As indicated earlier, R_α is half as smooth as F_α, but is only defined for positive values of its argument; because it is smooth however, it has a smooth extension ρ_α to negative values of η. Now let $h_\alpha(\eta, \mu)$ represent a family of functions, smoothly parameterized by μ, that satisfy

$$0 = h_\alpha(\eta, 0) \tag{21}$$

$$0 = \sum_\alpha q_\alpha \int_0^\infty h_\alpha\left(\frac{u^2}{2}, \mu\right) du \tag{22}$$

$$0 \neq \sum_\alpha \frac{q_\alpha^2}{m_\alpha} \int_0^\infty \frac{\partial^2 h_\alpha}{\partial \eta \partial \mu}\left(\frac{u^2}{2}, 0\right) du \tag{23}$$

Such families can always be found; for example, $h_\alpha(\eta, \mu) = \mu N_\alpha \exp(-\eta/T_\alpha)$ with N_α and T_α chosen properly will work.

Writing $r_\alpha(\eta, \mu) = \rho_\alpha(\eta) + h_\alpha(\eta, \mu)$, Eq. (18) becomes the one-parameter family of equations

$$\frac{d^2\phi}{dx^2} + \mathcal{H}(\phi, \mu) = 0 \tag{24}$$

with

$$\mathcal{H}(\phi, \mu) = 4\pi \sum_\alpha q_\alpha \int_0^\infty \left[\rho_\alpha\left(\frac{u^2}{2} + \frac{q_\alpha}{m_\alpha}\phi(x)\right) + h_\alpha\left(\frac{u^2}{2} + \frac{q_\alpha}{m_\alpha}\phi(x), \mu\right)\right] du \tag{25}$$

The function $\mathcal{H}$ is at least as smooth as ρ_α and h_α, provided that $\rho_\alpha(\eta)$, $h_\alpha(\eta)$ and their derivatives have suitable decay properties as $\eta \to \infty$.

Equation (22) implies that $\phi = 0$ is a solution of Eq. (24) for all values of the parameter μ. If we linearize Eq. (24) at $\phi = 0$, $\mu = 0$ we find

$$\frac{d^2\phi}{dx^2} + 4\pi \sum_\alpha \frac{q_\alpha^2}{m_\alpha} \int_0^\infty \frac{d\rho_\alpha}{d\eta}\left(\frac{u^2}{2}\right) du\, \phi \tag{26}$$

But, because $\rho_\alpha(\eta) = R_\alpha(\eta)$ for $\eta \geq 0$, Eq. (15) implies that this is equivalent to $d^2\phi/dx^2 + \kappa^2\phi = 0$. By the hypothesis $k^2 = \kappa^2 > 0$, this linear equation has $2\pi/k$ periodic solutions of the form $\phi = A\cos(kx) + B\sin(kx)$; this immediately suggests the existence of bifurcating solution branches tangent to the null space spanned by $\cos(kx)$ and $\sin(kx)$. A careful nonlinear analysis[1] using bifurcation theory and the equivariance of Eq. (24) under the $O(2)$ symmetry group has revealed that Eq. (24) indeed does have bifurcating families of solutions; one such family is described by

$$\phi = A\cos(kx) + \Psi(A, x) \tag{27}$$

$$\mu = M(A) \tag{28}$$

where $\Psi(A,x) \perp (\cos(kx), \sin(kx))$ and both Ψ and M are smooth functions of A with $\Psi(0,x) = 0$ and $M(0) = 0$. All other $2\pi/k$ periodic solutions of Eq. (24) near $\phi = 0$, $\mu = 0$ can be found from this family by phase shifts $x \mapsto x + \theta$ and spatial inversion $x \mapsto -x$.

Because the functions Ψ and M are smooth, Taylor's theorem justifies the expansions

$$\phi = A\cos(kx) + A\psi_1 + A^2\psi_2 + A^3\psi_3 + o(A^3) \tag{29}$$

$$\mu = A\mu_1 + A^2\mu_2 + o(A^2) \tag{30}$$

with ψ_1, ψ_2 and ψ_3 orthogonal to $(\cos(kx), \sin(kx))$. Substituting these expansions into the ODE and solving for each quantity yields

$$\psi_1(x) = 0 \tag{31}$$

$$\psi_2(x) = \frac{1}{4k^2}\mathcal{H}_{\phi\phi}(0,0)\left[-1 + \frac{1}{3}\cos(2kx)\right] \tag{32}$$

$$\psi_3(x) = \frac{1}{8k^2}\left[\frac{1}{24k^2}\left[\mathcal{H}_{\phi\phi}(0,0)\right]^2 + \frac{1}{24}\mathcal{H}_{\phi\phi\phi}(0,0)\right]\cos(3kx) \tag{33}$$

and

$$\mu_1 = 0 \tag{34}$$

$$\mu_2 = \frac{1}{\mathcal{H}_{\phi\mu}(0,0)}\left[\frac{5}{24k^2}\left[\mathcal{H}_{\phi\phi}(0,0)\right]^2 - \frac{1}{8}\mathcal{H}_{\phi\phi\phi}(0,0)\right] \tag{35}$$

where the subscripts on $\mathcal{H}$ indicate the derivatives of $\mathcal{H}$ with respect to those variables. Most importantly,

$$\mathcal{H}_{\phi\mu}(0,0) = 4\pi\sum_\alpha \frac{q_\alpha^2}{m_\alpha}\int_0^\infty \frac{\partial^2 h_\alpha}{\partial\eta\partial\mu}\left(\frac{u^2}{2}, 0\right) du \neq 0 \tag{36}$$

by Eq. (23), so μ_2 is well defined. Thus, the amplitude of the wave goes like $\sqrt{\mu}$, where μ measures the deviation of the functions h_α from zero. This is a classical pitchfork bifurcation like the one illustrated in Fig. 1. When μ_2 is positive μ is quadratic in A and the pitchfork opens to the right as shown in the figure. When it is negative μ again is quadratic in A, but the pitchfork opens to the left. When it is zero, the pitchfork is higher order, but the symmetries of the problem guarantee that $M(A) = M(-A)$, so the bifurcating branch will always appear only on one side of $\mu = 0$.

The one remaining task is to construct a family of distribution functions $f_\alpha(x,u,A)$ that go to $F_\alpha(u+V)$ as $A \to 0$. To do this, for each A, two functions $g_\alpha^\pm(\eta, A)$, defined on $\eta \in [\Phi_\alpha^{min}, \infty)$, must be chosen. These functions must satisfy

$$i)\ \ g_\alpha^+(\eta,A) + g_\alpha^-(\eta,A) = \rho_\alpha(\eta) + h_\alpha(\eta, M(A)), \quad \Phi_\alpha^{max} \le \eta$$

$$ii)\ \ g_\alpha^\pm(\eta,A) = \frac{1}{2}\left(\rho_\alpha(\eta) + h_\alpha(\eta, M(A))\right), \quad \Phi_\alpha^{min} \le \eta \le \Phi_\alpha^{max}$$

and

$$iii)\ \ 0 = 4\pi\sum_\alpha q_\alpha \int_{\Phi_\alpha^{max}}^\infty \left[g_\alpha^+(\eta,A) - g_\alpha^-(\eta,A)\right] d\eta\,.$$

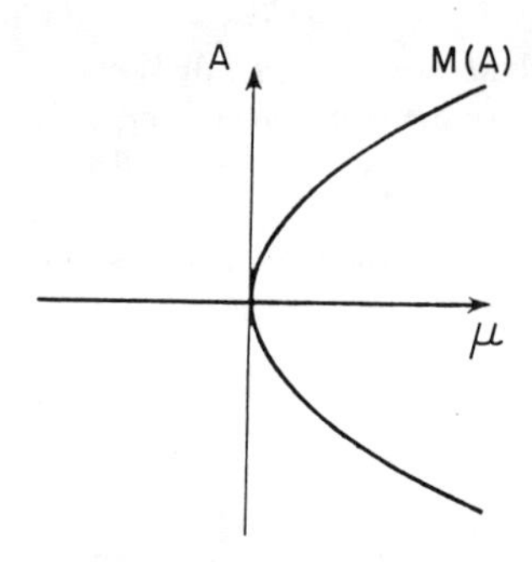

Fig. 1 The Pitchfork Bifurcation. $A = 0$ is a solution for all μ. The branch of bifurcating solutions is described by $\mu = M(A)$.

Since $\Phi_\alpha^{max} \geq 0$ for A small, ρ_α can be replaced with R_α in condition *i*. With $\delta_\alpha^\pm(\eta, A) = g_\alpha^\pm(\eta, A) - G_\alpha^\pm(\eta)$ this condition becomes

$$\delta_\alpha^+(\eta, A) + \delta_\alpha^-(\eta, A) = h_\alpha(\eta, M(A)), \quad \Phi_\alpha^{max} \leq \eta \tag{37}$$

while, in terms of $\delta_\alpha^\pm$, condition *iii* becomes

$$0 = 4\pi \sum_\alpha q_\alpha \left[\int_{\Phi_\alpha^{max}}^\infty [G_\alpha^+(\eta) - G_\alpha^-(\eta)] \, d\eta + \int_{\Phi_\alpha^{max}}^\infty [\delta_\alpha^+(\eta, A) - \delta_\alpha^-(\eta, A)] \, d\eta \right] \tag{38}$$

However, since the F_α describe a spatially uniform equilibrium of the one-dimensional Vlasov-Maxwell equations, they must satisfy the zero current condition; this implies that Eq. (38) is equivalent to

$$\sum_\alpha q_\alpha \int_0^{\Phi_\alpha^{max}} [G_\alpha^+(\eta) - G_\alpha^-(\eta)] \, d\eta = \sum_\alpha q_\alpha \int_{\Phi_\alpha^{max}}^\infty [\delta_\alpha^+(\eta, A) - \delta_\alpha^-(\eta, A)] \, d\eta \tag{39}$$

Functions $\delta_\alpha^\pm(\eta, A)$ that both satisfy Eqs. (37) and (39) and also go to zero as $A \to 0$ can always be chosen.

Suppose that small $\delta_\alpha^\pm$ have been so chosen. This fixes the functions $g_\alpha^\pm$ and then

$$f_\alpha(x, u, A) = \begin{cases} g_\alpha^+\left(\dfrac{u^2}{2} + \dfrac{q_\alpha}{m_\alpha}\phi(x), A\right), & u \geq 0 \\ g_\alpha^-\left(\dfrac{u^2}{2} + \dfrac{q_\alpha}{m_\alpha}\phi(x), A\right), & u \leq 0 \end{cases} \tag{40}$$

provides a family of distribution functions, depending on the amplitude A of the potential; together, these distribution functions and the potential provide the desired solution (in the wave frame). Then, with $U_\alpha(x) = \sqrt{2\Phi_\alpha^{max} - 2(q_\alpha/m_\alpha)\phi(x)}$,

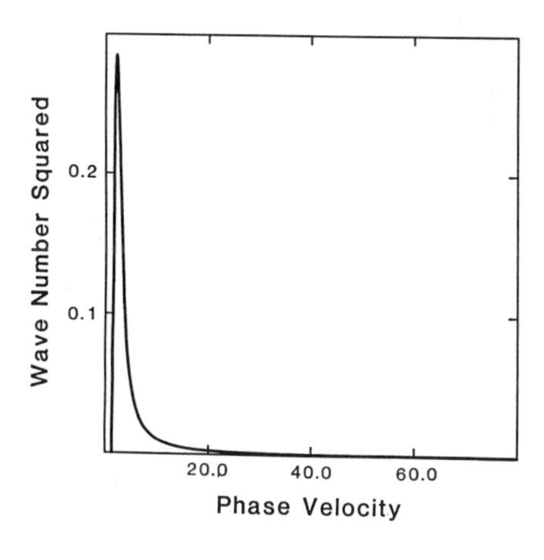

Fig. 2 κ^2 vs V, in units of inverse electron Debye length squared and electron thermal velocity respectively, for a thermal proton-electron plasma.

$$\int_0^{\pm\infty} |f_\alpha(x,u,A) - F_\alpha(u+V)|\,du$$

$$= \int_0^{U_\alpha(x)} \left| \frac{1}{2}\left[\rho_\alpha\left(\frac{u^2}{2} + \frac{q_\alpha}{m_\alpha}\phi(x)\right) + h_\alpha\left(\frac{u^2}{2} + \frac{q_\alpha}{m_\alpha}\phi(x), A\right)\right] - G_\alpha^\pm\left(\frac{u^2}{2}\right)\right| du$$

$$\int_{U_\alpha(x)}^{\infty} \left| G_\alpha^\pm\left(\frac{u^2}{2} + \frac{q_\alpha}{m_\alpha}\phi(x)\right) + \delta_\alpha^\pm\left(\frac{u^2}{2} + \frac{q_\alpha}{m_\alpha}\phi(x)\right) - G_\alpha^\pm\left(\frac{u^2}{2}\right)\right|$$

Both of these integrals go to zero as $A \to 0$; thus $f_\alpha \to F_\alpha$ in the sense that

$$\int_{-\infty}^{\infty} |f_\alpha(x,u,A) - F_\alpha(u+V)|\,du \longrightarrow 0 \tag{41}$$

as $A \to 0$.

There is no guarantee however that this family of distribution functions will be smooth in the amplitude. A formal expansion of $g_\alpha^\pm(u^2/2 + (q_\alpha/m_\alpha)\phi(x), A)$ in A yields

$$g_\alpha^\pm\left(\frac{u^2}{2} + \frac{q_\alpha}{m_\alpha}\phi(x), A\right) = G_\alpha^\pm\left(\frac{u^2}{2}\right) + A\cos(kx)\frac{q_\alpha}{m_\alpha}\frac{dG_\alpha^\pm}{d\eta}\left(\frac{u^2}{2}\right) + A\frac{\partial g_\alpha^\pm}{\partial A}\left(\frac{u^2}{2}, 0\right) + \ldots \tag{42}$$

But $dG_\alpha^\pm/d\eta|_{\eta=0}$ *need not exist!* When it does not, the amplitude expansion of f_α cannot be even first order in A at the phase velocity V ($u = 0$ in the wave frame), even though the expansion of ϕ is through third order; the amplitude expansion of f_α can be carried out to first order only in the very special case in which $F_\alpha'(V) = 0$, and to second order only when $F_\alpha'''(V) = 0$ as well, otherwise a direct amplitude expansion of the distribution function will be plagued by singularities at the phase velocity. This indeed proved to be the case in papers by

Simon and Rosenbluth[4] and Burnap, Micklavčič, Willis and Zweifel,[5] in which such expansions were attempted and led to ill-defined quantities.

4. The Dispersion Function For The Nonlinear Waves

A $2\pi/k$ periodic wave moving at the velocity V with respect to an observer gives the appearance of an electric field oscillating at the frequency $\omega = kV$. Substituting $V = \omega/k$ into $k^2 = \kappa^2$ and shifting the integration variable yields the following relation between k and ω

$$k^2 = 4\pi \sum_\alpha \frac{q_\alpha^2}{m_\alpha} \mathrm{P}\!\int \frac{F'_\alpha(u)}{u - \omega/k}\, du \tag{43}$$

This relation can be regarded as a dispersion relation for the nonlinear plasma waves constructed in Sec. 3, although it does not have the same significance as a dispersion relation for linear waves: the concept of group velocity does not have any physically useful meaning for nonlinear waves. Furthermore, although this is a dispersion relation for the waves constructed in Sec. 3, it is possible to construct other nonlinear longitudinal waves that do not satisfy Eq. (43).

This relation between wave number and frequency is different from the dispersion relation derived in the linear theory.[6] There, a dispersion relation is derived by analytically continuing certain integrals that appear in solving the linear initial value problem, subject to the hypotheses that the functions that describe the equilibrium and the initial perturbation are analytic functions on a wide strip around the real axis in the complex velocity plane. The resulting dispersion relation for undamped linear waves takes the form

$$k^2 = 4\pi \sum_\alpha \frac{q_\alpha^2}{m_\alpha} \mathrm{P}\!\int \frac{F'_\alpha(u)}{u - \omega/k}\, du - 4\pi^2 i \sum_\alpha \frac{q_\alpha^2}{m_\alpha} F'_\alpha(\omega/k) \tag{44}$$

Because of the addition of the imaginary term on the right hand side, this dispersion relation cannot be satisfied unless $0 = 4\pi \sum_\alpha (q_\alpha^2/m_\alpha) F'_\alpha(\omega/k)$, that is, $F(u) \equiv 4\pi \sum_\alpha (q_\alpha^2/m_\alpha) F_\alpha(u)$ must have an extremum at the phase velocity ω/k. Furthermore, for physically relevant—that is to say positive—equilibrium distribution functions, this dispersion relation for undamped linear waves cannot be satisfied when F has only a single extremum;[7,8] when ω/k is at that extremum, the integral on the right hand side is always negative.

For undamped nonlinear waves of a given phase velocity to exist near F_α it is necessary that $\kappa^2 \geq 0$. Thus a plot of κ^2 vs V immediately indicates over what range of phase velocities such waves can exist; indeed, in light of the explicit construction of small amplitude waves when $k^2 = \kappa^2$, such a plot reveals for what phase velocities such waves do exist. Figure 2 shows a numerically calculated plot of κ^2 as a function of V for a proton-electron plasma (mass ratio 1/1840) in which the ions and electrons have equal temperatures and mean velocities. For phase velocities below $1.35v_e^{th}$, κ^2 is negative and so undamped plasma waves cannot be supported. For phase velocities above this velocity, however, κ^2 is positive, and so there will be small amplitude waves with wave number κ and phase velocity $V > 1.35v_e^{th}$.

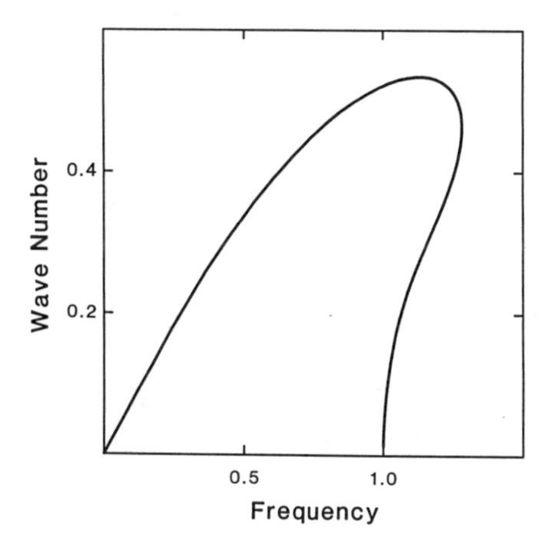

Fig. 3 Wave number vs frequency, in units of inverse electron Debye length and electron plasma frequency respectively, for a thermal proton-electron plasma.

The existence of longitudinal wave solutions with arbitrarily small amplitude electric fields in such a single-humped plasma directly contradicts the predictions of the standard linear analysis.[6,7] According to the linearized description, the electric field that results from any small amplitude disturbance in such a plasma must damp, albeit slowly for Langmuir and ion-acoustic waves. In contrast, the nonlinear analysis summarized here has shown that there are indeed small amplitude waves with undamped electric field in such plasmas.

From the graph of κ^2 vs V, the set of k, ω that satisfy the dispersion relation, Eq. (43), can be generated. The resulting plot of k vs ω for the thermal equilibrium plasma is presented in Fig. 3. This curve shows that for any wave number k there are two kindred frequencies ω, which correspond to a fast wave and a slow wave with identical wavelengths. Furthermore, for frequencies above the plasma frequency but below the cut-off frequency at about $1.3\omega_p$ there are two distinct wave numbers for each frequency, this time corresponding to two distinct wave motions, one fast and one slow, with identical frequency but different wavelengths. Thus, in this regime, a measurement of either wavelength or frequency alone can not identify the other. This is again at variance with the linear theory of longitudinal waves in such a plasma, which predicts (slowly damped) Langmuir and ion-acoustic waves which can be completely specified by their frequency. In fact, the dispersion curve in Fig. 3 shows that ion-acoustic and Langmuir waves are simply two ends of a continuum: in the low frequency, small wave number regime the dispersion curve is approximately linear, as traditionally predicted for ion-acoustic waves, while in the plasma frequency, small wave number regime the familiar parabola $\omega^2 = \omega_p^2 + c^2k^2$ is seen.

Acknowledgment

At the time this work was undertaken, James Paul Holloway was supported by NASA Grant NGT-50183 and J. J. Dorning was supported in part by US ONR Contract N00014-85-K-0382. This work is currently supported by NASA Grant NAGW-1696.

References

[1]Holloway, J. P., "Longitudinal Traveling Waves Bifurcating From Vlasov Plasma Equilibria," Ph.D. Dissertation, Engineering Physics, Univ. of Virginia, Charlottesville, VA, Jan. 1989.

[2]Kato, T., "Perturbation Theory for Linear Operators," 2nd corrected printing of the 2nd ed., Springer-Verlag, Berlin, 1984, pg. 196.

[3]Bernstein, I. B., Greene, J. M., and Kruskal, M. D., "Exact Nonlinear Plasma Oscillations," *Physical Review*, Vol. 108, 1957, pp. 546–550.

[4]Simon, A., and Rosenbluth, M. N., "Single-Mode Saturation of the Bump-on-the-Tail Instability: Immobile Ions," *Physics of Fluids*, Vol. 19, 1976, pp. 1567–1580.

[5]Burnap, C., Miklavčič, M., Willis, B. L., and Zweifel, P. F., "Single-Mode Saturation of a Linearly Unstable Plasma," *Physics of Fluids*, Vol. 28, 1985, pp. 110–115.

[6]Landau, L., "On the Vibrations of the Electronic Plasma," *Journal of Physics*, Vol. 10, 1946, pp. 25–34.

[7]Backus, G., "Linearized Plasma Oscillations in Arbitrary Electron Velocity Distributions," *Journal of Mathematical Physics*, Vol. 1, 1960, pp. 178–191.

[8]Jackson, J. D., "Longitudinal Plasma Oscillations," *Journal of Nuclear Energy, Part C*, Vol. 1, 1960, pp. 171–189.

Chapter 3. Atomic Oxygen Generation and Effects

Laboratory Simulations of Energetic Atom Interactions Occurring in Low Earth Orbit

G. E. Caledonia*
Physical Sciences, Inc., Andover, Massachusetts

Abstract

The Space Shuttle flights provided the first significant data base on the environment experienced by a large space structure operating in low Earth orbit (LEO). A number of interesting and unanticipated effects were observed. These include material erosion induced by ambient oxygen atoms, the visible Shuttle glow occurring above surfaces exposed to the ram flow, and large near= field perturbations and variability in the gaseous neutral and plasma environment about the Shuttle. These latter observations are coupled to the contaminants introduced by the Shuttle and their interaction with the ambient gases. The understanding of these phenomena is critical to the proper design and specification of future large LEO space structures such as a space station. This paper provides a brief overview of these observations and their phenomenological interpretation and then discusses laboratory approaches for their investigation. The emphasis of the paper will be on the state-of-the-art in the development of energetic (v ~ 8 km/s) oxygen atom sources and the variety of experiments presently being performed with such devices.

Introduction

The Space Shuttle flights provided the first opportunity to quantitatively examine the local environment found about a large space structure in low Earth orbit (LEO), 250-350 km. In essence, it was found that contaminants on Shuttle surfaces continually outgassed forming a contaminant cloud about the Shuttle. The interaction

---

Presented as an Invited Paper.

*Senior Vice President.

between the ambient environment and the Space Shuttle with its contaminant cloud occurs at the orbital velocity of 8 km/s and can produce deterious effects for Shuttle performance (see Ref. 1 for an overview). These include material erosion, Shuttle glows, and plasma enhancements.

This interaction can provide for structural dysfunction by material erosion as well as operational dysfunction through oxidation or coating phenomena. Furthermore, the contaminant cloud can provide a more difficult environment for external probes to operate in because of increased radiative backgrounds due to surface and "cloud" glows, enhanced plasmas and surface charging, and also direct deterioration of diagnostic equipment.

In the next section we will provide a brief overview of the phenomenology occurring in this ambient environment/contaminant cloud Shuttle interaction and review the physical data required to characterize it. This discussion will be followed by a description of laboratory beam techniques presently being developed to provide the required data and brief examples of typical measurements made on the Physical Sciences Inc. (PSI) energetic oxygen atom facility.

Contaminant Cloud

Of course the dominant source of the contaminant cloud is the space structure itself. Contaminant species are naturally introduced around the structure and on its surfaces during operational events such as thruster firings, water dumps, and other vents. Furthermore, particles will shake off of surfaces and outgassing will occur. The ambient hard uv flux will also act to enhance desorption and outgassing and indeed may interact with some species to provide polymerization on surfaces. Thus, the specification of the chemical form of these outgassed species is itself critical to determining their ultimate impact on space structure performance.

Ambient species, primarily O and N_2, but also lesser species such as N, O_2, and H, will impact space station surfaces at orbital velocities of 8 $km \cdot s^{-1}$. It has been found that in many materials this interaction produces material erosion. It is generally assumed that this erosion is the result of oxygen atom attack, and for many hydrocarbon materials mass loss is estimated to occur in 1 out of 10 impacts.[2] The reaction products of these interactions have not been measured but in many cases can be estimated from mechanistic arguments. Erosion species identification is, of course, critical for specification of subsequent reaction, evaluation of deposition tendency, and understanding of erosion-induced glows. It has been suggested in the past that limited key components could be protected from oxygen atom attack by the application of sacrificial coatings. The ultimate impact of these eroded materials on the local environment must be carefully evaluated before such applications.

Oxygen atom attack can also provide for functional deterioration in more insidious ways. For example, Leger and Visentine[3] have recently pointed out that moly-disulfide, a common lubricant, can oxidize under oxygen atom attack, becoming abrasive. Such a transformation would provide increased particle loading and decreased mobility for moving parts. Other materials, although they do not erode, will oxidize, resulting in changing thermal and radiative properties. Furthermore, possible synergistic effects on material erosion resulting from uv loading or surface charging remain to be evaluated.

The catalytic properties of various materials in high velocity interactions must also be evaluated. For example, knowledge of the surface accommodation coefficient for momentum is critical to specifying the local cloud density. Specifically, if the ambient species accommodate their momentum on the surface, they will then effuse away thermally, resulting in a higher local gas density than if they had scattered elastically from the surface. The momentum accommodation coefficient is a key parameter in contaminant cloud models. As another example, catalytic reactions of ambient species on surfaces have long been suggested[4] as possible sources for excited states, which could then either further interact or produce a surface glow. No data are presently available on such catalytic effects at orbital velocities for the various materials of importance to the space station.

The ambient gasses will also interact with outgassed species around the space structure. This interaction, initially occurring at orbital velocities but also of importance at lower velocities, will produce a scattering pattern that plays a role in defining the density profile and extent of the contaminant cloud. To the author's knowledge there are no measurements of the angular differential cross sections or momentum transfer resulting from such heavy body collisions. Furthermore, inelastic collisions will also occur producing radiation from direct excitation or chemiexcitation, as well as species transformation. The data base for such interactions is very sparse in the energy range of interest.

Last, the importance of positive ion reactions must be evaluated. Although ambient ion concentrations are typically small compared to neutral concentrations, there can be charge buildup around the space structure. One way this can occur is through reactions between ambient ions and contaminant neutrals. Many reactions of this type will move charge between species without significant momentum transfer. Thus, charge initially at rest in the Earth frame may be swept along with the Shuttle.[5] Such reactions can also produce excited species that can radiate, and new ionic species that are more likely to provide surface deposition. The efficiencies for ion neutralization on various space station materials remain to be evaluated. We note that enhanced ionization levels have been proposed as a source of Shuttle glow.[6]

Table 1 Required data for space station contamination level specification

Data Required		
Material behavior under uv Loading	**Ambient/surface interactions**	**Ionic interactions**
- Outgassing rates - Products - Surface effects - Particle formation	- Momentum transfer/ accommodation	- Surface neutralization efficiencies
Material "erosion" studies under energetic species impact	- Surface reactions e.g., fast $N_2 \rightarrow N + N \rightarrow N_2(A)$	- Ambient ion/ contaminant reactions
- Erosion rates	- Surface collision induced glows	- Ion velocity separation
- Passivation effects-nonlinear behavior	- Material dependence of all above	- Quasi-neutrality
- Species produced · State changes · Deposition	**Ambient/contaminant cloud interactions**	**Nonlinear effects**
- Surface property changes	- Differential scattering cross sections	
- Erosion-induced glows	- Inelastic collisions · Chemical reaction	
- Synergistic effects · uv Loading · Charged surfaces	· Radiative inducing e.g., $O + M \rightarrow O + M^*$; $O + AB \rightarrow OA^* + B$	

The various data requirements discussed above have been summarized in Table 1. Potential laboratory techniques for developing this data base are examined below.

Laboratory Studies

Development of the data base requires the use of state-of-the-art neutral and ionic beams exhibiting characteristic velocities of 8 $km \cdot s^{-1}$. A number of neutral oxygen atom beams have been under development in response to the Shuttle observations of significant material erosion (e.g., Refs. 7-15). These devices have recently been reviewed by Visentine[16] and the following discussion is an expansion of his effort.

Presently, extant oxygen atom sources may be broken into five types: thermal, high-temperature electrical discharges; ion beams, beam-surface interaction, and laser breakdown.

Thermal sources employ a variety of techniques to partially dissociate oxygen molecules without significantly altering the thermal content of the gas. Such techniques include plasma ashers, microwave discharges, etc. These sources tend to have low atomic fluxes in that the atoms have only thermal velocities, and can be plagued by significant concentrations of other reactive species, particularly metastable states. These devices are frequently used for material testing because of their wide availability and simplicity of operation. Nevertheless, their value as a simulation of the higher velocity space interaction remains to be demonstrated.

Table 2 Discharge-created neutral beams

Type	Technique	Source	Species	Energy eV	Status
Surfatron (microwave)	Freejet	UTIAS, Tennyson et al.	O, O_2, 98% He	< 3	Operational
Plasma torch (dc arc)	Supersonic nozzle	Aerospace, Arnold and Peplinski	O, O_2, 98% He	1-2	Operational
Plasma torch (dc arc)	Supersonic nozzle	ARI, Freeman	O (He, O_2)	1.3	Operational

Higher-velocity beams can be achieved by strongly heating gases as in a plasma torch. Here the concept is to highly excite a gas by rf, dc, or microwave sources and subsequently expand the gas through a freejet or hypersonic nozzle converting the sensible heat to velocity. The characteristics of several such sources are listed in Table 2. To maximize the expanded velocity helium is the gas of choice for excitation with a few percentage points of oxygen added downstream. The expanded beam is then composed of a mix of oxygen atoms and molecules dilute in helium. These are high-flux devices, allowing oxygen atom brightnesses of 10^{18} to 10^{19} atoms/s-sr but are limited to O atom energies below 3 eV, primarily by material limitations.

Perhaps the most popular type of beam under development is the ion source. Here positive or negative ions are created by either electron bombardment or rf excitation and then are electrostatically accelerated and focused to achieve the proper velocity, at which point the charge is stripped by various techniques such as charge exchange or grazing incidence surface neutralization. Such beams, which can readily achieve the appropriate velocity, however, are typically limited to low fluxes because of coulombic repulsion effects. For standard ion sources, achievable neutral fluxes as high as 10^{15} $cm^{-2} \cdot s^{-1}$ have been predicted but not yet demonstrated.

An overview of existing ion sources is provided in Table 3. Note that focusing limitations caused by coulombic repulsion can be eliminated by modifying the beam to be a neutral plasma, i.e., electron injection. Singh et al.[12] and Langer et al.[10] have adopted the technique and report higher fluxes; the beam of Singh et al. has not been neutralized after focusing, however, and remains a beam of O^+,e pairs rather than oxygen atoms. Indeed, ultimate beam neutralization appears to present a major complication for positive ion sources. Gaseous charge exchange is inefficient because of low cross sections, and surface neutralization produces dispersed beams. Indeed, the majority of positive ion sources are presently being operated as ion rather than neutral beam sources. The negative ion sources show more promise inasmuch as laser photodetachment can be utilized for the neutralization process. Furthermore, variable beam $O(^1D)$ to $O(^3P)$ concentrations can be achieved by adjusting the photodetaching laser wavelength appropriately.

As outlined in Table 4, several groups have utilized beamed energy/surface interactions to provide energetic oxygen atoms. The approach is to use beamed energy in various forms (electrons, ions, lasers) to promote oxygen

Table 3 Ion beam techniques

Type	Technique	Source	Species	Energy eV	Status
Electron bombardment	Charge	Martin Marrieta; Sjolander; MSFC Carruth	O^+, O_2^+	5	Ions only at present
Electron bombardment	X^+ surface neutralization	Vanderbilt Univ., Tolk and Albridge[a]	O	5-10	Low flux device
Electron bombardment	Charge exchange	LeRC, Hanks, and Rutledge	O^+, O, O_2	40-200	Ions only at present, adding uv
Electron bombardment	Neutral plasma	G.E., Singh	O^+, O_2^+, e	3-10	No neutralization plans
rf Discharge	Neutral plasma surface neutralization	Princeton Univ., Langer	O, O_2	~5	In construction
Hollow anode discharge	Charge exchange	USCLA, Munz	O^+, O_2^+		Ions only at present
Electron bombardment/N_2O	O^- Photodetachment	Boeing, Rempf	O	5	Proof of principle
Electron bombardment/N_2O	O^- Photodetachment	JPL, Chutjian	O	5	Proof of principle

[a]Also examines negative ion beams.

atom removal from surfaces. Ferrieri et al.[13] have examined ion sputtering off of metal pentoxides and Brinza[14] has utilized laser/surface breakdown as a sputter source. Of course, other elements present in the solid can track the resulting oxygen atoms. Brinza's approach of breaking down cryogenically deposited thin films of ozone is appealing in this respect. The technique of Outlaw et al.[11] can provide a high-purity oxygen atom beam. In their approach high-pressure and vacuum chambers are connected by a silver membrane. Oxygen molecules introduced into the

Table 4 Beam-surface interaction devices

Type	Technique	Source	Species	Energy eV	Status
Ion blowoff	Sputtering on thin films, e.g., VaO_5, TaO_5	Brookhaven, Ferrieri	0	2-18	Operational
Laser blowoff	Breakdown on thin films, e.g., O_3, ITO	JPL, Brinza	0	2-7	Operational
Electron-stimulated desorption	Surface dissociation/diffusion through Ag membrane	LaRC, Outlaw	0	5	Proof of principle, 10^{12} $cm^{-2} \cdot s^{-1}$

Table 5 Laser sustained discharge neutral beams

Type	Technique	Source	Species	Energy eV	Status
Laser discharge	CW breakdown	Los Alamos, Cross	0, O_2, He	< 4	Operational
Laser discharge	Pulsed breakdown	PSI, Caledonia and Krech	0, O_2	2-14	Operational

high-pressure chamber interact with the silver, dissociate, diffuse through the membrane, and ultimately are adsorbed on the vacuum-side silver surface. Electron-stimulated desorption is then used to free and energize the oxygen atoms. The predicted ultimate oxygen atom flux achievable by this technique is 10^{15} $cm^{-2} \cdot s^{-1}$. These devices are all operational, and their disadvantages are wide energy spread and diverging beams.

Perhaps the most mature technology for materials testing applications are the laser-sustained discharge sources,[8,9] e.g., where lasers are used to produce a high=

temperature plasma that is subsequently expanded in a freejet or supersonic nozzle to produce a high-velocity neutral beam. Such sources have been demonstrated to produce beams of the desired velocity of 8 $km \cdot s^{-1}$ at flux levels of 10^{17}-10^{18} $cm^{-2} \cdot s^{-1}$. To date they are the only O atom devices that exhibit both the appropriate energy for orbital simulation and high flux. The characteristics of the two extant devices are listed in Table 5. Both devices have been operated for long periods of time and are useful for aging studies.

Example Beam Experiments

In principle, many of the devices described above can be used to develop a data base for the phenomenology listed in Table 1. For purposes of illustration, several such measurements performed with the PSI pulsed laser oxygen atom source will be briefly described below.

The PSI source has been described in some detail elsewhere[8,17] and will only be briefly reviewed here. In operation a fast-acting valve is used to introduce a pulse of oxygen molecules into a previously evacuated supersonic nozzle. A pulsed CO_2 laser focused near the nozzle throat is used to break down this gas and form a high-temperature plasma. The plasma subsequently expands producing a high-velocity beam made up primarily of oxygen atoms. A schematic of the PSI system, as it is used for material erosion studies, is provided in Fig. 1. The laser beam enters from the left and the atomic beam propagates to the

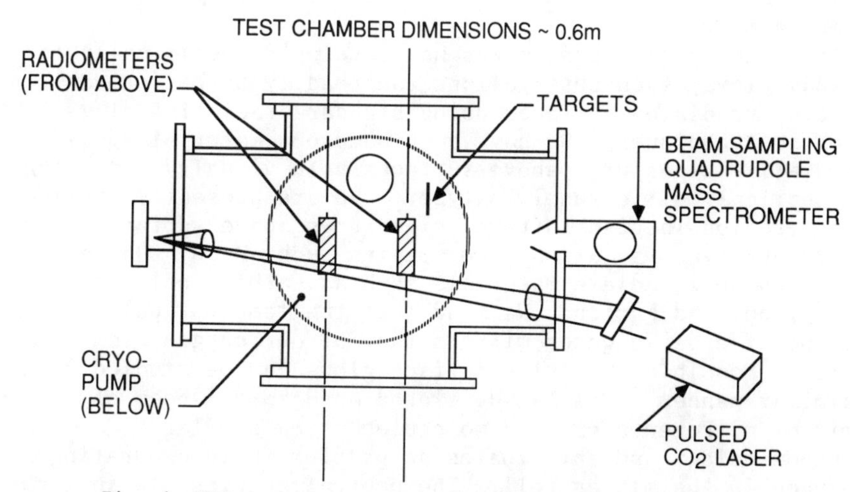

Fig. 1 PSI oxygen atom test facility under development.

right striking material targets as shown. The large expansion ratio allows samples as large as several hundred square centimeters to be irradiated. A mass spectrometer is available for beam characterization and radiative diagnostics are available to monitor both beam properties and radiation from the beam target interaction. A second mass spectrometer head will soon be installed to allow monitoring of erosion products.

A number of materials have already been studied with this device with total 8 $km \cdot s^{-1}$ O atom irradiation levels typical of those encountered during a few weeks operation at Shuttle altitudes, $\leq 10^{21}$ O atoms cm^{-2}. In general mass removal rates and surface properties have been found to be similar to those observed during Shuttle operation.[16]

A typical SEM sequence of an irradiated material, carbon fiber reinforced plastic, is shown in Fig. 2. Irradiation level was ~ 3×10^{20} oxygen atoms/cm^2. The top left panel of the figure contrasts virgin and irradiated portions of the material, and, then, with increasing magnification the remaining panels highlight the differing erosion patterns of the fiber and plastic portions of the composite. These ruglike erosion patterns are similar to those observed on materials irradiated on Space Shuttle flights.

As soon as the second mass spectrometer head is operational, this system will be capable of addressing many of the issues discussed in the previous sections. These include mass loss rates, erosion species identification, and surface property changes. Synergistic effects resulting from uv loading, heating cycles, stress, and flexing can also be investigated with modest system improvements.

Although the system was not developed specifically to study glows, such observations can readily be performed above irradiated surfaces using standard radiative diagnostic techniques. We have seen numerous material-specific radiative signatures above surfaces both visually and using an optical multichannel analyzer. We are presently examining erosion-induced infrared signatures above surfaces. We have observed radiation from species such as CO, CO_2, and OH when we irradiate materials such as carbon, polyethylene, and kapton. We find that the temporal pulse shape of erosive gases mirrors that of our oxygen atom beam. Possible catalytic surface glows can be studied in a similar manner. PSI has developed an 8 $km \cdot s^{-1}$ beam of a mix of nitrogen atoms and molecules using similar phenomenology and anticipates no problem in incorporating oxygen in the mix as well. The neutral species mix in such beams will be evaluated using the mass spectrometer.

Fig. 2 Scanning electron micrograph analysis of carbon fiber reinforced plastic. Irradiated by 3 x $10^{20}/cm^2$ 5-eV oxygen atoms. Top left contrasts virgin and irradiated materials. Remaining views emphasize erosion patterns at increased magnification as shown.

PSI has also developed a crossed-beam experiment to study infrared excitation resulting from energetic oxygen atom collisions with species such as CO, CO_2, and CH_4. A schematic of the device is shown in Fig. 3. Here a skimmed beam of fast oxygen atoms is crossed at right angles with a skimmed pulse (again using a fast pulsing valve) of thermal target molecules. The IR radiation produced by the interaction is monitored by a wide field of view, variable-

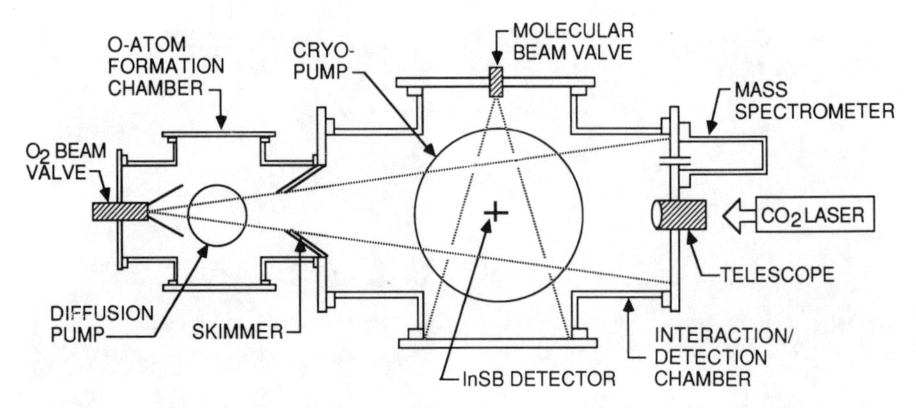

Fig. 3 Crossed beam facility.

filtered detector. This experiment design is challenging in that the measurement must be made under single collision conditions; i.e., both the target molecules and the oxygen atoms are required to experience only one collision in the interaction zone to ensure that the radiation is characteristic of fast atom impact. These measurements are in process, and sufficient signal is available to evaluate excitation cross sections. Similar techniques can be used to study visible excitation and chemical reaction in such systems.

Summary

A number of important quantities that must be evaluated in order to both understand and predict the contamination field about large LEO space structure have been enumerated. It has been shown that the recent development of energetic oxygen atom sources enables the laboratory evaluation of the majority of these quantities. A number of potential measurement techniques have been briefly reviewed.

Acknowledgments

The PSI work presented has been performed in collaboration with R. H. Krech, B. D. Green, K. Holtzclaw, B. Upschulte, M. Fraser, and A. Gelb. This research was suported by NASA under Contract NAS7-963 and by Physical Sciences Inc. internal funds.

References

[1]Green, B. D., Caledonia, G. E., and Wilkerson, D. T., "The Shuttle Environment: Gases, Particulates, and Glow," Journal of

Spacecraft and Rockets, Vol, 22, September-October 1985, pp. 500-511.

[2]Leger, L. J. and Visentine, J. T., "A Consideration of Atomic Oxygen Interactions with the Space Station," Journal of Spacecraft and Rockets, Vol. 23, January 1986, pp. 50-56.

[3]Leger, L. J. and Visentine, J. T., unpublished results, 1987.

[4]Green, B. D., Rawlins, W. T., and Marinelli, W. J., "Chemiluminescent Processes Occurring Above Shuttle Surfaces," Planetary and Space Science, Vol. 34, September 1986, pp. 879-887.

[5]Caledonia, G. E., Person, J. C., and Hastings, D. E., "The Interpretation of Space Shuttle Measurements of Ionic Species," Journal of Geophysical Research, Vol. 92, January 1987, pp. 273-281.

[6]Papadopoulos, K., "On the Shuttle Glow (The Plasma Alternative)," Radio Science, Vol. 19, 1974, pp. 571-577.

[7]Arnold, G. S. and Peplinski, D. R., "Reaction of Atomic Oxygen with Vitreous Carbon Laboratory and STS-5 Comparisons," AIAA Journal, Vol. 23, June 1985, pp. 976-977.

[8]Caledonia, G. E., Krech, R. H., and Green, B. D., "A High Flux Source of Energetic Oxygen Atoms for Material Degradation Studies," AIAA Journal, Vol. 25, January 1987, pp. 59-63.

[9]Cross, J. B. and Cremers, D. A., "High Kinetic Energy (1 to 10 eV) Laser Sustained Neutral Atom Beam Source," Nuclear Instruments & Methods, Vol. 813, 1986, p. 658.

[10]Langer, W. D., Coben, S. A., Manus, D. M., Motley, P. W., Oro, M., Paul, S. F., Roberts, D., and Selberg, H., "Detection of Surface Glow Related to Spacecraft Glow Phenomena," Geophysics Research Letters, Vol. 13, April 1986, pp. 377-386.

[11]Outlaw, R. A., Hoflund, G. B., and Corallo, G. R., "Electron-Stimulated Desorption of Atom: Oxygen From Polycrystalline Ag," Applied Surface Science, Vol. 28, March 1987, pp. 235-246.

[12]Singh, B., Amore, L. J., Saylor, W., and Racette, G., "Laboratory Simulation of Low-Earth Orbital Atomic Oxygen Interaction with Spacecraft Surfaces," AIAA Paper 85-0477, January 1985.

[13]Ferrieri, R. A. and Chu, Y. Y., "Ion Sputtering on Metal Oxides: A Source of Translationally Hot $O(^3P_j)$ Atoms for Chemical Studies Above 1 eV," Revision Scientific Instruments, Vol. 59, October 1988, pp. 2177-2184.

[14]Brinza, D. (ed.), Proceedings of the NASA Workshop on Atomic Oxygen Effects, JPL Publ. 67-14, 1987.

[15]Hoggatt, J. T., Hill, S. G., and Johnson, J. C. (eds.), "Materials for Space--The Gathering Momentun," 18th International SAMPE Technical Conference, Vol. 18, Society for the Advancement of Material and Process Engineering, Seattle WA, 1986.

[16]Visentine, J. T., Atomic Oxygen Effects Measurements for Shuttle Missions STS-8 and 41G, Vol. III, NASA Technical Memorandum 100459, September 1988.

[17]Caledonia, G. E. and Krech, R.H., "Energetic Oxygen Atom Material Degradation Studies," AIAA Paper 87-0105, 25th Aerospace Sciences Meeting, Reno, NV, Jan. 1987.

High-Energy/Intensity CW Atomic Oxygen Beam Source

J. B. Cross* and N. C. Blais†
Los Alamos National Laboratory, Los Alamos, New Mexico

Abstract

A high-intensity (10^{19} O-atoms/s-sr) continuous-wave (cw), high-energy (1-5 eV) source of neutral oxygen atoms (AO) has been developed, which produces a total fluence of 10^{22} AO/cm^2 at a distance of 15 cm from the source in less than 100 hr of continuous operation. The source employs a cw CO_2 laser-sustained discharge to form a high-temperature (15,000-20,000 K) plasma in the throat of a nozzle (0.3-mm-diam) using 3-8 atm of rare-gas/O_2 mixture. Visible and infrared photon flux levels of 1 W/cm^2 have been measured 15 cm downstream of the source, whereas vacuum ultraviolet fluxes are comparable to that measured in low Earth orbit (LEO). The reactions of atomic oxygen with Kapton, Teflon, silver, and various coatings have been studied. The oxidation of Kapton (reaction efficiency = 3×10^{-24} cm^3/atom ±50%) has an activation energy of 0.8 kcal/mole over the temperature range of 25-100°C at a beam energy of 1.5 eV and produces low-molecular-weight gas-phase reaction products (H_2O, NO, CO_2). Teflon reacts with ≈0.1-0.2 efficiency to that of Kapton at 25°C, and both surfaces show a rug-like texture after exposure to the AO beam. Angular scattering distribution measurements of O-atoms show a near-cosine distribution from reactive surfaces, indicating complete accommodation of the translational energy with the surface, whereas a nonreactive surface (nickel oxide) shows specularlike scattering with 50% accommodation of the translational energy with the surface. A technique for simple on-orbit chemical experiments using resistance measurements of coated-silver strips is described.

---

*Section Leader, Materials Chemistry, Chemical and Laser Sciences Division.

†Member of Professional Staff, Chemical and Laser Sciences Division.

Introduction

Operations in low Earth orbit (LEO) (100-500 km) must take into consideration the highly oxidative character of the environment. Partial pressures in the range of 10^{-6} to 10^{-7} Torr of atomic oxygen (AO) are present, which produce extensive oxidation of materials facing the direction of travel (ram direction). The ram oxidation is most severe not only because of the high flux (10^{15} O-atoms/s-cm^2) but also because of the high collision energy (5 eV) of oxygen atoms with the ram surfaces caused by the orbital velocity (8 km/s) of the spacecraft. Ground-based simulation of these conditions is being accomplished using a continuous-wave (cw) laser-sustained discharge source for the production of a 1-5 eV beam of O-atoms with a flux of up to 2 X 10^{17} O-atoms/s-cm^2 (200 X LEO flux).

O-Atom Plasma Source

The AO source[1] (Fig. 1) employs a cw plasma formed by focusing a high-power CO_2 laser beam to produce plasma temperatures of 15,000-20,000 K in a rare-gas/oxygen mixture. The cw CO_2 laser (10.6 μm) is used to sustain the spark-initiated plasma in the mixture which subsequently flows through the throat (0.3 mm in diameter) of a hydrodynamic expansion nozzle producing an atomic beam of neutral species. Stagnation pressures of 2-8 atm are employed depending on the rare gas; i.e., 2 atm for 50% O_2 in argon and 8 atm for 15% O_2 in helium. A 2.54-cm focal-length ZnSe lens is used to focus the laser beam to a 70-100-μm spot producing power densities of 10^7 W/cm^2, which sustains the plasma at a roughly 50% ionized condition. The lens is moved axially to position the plasma ball in the throat of the water-cooled nozzle. Continuous operation times of greater than 75 hr have been obtained producing fluences >10^{22} O-atoms/cm^2. The source is mounted in a molecular beam apparatus (Fig. 2), where the gas mixture is skimmed after exiting the nozzle and then collimated into a neutral atomic beam of rare gas and O-atoms. The facility consists of 1) the laser-sustained AO beam source, 2) three stages of differential pumping between source and a sample manipulator located 15 cm from the source, 3) a rotatable mass spectrometer with TOF capability for measurements of scattered particle angular and velocity distributions to determine drag coefficients and gas phase reaction products, and 4) a mass spectrometer calibration chamber and quadrupole mass spectrometer located 120 cm from the

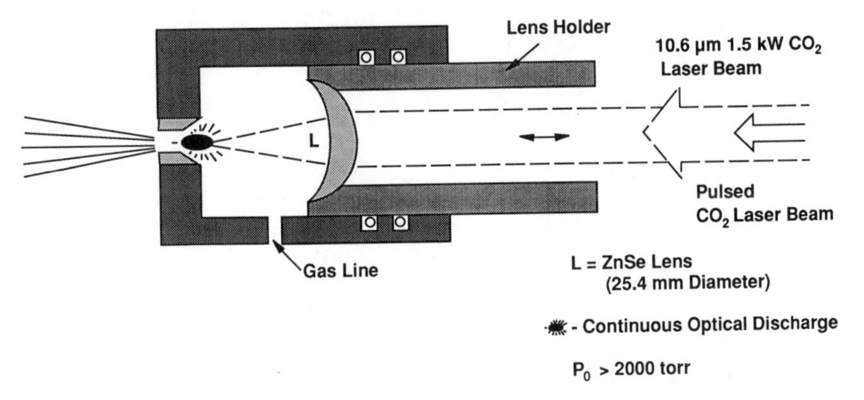

Fig. 1 Continuous-wave laser-sustained plasma neutral O-atom source.

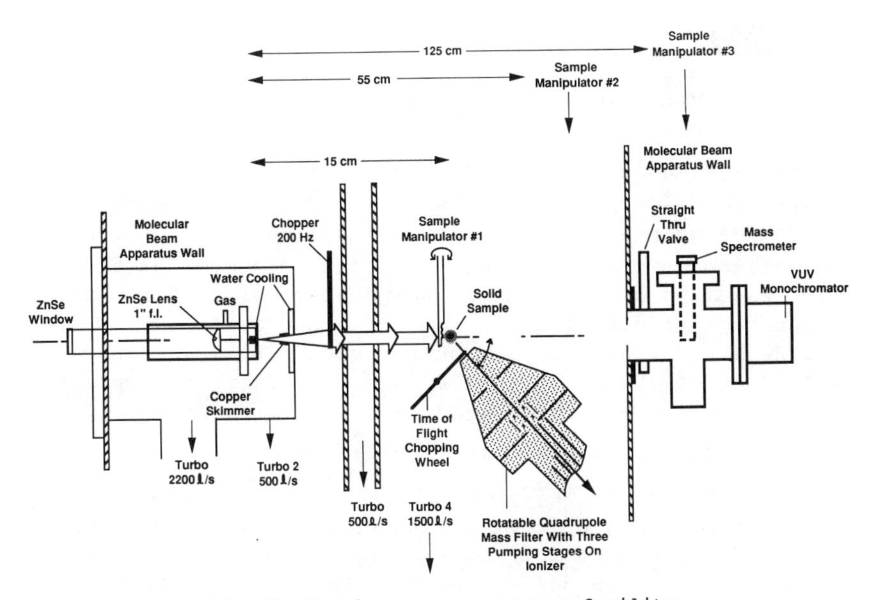

Fig. 2 Atomic oxygen exposure facility.

source, which is used for beam TOF measurements. At the sample manipulator position, 15 cm from the source, O-atom flux densities of 10^{17} AO/s-cm^2 are obtained whereas at the mass spectrometer calibration chamber position, flux densities of 10^{15} AO/s-cm^2 are recorded. A base pressure of 1 X 10^{-9} Torr is measured in the sample exposure chamber, which rises to 2 X 10^{-7} Torr when the AO beam is operating. Figure 3 shows AO energy distributions obtained from time-of-flight (TOF) analysis of the plasma source beam. The

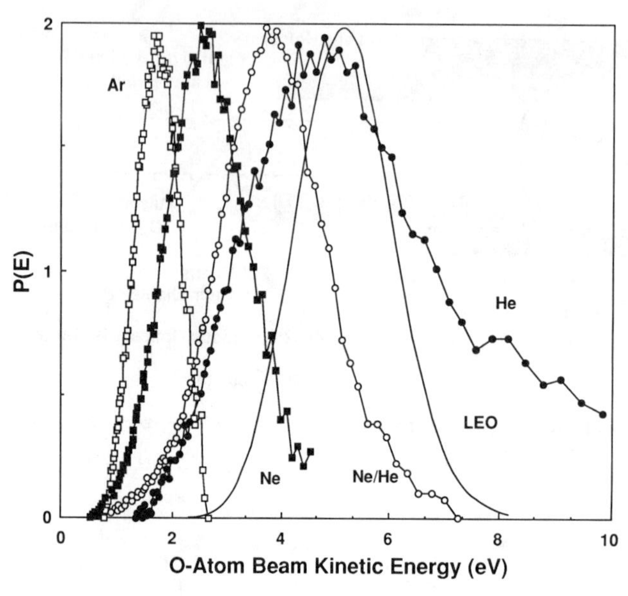

Fig. 3 Flux density kinetic energy distributions [P(E)].

rare-gas symbols indicate the oxygen/rare-gas composition; i.e., Ne designates a mixture of 50% Ne + 50% O_2, whereas Ne/He designates 25% O_2 + 25% Ne + 50% He. Also shown for comparison is the calculated in-orbit AO energy distribution assuming an atmospheric temperature of 1000 K. The percentage of dissociation for the mixtures Ar, Ne, Ne/He, and He are 85, 87, 96, and an estimated 98% at stagnation pressures of 1900, 2200, 4200, and 6000 Torr, respectively. The cw CO_2 laser power ranged from 1100 W for Ar/O_2 to 1600 W for He/O_2. Beam intensities range from 8 X 10^{16} O-atoms/s-cm^2 for an Ar/O_2 mixture to 2 X 10^{17} O-atoms/s-cm^2 at the sample manipulator position, which is 15 cm from the nozzle (see results for measurement methods). Long-term operation with the He/O_2 mixture at 6000 Torr requires additional mechanical pumping capacity on the source foreline--this is presently being added to the system. Figure 4 presents the plasma spectral distribution as measured with a 0.5-m vacuum monochrometer employing both CsTe solar blind and bialkali photomultiplier tubes (PMT). The distribution has been corrected for the wavelength dependence of the PMT's quantum efficiency. Since power measurements taken at the sample manipulator position (Fig. 2) indicated a heating power density of ≈ 1 W/cm^2 compared with 0.14 W/cm^2 for solar radiation, the solar spectrum was normalized to 0.14 that of the plasma spectrum at 8000 Å for purposes of comparison. Vacuum ultraviolet (VUV) intensities com-

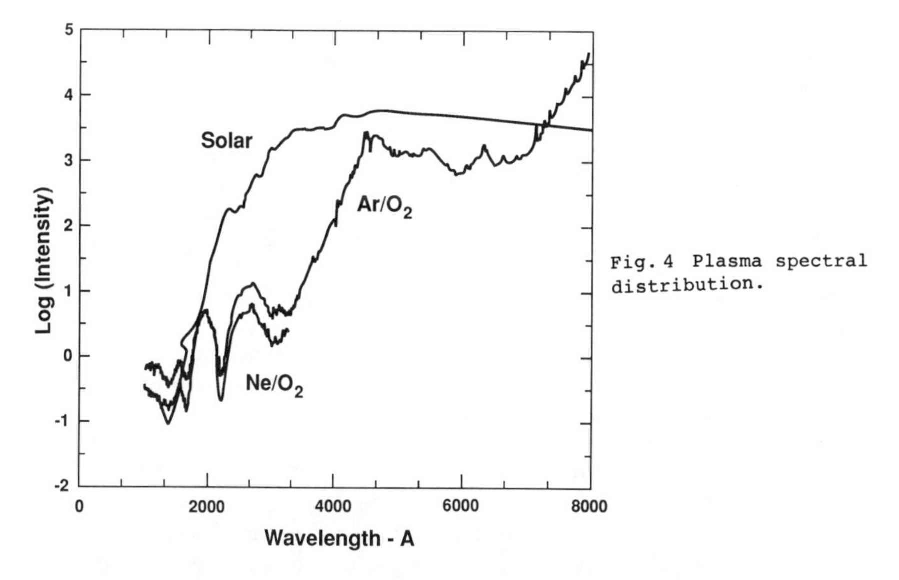

Fig. 4 Plasma spectral distribution.

parable to that of solar radiation are observed with a Ne/O_2 mixture, whereas Ar/O_2 produces VUV intensities a factor of 3 higher; thus, polymer materials can be investigated under the same VUV conditions that exist in orbit. Future work will provide a high transmission velocity filter to eliminate photons from the AO beam in order to investigate the mechanisms of VUV/AO polymer degradation.

Results

Beam Flux Calibration

Samples are normally exposed to the beam at the sample manipulator position 15 cm from the nozzle, where the O-

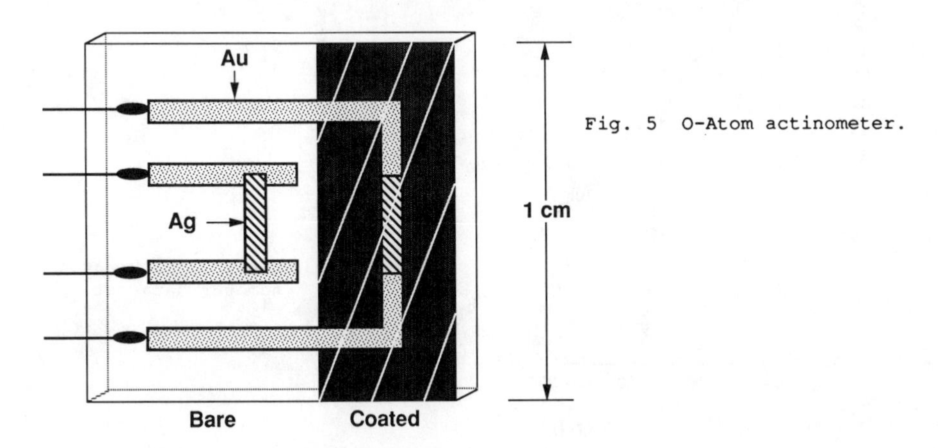

Fig. 5 O-Atom actinometer.

atom flux is ≈10^{17}/s-cm^2 and beam calibration is performed at the mass spectrometer calibration position 120 cm from the nozzle where the flux is approximately equal to orbital flux (10^{15}/s-cm^2). A simple technique[2] was developed to perform beam calibrations and to evaluate reaction rates of coatings and conductive materials. A thin strip of conductive material is deposited on a nonconducting and nonreactive substrate (Fig. 5), and a resistance measurement is made as a function of time of exposure to the AO beam to determine the rate of material loss or conversion to a nonconducting oxide. In the case of silver, the conducting metal is transformed into its nonconducting oxide, thus increasing the device's resistance, whereas carbon, if deposited in place of the silver, forms volatile oxides, which alters the film's cross- sectional area. The time history of a 250-Å-thick bare silver film, when exposed to the O-atom beam (2.5 eV) 120 cm from the nozzle is shown in Fig. 6 and represents an oxidation rate of roughly 1.7 monolayers/s or a beam flux of approximately 2 x 10^{15} O-atoms/s-cm^2. The oxidation rate and beam flux were calculated using a number of assumptions: 1) the sticking

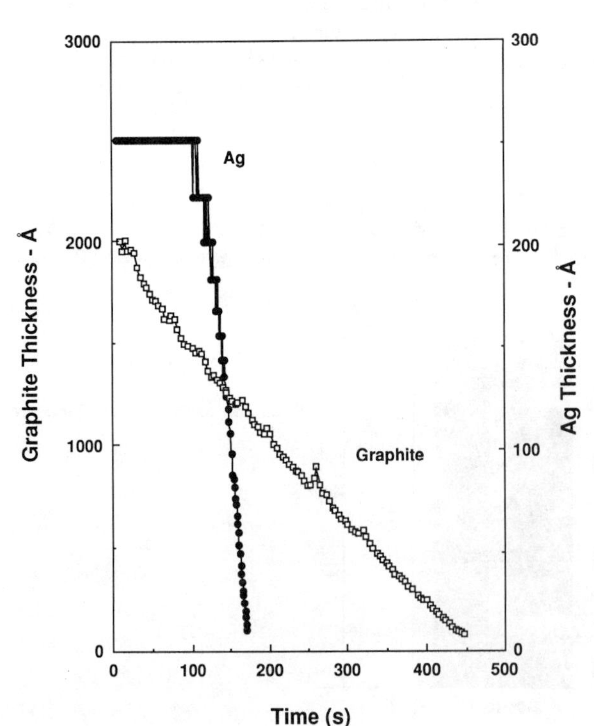

Fig.6 Typical actinometer data set.

probability of O-atoms on bare polycrystalline silver is unity over the entire range of oxide formation; 2) the nearest-neighbor distance (NND) of atoms in polycrystalline silver is that of silver-face-centered cubic structure, 2.89 Å; and 3) the O_2 molecules in the beam (15% using Ar/O_2 mixture) do not react with silver to form silver oxide. The absolute flux and fluence values, therefore, may have 10-15% errors, and additional work is required to ascertain more accurate values (5%). The silver thickness (th) as a function of time (t) is calculated from the resistance (R) measurement as follows:

$$th(t)=th(0)\cdot R(0)/R(t)$$

where th(0) is the original thin-film thickness (250 Å) and R(0) the film resistance at t = 0. The silver oxidation in Fig. 6 was performed 120 cm from the beam source, and the AO flux F(120) calculated from the plot is

$$\begin{aligned} F(120) &= (\Delta th/\Delta t)(1/NND)(\#Ag\ atoms/cm^2) \\ &= (250\ \text{Å}/50\ s)(1/2.89\ \text{Å})(1.2 \times 10^{15}) \\ &= 2.1 \times 10^{15}\ AO/s\text{-}cm^2 \end{aligned}$$

The flux 15 cm from the nozzle is then

$$\begin{aligned} F(15) &= F(120)(120\ cm/15\ cm)^2 \\ &= 1.3 \times 10^{17}\ AO/s\text{-}cm^2 \end{aligned}$$

The two primary assumptions in this AO estimate, 1) nonreactivity of O_2 at high kinetic energy and 2) a unity sticking coefficient of AO on Ag_2O/Ag, need to be investigated further before this technique can be used as a measure of absolute AO fluxes. Completion of this effort will then provide the basis for accurate absolute measurements of AO flux not only in the laboratory but also in orbit. In addition to these uses, the reactivity of coating can be determined by measuring the time required to burn through a known thickness of over-coat. The burnthrough is detected by the subsequent oxidation of the silver.

As a check on this silver actinometer technique, absolute partial $[C_i]$ number density measurements were made in the mass spectrometer calibration chamber (Fig. 2) after thermal equilibration of the beam in the chamber. An orifice of known diameter (1.270 cm) operating under effusive flow conditions was employed to pump on the chamber having a known temperature T(ch). The beam composition was measured using lock-in detection of the modulated beam and a residual gas analyzer corrected for known relative ionization cross sections. From these data, the partial flux density of each component of the beam was determined from

$$F_i = 1/4[C_i] \langle V_i \rangle$$

where $\langle V_i \rangle = [8kT(ch)/\pi M_i]^{1/2}$ is the average velocity, k is the Boltzmann constant, and M_i is the atomic mass of the ith species. This technique produces AO flux values to within 80% of the silver actinometer technique.

Reaction Rate Dependence on Surface Temperature and Beam Energy

Investigations were performed to determine the relative importance of surface temperature (activation energy) and beam translational energy on the reaction rate of AO with sputter-deposited graphite. Figure 6 shows a typical data set, where, first, a silver actinometer is used to obtain a beam flux calibration at a distance of 120 cm, and then a sputter-deposited graphite actinometer is exposed at a distance of 15 cm from the AO source. The 250-Å bare silver strip was exposed to the AO beam at the position 120 cm from the source and required over 100 s before the silver began to react. This effect is caused by the burning off

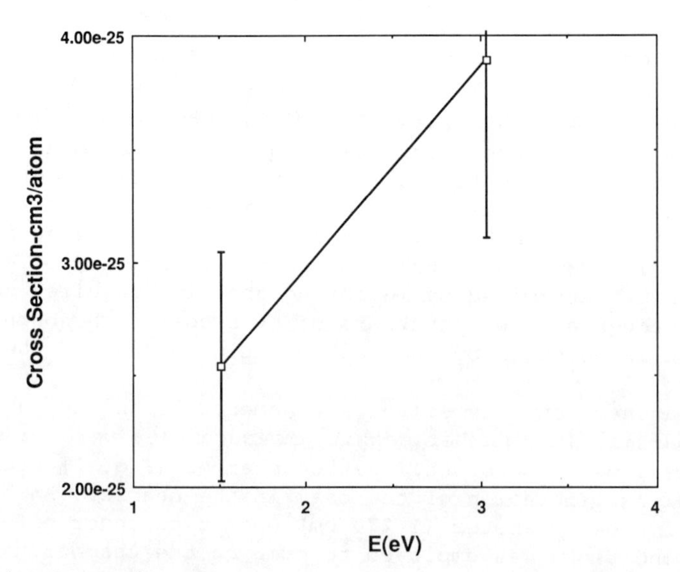

Fig. 7 Atomic oxygen beam energy dependence on reaction rate.

of a hydrocarbon overlayer, which was produced in the device fabrication. A large number of bare silver actinometers have been exposed, and, in all cases, the lag or burnoff time correlates with the silver oxidation rate or AO flux. The slope of the silver oxidation plot represents a flux of $\approx 2 \times 10^{15}$ O-atoms/s-cm^2 at a distance of 120 cm or a flux of 1.3×10^{17} at the 15-cm sample manipulator distance. The 2000-Å sputter-deposited graphite film was oxidized at the sample manipulator position 15 cm from the nozzle. The rate is 4.4 Å/s at a flux level of 1.3×10^{17} O-atoms/s-cm^2. A reaction efficiency (RE) is obtained having units of cm^3/AO

$$\begin{aligned} \text{RE} &= \text{Graphite reaction rate/AO flux} \\ &= [4.4 \times 10^{-8}\ \text{cm/ s}]/[1.3 \times 10^{17}\ \text{AO/s-cm}^2] \end{aligned}$$

$$\text{RE(Graphite at 2.5 eV)} = 3.4 \times 10^{-25}\ \text{cm}^3/\text{AO}$$

Figure 7 shows the reaction efficiency variation of sputter-deposited graphite with AO over the kinetic energy range of 1.5-3 eV at a surface temperature of 35°C. The beam energy was varied by seeding oxygen with various rare gases (see Fig. 3). The error bars represent the worst case of the still yet unknowns in the experiment; i.e., whether the reactions of O_2 molecules present (15%) in the beam with silver or graphite have a translational energy barrier between 2 and 4 eV. The data represent a 10-30% change in reaction efficiency between 1.5 and 3 eV. Similar investigations varying the surface temperature at constant beam energy show an activation energy of 800 cal/mole for graphite oxidation in agreement with flight experiments.[3] These results indicate that graphite-based materials are affected primarily by high kinetic AO but that high-temperature surfaces will suffer additional degradation in LEO caused by the thermal energy component of the reaction.

Angular Scattering Distribution Measurements

AO angular scattering distributions from reactive and nonreactive materials were measured at a beam kinetic energy of 1.5 eV using the rotatable mass spectrometer (Fig. 2). These distributions indicated that the reactive surface produces complete accommodation of the AO, with the AO leaving the surface in a cosine distribution having forgotten its initial velocity vector (Fig. 8). The nonreactive surface, however, produces specularlike scattering, with the AO leaving having remembered its initial

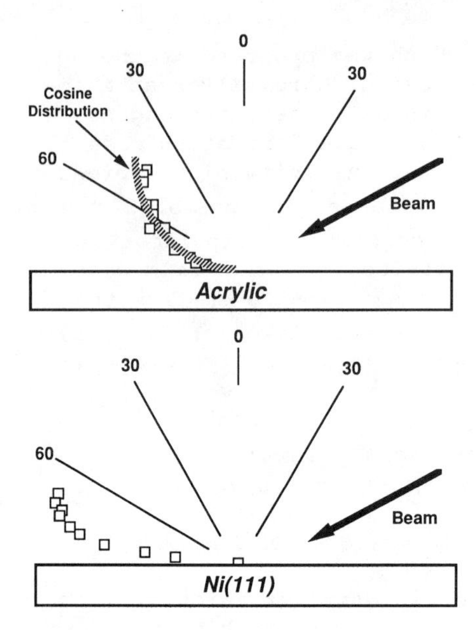

Fig. 8 AO angular scattering distributions.

velocity vector. Separate TOF experiments[1] (not shown here) indicate that approximately 50% of the initial energy is lost to the nonreactive surface, whereas the reactive surface showed complete accommodation. The polyethylene gas-phase products were found to be H_2O, CO, and CO_2, all of which showed accommodation to the material surface temperature (393 K). These results have important implications for long-term operation of spacecraft in the LEO environment in terms of spacecraft drag and contamination. Nonreactive surfaces reduce spacecraft drag compared to reactive ones while simultaneously acting as mirrors for AO, possibly creating higher than normal intensity AO fluxes on nonram surfaces that are shielded from direct AO attack. The CO_2 and CO reaction products emitted in a cosine distribution from surfaces may pose a contamination problem on other parts of the craft.

Accelerated Materials Testing

Spacecraft operating in LEO for extended periods of time (20 yr) will experience total fluences[4] of 10^{22}-10^{23} O-atoms/cm^2. Materials certification studies will be needed to ensure reliable operation of spacecraft over their operational lifetime and will require exposures of selected materials to the same total fluence that the craft will experience in a period of time shorter than real time. The

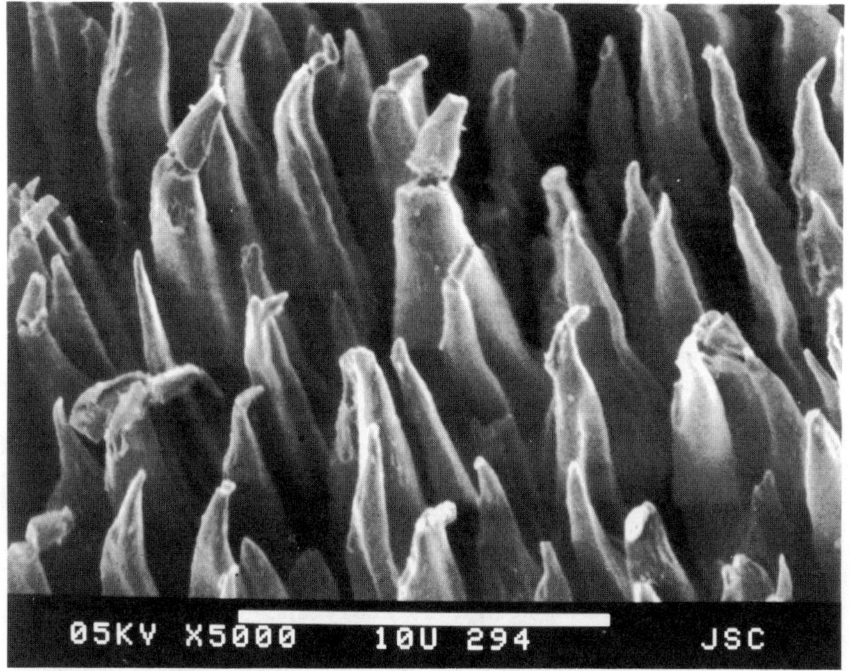

Fig. 9 SEM photograph of Teflon.

laser-sustained AO plasma source described here is capable of performing such studies since in 100 hr of operation a total fluence of 3.6 X 10^{22} O-atoms/cm^2 can be obtained at the 1-5 eV energy level. The SEM photograph in Fig. 9 shows the resulting Teflon surface after a 76-hr exposure to the AO beam (total fluence 2 X 10^{22} O-atoms/cm^2). In addition to the ability to perform such exposures, an understanding of the surface reaction mechanisms is needed to interpret the results. For example, the increased AO flux needed for accelerated testing may provide a mechanism for enhanced AO recombination on surfaces before reaction with the surface can occur, thus giving a lower materials reaction probability than actually occurs in orbit.

Conclusion

A continuous-wave source of 1-5 eV atomic oxygen producing 100 times (10^{17}/s-cm^2) low Earth orbit (LEO) flux has been demonstrated and used to investigate the relative importance of atomic oxygen (AO) kinetic energy and surface temperature on the reaction with sputter-deposited graphite. Oxidation of graphite at 25°C is dominated by AO kinetic energy. The source produces a vacuum ultraviolet (VUV) flux that is roughly equal to the solar flux in LEO and has demonstrated the ability to perform accelerated testing as evidenced by the reaction observed on Teflon and Kapton. Interpretation of these results for use in predicting long-term oxidation effects on spacecraft surfaces will require a better understanding of the AO/VUV reaction mechanisms.

Angular and energy transfer distributions, which can be used to determine spacecraft drag coefficients as well as subsequent contamination by gas-phase reaction products, have been measured. Reactive surfaces produce more drag or energy accommodation and contamination than do nonreactive ones, but a nonreactive (nickel oxide) surface can act as a mirror scattering AO with considerable kinetic energy onto spacecraft surfaces that normally would be shielded from ram AO attack.

A very simple device (actinometer) for measuring thin-film and coating reaction rates has been demonstrated, which can be used for both laboratory and orbital flight experiments. Now that orbital energies and fluxes of AO have been obtained in the laboratory, one of the primary concerns will be to obtain reaction rate data in long-exposure active orbital experiments using highly characterized surfaces and conditions to compare with laboratory data. This undertaking would provide several benefits: 1) determine whether the existing facilities are really able to reproduce orbital rates; 2) determine if other effects, such as low intensity plasmas, affect the AO reaction rates; and 3) give confidence that accelerated testing methods actually can be used to predict long-term exposure effects in LEO.

Acknowledgments

The authors wish to acknowledge discussions with Dr. Lubert Leger, James T. Visentine, and Dr. Steve Koontz of NASA Johnson Space Center. They also wish to thank Dr. Steve Koontz for SEM and profilometer analysis of the Kapton and Teflon samples and to acknowledge funding for this work from NASA Johnson Space Center, McDonnell Douglas Corporation, the Air Force Geophysics Laboratory, and the Strategic Defense Initiatives Office.

References

[1]Cross, J. B., Cremers, L. H., Spangler, L. H., Hoffbauer, M.A., and Archuleta, F.A., "CO_2 Laser Sustained CW Discharge Atomic Beam

Source," Proceedings of the 15th International Symposium on Rarefied Gas Dynamics, Vol. 1, p. 657, Grado, Italy, June 1986.

[2]Cross, J. B., Lan, E. H., and Smith, C. A., "A Technique to Evaluate Coatings for Atomic Oxygen Resistance," 33rd International SAMPE Symposium and Exhibition, p. 126, Anaheim, CA, March 1988.

[3]Gregory, J. C., Polymer Preprints, Vol. 28, No. 2, 1987, p. 459.

[4]Leger, L., Visentine, J., and Santos-Mason, B., "Selected Materials Issues Associated with Space Station," 18th International SAMPE Technical Conference, Vol. 18, p. 1015, Seattle, WA, Oct. 1986.

Development of a Low-Power, High Velocity Atomic Oxygen Source

J. P. W. Stark* and M. A. Kinnersley†
University of Southampton, Southampton, England, United Kingdom

Abstract

The determination of gas/surface interactions appropriate to space vehicle applications requires laboratory-based measurements using a high-velocity (~8 km/s) neutral atomic oxygen source. The development of such a source has been an active area of research in a number of laboratories for several years, with a variety of techniques being investigated. The type of source under development at Southampton University is based on the seeded arcjet principle, where a low-molecular-weight carrier gas (helium) is used in conjunction with an appropriate seed gas. The source power of this system under typical operating conditions is less than 3 kW, with a beam velocity approaching 4.5 km/s and a source flux greater than 10^{14} oxygen atoms/cm^2/s. In this paper results are presented concerning the use of several seed gases in order to identify different reactions that may be useful in the efficient production of atomic oxygen. A variety of gases have been studied separately and in conjunction with other gases, including nitrous oxide both as sole source of atomic oxygen and jointly with arc-produced atomic nitrogen. The chemically simpler reactions of the direct dissociation of molecular oxygen is also reported. Our results show that a severe degree of chemical nonequilibrium exists in the source. At least two thermal components are required to explain these results.

*Senior Lecturer, Department of Aeronautics and Astronautics.
†Research Assistant, Department of Aeronautics and Astronautics.

I. Introduction

A fundamental problem of rarefied gasdynamics is that of the influence of a boundary within a flowfield on the flow itself. For analytic tractability, assumptions of diffuse or specular reflection are assumed; thus, a solution may be elegant. Experience of high-velocity rarefied flows on space vehicles clearly indicates that such simple assumptions are not realistic because the nature of spacecraft material erosion indicates that at a minimum the components of flow species may change due to the boundary interactions. Space vehicles provide us with one of the most interesting aspects of rarefied gas flows; however, all of these flows are at high velocity. For example impact between atmospheric species and the vehicle occurs at nearly 8 km/s, and impact between exhaust plumes and surfaces are generally at velocities exceeding 2 or 3 km/s. Therefore, to provide experimental data on the nature of gas/surface interactions appropriate to space vehicles, it is necessary to produce a high velocity source that can produce a variety of species, together with an appropriate measurement system for determining the nature of the species interaction with surfaces.

Two particular aspects of space vehicle aerodynamics highlight the need for facilities to examine these high-velocity flows. The first one, high-velocity atmospheric atomic oxygen flows impacting on space vehicle surfaces, is particularly relevant to the selection of materials for spacecraft. A second one, shuttle glow, is of greater relevance to the users of space vehicles who are concerned with the radiation environment (particularly at near infrared and optical wavelengths) surrounding their instrument space platform. For the first of these aspects a pure atomic oxygen source is desirable, however, for the second it is probable that other species play a role in the radiation observed.[1] A source that is appropriate for simulating the space environment to investigate these phenomena should therefore operate with O, O_2, N, N_2, NO and He, ideally with the capability to operate with single species, or with gas mixtures. (Although most degradation of surfaces is most likely caused by O, synergistic effects should not be overlooked; in particular, material degradation observed on the Solar Maximum Mission has shown that material defects arising from micrometeorite impact tends to be associated with enhanced degradation.)

In this paper we report the development of a seeded arcjet source that is capable of producing both single-component species flows and also gas mixtures. Using helium as a seed gas leading to flow mixtures of atomic oxygen and helium, velocities approaching 4.5 km/s have been achieved, with source fluxes of atomic oxygen in excess of 10^{14} atoms/cm^2/s. Similar velocities have also been achieved with flows containing molecular nitrogen (and its oxides) together with oxygen and helium.

An additional feature of the source is the instrumentation. The determination of a gas/surface interaction requires evaluation of the re-emitted species velocity distribution function. This is achieved by the use of the combination of a cylindrical mirror analyzer and mass spectrometer. This combination permits the unambiguous evaluation of both re-emitted species and their velocity spectrum.

Section II of this paper describes the source and its instrumentation. Section III provides some details of the source operating characteristics that have been achieved. These results are analyzed in Sec. IV with regard to arcjet conditions and the nature of equilibrium within the flow.

II. Experimental Apparatus

The experimental facility uses a continuum beam source generation as devised by Kantrowitz and Grey. Production of energetic atomic oxygen in the fraction of electron volts to electron volt range is achieved with a conventional arc heated near atmospheric pressure source with the necessary differential pumping required for beam formation.

The overall facility consists of two vacuum chambers (Fig. 1). The first is used for the nozzle expansion and subsequent skimming to form the molecular beam. This in essence requires high pumping speeds for maintenance of low pressures in the 10^{-2} to 10^{-3} mbar region and uses three large vapor booster pumps (Edwards 18B3). The chamber is a 0.5-m-diam, 1-m-long cylindrical mild steel vessel using nitrile O rings and is capable of an ultimate pressure of 10^{-4} mbar. The second "target" chamber is a stainless steel system with both conventional and metal seals, pumped by a liquid nitrogen-trapped diffusion pump (Varian VHS6) and can achieve an ultimate pressure of 10^{-7} mbar. Both pump systems are able to use conventional diffusion and rotary pump oils. This is due to the presence of large amount of helium during the atomic oxygen beam runs and thus negates the need for the more costly poly perfluoroethers more commonly used for oxygen pumping. In addition, there are extensive mechanical motion and electrical feedthroughs for remote target movement and beam monitoring.

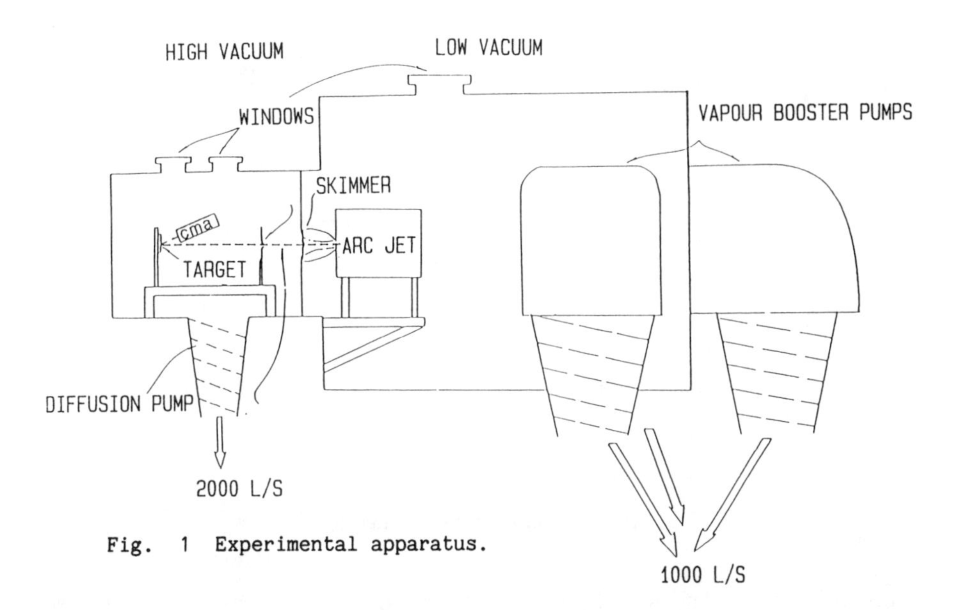

Fig. 1 Experimental apparatus.

The beam source is a Knuth[2] type arc-heated source operating over a power range of 1.5 - 3 kW. The heater comprises three sections: the nozzle, body, and cathode, all of which have high-pressure water cooling. The cathode is manufactured from 1/4-in. thoriated tungsten rod and is silver-soldered to the same size copper pipe for electrical and cooling services. The body houses a copper anode of annular shape and is connected to the positive terminal of a conventional 8-kW welding supply via water-cooled power cables. The nozzle serves both to stabilize the arc-heated gas flow and to provide a means of introducing the heavier seed gas to be accelerated. A schematic can be seen in Fig. 2.

High-energy heavy-species particle generation is accomplished via the "seeding" technique,[3] which uses a light carrier gas (He) to provide the necessary high-velocity freejet expansion for beam formation. The heavy gas is introduced as a minor constituent downstream of the arc (so as not to oxidize the electrodes). The heavy gas expands close to the light gas velocity, with some velocity slippage occurring. With beam enrichment, a high percentage of the beam flow can consist of the heavy species.

Atomic Oxygen Generation Techniques

The conventional method of generating atomic oxygen in arc heaters utilizes molecular oxygen as the minor

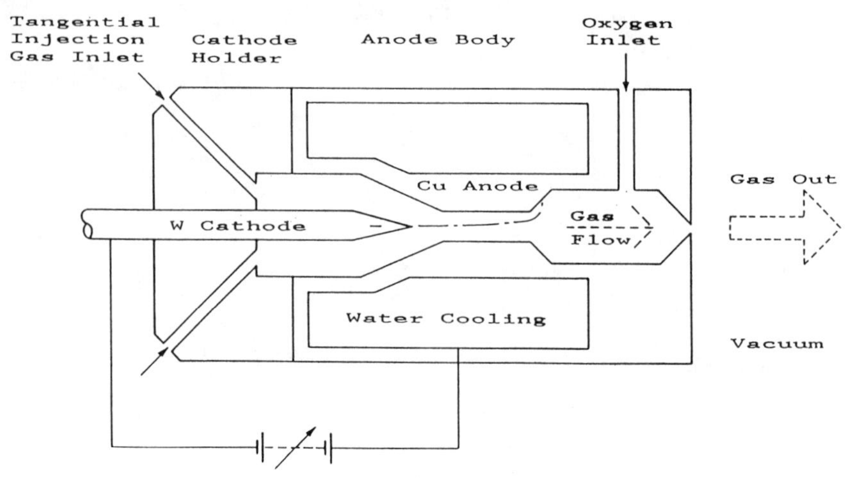

Fig. 2 Arc heater schematic.

constituent injected downstream of a high-temperature helium flow. This method has been used in the Southampton source, but recently other seed gases have been investigated with one aim being to increase the O_1 yield in the source. The injection of O_2 as source gas of O_1 has been used in various arc-heated facilities,[4,5] and, although producing high flux levels, is seen not to be fully dissociated in a pure helium flow. In fact, Silver et al.[5] added argon gas at very high arc power levels to realize full dissociation. This is by virtue of argon's lower themal diffusivity and hence higher nozzle exit temperatures. The disadvantages of operating with even a small percentage of argon is the resultant increase in the bulk mixture molecular weight and hence decrease in attainable exit speed. This effect can clearly be seen in Ref. 5. This is where the direct molecular oxygen route may suffer in comparison to chemical routes such as nitrous oxide/nitrogen.

Instrumentation - flow Characteristics

The necessary information needed to adequately characterize a molecular beam requires both knowledge of individual species concentration and species velocities. Absolute values of fluxes, although desirable, are hard to achieve with any certainty, due to the extensive calibration procedures needed in the number-density-measuring device. With mass spectroscopy, care has to be taken in subtracting the contribution of O_1 signal from

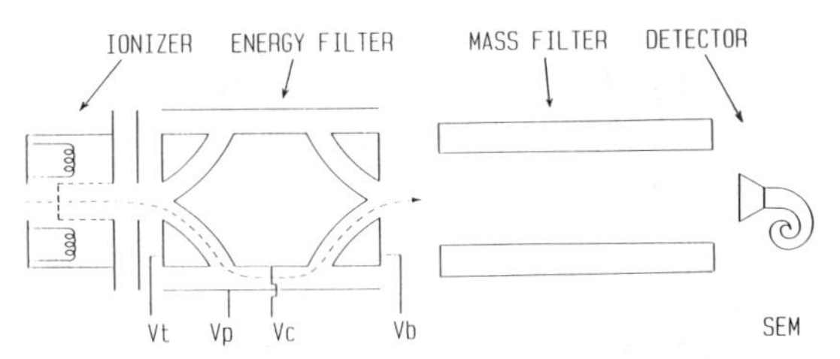

Fig. 3 CMA/mass filter combination.

cracked parent ions. Furthermore, the corrections used to account for differing collision cross sections in an electron bombardment detector will also contribute to the overall error.[6] Other effects such as O_1 recombination on the ionizer inlet will also add to the problem. Thus, extreme care has to be taken when basing flux measurements solely on mass spectroscopic measurements, and indeeed an article citing this problem for spacecraft-borne mass spectrometers confirms these fears.[7]

Measurements on the molecular beam were accomplised by two methods. In the first instance, flux values were based on mass spectroscopic measurements, accompanied by the standard correction formula of Sibener et al.[8] An alternative method involved the use of velocity discrimation to decrease the uncertainty in discriminating between beam and thermal background. This made use of a cylindrical mirror analyzer (CMA) (VG Quadrupoles) electrostatic energy filter located on the front end of the mass spectrometer with a low-emission ionizer (to minimize space charge effects). A schematic of both can be seen in Fig. 3. The CMA is able to scan energy with intensity at a constant filter transmission, irrespective of filter

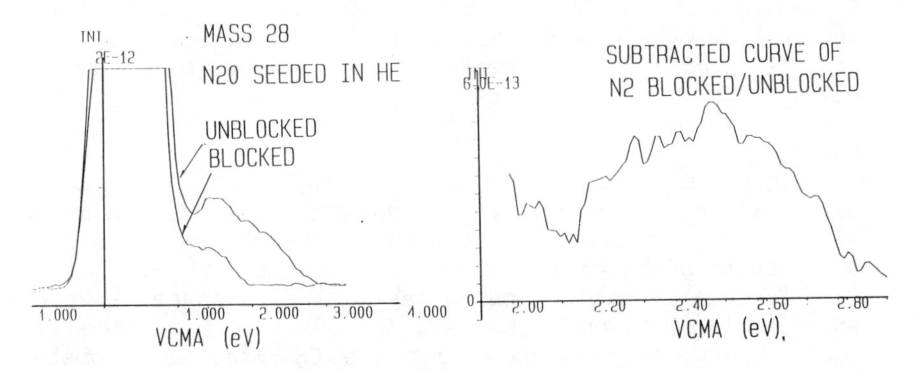

Fig. 4 Nitrogen blocked/unblocked beam signal.

energy. Hence, resultant scans will be a convolution of the measured energy distribution and CMA response. Typical results for the CMA/mass analyzer combination provide both mass and energy filtering, with a typical graph shown in Fig. 4, with a beam flag on and off (blocking and unblocking the beam). This particular scan shows nitrogen molecules accelerated in a helium beam. Note the large thermal background peak in comparison with the 1.7 eV beam peak. Of interest is the difference in thermal background in both blocked and unblocked cases. This would cause an underestimation of beam dissociation if a mass spectrometer was used in isolation.

III. Source Performance

The arc heater at Southampton differs from earlier arc-heated O_1 generation techniques in both gas flow rate and arc power. Helium is the working gas in the Southampton source, operating with a helium throughput of approximately 10-20 mbar. l/s. This scales as a factor of about 1/50 the flow speed of other sources with a consequent drop in necessary pumping speed to maintain an adequate background vacuum. Arc power is in the range of 1.5-3 kW and thus reduces the costs of the dc power supply. With both the gas flow speeds and arc power scaled down, it is still possible to produce helium average nozzle exit temperatures in the 1500° K range. This produces a beam velocity of ~4 km/s, which corresponds to about 1 eV atomic oxygen with the helium carrier traveling at an energy of approximately 0.3 eV. The helium energy can be seen in Fig. 5, via a CMA scan. Intensitites of a pure helium beam 0.5 m downstream of the source have been estimated at 1×10^{17} at cm^2/s.

Production of atomic oxygen via molecular oxygen and nitrous oxide routes provide approximate values of O_1 fluxes of 0.5×10^{15} at cm^2/s at 0.1 m downstream of the source. This has been estimated by a combination of both mass spectroscopic/CMA measurements and by the rate of degradation of a carbon target held in the beam and then comparing this result with orbital and other experimental data.

Improvements to this figure for this particular source may well rely on either operating with helium/argon mixtures or via the nitrous oxide route, hence increasing O_1 production via greater dissociation.

Typical relative percentages and average species energies of the seeded N_2O and O_2 routes are provided in Table 1 with the raw data from the CMA/mass spectrometer combination seen in Fig. 4.

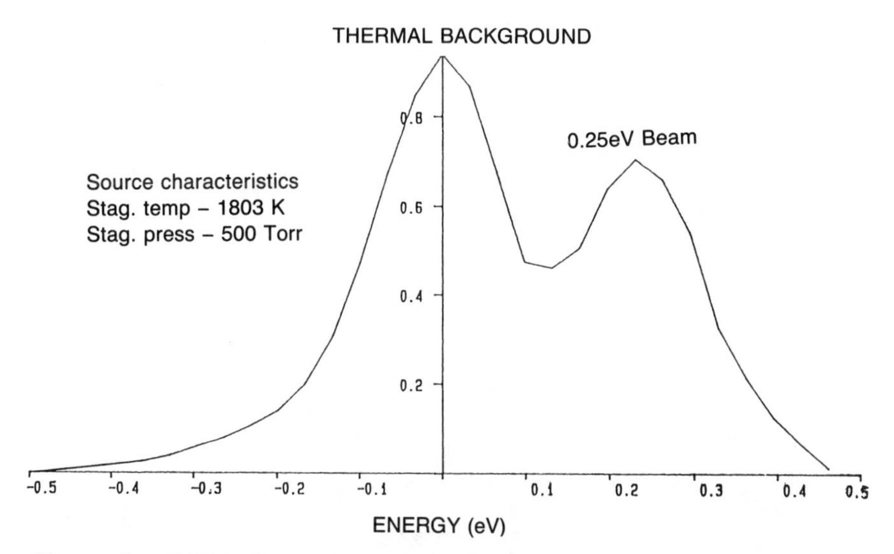

Fig. 5 Helium beam energy distribution measured by CMA/mass filter combination.

Table 1 Experimental data summary

	N_2O		O_2	
	Relative Intensity (±20%)	Energy (±0.1 eV)	Relative Intensity (±20%)	Energy (±0.1eV)
$(O_1)16$	37	1.03	32	1.05
$(N_2)28$	100	1.55	-	-
$(NO)30$	56	1.59	-	-
$(O_2)32$	32	1.76	100	1.90

Further gains in O_1 production using the nitrous oxide method have also been achieved by injection of molecular nitrogen through the arc. Atomic nitrogen is produced, which, when mixed with the N_2O reactive products downstream of the arc, promotes the reaction

$$NO + N \rightleftharpoons N_2 + O$$

Hence the amount of reactive NO is reduced with a consequent increase in the desired O_1. This effect can be seen qualititively in Fig. 6.

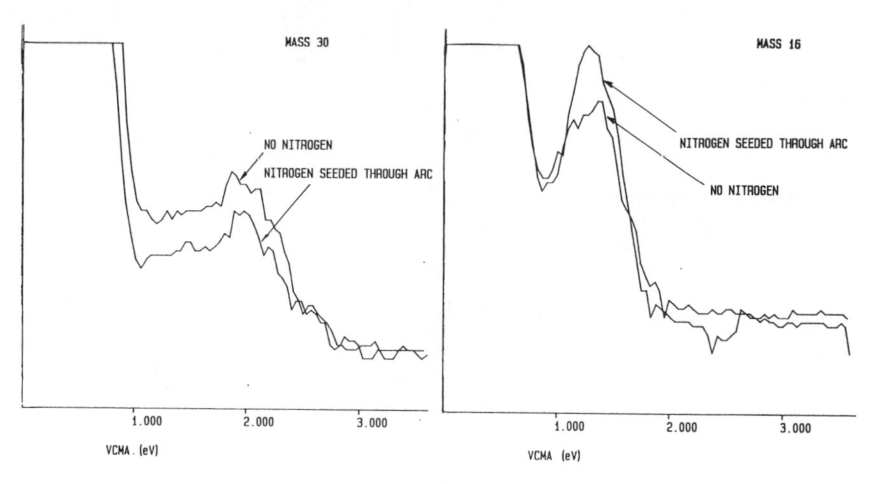

Fig. 6 Nitrous oxide seeded in helium.

IV. Discussion

The results provided in the previous section provide information on both the chemical state of the species in the molecular beam source and also the translational temperature of the beam via independent measurements. It is appropriate therefore to consider the consistency of these measurements. The results may be considered separately for the pure helium, the helium plus nitrous oxide, and helium plus oxygen.

Helium Source

Use of a cylindrical mirror analyzer indicates that the pure helium arcjet produces species with a translational energy of ~0.3 eV (depending on current). The equivalent source stagnation temperature is ~1500°K. This result may be compared to the temperature expected for an arcjet operating under the conditions indicated in the previous section. The work of Incropera and Murrer[9] studies fully developed constricted arcs of helium and nitrogen at near atmospheric pressure with no swirl. Their results show that there is a great deal of thermochemical nonequilibrium in such arcs. Through spectroscopic techniques they determined that the excitation temperature (characteristic of the electron energy) demonstrates a flat, radial profile through the arc, with a value of Te ~ 13,000°K. Clearly the high thermal diffusivity of helium plays a significant role in this profile. For helium arcs at these relatively low currents, Lukens[10] determined , that the mean helium atom temperature is however much lower, being of the order 10^3°K, a finding that corresponds to both our work and that of Incropera and Murrer.[9] It is worth noting that, for pure nitrogen arcs operating under these conditions, a similar excitation temperature to those noted in helium occurs on the centreline[9]; off axis this temperature falls by a factor of ~2. However, in nitrogen arcs the atom translational temperature is significantly higher, with values determined by molecular dissociation being of order 6000°K.

In summary, we see therefore that the translational temperature of the arc-heated helium atoms does not appear to fall significantly if we assume that the results of Ref. 10 are relevant to our data.

Helium Arc Seeded with N_2O

The results presented for the N_2O seeded flow provide an interesting alternative method for the determination of temperature produced in the arcjet. The CMA data shows that the heavy species velocity in the beam is not significantly below the value of He, lying close to 4 km/s. Thus, the equivalent stagnation temperature of the gas mixture is still of order 1500°K

An alternative method of temperature estimation may be made by modeling the dissociation reactions of N_2O. Baulch[11] provides a list of some 42 reactions associated with this dissociation and containing nitrogen and oxygen. These need to be solved similtaneously together with the direct dissociation of molecular oxygen by atomic collisions. Radiative dissociation has been omitted from the present analysis, however, since the influence of such effects is likely to be minor under the present conditions.[12]

Using the rate data for the reactions given by Baulch, calculations were performed to determine the species concentration as a function of time on the basis that the N_2O is heated by atomic collisions with helium at some specified arc temperature. A representative curve of the N_2O dissociation is shown in Fig. 7 for the case of helium at a temperature of 2800°K. Several features should be noted. Under the prevailing conditions the reactions appear to go to completion on a time scale much less than the residence time of the injected gas in the plenum chamber (~5ms); thus, dissociation takes place rapidly. However, it is also clear from Fig. 7 that some significant uncertainty exists with regard to the relative species

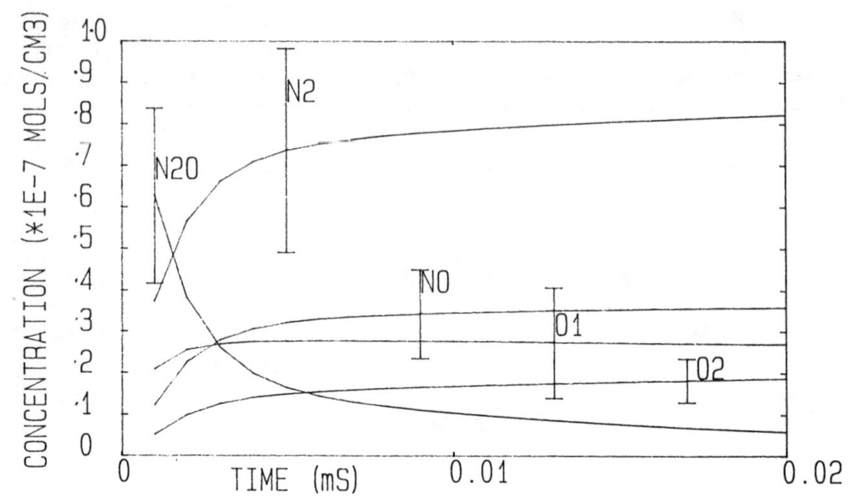

Fig. 7 Four percent nitrous oxide seeded in helium.

concentration ratios due to the uncertainties noted by Baulch, relating to the kinetic data. The 1σ uncertainties shown in Fig. 7 were determined by performing Monte Carlo calculations wherein the kinetic rate data noted in Ref. 11 was allowed to vary by the suggested uncertainties in Ref. 11.

In Fig. 8, species concentration ratios, relative to the concentration of nitrogen, are shown as a function of temperature. For an excitation temperature of 1500°K, relatively little dissociation of N_2O occurs at all. From this data it is apparent that at 1500°K the dissociation of N_2O is negligible. (It should be noted that the ratios noted at this temperature shown in the figure are somewhat distorted due to the low molecular nitrogen concentration.) Results obtained at higher temperatures, however, do not produce the high relative concentration levels of either N_2O or NO.

Before consulting these results in any futher detail, the possibility of relaxation should be considered. Since the arcjet is vortex-stabilized, swirl will be evident in the flow. The high wall velocity in the plenum chamber due to its high operating temperature will lead to a high wall heat flux and thus a rapid fall in gas kinetic temperature. Figure 9 shows the chemical relaxation process from 4000 to 1500°K. After 5 ms it is clear that the equilibrium conditions for a 1500°K gas mixture are not achieved, with still negligible concentrations of N_2O being formed. Indeed, the chemical relaxation takes place slowly being much larger than the gas residence time in the plenum

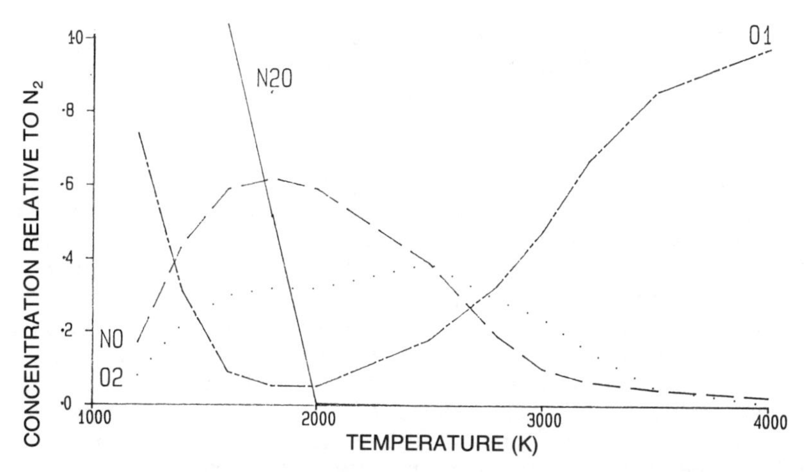

Fig. 8 Dissociated N_2O equilibrium concentration.

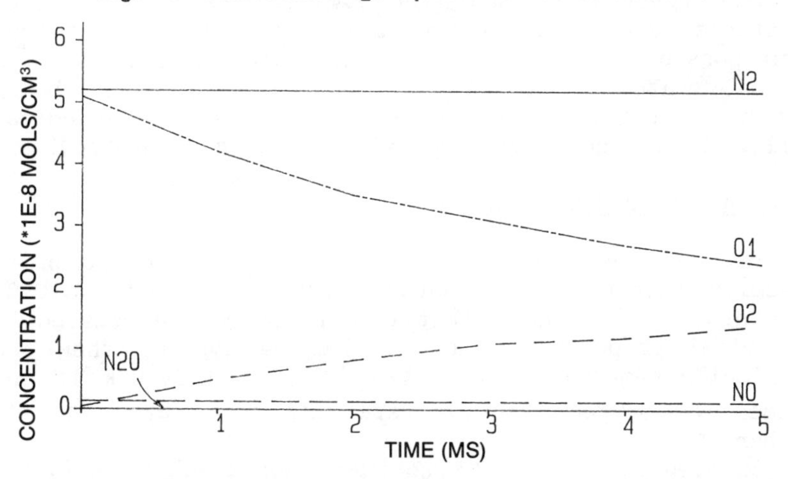

Fig. 9 N_2O equilibrium mixture at 4000°K cooled to 1500°K.

chamber. Thus, the equilibrium values predicted in Fig. 8 provide a good approximation to the chemical state, even if translational temperature cooling takes place rapidly.

Although it is clear from the data that no single temperature (certainly as low as 1500°K) is able to define the chemical state of the seed gas, it is also apparent that, in order to obtain the high levels of atomic oxygen formation in the source, a thermal component of at least ~3000°K is required. This seems to require helium heavy particle temperatures a factor ~3 larger than that proposed by Lukens[10] to explain his results. Two possible features could be influential in accounting for these differences.

First, the earlier cited work was concerned with fully developed arcs with no swirl. Under the conditions used in the source reported herein the arc is rather short (less then ~4 mm). The models of Watson and Pegot[13] show that in the early part of constricted arcs the conditions are indeed different from the fully developed conditions. Indeed, the centerline enthalpy can demonstrate very high predicted values in the entrance region of an arc. This entrance region is also sensitive to the assumptions of flow properties.

An alternative possibility arises frorm a lack of knowledge concerning the reaction rates. The data of Baulch provides reaction rates related to ground state reactions. Although uncertainty is associated with these rates (see Fig. 7), additional uncertainty surrounds the rates of reaction between metastable helium (molecules and atoms) and other reactants (see, for example, Ref. 14). Such metastable states will be produced in the arc, and it thus seems possible that this may contribute to the observed chemical state of the source. Further results are being pursued to resolve the source of the lack of chemical equilibrium in the complex gas mixture being generated.

Helium Arc Seeded with Oxygen

The results obtained from the direct dissociation of molecular oxygen to form atomic oxygen yielded an O/O_2 ratio of 0.32. Assuming that this ratio may be related to a chemical temperature for the $O/O_2/He$ system, then the equilibrium temperature is between 2850 and 2900°K if one assumes the oxygen reaction rate data given by Camac and Vaughan.[15]

In view of the preceding discussion on N_2O reactions, the concept of chemical temperature for the seed gas mixture is hazardous; however, it is interesting to note that on the basis of the O/O_2 ratio for the N_2O seed results, the chemical equilibrium temperature lies (1σ) within the range of 2650 and 2890°K with a best-fit value of 2810°K.

V. Conclusions

The development of a low-power, high-fluence source of atomic oxygen has been undertaken. Two routes to the production of atomic oxygen have been investigated: direct dissociation of molecular oxygen by arc-heated helium and through the chemical dissociation of N_2O by arc-heated helium. Both techniques produce an atomic oxygen flux of

~0.5 × 10^{15} atoms/cm^2/s. The latter route produces the additional constituents of N_2O and NO. With the addition of nitrogen through the arc, these components may be materially modified, making the source appropriate for the study of phenomenon such as Shuttle glow.

When considering the chemical nature of the gas, it is evident that a severe degree of chemical nonequilibrium appears to exist in the source. The evidence suggest a two-component thermal structure to the gas, with one component approaching 3000°K, and the other component approximately half this value.

Acknowledgments

Partial support for this work came from the Science and Engineering Research Council of the United Kingdom. We would like to thank Ian Smart for his assistance in the design and construction of the arc heater.

References

[1]Torr, M., "Optical Emissions Induced by Spacecraft-Atmospheric Interactions," Geophysical Research Letters, Vol. 10, 1983, pp. 114-117.

[2]Young, W. S., Rogers, W. E., and Knuth, E. L., "An Arc Heater for Supersonic Molecular Beams," Review of Scientific Instruments, Vol. 40, No. 10 - Oct 1969, pp. 1346-1347.

[3]Abuaf, N., Anderson, J. B., Andres, R. P., Fenn, J. B., and Marsden, D. G. H., "Molecular Beams with Energies Above 1eV," Science Vol. 155, Feb. 1967, pp. 997-999.

[4]Arnold, G. S., and Peplinski, D. R., "A Facility for Investigating Interactions of Energetic Atomic Oxygen with Solids," NASA CP-2340, Oct. 1984, pp 150-168.

[5]Silver, J., Freedman, C., Kolb, C., Rahbee, A., and Dolan, C., "Supersonic Nozzle Beam Source of Atomic Oxygen Produced by Electric Discharge Heating," Review of Scientific Instrumentation, Vol. 40, Nov. 1982, pp 1714-1718.

[6]Keiffer, L. J., and Dunn, G. H., "Electron Impact Ionization Cross-section Data for Atoms, Atomic Ions and Diatomic Molecules: I: Experimental Data," Review of Modern Physics, Vol. 38, No. 1, Jan. 1966, pp. 1-35.

[7]Ballenthin, J. O., and Nier, A. O., "Molecular Beam Facility for Studying Mass Spectrometer Performance," Review of Scientific Instrumentation, Vol. 38., No. 7, July 1981, pp 1016-1028.

[8]Sibener, S. J., Buss, R. J., Ng, C. K., and Lee, Y. T., "Development of a Supersonic O(3Pj), O(1D2) Atomic Oxygen Nozzle

Beam Source", Review of Scientific Instrumentation, Vol. 51, No. 2, Feb 1980, pp. 167-182.

[9]Incropera, F. P., and Murrer, E. S., "Spectroscopic Measurements for Atmospheric Nitrogen and Helium Arcs", Journal of Quantitative Spectroscopy & Radiative Heat Transfer, Vol. 12, No. 10, Nov. 1972, pp. 1369-1377.

[10]Lukens, L. A., Ph.D. Disertation, "An Experimental Investigation of Electric Field Intensity and Wall Heat Transfer for the Heating Region of a Constricted Arc Plasma," School of Mechanical Engineering, Purdue Univ., Lafayette, IN, 1971.

[11]Baulch, D. L., Drysdale, D. D., Horne, D. G., and Lloyd, A. C., Evaluation of Kinetic Data for High Temperature Reactions, Butterworths, London, 1973, Vol. 2.

[12]Giannaris, R. J, and Incropera, F. P., "Radiative and Collisional Effects in a cylindrically Confined Plasma - I: Optically Thin Considerations," Journal of Quantatice Spectroscopy and Radiative Heat Transfer, Vol. 13, No. 2, Feb. 1973, pp. 167-181.

[13]Watson, V. R., and Pegot, E. B., "Numerical Calculations for the Characteristics of a Gas Flowing Axially Through a Constricted Arc," NASA TN D-4042, 1967.

[14]Pouvesele, J. M., Stevefelt, J., Lee, F. W., Jahani, H. R., Gylys, V. T., and Collins, C. B., "Reactivity of Metastable Helium Molecules in Atmospheric Pressure after glows," Journal of Chemical Physics, Vol. 83, 1985, pp. 2836-2839.

[15]Camac, M. and Vaughan, A., "O_2 Dissociation Rates in O_2-Ar Mixtures," Journal of Chemical Physics, Vol. 34, No.2, Feb. 1961, pp. 460-470.

Options for Generating Greater Than 5-eV Atmospheric Species

H. O. Moser*
Kernforschungszentrum Karlsruhe, Karlsruhe, Federal Republic of Germany
and
A. Schemppt†
University of Frankfurt, Frankfurt, Federal Republic of Germany

Abstract

The acceleration of cluster ion beams of species, including those occurring in the Earth's atmosphere with the aim to produce intense, energetic particle beams for interaction and simulation studies, is discussed. Results on beams formed with a dc accelerator and design values of a radio-frequency quadrupole cluster ion accelerator show that, in either case, the flux densities and velocities satisfy the requirements comfortably. If the cluster ion beams can be converted effectively into atomic or molecular beams, then such facilities will be very useful to study interactions between fast particles and various targets, and to simulate spacecraft flight in low-density gas atmospheres.

Introduction

The need of data on the interaction of spacecraft with low-density gas atmospheres arises in various spaceflight situations, including satellite launching, Space Shuttle flight, and missions to other planets or to comet tails.[1] Using the case of a spacecraft in the Earth's atmosphere as an example,[2] this need translates into the necessity to investigate the interaction of gas species occurring in the Earth's atmosphere, such as atomic oxygen or molecular nitrogen, at energies greater than about 5 eV with solid surfaces. For example, an oxygen atom at a speed of 7.7 km/s has a kinetic energy of 5 eV. Figure 1 shows the flux

*Senior Physicist, Institut für Mikrostrukturtechnik; presently at Brookhaven National Laboratory.

†Senior Physicist, Institut für Angewandte Physik.

densities occurring in the Earth's atmosphere for a body with a speed of 10 km/s. Usually, the information about the interaction between spacecraft and gas atmosphere is extracted from orbital data[3] or inferred from laboratory simulation.[4] Recently, orbiting experimental facilities[5] and tethered satellites[1] have been proposed. Tracking as well as performing experiments on orbiting facilities or tethered satellites clearly provides for a real atmosphere. However, these methods do not easily allow to vary parameters separately and are rather expensive. Laboratory simulation allows, in principle, selection of single-flow components, provided they can be generated at all, and variation of parameters of both gas flow and its composition, and the shape and surface of the spacecraft model. Unfortunately, appropriate sources closely modeling the flow conditions in space are not readily available.

The following discussion of options for producing molecular beams of atmospheric species at kinetic energies in excess of 5 eV is based on results obtained with

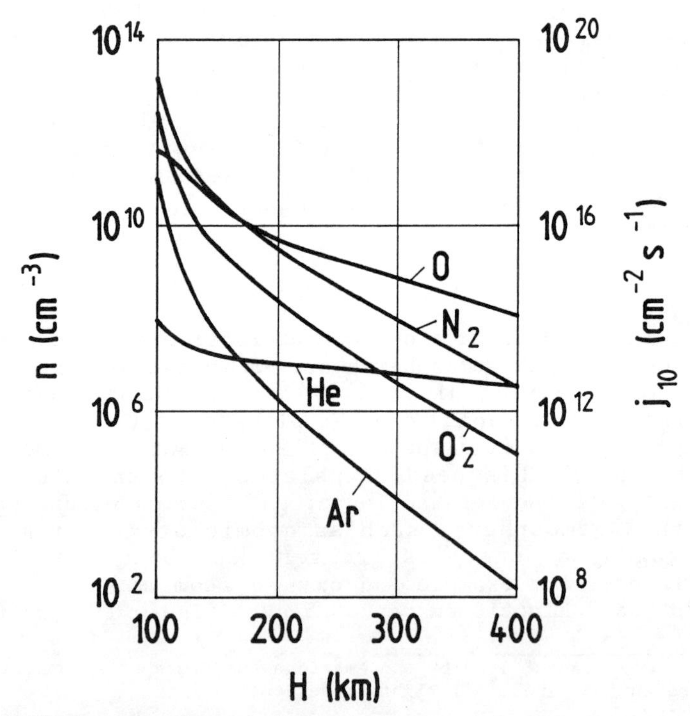

Fig. 1 Number density n of the main atmospheric species and a typical flux density j_{10} corresponding to a speed of 10 km/s vs altitude H (adapted from Ref. 2.) The order of magnitude of the accelerator output flux density is 10^{19} cm^{-2} s^{-1}.

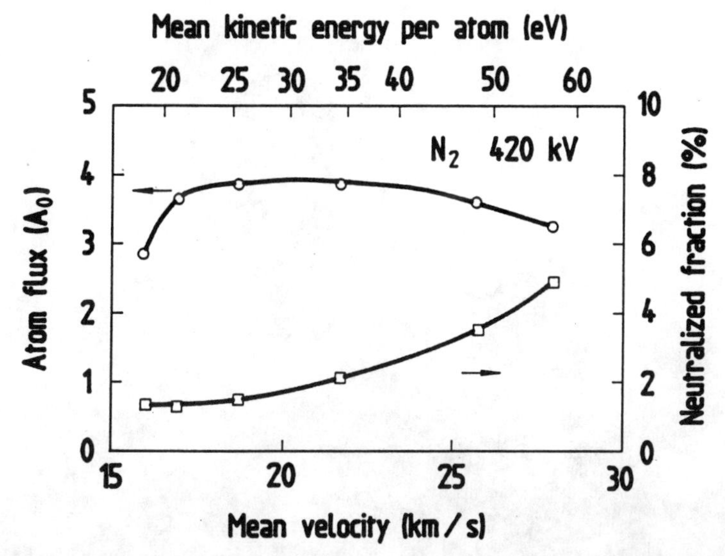

Fig. 2 Measured atom flux in units of equivalent amperes A_o ($\triangleq$ 6.25 x 10^{18} s^{-1}) and neutralized fraction of atom flux vs mean velocity of nitrogen cluster ions accelerated with 420 kV (flux density of the order of 1 A_o/cm^2). Obviously, a dc cluster ion accelerator can produce an intense cluster ion beam with only a small extent of neutrals over a large range of velocities, or energies per atom, respectively.

electrostatically accelerated cluster ion beams of nitrogen and hydrogen,[6–8] and on planned work aiming at the acceleration of cluster ions by means of a radio-frequency quadrupole (RFQ) accelerator.[9,10] Both methods of acceleration will be compared.

Dc Cluster Ion Acceleration

The electrostatic acceleration of intense beams of cluster ions is well established.[11] A typical result concerning the beam flux is represented in Fig. 2. It was obtained with the Kernforschungszentrum Karlsruhe (KfK) 1-MV cluster ion accelerator (Fig. 3). The average velocities achievable with 100-keV to 1-MeV rectifier cascades extend significantly beyond the lower limit corresponding to 5 eV. Since these beams have a comparably broad velocity spectrum of typically 40% relative bandwidth (Fig. 4), a smaller velocity interval eventually must be cut out of the total spectrum. It has been estimated that this could be done by means of a mechanical velocity selector of the multiple slotted-disk type.[8] As long as the cluster ion beam remains ionized (see Fig. 2), momentum

Fig. 3 KfK 1 MV dc cluster ion accelerator.

analysis with magnetic fields produced from superconducting dipole magnets is an attractive alternative. As an example, a state-of-the-art superconducting dipole of 1 m in length and 4 T magnetic flux density would produce a deflection of 10 cm for 100-keV particles at 10 km/s, allowing to cut out a 10% velocity band with a 1-cm-wide aperture. The components of a facility based on dc cluster acceleration are depicted schematically in Fig. 5.

Cluster Ion Acceleration by an RFQ

The radio-frequency quadrupole is a linear resonance accelerator with simultaneous strong transverse focusing.[12] Figure 6 shows a photograph of a four-rod RFQ developed at the University of Frankfurt.[13] Among the main advantages of an RFQ compared with an electrostatic accelerator is the considerably narrower velocity spectrum that can be expected from the longitudinal phase focusing. For comparable cases, a reduction of about an order of magnitude has been estimated.[9] As a consequence, an RFQ accelerator directly produces a useful velocity spectrum and, thus, can dispense with a subsequent velocity analyzer. In addition, the flux density limit caused by intrabeam collisions[7] is expected to be significantly higher. Table 1 shows some parameters of an RFQ designed to accelerate cluster ions of mass 3.2×10^5 u from 0.06 to

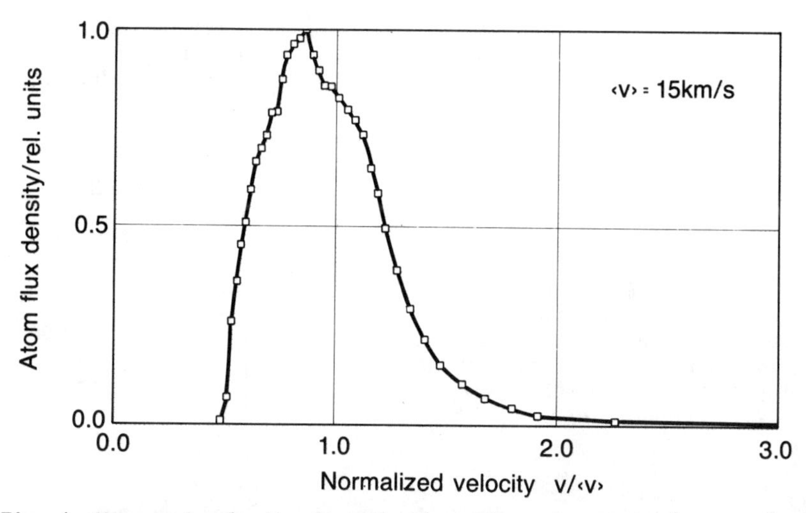

Fig. 4 Measured velocity distribution of a nitrogen cluster ion beam accelerated with the KfK cluster ion accelerator. Slower beams can be produced easily by reducing the acceleration voltage.

Table 1 Main design parameters of an RFQ producing a mean velocity of 10.7 km/s

Initial kinetic energy per mass unit u, eV	0.06
Final kinetic energy per mass unit u, eV	0.6
Maximum total kinetic energy, keV	192
Maximum ion mass, u	3.2×10^5
Length of structure, m	0.5
Clear aperture between electrodes, mm	6
Maximum modulation of electrodes	2
Diameter of vacuum chamber, m	0.5
Frequency, kHz	250
Peak voltage, kV	30
Initial velocity acceptance, %	5
Final velocity spread, %	5
Transmission, %	50
Maximum electric current at maximum mass, μA	100
Maximum oxygen atom current at maximum mass, A_o	2
Maximum oxygen atom flux density at maximum mass, A_o/cm^2	2.5

0.6 eV/u. This corresponds, for example, to cluster ions with 10^4 oxygen molecules, accelerated from 0.96 to 9.6 eV per oxygen atom. The initial energy of 19.2 keV per cluster ion comes from the extraction voltage. With an initial velocity acceptance of 5% (see Table 1), about one-eighth of the total extracted beam can be accelerated. A block diagram of an RFQ-based beam facility is shown in

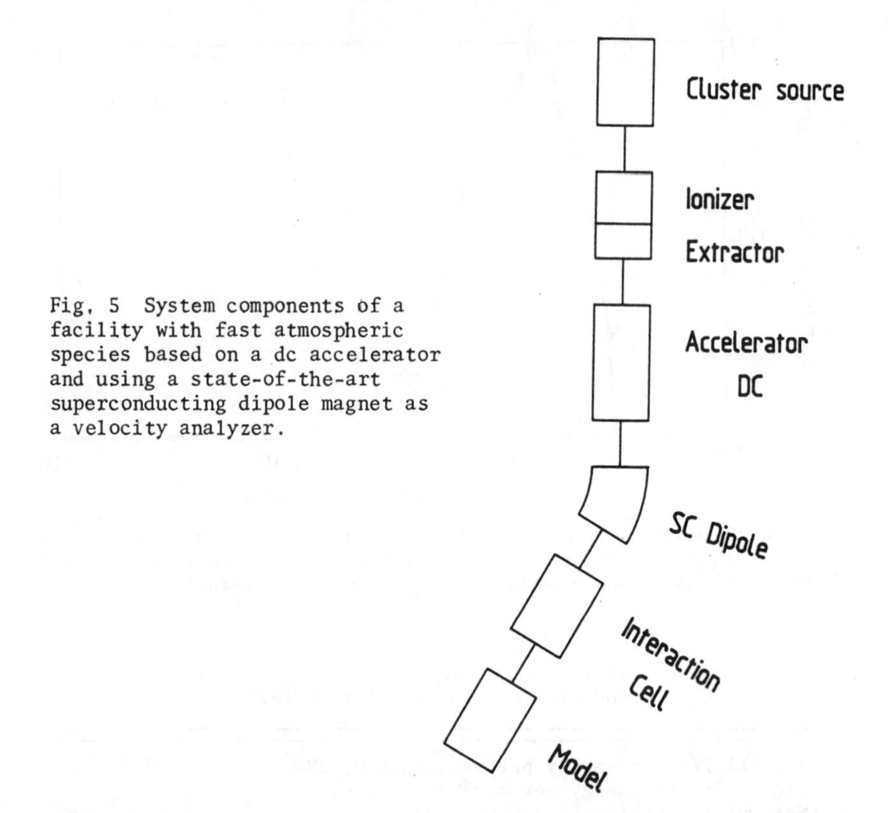

Fig. 5 System components of a facility with fast atmospheric species based on a dc accelerator and using a state-of-the-art superconducting dipole magnet as a velocity analyzer.

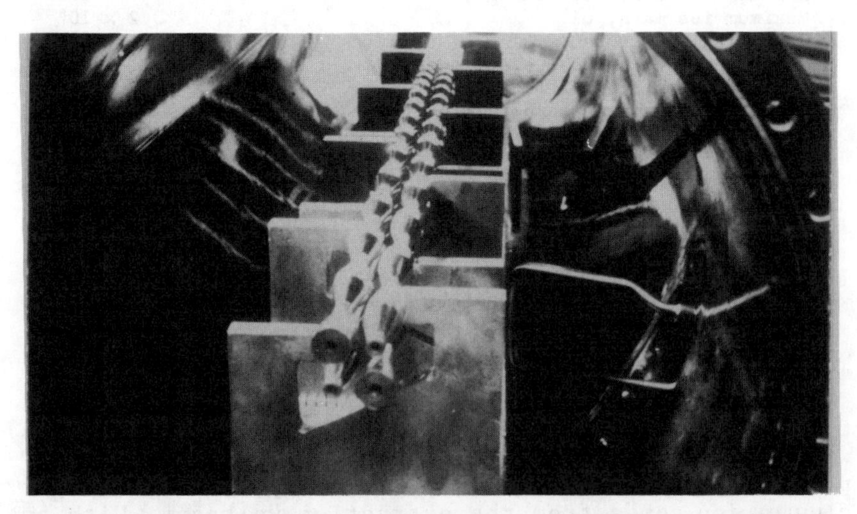

Fig. 6 Four-rod RFQ structure. As in a quadrupole mass filter, a radio-frequency voltage is applied to pairs of opposite rods. The longitudinal diameter modulation of the rods creates an accelerating field component.

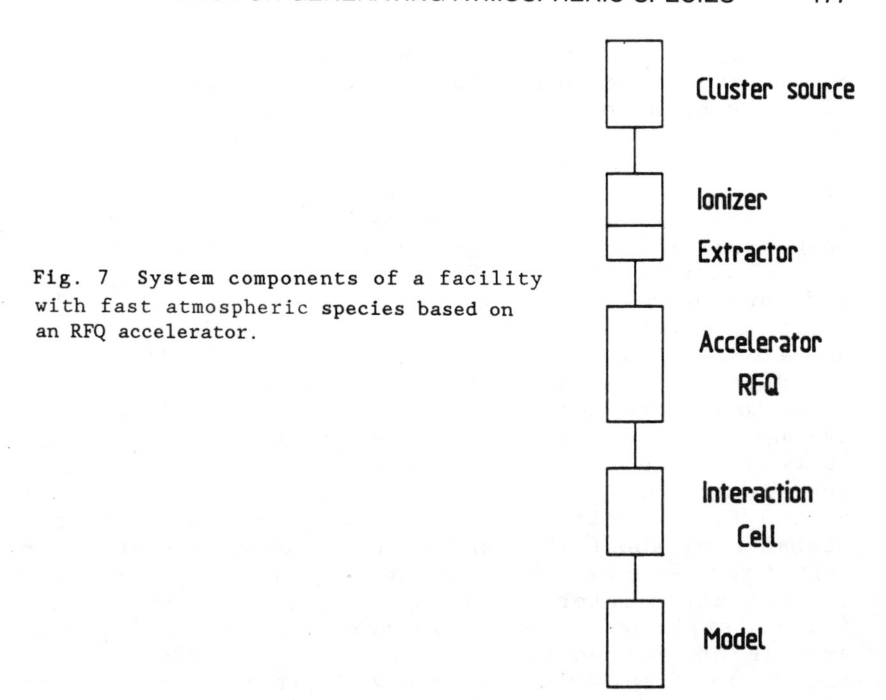

Fig. 7 System components of a facility with fast atmospheric species based on an RFQ accelerator.

Fig. 7. It differs from Fig. 5 by the lack of a velocity analyzer and, of course, the other accelerator type. Further important advantages of the RFQ, particularly when the beam energy exceeds about 400 keV and extends into the million-electron-volt range, are its much better accessibility and considerably lower cost. The periodic time structure of the beam from an RFQ, in occurrence, a bunch train with 250 kHz repetition rate, may be useful for the study of time-dependent processes. The first project to accelerate hydrogen clusters in the mass range from 1 to 50 u up to an energy of 5 MeV is well under way.[9,10] Operation of this RFQ-based accelerator is expected to start by the end of 1989.

For both cases of accelerators discussed so far, it is conceivable to merge individual beams of different species at the entrance of the accelerator and to accelerate them simultaneously to the same speed. This feature would allow to come closer to a real simulation and to study synergistic effects.

Disintegration and Dissociation

The remaining question is how to transform the cluster ion beams into atomic or molecular beams of the desired species. Here, we must speculate in order to find out in which direction further experimental work should go. First, according to Ref. 8, the mere disintegration of clusters of atoms or molecules into atoms or molecules, respectively, seems readily feasible using the interaction with a state-of-the-art gas target but might need downstream velocity selection. This is one of the questions requiring experimental investigation. Second, photodissociation/disintegration, in particular to form atomic oxygen, requires fairly high photon flux densities and long beam-beam interaction spaces. For example, assuming a speed of 7.7 km/s, an interaction length of 1 m, and a photodissociation cross section of 10^{-18} cm^2, we estimate a photon flux density of 7.7×10^{21} cm^{-2} s^{-1} in order to achieve photodissociation of each molecule, on average. Synchrotron radiation from wigglers or undulators seems to be the most advanced means to provide radiation in the spectral region needed. Present state wigglers have photon flux densities of 10^{17} to 10^{18} cm^{-2} s^{-1} at a distance of about 10 m and in the vacuum-ultraviolet to soft x-ray region of the spectrum. In the long term, the free-electron laser may become a light source appropriate for this photodissociation task. Meanwhile, the experiments needed to determine the dissociation yield of cluster ions seem quite feasible with present-day means and a moderate effort using thermal-energy cluster beams, and either pulsed excimer lasers or synchrotron radiation from a wiggler/undulator beam line.

Conclusion

From the accelerator point of view, the options discussed seem fairly realistic. A facility based on a radio-frequency quadrupole (RFQ) appears more favorable because of some of the features intrinsic to an RFQ, such as the narrow velocity spectrum at the exit, accessibility, and low cost. Further experimental work, in particular on the photodissociation of oxygen clusters, must be carried out before the feasibility of the whole scheme can be assessed.

References

[1]Proceedings of the 15th International Symposium on Rarefied Gas Dynamics, edited by V. Boffi and C. Cercignani, Teubner, Stuttgart, Federal Republic of Germany, 1986, with references to previous proceedings.

[2]Koppenwallner, G., "Free Molecular Aerodynamics for Satellite Application," DFVLR-FB 82-08, Göttingen, Federal Republic of Germany.

[3]See, for example, Kovalevsky, J., "Recent Progress in the Dynamics of Artificial Satellites and Its Application," Space Research X, North Holland, Amsterdam, the Netherlands, 1970, pp. 1-16.

[4]See, for example, Böttcher, R.-D., Koppenwallner, G., and Legge, H., "Investigations of Satellite Aerodynamics," ESA Journal, Vol. 4, 1980, pp. 357-370.

[5]Bütefisch, K. and Koppenwallner, G., "Orbitaler Strömungs-versuchsstand (OSV) or an Orbital Research Facility for Rarefied, Reactive, and Plasma Flows," Rarefied Gas Dynamics, edited by R. Campargue, Commissariat à l'Energie Atomique, Paris, 1979, pp. 521-543.

[6]Moser, H. O., Falter, H. D., Hagena, O. F., Henkes, P. R. W., and Klingelhöfer, R., "Cluster Ion Acceleration as a Means of Producing Multiampere Particle Beams in the Energy Range of 1 eV to 1 keV/Atom," Surface Science, Vol. 106, 1981, pp. 569-575.

[7]Moser, H. O., "On the Physics of Intense, Electrostatically Accelerated Cluster-Ion Beams," KfK 4068, Kernforschungszentrum Karlsruhe, Karlsruhe, Federal Republic of Germany, 1986.

[8]Moser, H. O., "Investigation of the Influence of Low Density Gas Atmospheres on Spacecraft by Means of Accelerated Cluster Ion Beams," Zeitschrift für Flugwissenschaft und Weltraumforschung, Vol. 11, 1987, pp. 291-294.

[9]Moser, H. O. and Schempp, A., "Cluster Ion Acceleration with the Radiofrequency Quadrupole," Nuclear Instruments & Methods, Vol. B24/25, 1987, pp. 759-762.

[10]Moser, H. O. and Schempp, A., "Upgrading the Lyon Cluster Ion Accelerator by a Radiofrequency Quadrupole," KfK 4201, Kernforschungszentrum Karlsruhe, Karlsruhe, Federal Republic of Germany, 1987.

[11]Becker, E. W. et al., "Construction and Test of a High Power Injector of Hydrogen Cluster Ions," Fusion Technology, Pergamon Press, Oxford, 1979, pp. 331-337.

[12]Kapchinskii, I. M. and Teplyakov, V. A., "Linear Ion Accelerator with Spatially Homogeneous Strong Focusing," Pribory; Tekhnika Eksperimenta [Instruments and Experimental Techniques (USSR)], Vol. 2, 1970, p. 19.

[13]Schempp, A. et al., "Four-Rod-Lambda/2-RFQ for Light Ion Acceleration," Nuclear Instruments & Methods, Vol. B10/11, 1985, pp. 831-834.

Laboratory Results for 5-eV Oxygen Atoms on Selected Spacecraft Materials

Gary W. Sjolander* and Joseph F. Froechtenigt†
Martin Marietta Corporation, Denver, Colorado

Abstract

Ground-based simulation of the atomic oxygen (AO) environment as it interacts with spacecraft surfaces is a vital part of correctly selecting materials that will be able to withstand tens of years in low-Earth-orbit. The Martin Marietta Corporation has built a laboratory facility called OMEGA (Oxidation/Materials Erosion and Glow Analysis) that has started to test spacecraft materials to characterize mass loss rates and corresponding degradation effects and to understand mechanisms and influencing parameters for both oxidation and glow. Initial laboratory testing with AO^+ shows close agreement with AO materials testing in space. The laboratory is presently testing materials with a 5-eV ion beam; however, work is progressing on ion beam neutralization in order to more closely simulate the low Earth environment. This paper presents initial qualitative results on silver and Kapton and quantitative ion chemistry reaction rates on copper and various carbon samples. These results on carbon are then compared with neutral oxygen reaction rates derived from Space Shuttle flight experiments. The laboratory experiments for 5-eV oxygen ions on carbon consistently produce higher reaction rates than observed on orbit for neutral oxygen chemistry.

Introduction

Many new generation spacecraft such as Space Station, Space Telescope, and Tethered Satellite will be highly exposed to the atomic oxygen (AO) environment either through long duration in low-Earth-orbit (200-700 km) or high flux

* Senior Group Engineer, Martin Marietta Strategic Systems.
† Staff Engineer, Martin Marietta Space Systems.

densities for shorter periods at lower altitudes (100-200 km). Space Shuttle flight data have shown that physical properties of metallic and nonmetallic materials are degraded by exposure to this environment. This includes visible surface damage,[1] severe loss of mass,[2,3] and degradation of thermo-optical properties, such as solar absorptivity and hemispheric emissivity. These data also indicate that certain surface materials or outgassing products visually glow on exposure to the ambient orbital environment.[4] Interaction of the ambient environment, which is predominantly AO, with spacecraft materials is believed to cause material degradation and glow phenomenon processes.

Simulation goals for laboratory AO testing were derived from the spacecraft-induced environment while orbiting through the AO atmosphere. Specifically, the AO number density in the orbital range of interest, between 250 and 650 km, varies inversely from 10^9 to 10^6 cm^{-3}, respectively. The corresponding AO flux density to spacecraft surfaces facing into the "AO wind" ranges from 10^{15} to 10^{12} cm^{-2}-s^{-1}. The AO wind is induced by the orbital speed of the spacecraft, which is approximately 7.8 km-s^{-1}. For oxygen atoms this collision velocity corresponds to a kinetic energy of approximately 5 eV. This is the nominal energy goal that is most often quoted for ground-based simulation of AO-surface interactions. A closer investigation of the actual on orbit AO-surface reaction energy shows that, when the corotation of the atmosphere and the Earth and variations in energy due to changes in orbital altitude and solar activity are considered, the interaction energy is closer to 4.2 ± 0.5 eV, which is an energy more easily obtained by supersonic expansion techniques. Preliminary testing with our ion beam apparatus and testing at other laboratories[5] have verified that AO causes material degradation.

Apparatus

The simulation approach taken to produce a 5-eV beam was by the acceleration/deceleration of AO ions produced in an electron bombardment plasma source.[6] A diagram of the ion beam apparatus is shown in Fig. 1. The main components are the ion source, acceleration and focusing lens, mass analyzer, decelerator, and exposure chamber (not illustrated). The ion source generated mostly oxygen ions; however, trace amounts of molecular oxygen, nitrogen, and water vapor ion are also produced. The magnetic mass analyzer is used to select only the AO ion species that produced a pure AO beam on the material target.

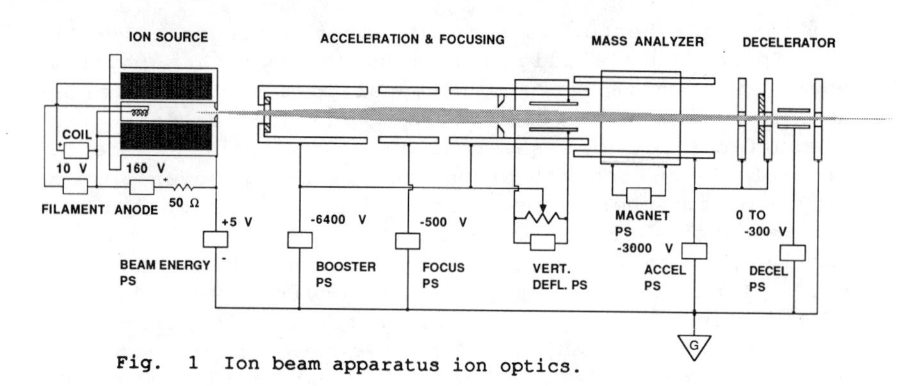

Fig. 1 Ion beam apparatus ion optics.

Table 1 OMEGA performance status

	Goals	Status
1	Beam energy ~ 5 eV	5 ± 0.8 eV
2	Flux ~ 10^{15} atoms cm^{-2}-s^{-1}	5 x 10^{14} @ 5 eV 0.2-in. diam
3	Fluence ~ 10^{18} atom cm^{-2} over 8 h time	> 10^{19} @ 5 eV 8 hr run, 0.2 in. diam
4	Neutral AO beam	Neutral plasma
5	Uniform exposure ~ 0.5 in.	0.5 inch diam with reduced flux density

Energy analysis of the ion beam has demonstrated that the beam energy can be adjusted from 3 eV to several hundred electron volts with an energy resolution of ± 0.8 eV (half-maximum) at the exit of the decelerating lens. However, there are indications that as the beam continues for several centimeters beyond the decelerator exit, the spread in ion energy becomes broader and somewhat skewed toward higher energies (+5/-2 eV), which is a result of beam space charge limitations.

Neutralization of atomic oxygen ions will be accomplished by gas charge exchange using water vapor or molecular oxygen; however, this part of the system is still in the testing phase. A major obstacle in meeting a flux density goal of 10^{15} cm^{-2}-s^{-1} for a neutral species is the high charge density in the 5-eV ion beam that leads to severe space charge spreading of the beam. At the present time, there is a factor of 10 loss in ion beam current as it passes through the gas charge exchange hardware. Improving the ion beam transmission will continue to be the focus of the development activity. Currently, neutralization is accomplished by electron injection into the beam, which creates a neutral plasma that reduces space charge built up on nonconducting materials. In summary, our ion beam system is basically on line. Operational parameters are summarized against the goals in Table 1.

AO-Surface Chemistry Test Results

Initial quantitive material testing included continuous monitoring of the ion current to a well defined area (1.27 cm^2) of the target that allows calculation of the AO fluence. Reaction rate calculations were made by combining fluence measurements with off-line measurements of volume loss using profiling Auger techniques. For two of the bulk carbon samples that were tested, reaction rates were determined by mass change measurements, which is described in the section on Carbon results.

The ion beam apparatus was used to investigate atomic oxygen ion reaction at 5 eV with silver, Kapton, copper, and carbon surfaces to establish a performance data base for comparison with neutral AO reactions experiments in space. The initial results include both quantitive and qualitive data that is summarized as follows.

Silver

The silver samples were prepared by vacuum evaporation of silver onto a glass substrate. The unexposed film thickness ranged between 600 and 900 Å and had a roughness of ±100 Å. Exposure of the silver film to the oxygen ion beam converted the highly optically reflecting silver to transparent silver oxide, which was similar to the results obtained for the EOIM-2 experiment.[2] The total exposure for our experiment was ~10^{19} cm^{-2}, which oversaturated the exposure needed to convert all the silver film to silver oxide by a factor of 30. The image of the exposed area showed that the ion beam had a complex structure that included rings and a hollow core (absence of ions). This complex beam structure is not surprising in light of the highly space charged limited nature of the beam. Silver is an important material for space application, and it should continue to be studied.

Kapton

The testing of Kapton in the ion beam apparatus proved to be difficult since Kapton is a dielectric material. As the beam of oxygen ions struck the target, the surface charged up to a potential equal to the ion beam energy (5 eV). This surface potential deflected most of the ions away from the surface, which resulted in a low current to the target (~ 10 nA). For an 8-h exposure, the fluence was no greater than 2 x 10^{15} atoms, which produced no noticeable damage. In a subsequent experiment, the beam energy was increased to 300 eV to improve the fluence level in a reasonable time. The analysis of the surface using scanning electron micrography (SEM) revealed similar morphology changes to that of AO damage to Kapton that was flown on EOIM-2.[1]

Copper

Oxygen-free high conductivity (OFHC) copper was used as the target material. The visual appearance of the exposed area was the reddish color associated with Cu_2O. The sample was analyzed using auger electron spectrography and argon sputtering for depth profiling of elemental composition. The total fluence was ~10^{18} oxygen ions-cm^{-2}, and the Auger analysis revealed detectable oxygen to a depth of 200 Å. From these data it is estimated that 4% of the oxygen ions reacted with the copper. This is a factor of ~1000 higher than the reaction rate reported for neutral oxygen atoms on copper for the STS-8, EOIM-2 experiment.[2] This difference is probably due to real differences between neutral and ion chemistry, but it could equally be due to differences in surface preparation and subsequent analysis.

Carbon

Three carbon samples were tested in the ion beam apparatus. One sample was a carbon film, whereas the other two samples were bulk graphite samples, which were provided by the NASA Lewis Research Center (LeRC) as part of their Atomic Oxygen Effects Test Program.[7]

The carbon film sample was prepared by vacuum depositing carbon vapor onto a glass substrate. The film thickness ranged from 600 to 1000 Å with a roughness of ~100 Å. Auger analysis combined with argon sputtering and fluence measurements yield a calculated erosion rate of 2.1 x 10^{-24} cm^3/AO^+. This compared well with the value for vitreous carbon flown on EOIM-2 that was reported as 1.3 x 10^{-24} cm^3/AO, and with carbon composites with reported erosion rates of 2.5 x 10^{-24} cm^3/AO.[2,3] The Auger analysis also revealed that oxygen had diffused into the bulk of the carbon film. Although diffusion had been seen in copper and silver, it was expected that all the oxygen that reacted with the carbon surface would leave the surface as CO or CO_2. With oxygen dissolved in carbon, it is possible that there could be latent effects that would lead to further erosion.

Carbon-oxygen ion reaction efficiences on highly oriented pyrolytic graphic (HOPG) and pyrolytic graphite, which were provided by the LeRC, were measured in our laboratory to an accuracy of 15%. The reaction effeciency R for the HOPG and pyrolytic graphite were determined by measuring mass loss and monitoring the integrated current as expressed by

$$R = \frac{\Delta mq}{\rho \int_0^T i(t)\, dt}$$

where q is the electronic charge, i the current, m mass, T the total exposure time, and ρ the sample density. The advantage of detemining reaction efficiency by mass loss is that the ion beam does not need to be uniform and the exposured area does not enter into the calculation. Both the mass loss and current can be measured to within a few percentage points.

Results for oxygen-carbon reaction efficiencies for our three samples, along with other reported results for oxygen-carbon reaction efficiencies,[2,3,8] are summarized in Table 2. As is shown in Table 2, the range of reaction efficiencies span greater than three orders of magnitude, which represents combinated differences of reaction rates and experimental technique. In the OMEGA system, the oxygen

Table 2 Atomic oxygen reaction efficiencies for various carbons

Carbon form	Reaction efficienty, 10^{-24} cm^3	Ref. no.
Diamond (STS-8)	0.022	2
Carbon (STS-4)	0.043	8
Vitreous (STS-5)	0.23	8
Amorphous (STS-8)	0.23	8
Film (STS-8)	0.36	8
Basal oriented graphite (STS-8)	0.63	2
Glassy (STS-8)	0.8	3
Molded graphite (STS-8)	0.85	3
Carbon (STS-3)	1.0	8
Glassy (STS-8)	1.3	2
Single crystal graphite (STS-8)	1.4	2
Film (OMEGA)[a]	2.1	This paper
HOPG (OMEGA)	9.2	This paper
Pyrolytic graphite (OMEGA)	29-36	This paper

[a]OMEGA- The Martin Marietta AO facility. Species: AO ions.

atom species is ionized, which differed from the actual space environment. The ion reaction effeciency appears to be systematically higher than the results reported for neutral reaction in space, which is consistent with other laboratory results.[9]

Conclusion

The AO ion beam apparatus has been demonstrated to be a useful tool in the study of AO reaction on conduction surfaces. The testing of dielectric materials may provide useful information, but real-time testing awaits improvements in the neutralization technology. Pure beam systems, like OMEGA, need to be fully developed and characterized in order to obtain reliable data for the selection of spacecraft material that will survive the low-Earth-orbit oxygen environment.

References

[1]Visentine, J. T., Leger, L. J., Kuminecz, J. F., and Spiker, I. K., "STS-8 Atomic Oxygen Effects Experiment," AIAA Paper 85-0415, 1985.

[2]Peters, P. N., Gregory, J. C., and Swann, J. T., "Effects on Optical Systems from Interactions with Oxygen Atoms in Low Earth Orbits," Applied Optics, Vol. 25, No. 8, April 1986, pp. 1290-1298.

[3]Lee, A. L. and Rhoads, G. D., "Prediction of Thermal Control Surface Degradation Due to Atomic Oxygen Interaction," AIAA Paper 85-1065, 1985.

[4]Mende, S. B., "Experimental Measurement of Shuttle Glow," AIAA Paper 84-0550, 1984.

[5]Visentine, J. T. and Leger, L. J., "Atomic Oxygen Effects Experiments: Current Status and Future Directions," Lyndon B. Johnson Space Center, NASA TM 100459, Vol. 3, Sept., 1988.

[6]Sjolander, G. W. and Bareiss, L. E., "Martin Marietta Atomic Oxygen Beam Facility," Proceedings of the 18th International SAMPE Technical Conference, Society for the Advancement of Material and Process Engineering, Covina, CA. 1986.

[7]Banks, B. A., private communication, NASA Lewis Research Center, Feb. 10, 1988.

[8]Bareiss, L. E., Payton, R. M., and Papzian, H. A., "Shuttle/Spacelab Contamination Environment and Effects Handbook," Martin Marietta Corporation, MCR-85-583, Dec., 1986.

[9]Banks, B. A,. private communication, NASA Lewis Research Center, Dec. 7, 1988.

Chapter 4. Plumes

Modeling Free Molecular Plume Flow and Impingement by an Ellipsoidal Distribution Function

Hubert Legge*
DFVLR, Göttingen, Federal Republic of Germany

Abstract

Modeling frozen plume flow is outlined using a freezing surface and an ellipsoidal distribution function beyond a freezing surface. Formulas are given for the number flux, pressure, shear stress, and heat-transfer on a surface element in free molecular flow with an ellipsoidal distribution function. These formulas are applied to frozen plume flow. Whereas the relevant coefficients as pressure coefficient are larger, the actual values as the pressure itself are smaller in frozen flow. To prove the accuracy of the modeling, it is applied to the totally frozen Knudsen effusion, for which the flow quantities are given. The approximation is compared to an exact calculation of the number flux, pressure, shear stress, and heat-transfer on a surface element on the axis of the free molecular effusion. For the given example the accuracy is better than 2.5% and will even be better if the flow freezes at large speed ratios in a jet.

Nomenclature

c_e = heat-transfer coefficient
c_j = thermal molecular velocity
c_p = pressure coefficient
c_τ = shear stress coefficient
d = orifice diameter (see Fig. 7)
$\dot{e}$ = heat flux
E = average energy per molecule
f = distribution function
k = Boltzmann constant

---

*Research Scientist, Fluid Mechanics Department.

m = mass of a molecule
p = pressure
P = freezing parameter [see Eq. (1)]
Q = impingement quantity
r = distance from jet source
R = specific gas constant
s = distance along streamline
S = speed ratio $S = u/(2RT)^{1/2}$
T = temperature
u = flow velocity
x = distance from orifice on jet axis
α = angle (see Fig. 1)
θ = angle in Knudsen effusion (see Fig. 7)
ν = collision frequency
ξ_j = molecular velocity
ρ = freestream density
σ_E = accommodation coefficient for energy transfer
σ_n = accommodation coefficient for normal momentum transfer
σ_τ = accommodation coefficient for tangential momentum transfer
τ = shear stress
φ = angle (see Fig. 1)

Subscripts

ell = ellipsoidal
f = frozen
i = incident on surface element
j = coordinate axis
M = Maxwellian
r = reflected from surface element
rot = rotational
vib = vibrational
w = reflected at complete accommodation
0 = stagnation chamber condition
$\perp$ = perpendicular (to flow velocity)
$\parallel$ = parallel (to flow velocity)

Introduction

Freejets[1] from sonic orifices are well suited to generate flows with high Mach numbers and low densities. They are used in molecular beam systems and in experiments to determine force and heat-transfer on small models up to free molecular flow. At these rarefied conditions even the translational degrees of freedom of the jet gas can be frozen. Similarly to the sonic orifice freejet, the plume from a thruster can be frozen when it impinges free molecularly on spacecraft surfaces. The question is, how does freezing influence the number flux, force, and heat-transfer on bodies in free molecular flow?

In the transition regime of a jet flow an ellipsoidal distribution function seems a good approximation to the actual one as shown by experiment[2] and theory.[3] In the following discussion the modeling of the rarefied flow by a freezing surface and an ellipsoidal distribution function is described. The number flux, pressure and shear stress coefficients, and heat-transfer characteristics on a surface element in free molecular flow are derived for an ellipsoidal distribution function. A comparison of an exact calculation for a surface element in the totally frozen Knudsen effusion from an orifice and the approximation by an appropriate ellipsoidal distribution function are given.

Modeling Frozen Plume Flow

To model frozen jet flow[4] for engineering estimations of plume impingement effects on spacecraft, a freezing surface is assumed which is described by Bird's[5] freezing parameter

$$P = \frac{u}{\rho \cdot v} \left| \frac{d\rho}{ds} \right| = \text{const} \tag{1}$$

Up to the freezing surface the flow can be calculated by a plume model[4] or freejet equations.[1] Beyond the freezing surface the flow[6] and impingement quantities can be estimated by an integration of the f_M on the freezing surface over the solid angle, which is spanned by the line of sight from the inspection point to the freezing surface as indicated in Fig. 1. If $\phi(\xi_j)$ is some quantity of a molecule with velocity ξ_j and $f(\xi_j)$ the distribution function, the moments

$$<\phi(\xi_j)> = \int_{-\infty}^{+\infty}\int_{-\infty}^{+\infty}\int_{-\infty}^{+\infty} \phi(\xi_j) \cdot f(\xi_j)\, d\xi_1\, d\xi_2\, d\xi_3 \tag{2}$$

define the observable quantities as density $\rho = <m>$, velocity $u_j = (1/n) <\xi_j>$ etc., where n is the number density. For any distribution function we can define temperatures in the direction of

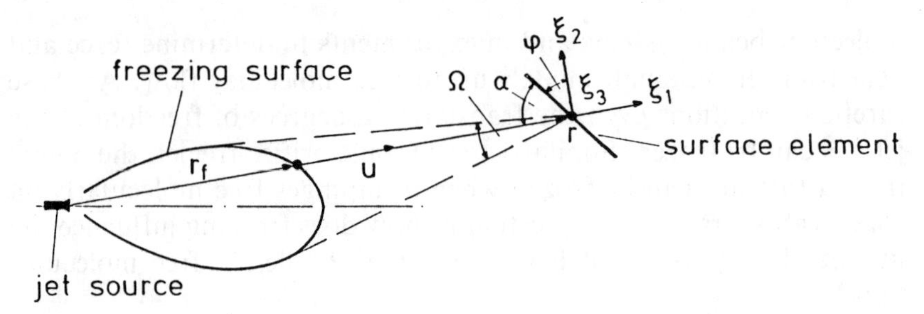

Fig. 1 Surface element in free molecular jet flow.

the coordinate axis j:

$$T_j = \frac{2}{nk} < \frac{1}{2} mc_j^2 > = \frac{1}{\rho R} p_j \tag{3}$$

with the thermal velocity

$$c_j = \xi_j - u_j \tag{4}$$

and the bulk velocity

$$u_j = \frac{1}{n} < \xi_j > \tag{5}$$

As the stress tensor σ_{ik} is symmetric for any distribution function, one can choose a coordinate system, so that the components out of the diagonal become zero. This means that using the equation

$$T_j = -\sigma_{jj}/\rho R = p_j/\rho R \tag{6}$$

an ellipsoidal distribution function

$$f = \frac{n}{\sqrt{T_1 \cdot T_2 \cdot T_3}\,(2\pi R)^{3/2}} \cdot \exp\left(- \frac{(\xi_1 - u_1)^2}{2RT_1} - \frac{(\xi_2 - u_2)^2}{2RT_2} - \frac{(\xi_3 - u_3)^2}{2RT_3} \right) \tag{7}$$

can be given resulting in the correct density, temperatures, pressures, and stress tensor and combinations of these quantities, which could be known from experiment, for example. The heat flux based on the thermal velocity alone

$$\dot{e}_j = m/2 < c_j^2 c_j > \tag{8}$$

is always zero for an ellipsoidal distribution function and cannot be fitted to reality if $\dot{e}_j \neq 0$.

The flow quantities at the inspection point in the plume, i.e., the surface element, can be judged from experiment and theory and have the following general behavior. The velocity of a molecule is constant for $r > r_f$, where r denotes the distance from the nozzle exit and subscript f the freezing surface. Therefore, the bulk velocity u is also constant beyond the freezing surface, if freezing occurs at hypersonic S and minor geometry effects are neglected. The number density n on a streamline, i.e., along r, decays (as in the continuum case) with $n \sim r^{-2}$. The T_2 and T_3 are assumed to be equal if they are based on the thermal velocity perpendicular to u and r (which is strictly correct only on the centerline). The T_2 and T_3 are then called perpendicular temperature $T_{\perp}$. It decays with increasing distance:

$$T_{\perp} = T(r_f) \cdot \left(\frac{r_f}{r} \right)^n \qquad (n \approx 1.5) \tag{9}$$

The T_1 is called parallel temperature and stays constant beyond the freezing point:

$$T_{\parallel} = T(r_f) \,. \tag{10}$$

Thereby all quantities, n, u, and T_j, can be expressed in Eq. (7) as functions of r and θ.

Forces and Heat Transfer

Let us now assume that we have an ellipsoidal distribution function of the form

$$f_{ell} = \frac{n}{\sqrt{T_{\parallel}}\, T_{\perp} (2\pi R)^{3/2}} \cdot \exp\left(-\frac{(\xi_1 - u)^2}{2RT_{\parallel}} - \frac{\xi_2^{\,2}}{2RT_{\perp}} - \frac{\xi_3^{\,2}}{2RT_{\perp}} \right) \tag{11}$$

with

$$p_{\perp} = \rho R T_{\perp}, \quad p_{\parallel} = \rho R T_{\parallel}, \quad p = (2p_{\perp} + p_{\parallel})/3 = \frac{\rho R}{3}(2T_{\perp} + T_{\parallel}) = \rho R T$$

The free molecular quantities transferred to a surface element are evaluated similarly to the procedure given by Schaaf and

Chambre[7] for a Maxwellian distribution function. The transferred quantities are calculated separately for the incident molecules, subscript i, and the reflected quantities at complete accommodation, subscript w. With the usual definition[7] of σ, p, τ, and $\dot{e}$ are given by

$$p = p_i + p_r = (2 - \sigma_n)p_i + \sigma_n p_w \tag{12}$$

$$\tau = \tau_i - \tau_r = \sigma_\tau \cdot \tau_i \tag{13}$$

$$\dot{e} = \dot{e}_i - \dot{e}_r = \sigma_E(\dot{e}_i - \dot{e}_w) \tag{14}$$

where the subscript r denominates the actually reflected quantity. Assuming for the particle flux $\dot{n}_i = \dot{n}_r$ it is[7]

$$p_w = \frac{1}{2} m\sqrt{2\pi RT_w} \cdot \dot{n}_i \tag{15}$$

and

$$\dot{e}_w = (E_{wrot} + E_{wvib} + 2kT_w) \cdot \dot{n}_i \tag{16}$$

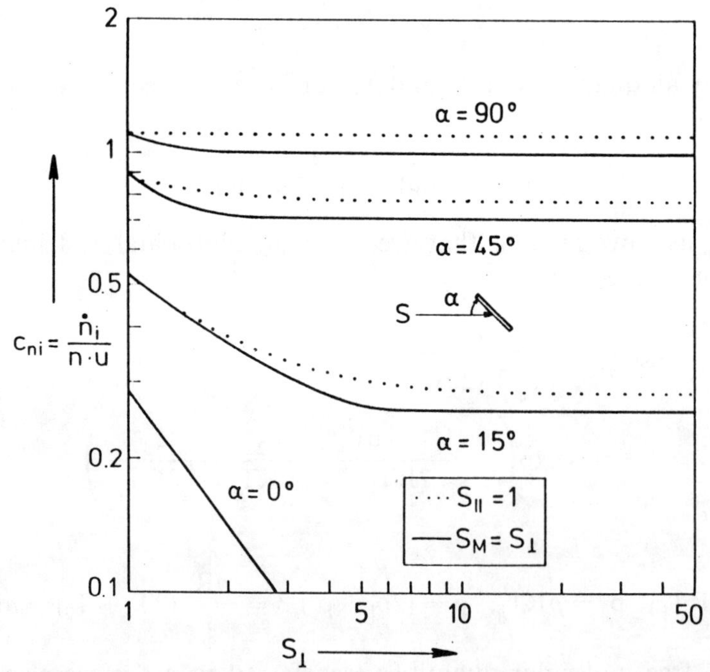

Fig. 2 Number flux on a surface element in free molecular flow with an ellipsoidal distribution function.

To determine the incoming quantities the ellipsoidal distribution function Eq. (1) is written for a coordinate system turned by φ = 90 deg - α around the x_3 axis, which is laid into the surface:

$$f_{ell} = \frac{n}{\sqrt{T_{\|}}\,T_{\perp}\,(2\pi R)^{3/2}} \cdot \exp\left(-\left[\frac{((\xi'_1 - u\sin\alpha)\cos\varphi - (\xi'_2 + u\cos\alpha)\sin\varphi)^2}{2RT_{\|}} + \frac{((\xi'_1 - u\sin\alpha)\sin\varphi + (\xi'_2 + u\cos\alpha)\cos\varphi)^2}{2RT_{\perp}} + \frac{\xi'^2_3}{2RT_{\perp}} \right]\right) \tag{17}$$

With the definitions

$$S_{\perp} = \frac{u}{\sqrt{2RT_{\perp}}} \quad \text{perpendicular speed ratio}$$

$$S_{\|} = \frac{u}{\sqrt{2RT_{\|}}} \quad \text{parallel speed ratio}$$

$$S_{ell} = \frac{1}{\sqrt{\sin^2\alpha/S^2_{\|} + \cos^2\alpha/S^2_{\perp}}} \quad \text{ellipsoidal speed ratio}$$

$$S' = S_{ell} \cdot \sin\alpha, \quad S_w = \frac{u}{\sqrt{2RT_w}}$$

With E_{irot}, E_{ivib} we obtain (after some calculations),

$$\begin{aligned} \dot{n}_i &= \int_{-\infty}^{+\infty}\int_{-\infty}^{+\infty}\int_{0}^{\infty} \xi'_1 f_{ell}\, d\xi'_1\, d\xi'_2\, d\xi'_3 \\ &= \frac{n}{2\sqrt{\pi}}\frac{u}{S_{ell}}\left[\exp(-S'^2) + \sqrt{\pi}\,S'(1 + \mathrm{erf}(S'))\right] \end{aligned} \tag{18}$$

$$p_i = \frac{1}{2}\rho u^2 \cdot \frac{1}{\sqrt{\pi}\,S_{ell}^{\ 2}}\Big[S'\exp(-S'^2) + \sqrt{\pi}\left(\frac{1}{2}+S'^2\right)(1+\mathrm{erf}(S'))\Big] \tag{19}$$

$$\begin{aligned}\tau_i = \frac{1}{2}\rho u^2 \cdot \Bigg\{ & \frac{S_{ell}}{\sqrt{\pi}}\frac{\cos\alpha}{S_{|}^{\ 2}}\Big[\exp(-S'^2) + \sqrt{\pi}\,S'(1+\mathrm{erf}(S'))\Big] \\ & + \frac{1}{\sqrt{\pi}}\left[\left(\frac{1}{S_{\|}^{\ 2}} - \frac{1}{S_{|}^{\ 2}}\right)\sin\alpha\cdot\cos\alpha\right] \\ & \cdot\Big[S'\exp(-S'^2) + \sqrt{\pi}\left(\frac{1}{2}+S'^2\right)(1+\mathrm{erf}(S'))\Big]\Bigg\}\end{aligned} \tag{20}$$

$$\begin{aligned}\dot{e}_i = \frac{\rho}{2}u^3\frac{1}{\sqrt{\pi}}\cdot S_{ell}\cdot\Bigg\{ & \frac{1}{2}\left(\frac{\cos^2\alpha}{S_{|}^4} + \frac{\sin^2\alpha}{S_{\|}^4}\right) \\ & \cdot\left[e^{-S'^2}(1+S'^2) + S'\sqrt{\pi}\,(1+\mathrm{erf}(S'))\left(S'^2+\frac{3}{2}\right)\right] \\ & + S'\cos^2\alpha\frac{1}{S_{|}^2}\left(\frac{1}{S_{\|}^2} - \frac{1}{S_{|}^2}\right) \\ & \cdot\left[e^{-S'^2}\cdot S' + \sqrt{\pi}\,(1+\mathrm{erf}(S'))\left(S'^2+\frac{1}{2}\right)\right] \\ & + \frac{1}{4}\frac{1}{S_{|}^2}\left(\frac{1}{S_{ell}} + 2S_{ell}^2\cos^2\alpha\cdot\frac{1}{S_{|}^2} + \frac{1}{S_{\|}^2}\right) \\ & \cdot\left[e^{-S'^2} + \sqrt{\pi}\,(1+\mathrm{erf}(S'))\cdot S'\right]\Bigg\} + (E_{irot}+E_{ivib})\dot{n}_i\end{aligned} \tag{21}$$

with

$$\mathrm{erf}(t) = \frac{2}{\sqrt{\pi}}\int_0^t \exp(-z^2)dz$$

If $S_{\|} = S_{|} = S_M = u/[(2RT)^{1/2}]$, the Maxwellian distribution function and the corresponding formulas are obtained.

The equations used to calculate $\dot{n}_i$, p_i, and p_w (but not τ_i and $\dot{e}_i$) for an ellipsoidal distribution function can be obtained from the ones for a Maxwellian by replacing S_M by S_{ell}. In addition, it is $S_{ell} = S_{|}$ for $\alpha = 0$ deg and $S_{ell} = S_{\|}$ for $\alpha = 90$ deg.

Example Calculations with an Ellipsoidal Distribution Function

Some example calculations for the coefficients

$$c_{ni} = \frac{\dot{n}_i}{n \cdot u}, \quad c_{pi} = \frac{p_i}{\frac{1}{2}\rho u^2}, \quad c_{\tau i} = \frac{\tau_i}{\frac{1}{2}\rho u^2}, \quad c_{ei} = \frac{\dot{e}_i}{\frac{1}{2}\rho u^3}$$

are given in Figs. 2-5 to show the general behavior and the differences against a Maxwellian distribution, represented by the full lines with $S_M = S_\perp$. For the dotted and dashed lines the parallel speed ratio is frozen at $S_\parallel = 1$ or $S_\parallel = 4$, respectively.

In Figs. 2-5 all coefficients for frozen flow are larger than the ones for the corresponding Maxwellian with $S_M = S_\perp$. The differences between the coefficients for $S_\parallel = \text{const}$ and the Maxwellian are usually smaller the larger the $S_\parallel$ and the smaller the difference between $S_\perp$ and $S_\parallel$. For $S_\parallel \geq 4$ the differences in c_{ni} could not be distinguished in Fig. 2.

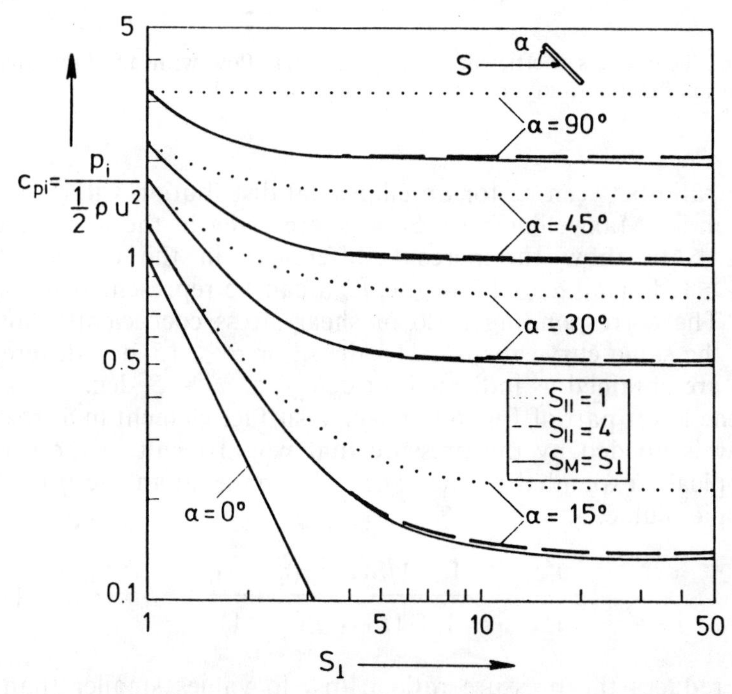

Fig. 3 Pressure coefficient in free molecular flow with an ellipsoidal distribution function (only incoming molecules or $\sigma_n = 1$, $T_w = 0$).

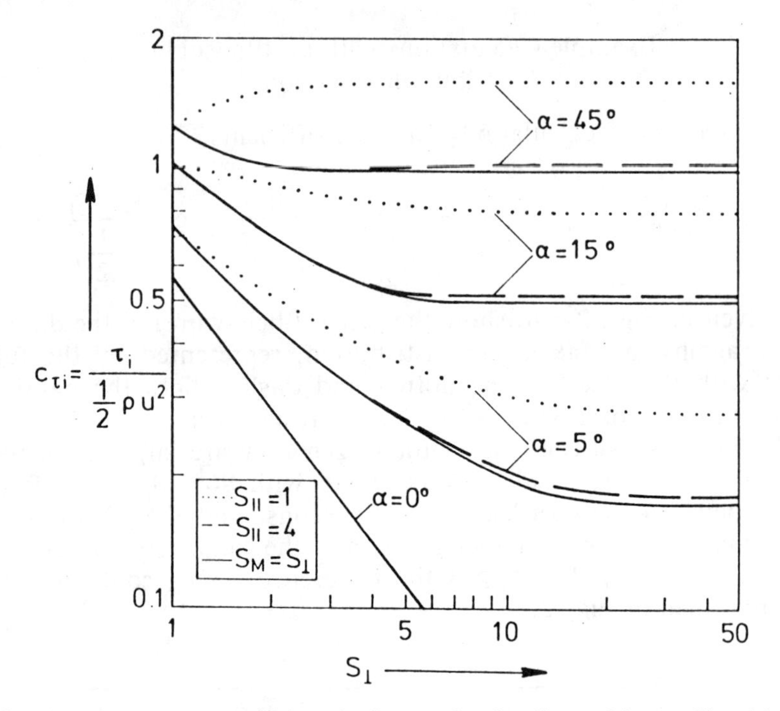

Fig. 4 Shear stress coefficient in free molecular flow with an ellipsoidal distribution function (only incoming molecules or $\sigma_\tau = 1$).

The ratio between c_{pi} for an ellipsoidal distribution with $S_\| = 4$, $S_| \geq 4$ and a Maxwellian with $S_M = S_|$ are given in the upper part of Fig. 6 to show the percent differences in the coefficients, $c_{pi}/c_{piM} \geq 1$. For all $\alpha \geq 15$ deg, c_{pi}/c_{piM} can be represented by one curve. The corresponding ratio of shear stress coefficients shows nearly the same curve for $\alpha \geq 15$ deg. For $\alpha \leq 15$ deg different results are obtained as indicated for c_{pi}/c_{piM} at $\alpha = 5$ deg.

In the lower part of Fig. 6 the p_i on a surface element in a frozen jet flow is divided by the pressure that would result in the same isentropically expanded plume. The u_f is smaller than the u in the isentropic plume:

$$\frac{u_f^2}{u^2} = \frac{1 + 1/S_M^2 \cdot \kappa/(\kappa + 1)}{1 + 1/S_\|^2 \cdot \kappa/(\kappa - 1)} \tag{22}$$

which reduces the pressure ratio p_{if}/p_{iM} to values smaller than 1 though $c_{pi}/c_{piM} > 1$. Similar curves are obtained for the number flux and the energy flux. The largest and smallest values are obtained

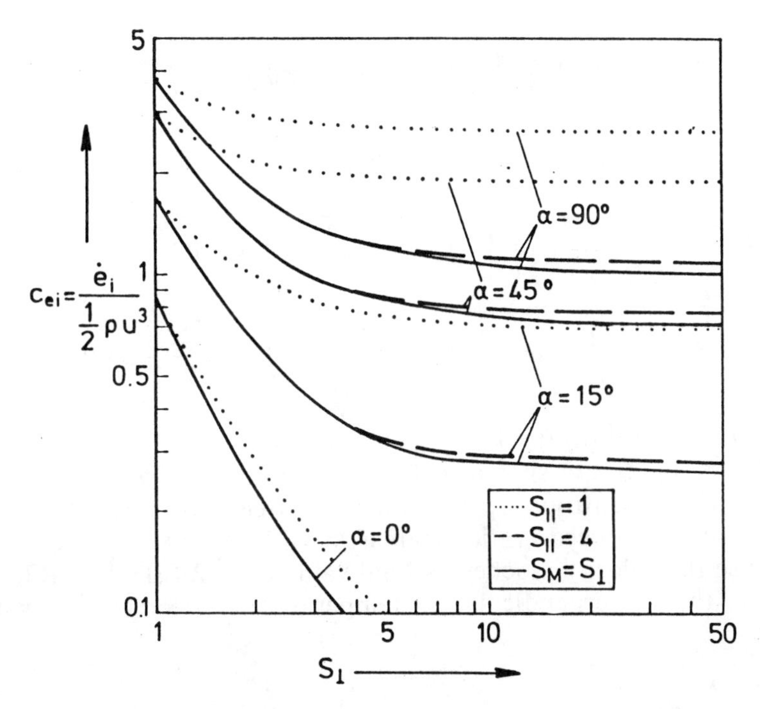

Fig. 5 Heat-transfer coefficient in free molecular flow with an ellipsoidal distribution function (only incoming molecules or $\sigma_E = 1$, $T_w = 0$).

for c_{ei}/c_{eiM} and $\dot{e}_{if}/\dot{e}_{iM}$, which amount to 1.09 and 0.87, respectively, for $S_\parallel = 4$, $S_\perp \rightarrow \infty$, and $S_M \rightarrow \infty$. These differences decrease rapidly when the jet freezes at larger speed ratios $S_\parallel > 4$.

Application of Ellipsoidal Approximations to Knudsen Effusion

To judge the accuracy of the pressure and shear stress determination by an ellipsoidal distribution function in a frozen jet flow, the extreme case of a rarefied jet, a Knudsen effusion, i.e., the free molecular flow from a large stagnation chamber through an orifice into vacuum, is considered in the following. The geometry is shown in the insert of Fig. 7. On the x axis it is easy[8] to calculate the quantities in the flow:

$$n = n_0 \sin^2(\theta/2) \tag{23}$$

$$u = 2\sqrt{(2RT_0)/\pi}\, \cos^2(\theta/2) \tag{24}$$

$$T_{\|} = T_0\left(\left(1 - \frac{2}{\pi}\right)4\cos^4\frac{\theta}{2} - \cos\theta\right) \tag{25}$$

$$T_{\perp} = T_0\left(\sin^2\frac{\theta}{2} + \frac{1}{2}\sin^2\theta\right) \tag{26}$$

$$S_{\perp} = \frac{2}{\sqrt{\pi}}\cos^2\frac{\theta}{2} \Big/ \sqrt{\sin^2\frac{\theta}{2} + \frac{1}{2}\sin^2\theta} \tag{27}$$

$$S_{\|} = \frac{2}{\sqrt{\pi}}\cos^2\frac{\theta}{2} \Big/ \sqrt{4\left(1 - \frac{2}{\pi}\right)\cos^4\frac{\theta}{2} - \cos\theta} \tag{28}$$

with $\theta = \arctan[(d/2)/x]$.

The $S_{\perp}$ and $S_{\|}$ are shown in Fig. 7 as function of x/d and compared to the isentropic speed ratio[1] in a freejet. For $x/d \gg 1$ the perpendicular speed ratio increases with $S_{\perp} \approx [8/(3\pi)^{1/2}] \cdot (x/d)$ because the velocity reaches its final value $u = [2/(\pi)^{1/2}] \cdot (2RT_0)^{1/2}$ and the perpendicular temperature decreases with

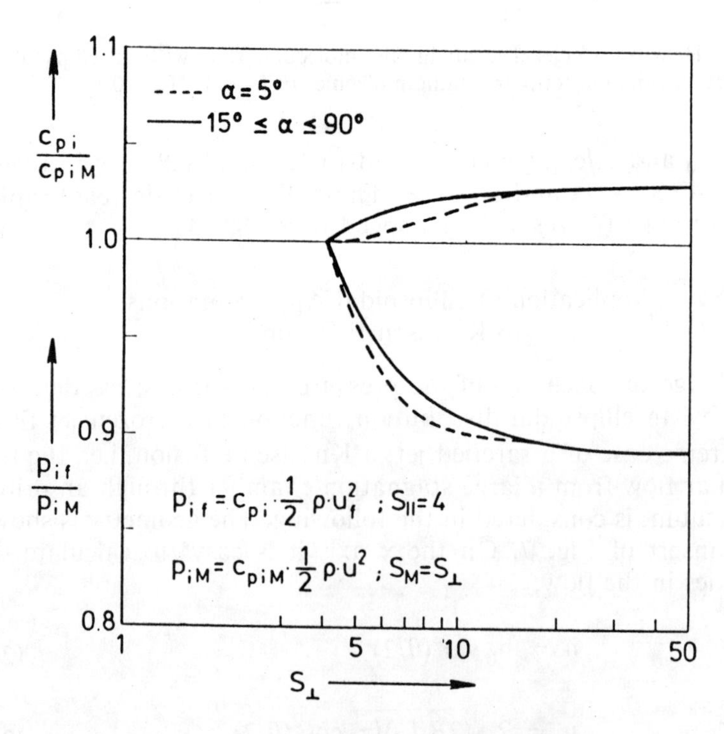

Fig. 6 Influence of freezing on surface pressure.

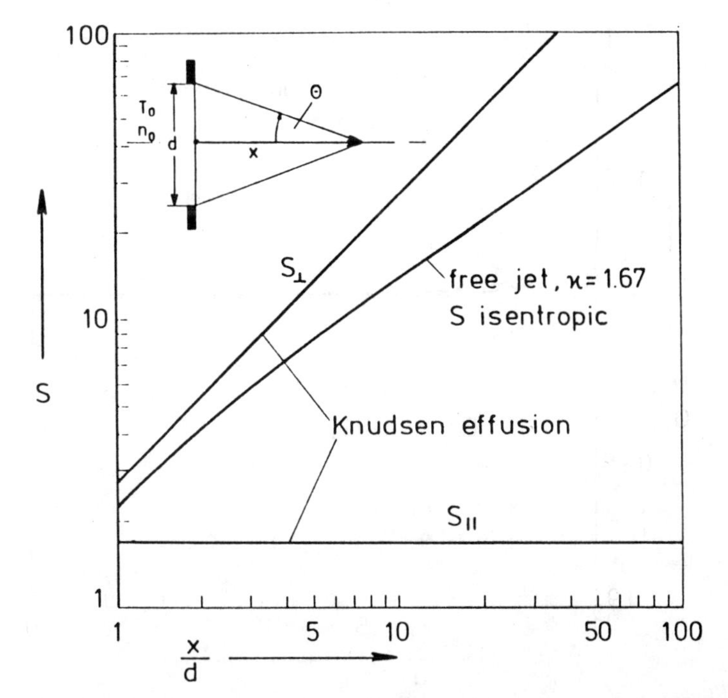

Fig. 7 Speed ratio in a Knudsen effusion and in an isentropic freejet.

$T_\perp \approx T_0 \cdot (3/16) \cdot (x/d)^{-2}$. The $T_\parallel$ is nearly constant. The physical reason is that the possible molecular velocity components perpendicular to the x axis become smaller within the lines of sight with decreasing θ, whereas in x direction half of the Maxwellian distribution function is still present.

Equations (23-28) can now be used in Eqs. (18-21) to obtain approximate values Q_{ell} for the quantities $\dot{n}_i$, p_i, τ_i, and $\dot{e}_i$ on a surface element on the axis of a Knudsen effusion. On the other side the exact values Q_{Kn} are obtained for $\alpha > \theta$ (see Fig. 8) by

$$\dot{n}_i = \frac{n_0}{2}\sqrt{\frac{2RT_0}{\pi}}\,\sin^2\theta \cdot \sin\alpha$$

$$p_i = \frac{p_0}{2}\left\{\left(1 - \frac{3}{2}\cos\theta + \frac{1}{2}\cos^3\theta\right)\cos^2\alpha + \left(1 - \cos^3\theta\right)\sin^2\alpha\right\}$$

$$\tau_i = \frac{p_0}{2}\cdot\frac{3}{2}\left(\cos\theta - \cos^3\theta\right)\cos\alpha \cdot \sin\alpha$$

$$\dot{e}_i = \frac{p_0\sqrt{2RT_0}}{\sqrt{\pi}} \cdot \sin^2\theta \cdot \sin\alpha$$

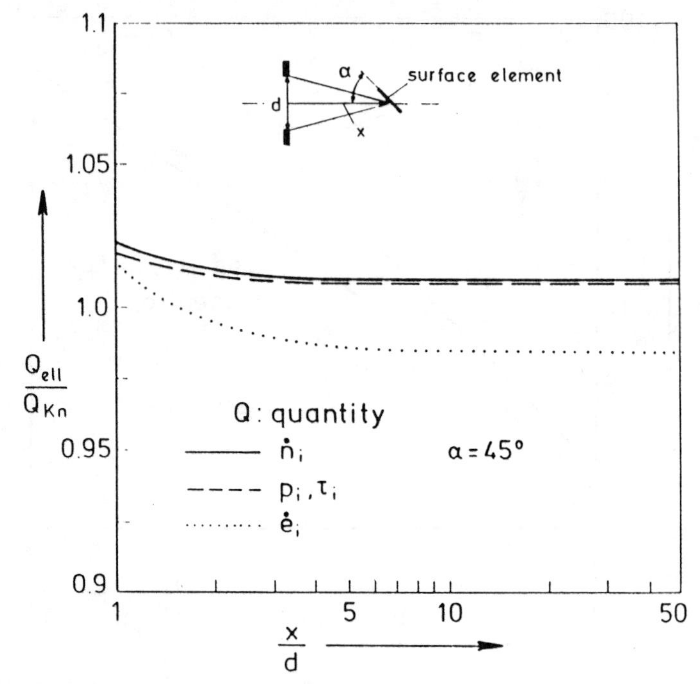

Fig. 8 Number flux $\dot{n}_i$, pressure p_i, shear stress τ_i, and energy flux $\dot{e}_i$ due to incoming molecules on a surface element on the axis of an Knudsen effusion. Comparison of approximations by an ellipsoidal distribution (Q_{ell}) and exact calculation (Q_{Kn}).

Figure 8 shows the ratio between the ellipsoidal approximation Q_{ell} and the exact calculation Q_{Kn}. The approximations have an error smaller than 2.5% for the shown example even though the Knudsen effusion is the extreme case of a frozen jet flow. It can therefore be expected that the approximations for $\dot{n}_i$, p_i, τ_i, and $\dot{e}_i$ are very accurate if the jet freezes at large speed ratios and the flow properties n, u, $T_\perp$, $T_\parallel$ are well known. For engineering estimations of forces and heat-transfer in plume impingement problems, the influence of freezing can usually be neglected as far as the distribution function is concerned.

References

[1]Ashkenas, H., and Sherman, F. S., "The Structure and Utilization of Supersonic Free Jets in Low-density Wind Tunnels," Rarefied Gas Dynamics, edited by J. H. de Leeuw, Academic, New York, 1966, Vol. II, pp. 84-105.

[2]Muntz, E. P., "Measurements of Anisotropic Velocity Distribution Functions in Rapid Radial Expansions," Rarefied Gas Dynamics, edited by C. L. Brundin, Academic, New York, 1967, Vol. 2, Suppl. 4, pp. 1257-1286.

[3]Willis, D. R., Hamel, B. B., and Lin, J. T., "Development of the Distribution Function on the Centerline of a Free Jet Expansion," Physics of Fluids, Vol. 15, No. 4, 1972, pp. 573-580.

[4]Legge, H., and Boettcher, R.-D., "Modelling Control Thruster Plume Flow and Impingement," Rarefied Gas Dynamics, edited by O. M. Belotserkowskii, M. N. Kogan, S. S. Kutateladze, and A. K. Rebrov, Plenum, New York, 1985, Vol. 2, pp. 983-992.

[5]Bird, G. A., "Breakdown of Continuum Flow in Free Jets and Rocket Plumes," Progress in Astronautics and Aeronautics, Rarefied Gas Dynamics, Vol. 74, P. 2, edited by S. S. Fisher, AIAA, New York, 1981, pp. 681-694.

[6]Allen, G. A., Koppenwallner, G., and Leners, K. H., "A Study of Species Separation in Free Jet Expansions," Rarefied Gas Dynamics, edited by V. Boffi, C. Cercignani, and B. G. Teubner, Stuttgart, FRG, 1986, Vol. 2, pp. 66-75.

[7]Schaaf, S. A., and Chambre, P. L., "Flow of Rarefied Gases," High-speed Aerodynamics and Jet Propulsion, edited by H. W. Emmons, Princeton Univ. Press, Princeton, NJ, 1958, Vol. III, Sec. H, pp. 687-739.

[8]Legge, H., "Entwurfsgrundlagen und kurze Baubeschreibung der dritten Meßstrecke des Hypersonischen Windkanals für kleine Gasdichten der DFVLR-AVA in Göttingen," Deutsche Luft- und Raumfahrt, DLR-FB 71-84, 1971.

Plume Shape Optimization of Small Attitude Control Thrusters Concerning Impingement and Thrust

K. W. Naumann*
Franco-German Research Institute of Saint-Louis (ISL), Saint-Louis, France

Abstract

This paper presents a comparative description of plumes emanating from different nozzles, similar to those of typical small hydrazine thrusters. The objective is to point out the effect of nozzle design on freejet expansion. The investigation shows that there exists an optimum nozzle wall angle, which causes the minimal spreading plume. With increasing nozzle size and stagnation pressure, the plume spreads significantly wider, and the optimum nozzle wall angle is marked more clearly, tending to somewhat lower values of that angle. In consideration of thrust efficiency and minimization of impingement effects, the use of smaller thrusters with greater wall angles is favorable.

Introduction

Small thrusters are used mostly for the attitude or orbit control of satellites. Without any back pressure acting, the plume spreads very widely. Hence, it is impossible to entirely avoid detrimental interactions between the freejet and surfaces of the satellite (see Fig. 1). However, the designer of the satellite has the possibility to minimize those effects by installing thrusters with a narrow spreading plume. This measure is not bound to inevitably cause loss in thrust efficiency. It makes use of the fact that the deflection of the plume streamlines in axially symmetric Prandtl-Meyer expansion is subject to rarefaction effects.[1,2] The process is named "freezing of flow deflection" and causes the streamlines

---

*Scientist, Department for Hypersonic Aerodynamics.

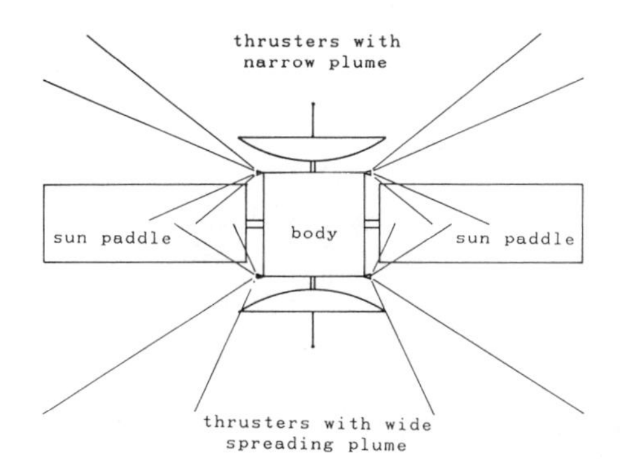

Detrimental interactions between expanding freejet and impinged surfaces

- parasitic (secondary) forces and moments
- thermal loads
- contamination effects
- chemical reactions

Fig. 1 Satellite and its thrusters

Table 1 Gas data (catalytically decomposed hydrazine)

Mole mass, kg/kmole	14.5
γ	1.24
T_0, K	1345
μ_0, kg/ms	4466×10^{-8}

not to entirely exploit the deflection capability given from continuum considerations. In axially symmetric expansion to vacuum, the deflection capability depends on the state parameters in the nozzle opening and the nozzle size, all being at the designer's disposal.

Thruster Data and Nozzle Flow

The shape and size of the conical nozzles examined are shown in Fig. 2, and the gas data are given in Table 1. The area ratio A_E/A^* is always 50 and produces the friction-free opening Mach number $Ma_E = 4.66$. The hydrazine is taken without water, being catalytically decomposed.[3] The nozzle Reynolds numbers range from 854 to 22540, which means that all nozzles have a laminar boundary layer.[4]

The designations for the nozzle flow are also given in Fig. 2, and Fig. 3 shows the relationships used for the calculations. The nozzle flow is subdivided into two regions: the isentropic core and the boundary layer (BL). We calculate the BL flow using an experimentally proven[2] approximation with a linear Mach number slope across the BL. The flow is considered to be adiabatic, with $T_0 = T_W$ and constant pressure across the BL. The Mach number in the

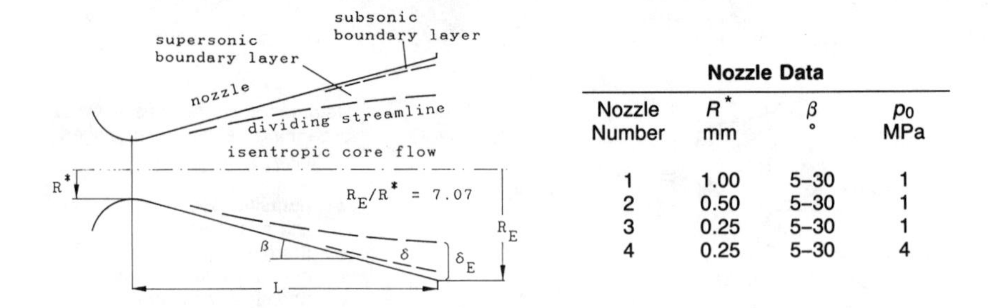

Nozzle Data

Nozzle Number	R^* mm	β °	p_0 MPa
1	1.00	5–30	1
2	0.50	5–30	1
3	0.25	5–30	1
4	0.25	5–30	4

Fig. 2 Nozzle data and designations for the nozzle flow.

$$\delta_E/L_E = 5.4 \cdot Re_{L,0}^{-1/2}$$

$$Re_{L,0} = \frac{\rho_E \cdot W_E \cdot L_E}{\mu_0}$$

$$Ma_{BL} = Ma_C \cdot \frac{R_E - y}{\delta}; \; (R_E - \delta) < y < R_E$$

$$T_{0,BL} = T_{0,C} = T_W$$

Fig. 3 Relationships for nozzle flow calculation.

core flow is the effective value, determined with the BL displacement effect. Table 2 gives a synopsis of the nozzle flow data.

Plume Flow Calculation

The plume calculation uses the principle of "freezing of the streamline deflection" in axially symmetric expansion to vacuum. This is presented in detail in Refs. 1 and 2, and Fig. 4 shows the designations. The essential formulas are listed in the Appendix.

The causal process is the disappearance of the streamline curvature because of rarefaction effects. It occurs together with the appearance of nonequilibria in the translational degrees of freedom, at approximately the same "continuum breakdown place." We name it the freezing point (FP) and assume, that there the freezing process is contracted as a sudden event. All properties receive the index F at this point. Beyond the FP, the streamline direction Θ_F remains unchanged. All other properties change according to the laws describing free molecular expansion.

Table 2 Nozzle flow data

Nozzle	R^*, mm	β, deg	L, mm	Re_L,	$Ma_{C,E}$,	δ_E/R_E,	p_0, MPa
1a	1	5	69.4	22543	4.22	0.353	1
1b	1	10	34.4	11185	4.36	0.249	1
1c	1	15	22.7	7360	4.41	0.202	1
1d	1	20	16.7	5419	4.45	0.173	1
1e	1	30	10.5	3416	4.49	0.137	1
2a	0.5	5	34.7	11272	3.97	0.499	1
2b	0.5	10	17.2	5593	4.22	0.351	1
2c	0.5	15	11.3	3680	4.30	0.285	1
2d	0.5	20	8.4	2710	4.36	0.246	1
2e	0.5	30	5.3	1708	4.42	0.194	1
3a	0.25	5	17.3	5636	3.58	0.706	1
3b	0.25	10	8.6	2796	3.97	0.497	1
3c	0.25	15	5.7	1840	4.13	0.403	1
3d	0.25	20	4.2	1355	4.23	0.346	1
3e	0.25	30	2.6	854	4.32	0.275	1
4a	0.25	5	17.3	22543	4.22	0.352	4
4b	0.25	10	8.6	11185	4.36	0.249	4
4c	0.25	15	5.7	7360	4.41	0.202	4
4d	0.25	20	4.2	5419	4.45	0.173	4
4e	0.25	30	2.6	3416	4.49	0.135	4

In contrast to other freezing effects, the vanishing of the streamline deflection at finite Mach numbers affects the macroscopic jet configuration.

The mathematical treatment of this phenomenon may be done analytically, using the well-known freezing parameter P_F derived by Bird.[5] The suitable formulation of this parameter leads to the formula[1] (see the Appendix)

$$P = (\pi\gamma/2)^{1/2} Ma_F \frac{\lambda_F(k_\mu; R; \gamma; p_0; T_0; Ma_{C,E}; Ma_E; Ma_F)}{r(\gamma; y/R_E; R_E; \Theta_E; Ma_E; Ma_F)} \tag{1}$$

where all of the parameters with the exception of Ma_F can be taken as known. The experiments in ref. 2 give the value of $P_F = 0.06$ at FP, which is in good agreement with Bird's[5,6] quantity for the appearence of nonequilibria in the translational degrees of freedom.

With Ma_F determined, all other state parameters at FP can be evaluated from the values at the nozzle opening. This holds to the same extent for each streamline emanating from the core or the supersonic BL. In the subsonic BL, i.e., in the immediate vicinity of the nozzle edge, the flow conditions become singular, making analytical considerations difficult. As the subsonic BL carries only some percentage of the total mass flux with a minimum of kinetic energy, we may neglect it. Similar to Boettcher and Legge,[7] we take the $Ma_{BL,E} = 1$ streamline as being the outermost streamline considered, called the "limiting streamline."

For the plume geometry, the deflection angles are most important. From Ma_E and Ma_F, we get the deflection capability of each streamline in the absence of any influences due to nozzle geometry or BL structure, i.e.,

$$\Theta'_{F,lim} = \nu(Ma_F) - \nu(Ma_E) \tag{2}$$

The maximum deflection potential of the actual streamline leaving the nozzle with Θ_E is

$$\Theta_{F,lim} = \nu(Ma_F) - \nu(Ma_E) + \Theta_E \tag{3}$$

and the freezing angle actually attained may be approximated by[2]

$$\Theta_F = \Theta_{F,lim} \cdot y/R_E \tag{4}$$

Figure 5 shows these values in relation to the maximum continuum deflection potential for the streamlines emanating from the core or the supersonic BL of one of the medium-sized nozzles considered. We see that $\Theta_{F,lim}$ as well as Θ_F are significantly smaller than the deflection capability in continuum Prandtl-Meyer expansion. Notice that the maximum value for $\Theta'_{F,lim}$ as well as $\Theta_{F,lim}$ does not belong to the limiting streamline!

The other significant streamline in the expanding freejet is the "dividing streamline" with $\Theta = \Theta_0$. It separates the core expansion flow from that emanating from the BL and encloses the region where the full dynamic pressure load hits the impinged structures. We must keep in mind that $\Theta_{F,0}$ depends strongly on the BL thickness. Accordingly, it gives only restricted information relating to plume spreading. For this, $\Theta_{F,1}$ is the more expressive parameter.

Comparison of the Plume Flowfields

Figures 6-8 show the deflection capability $\Theta_{F,lim}$ and the streamline angle Θ_F actually attained for the maximum deflection streamline, the dividing streamline, and the limiting streamline.

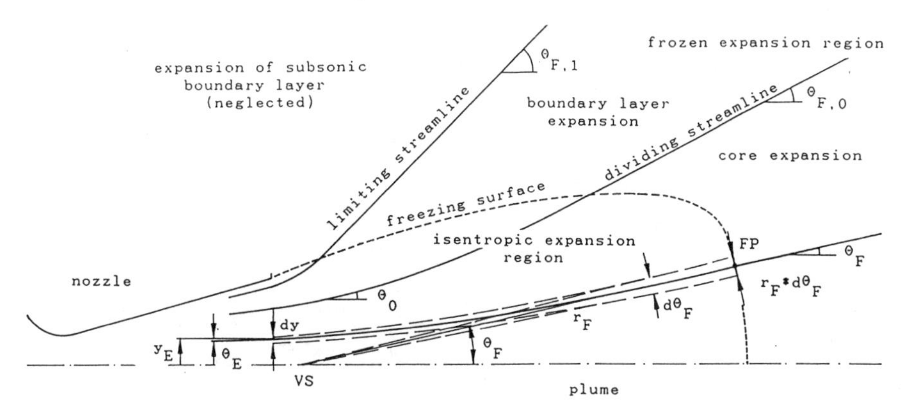

Fig. 4 Designations for plume flow calculation.

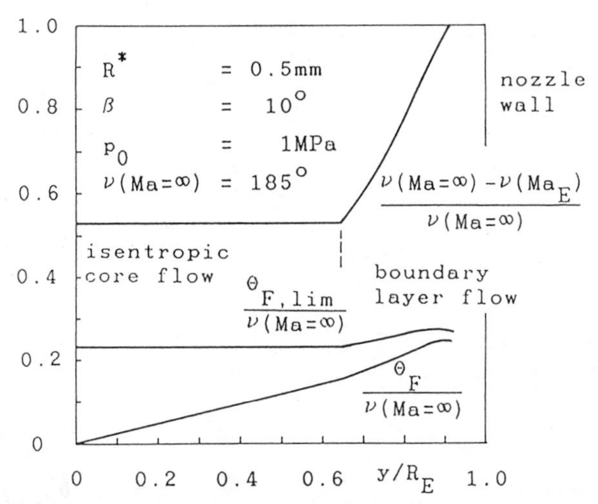

Fig. 5 $\theta_{F,lim}$, $\theta'_{F,lim}$, and $\nu(Ma = \infty)-\nu(Ma_E)$ compared with $\nu(Ma = \infty)$ for nozzle 2b.

We see that for each R^*/R_E there exists an optimum nozzle angle (and length) that produces the narrowest plume. This is caused by the BL displacement effect: the longer nozzle with smaller β has the lower Ma number, which increases $\theta'_{F,lim}$. On the other hand, smaller β means lower θ_E. At the optimum nozzle angle, the gradients of both effects balance each other.

The comparison of different nozzle sizes at equal pressure shows a significant reduction of the expansion angles for smaller nozzles. The optimum angle is shifted to greater β, and the optimum itself is less pronounced.

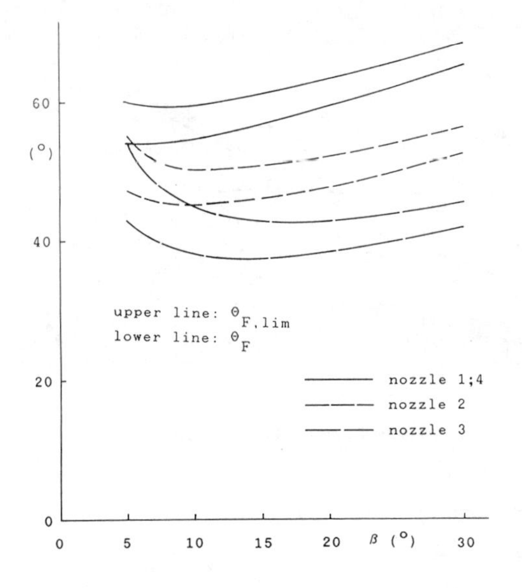

Fig. 6 $\Theta_{F,lim,max}$ and $\Theta_{F,max}$.

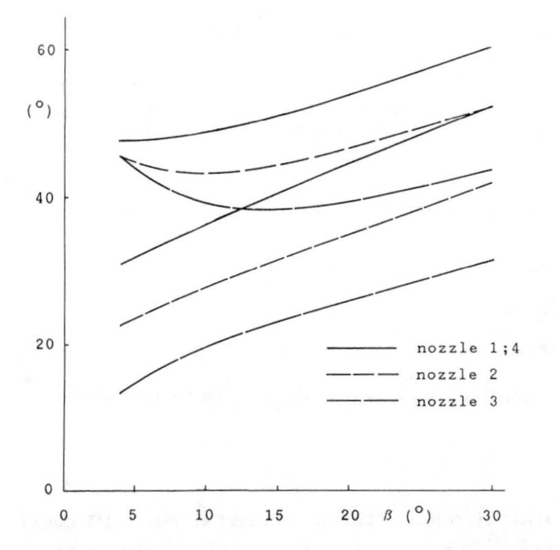

Fig. 7 $\Theta_{F,lim,C}$ (= $\Theta_{F,lim,0}$) and $\Theta_{F,0}$.

Reduction of the expansion angles is caused from the nozzle size parameter in the denominator of Eq. (1), indicating the smaller reference length in the freezing process. It causes the freezing process to occur at higher density, i.e., lower Mach number.

The comparison of two nozzles, Nos. 2 and 4, with the same thrust, shows that the larger nozzle produces a significantly smaller plume. Notice that p_0 in the

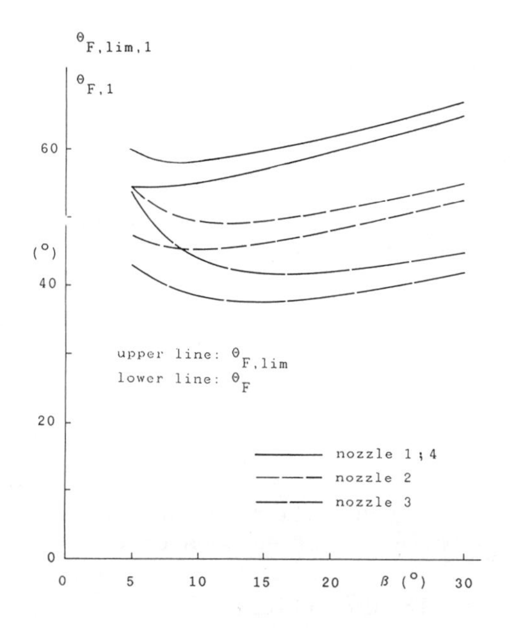

Fig. 8 $\theta_{F,lim,1}$ and $\theta_{F,1}$.

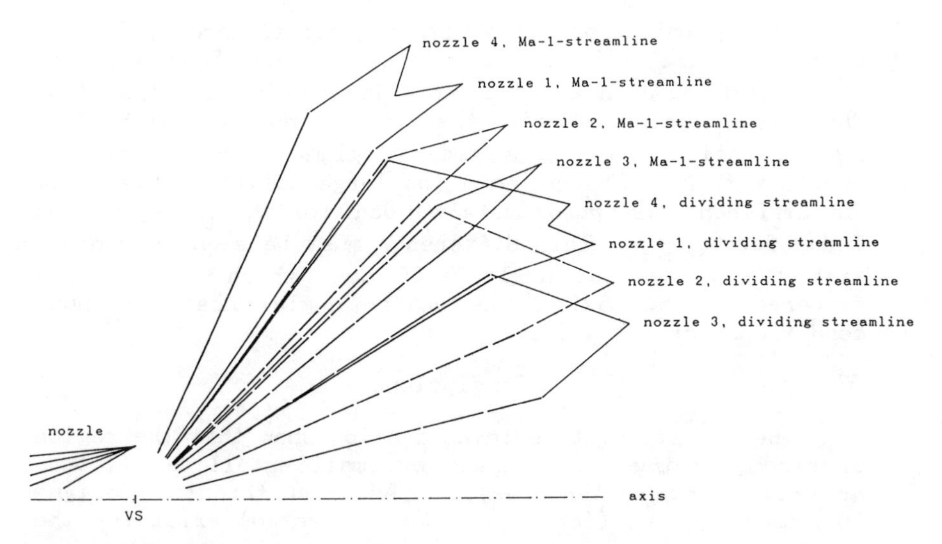

Fig. 9 Visual presentation of the minimum and maximum values for $\theta_{F,0}$ and $\theta_{F,1}$.

numerator of Eq. (1), being proportional to the thrust, overbears the corresponding (square root) change of the nozzle length parameter, say R^*, in the denominator. Comparison of the nozzles 1 and 4 with the same ratio p_0/R^* yields the same values. However, the nominal thrust of nozzle 1 is 4 times that of nozzle 4.

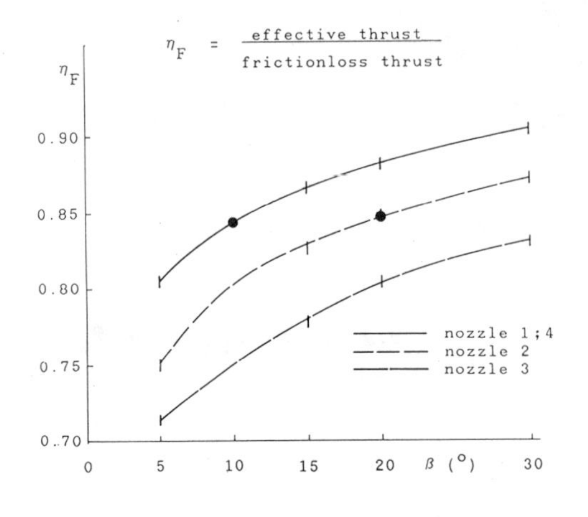

Fig. 10 Thrust efficiency factor.

Figure 9 gives a visual presentation of the minimum and maximum $\Theta_{F,1}$ and $\Theta_{F,0}$ for all nozzles considered.

Comparison of the Thrust Efficiency

For optimization, also with regards to thrust, Fig. 10 shows the nozzle thrust factors η_F varying from 0.714 to 0.905. Looking at nozzles 2 and 4, it can be seen that, for the same η_F being 0.845, the larger nozzle, with β = 20 deg, still produces a smaller plume than the smaller nozzle with β = 10 deg (note the large points in Fig. 10). The difference is approximately 8 deg for $\Theta_{F,lim,max}$ and 7 deg for $\Theta_{F,max}$. This difference must be seen in context with the plume flow angle < 65 deg, which is very low compared to the maximum deflection angle under continuum conditions for γ = 1.24.

Conclusions

The results of this investigation show that the region of strong impingement, caused by small satellite attitude or orbit control thrusters, depends on thruster design. Optimum solutions (for narrow plumes wanted) exist for the following:

- given p_0 and R^* ⇒ optimal β and length;
- given thrust ⇒ optimal R^*, β, and p_0;
- given p_0 ⇒ optimal R^*, β, and nozzle number.

Certainly, it is neither simple nor very cost-effective to install many small thrusters in place of a large one. The mass and the number of valves increase, and the control becomes more complicated.

The smarter method for narrow plumes is to install thrusters as small as possible and to extend the firing periods as far as possible. This measure is entirely positive, saving structural mass in addition to smaller impingement.

Appendix: Formulmas for Plume Flow Calculation

The well-known freezing parameter derived by Bird is written as

$$P = \frac{W\lambda}{c} \left| \frac{d\rho}{\rho ds} \right|$$

W is the speed, c the average thermal velocity with $c = (8RT/\pi)^{1/2}$, and λ the free path length, given by the Maxwell equation as

$$\lambda = \frac{16}{5} \frac{\mu}{\rho(2\pi RT)^{1/2}}$$

According to Legge,[8] the viscosity in the low-temperature range extending to 150 K can be written as $\mu = k_\mu T$.

Putting the isentropic relations for ρ and p and for the effctive stagnation pressure at the end of the nozzle,

$$p_{0,eff} = p_0 \frac{\left[2 + (\gamma-1)Ma_E^{\ 2}\right]^{\gamma/(\gamma-1)}}{\left[2 + (\gamma-1)Ma_{C,E}^{\ 2}\right]^{\gamma/(\gamma-1)}} = p_0 f_1 (Ma_E ; Ma_{C,E})$$

we obtain

$$\lambda = \frac{16\ k_\mu}{5\ p_0} (RT_0^{\ 3}/2\pi)^{1/2} \frac{(1 + (\gamma-1)Ma^2/2)^{(3-\gamma)/2(\gamma-1)}}{f_1 (Ma_E ; Ma_{C,E})}$$

In the freejet regions beyond FP, where the streamlines form a radial pattern, for the density gradient term we may use $|d\rho/\rho ds| = 2/r$. Treating the surface through which the freejet flows at FP as the generator surface of a spherical layer (see Fig. 4), we get

$$A_F = 2\pi rh = 2\pi r^2 d\Theta_F \sin\Theta_F$$

and for the area of the stream annulus at the nozzle aperture

$$A_E = 2\pi y dy \cos\Theta_E$$

The continuum relations are valid between A_E and A_F, and the area ratio A_F/A_E can be described by the function $f_2(Ma_F;Ma_E)$. The distance to the freezing point is then given by

$$r_F^{\,2} = \frac{y\cos\Theta_E}{\sin\Theta_F}\,\frac{dy}{d\Theta_F}\,f_2(Ma_F;Ma_E)$$

With the experimentally obtained relation $\Theta_F/\Theta_{F,lim} \cong y/R_E$ the differential of Θ_F is

$$d\Theta_F/dy = \Theta_{F,lim}/R_E + (y/R_E)(d\Theta_{F,lim}/dy)$$

As a first approximation, we may assume that $\Theta_{F,lim}$ does not vary very much over the nozzle outlet. Then only the term $\Theta_{F,lim}/R_E$ remains in the expression. The initially neglegted term $(y/R_E)(d\Theta_{F,lim}/dy)$ can be included in an iterative process in order to increase the accuracy of the calculation. We obtain

$$r_F = \frac{R_E\;(f_2 Ma_F;Ma_E)^{1/2}}{\left[\sin(\Theta_{F,lim}y/R_E)\Theta_{F,lim}/(\cos\Theta_E\;y/R_E)\right]^{1/2}}$$

where $\Theta_{F,lim}$ is also a function of Ma_E, Ma_F and Θ_E.

For FP, therefore, the freezing parameter is

$$P = (\pi\gamma/2)^{1/2}\,Ma_F\,\frac{\lambda_F(k_\mu;R;\gamma;p_0;T_0;Ma_{C,E};Ma_E;Ma_F)}{r(\gamma;y/R_E;R_E;\Theta_E;Ma_E;Ma_F)}$$

where all of the parameters with the exception of Ma_F can be taken as known.

References

[1] Naumann, K. W., "The Freezing of Flow Deflection in Prandtl-Meyer Expansion to Vacuum, "Proceedings of the 15th International Symposium on Rarefied Gas Dynamics, edited by V. Boffi and C Cercignani, Grado, Vol. 2, B. G. Teubner, Stuttgart, 1986, pp. 524-533.

[2] Naumann, K. W., "The Highly Underexpanded Exhaust Plume from Nozzles with Boundary Layer," DFVLR-AVA, Goettingen, FRG, DFVLR-FB 85-10, Jan. 1985 (German), ESA-TT 929, Sept. 1985 (English).

[3]Genovese, J. E., "Rapid Estimation of Hydrazine Exhaust Plume Interaction," AIAA Paper 78-1091, July 1978.

[4]Boynton, F. P., "Exhaust Plumes from Nozzles with Wall Boundary Layer," Journal of Spacecraft and Rockets, Vol. 5, Oct. 1968, pp. 1143-1147.

[5]Bird, G. A., Molecular Gas Dynamics, Clarendon Press, Oxford, 1976.

[6]Bird, G. A., "The Nozzle Lip Problem," Proceedings of the 9th International Symposium on Rarefied Gas Dynamics, Vol. 1, Paper A 22, Göttingen, FRG, July 1974.

[7]Boettcher, R.-D. and Legge, H., "A Study of Rocket Exhaust Plumes and Impingement Effects on Spacecraft Surfaces. II. Plume Profile Analysis. Part 1: Continuum Plume Modelling," DFVLR Internal Rept. DFVLR-AVA, Goettingen, FRG, 1980.

[8]Legge, H., Private communication, 1982.

Backscatter Contamination Analysis

B. C. Moore,* T. S. Mogstad,† S. L. Huston,‡
and J. L. Nardacci Jr.§
McDonnell Douglas Space Systems Company,
Huntington Beach, California

Abstract

A numerical procedure has been developed to compute the return flux due to backscattering of ambient molecules off of molecules emitted from a spacecraft. This analysis procedure computes the contamination of a particular surface from a selected point in space, given the local densities and velocity distributions of the two groups of molecules. This analysis includes the Maxwellian distribution of the ambient velocity and a selected emitted velocity distribution. It also includes the attenuation of both the emitted and backscattered molecules in passing through the ambient gas. This procedure has been used to determine the conditions behind an orbiting molecular wake shield. The effects of different emitted gas species, ambient and emitted gas temperatures, distance from emission source, and shield size are considered. These calculations show that N_2 emissions are more important than H_2O emissions for scattering of ambient molecules, and that pressures in the range of 10^{-11} to 10^{-10} torr are possible behind such a shield.

Introduction

Surfaces on orbiting spacecraft are frequently subject to contamination from various sources on the the space-

---

*Consultant.
†Engineer/Scientist, Advanced Technology Center.
‡Technical Specialist, Advanced Technology Center.
§Summer Intern; currently with University of California, Irvine.

craft, including vents, thruster plumes, and outgassing. The direct impingement of contaminants from these sources is a familiar problem in rarefied gas dynamics and is commonly accounted for in spacecraft design.[1] However, other sources of contamination result from the interaction of the spacecraft with its ambient environment. Obviously, the ambient molecules themselves may impinge directly on the spacecraft surface. Scattering is another source of contamination. Molecules emitted from the spacecraft can be scattered back to the spacecraft by collisions either with themselves or with the ambient gas.[2-4] In addition, an atmospheric molecule that collides with a molecule leaving the spacecraft with a thermal velocity can be backscattered towards the vehicle's wake if it is less massive than the thermal molecule.[5]

It is becoming increasingly important to be able to predict and control vacuum and contamination levels on spacecraft surfaces. High-performance sensor systems, for example, are extremely sensitive to contamination of optics. For space material processing experiments, such as molecular beam epitaxy, ultrapurification processes, and coating of large mirrors, vacuums as low as 10^{-12} to 10^{-14} torr are desired. These and other requirements have led to efforts to control contamination and to achieve ultravacuum conditions in space.

One such application is the orbiting molecular wake shield.[5] At 200-350 km altitude, atmospheric pressure ranges from 10^{-7} to 10^{-9} torr when measured at rest relative to the atmosphere. However, a spacecraft moves through this atmosphere at about 7.8 km/s. The molecular flux in the ram direction is about 10^{15} molecules/cm^2-s at 250 km, equivalent to 3×10^{-6} torr in an Earth-based vacuum chamber with 300 ^{o}K wall temperature.[5] Since the spacecraft travels several times faster than the average speed of lighter molecules such as hydrogen and helium and many times faster than heavier species such as oxygen and nitrogen, only molecules in the high-speed tail of the Maxwell-Boltzmann distribution can catch up with the spacecraft. Theoretically, pressures less than 10^{-14} torr are possible for H and He on the wake side of an orbiting spacecraft, and oxygen and heavier species would be almost completely absent.[5] However, these pressures will be increased by backscattering.

In this paper we describe the conditions immediately behind an orbiting molecular wake shield using a numerical procedure that is both accurate and computationally efficient. An axisymmetric geometry was assumed. This numerical approach permits the investigation of a large number of environmental, design, and operational parameters, and it permits an analysis of cases with complex gas expansion and attenuation patterns. The analysis could also be applied to any spacecraft where surface contamination is a concern, particularly contamination from waste dumps, rocket plumes, and power conversion system effluents.

Problem Description

There are many parameters to trade off in the design of a wake shield. Operational altitude (i.e., ambient density), distance between the shield and the host vehicle, and the shield size can all be varied to provide better vacuums. We also want to know what effluent fluxes are tolerable.

The configuration is shown schematically in Fig. 1; the wake shield is connected to the host vehicle via a boom extending in the positive z direction, i.e., antiparallel to the velocity vector of the Orbiter in the fixed reference frame. The wake shield, located symmetrically about the z axis, is assumed circular and planar in the x-y plane The effluents are released by the cabin of the

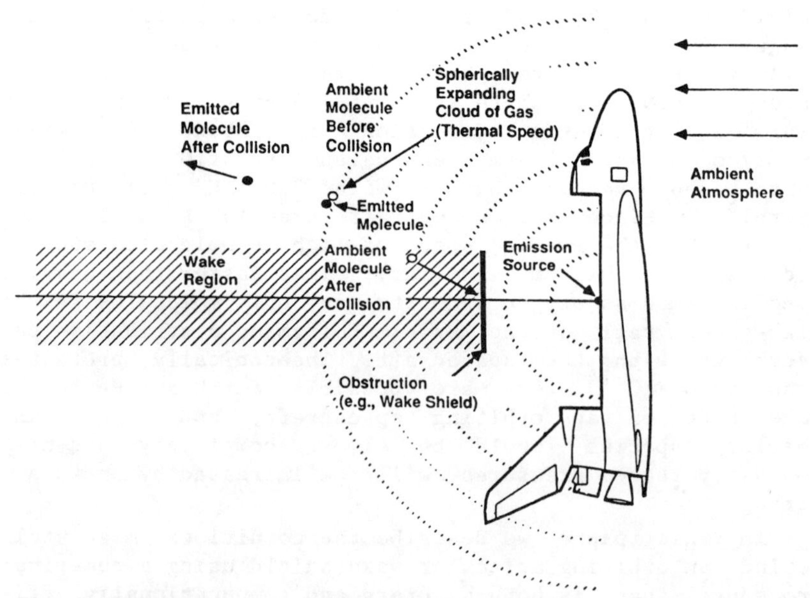

Fig. 1 Backscattering of ambient molecules off of a spacecraft outgas cloud.

host vehicle; thus, we have not included outgassing from the shield surface, molecules from the process being carried out behind the shield, or contaminants due to thruster activity, although such sources could be considered. The emission source is modeled as a point source located at the base of the boom.

For the ambient conditions we have adopted as a starting point a baseline case whose features are as follows. The orbital altitude is 200 km, and the Orbiter speed is 7775 m/s. At this altitude, the mean temperature of the ambient atmosphere (predominantly atomic oxygen) is about 900 ^{o}K, and the ambient density is typically $5x10^{9}$ cm^{-3}. Both the ambient temperature and density can vary significantly, sometimes nearly 100% depending on solar activity, season, local time, etc. The analysis considers two different species of emitted gas, N_2 and H_2O, which are typical of a cabin leak from the Shuttle Orbiter. We have considered a range of emission temperatures typical of those at which gases might be vented from the orbiter, although the baseline temperature is 300 ^{o}K. The cabin emission leak rate is taken to be 1 kg/day, which is within the range reported in the literature.[12,13] The wake shield radius is 100 cm, and the distance between the emission source and the shield (the boom length) is 1500 cm for the baseline case. In these analyses, the effect of the Orbiter on the ambient and emitted flowfields is neglected.

Approach

Both analytical and numerical techniques exist in the literature for computing contamination due to various sources. Harvey[3] and Scialdone[4] present analytical solutions for spacecraft self-contamination due to scattering of emitted molecules. Hueser and Brock[6] present a solution for the environment within a hemispherical wake shield oriented such that its opening faces aft; this solution, however, does not include the effects of scattered molecules. For more complicated geometries, Monte Carlo simulations are generally employed.[7] These techniques can now account for very general geometries and can include contributions from many sources; however, they tend to be time-consuming and expensive to set up and run. They are also rather inappropriate at conceptual design stages where many configurations must be analyzed.

In this case conventional Monte Carlo techniques worked well in defining the outgas cloud region (see Fig. 2, which shows contours of constant density in this re-

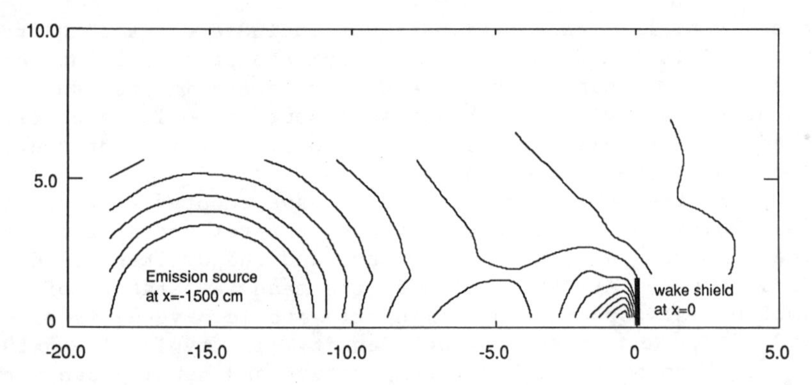

Fig. 2 Direct simulation Monte Carlo prediction of the flowfield near an orbiting wake shield, showing contours of constant density.

gion). Because of the large difference in density between the outgas cloud and the wake region, however, the Monte Carlo method yielded poor statistics in the extremely rarefied region behind the shield; thus, a different approach was required. In addition, we desired a fast screening tool in order to perform trade studies to optimize the design of the shield. To achieve these goals, a simplified approximation for the cloud region was used in order to increase the accuracy behind the shield and to increase computational speed.

In this analysis the gas emitted from the spacecraft is assumed to expand uniformly in all directions, forming a spherical cloud. The emitted molecules are assumed to have a uniform velocity given by the thermal velocity. The emitted molecules do not collide with each other, and the cloud density n_e is given by

$$n_e \propto \frac{1}{r^2} \exp\left(-\frac{r}{\lambda}\right)$$

where n_e is the local density of the emitted molecules, r is the distance from the center of the cloud of emitted molecules, and λ is the mean free path of the emitted molecules. The exponential term accounts for the emitted molecules being swept away by the ambient gas stream.

The ambient gas is assumed to have a Maxwellian velocity distribution and at any point in space is divided into discrete directional and velocity groups. Within each group all the molecules are approximated as having the same direction of motion and the same speed. The collisions between ambient and emitted molecules are transformed into a center of mass coordinate system; in this system scattering is uniform in all directions and the return flux to the target from a given point in space can be determined from simple geometric considerations.[8,9] The total return to the target is determined by summing the returns from all velocity groups over all space. Collision cross sections were taken from the literature[10,11] for elastic collisions at 5 eV relative velocity. Although the collision cross sections will vary with the relative velocity between the two molecules, in this analysis they are assumed to be invariant with the relative velocity.

Results

The analysis considers the effects of several parameter variations on the pressures at the target. The backscattered return flux has been computed for a range of ambient and emitted temperatures and ambient densities, and the equivalent pressure is determined from the return flux. The effects of varying the boom length and the radius of the wake shield have also been investigated. The results of the analysis are shown in Figs. 3-6. Figure 3 shows computed pressures at the shield as a function of emitted temperature for the two gases examined. It can be seen that the pressure at the target depends strongly on the emitted gas; the pressure is several orders of magnitude less for H_2O emissions than for N_2. For the baseline case, the pressure due to N_2 emissions is 1.01×10^{-10} torr, whereas the pressure due to water emission is only 2.54×10^{-14} torr. The H_2O molecules are

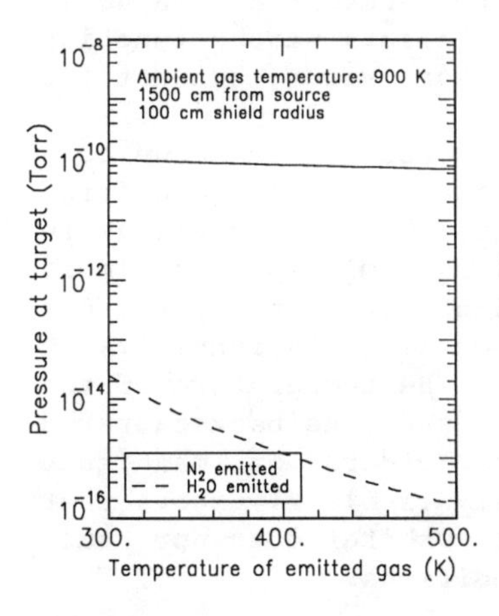

Fig. 3 Effect of temperature of emitted gas on equivalent pressure at target.

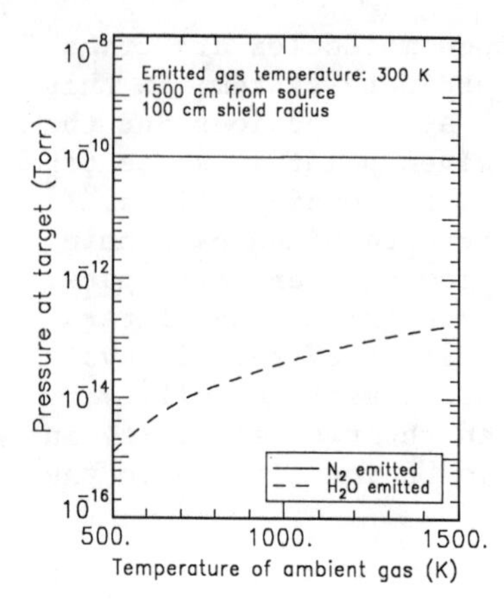

Fig. 4 Effect of ambient temperature on pressure at target.

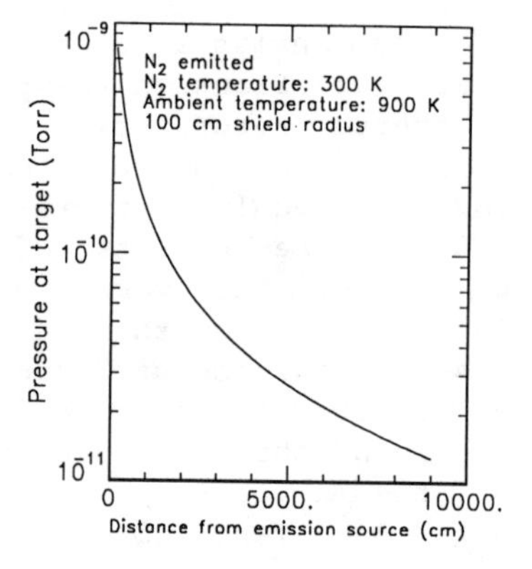

Fig. 5 Equivalent pressure at target as a function of distance from emission source (boom length).

so light that only collisions at high relative velocities have energy transfers sufficient to backscatter the ambient oxygen atoms. The pressure is also a strong function of the emitted gas temperature for water, but not for nitrogen.

In Fig. 4 the backscattered pressure is shown as a function of the ambient temperature for a fixed emitted gas temperature, and we recognize the same trends as in Fig. 3. The pressure due to N_2 emissions is almost invariant with respect to the ambient temperature, whereas the pressure due to H_2O emissions is quite sensitive to changes in ambient temperature. The temperature of the emitted N_2 and H_2O is 300 ^{o}K, and the backscattered pressure due to N_2 is several orders of magnitude greater than for H_2O. The analysis clearly shows that water vapor outgassing from the Orbiter does not cause much backscattering under these conditions.

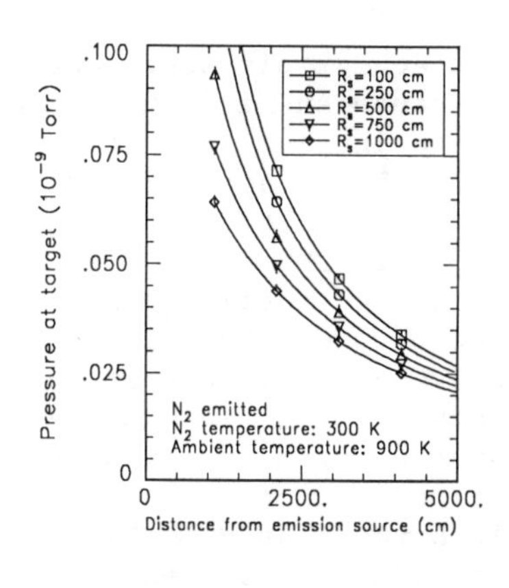

Fig. 6 Equivalent pressure at target as a function of distance from emission source and shield radius. Note linear axes.

The backscattered pressure is displayed as a function of the distance between the emission source and the shield in Fig. 5. As one would expect, the pressure depends quite strongly on boom length out to about 2000-3000 cm from the source but becomes much less sensitive at greater distances. Note that the longer distances are much greater than those practical with the Shuttle remote manipulator system (RMS); these conditions therefore would be more typical of a tethered or free-flying configuration.

Figure 6 shows the combined effects of shield radius and distance from the emission source. The pressure is a relatively weak function of shield radius, decreasing roughly linearly with increasing radius. At 2000 cm from the source, for example, increasing the shield radius from 100 to 1000 cm results in a less than 50% decrease in pressure. Again note that, beyond about 2500 cm, the pressure at the target is relatively insensitive to both shield size and distance from the source, indicating that the most effective means of controlling the pressure behind the shield may be to increase the orbital altitude, if pressures below about 10^{-10} torr are required.

Summary and Conclusions

A procedure has been developed for calculating contamination levels and pressures due to backscattered gas. This procedure is extremely flexible: any altitude and spacecraft velocity and any combination of ambient and emitted gas species, temperatures, and density distributions can be used.

This technique has been used to predict the backscattered return flux from atomic oxygen that would be expected for a molecular wake shield orbiting at 200 km altitude. It was determined that molecular nitrogen leaking from the spacecraft cabin could cause sufficient backscattering of ambient atomic oxygen to increase the effective pressure behind the wake shield to approximately 10^{-11} to 10^{-10} torr, for conditions representative of a wake shield deployed from the Shuttle RMS boom. Water vapor leaking from the cabin has a much smaller effect. These calculations also showed that beyond, about 2500 cm from the emission source, the equivalent pressure at the target is relatively insensitive to variations in the size of the wake shield or its distance from the emission source. Thus, placing the wake shield on a tether or operating it as a free flyer would have relatively little utility. The most effective way to reduce pressures below the levels determined here may be to increase the orbital altitude (although this parameter was not investigated in this study).

Acknowledgments

This work was supported by McDonnell Douglas Space Systems Company independent research and development. The authors wish to express their sincere thanks to Prof. J. P. Tonnies for his helpful suggestions and assistance in providing the latest data on collision cross sections.

References

[1]Lengrand, J. C., "Plume Impingement Upon Spacecraft Surfaces," Proceedings of the 14th International Symposium on Rarefied Gas Dynamics, 1984, pp. 217-228.

[2]Bird, G. A., "Spacecraft Outgas Ambient Flow Interaction," Journal of Spacecraft and Rockets, Vol. 18, Jan.-Feb. 1981, pp. 31-35.

[3]Harvey, R. L., "Spacecraft Neutral Self-Contamination by Molecular Outgassing," Journal of Spacecraft and Rockets, Vol. 13, May 1976, pp. 301-305.

[4]Scialdone, J. J., "Self-Contamination and Environment of an Orbiting Satellite," Journal of Vacuum Science and Technology, Vol. 9, February 1972, pp. 1007-1015.

[5]Naumann, R. J., "A More Perfect Vacuum," Aerospace America, Vol. 25, March 1987, pp. 44-47.

[6]Hueser, J. E., and Brock, F. J., "Theoretical Analysis of the Density within and Orbiting Molecular Shield," Journal of Vacuum Science and Technology, Vol. 13, May-June 1976, pp. 702-710.

[7]Bird, G. A., "Simulation of Multi-Dimensional and Chemically Reacting Flows," Proceedings of the 11th International Symposium on Rarefied Gas Dynamics, 1978, pp. 365-388.

[8]Warnock, T. T., and Bernstein, R. B., "Transformation Relationships from Center-of-Mass Cross Section and Excitation Functions to Observable Angular and Velocity Distributions of Scattered Flux," Journal of Chemical Physics, Vol. 49, 1968, pp. 1878-1886.

[9]Present, R. D., Kinetic Theory of Gases, McGraw Hill, New York, 1958.

[10]Leonas, V. B., "Studies of Short-Range Intermolecular Forces," Soviet Physics Uspekhi, Vol. 15, 1973, pp. 266-281.

[11]Foreman, P. B., Lees, A. B., and Rol, P. K., "Repulsive Potentials for the Interaction of Oxygen Atoms with the Noble Gases and Atmospheric Molecules," Journal of Chemical Physics, Vol. 12, 1976, pp. 213-224.

[12]Rantanen, R. O., and Jensen, D. A. S., "Orbiter Payload Contamination Control Assessment Support," MCR-77-107, Martin Marietta Aerospace, 1977.

[13]Cole, R. J., private communication, McDonnell Douglas Astronautics Co., Huntsville, AL, 1987.

Thruster Plume Impingement Forces Measured in a Vacuum Chamber and Conversion to Real Flight Conditions

Arthur W. Rogers*
Hughes Aircraft Company, El Segundo, California
Jean Allègre† and Michel Raffin†
Société d'Études et de Services pour Souffleries et Installations Aérothermodynamiques (SESSIA), Levallois-Perret, France
and
Jean-Claude Lengrand‡
Laboratoire d'Aérothermique du Centre National de la Recherche Scientifique, Meudon, France

Abstract

Measurements of forces caused by thruster plume impingement on two surface configurations were made in a vacuum chamber. One configuration was a scale model simulation of the axial thruster firing inside a solar cell cylinder of the INTELSAT VI spacecraft. The other simpler arrangement enabled plume impingement on flat surfaces parallel to or inclined away from the nozzle axis. A special balance was developed to measure the small impingement forces. Plume impingement induced a shear pressure along the internal cylindrical wall, resulting in thrust loss. The total force exerted by the plume results from integration of tangential and normal pressures over impinged surfaces. An interaction model is proposed, based on relative contributions of friction and normal forces, and was used to estimate real flight forces from model simulation measurements.

Introduction

Two main maneuver thrusters are mounted on the Hughes INTELSAT VI spacecraft. The radial thruster controls the

* Senior Scientist.
† Research Engineer.
‡ Research Scientist.

satellite rotation around its longitudinal axis. Plume impingement heat transfer rates on spacecraft surfaces around this submerged thruster nozzle were the subject of a previous paper.[1]

The second main maneuver thruster, the axial thruster, with its axis inclined 8 deg toward the satellite axis, controls the north-south station keeping. The plume from the axial thruster impinges on the surrounding surfaces, inducing wall pressure and friction and consequent torques and thrust losses. Because of the complexity of the flow interactions, plumes and their impingement effects on spacecraft are investigated experimentally in a vacuum chamber on reduced size test models and converted to real flight forces by a proposed interaction model.

Thruster Plume Impingement Configurations

For the axial thruster, Table 1 lists flow conditions, gas properties, nozzle geometry, and scale values for full scale and model configurations.

A number of requirements must be accounted for when the simulation conditions are determined:

1) Nozzle mass-flow rate must be kept within the pumping capacity of the facility.

2) Under space conditions, the plume background pressure is zero, which cannot be simulated in the testing facility. However, pressure inside the vacuum chamber was sufficiently low so that the impinged region was well within the plume first expansion cell. This reduced the influence of the background pressure on the plume-surface interaction.

3) Parameters I and δ^*/r_e, which characterize the angular density distribution within the plume and the relative boundary-layer displacement thickness in the nozzle exit plane, respectively, were nearly identical for both real flight and simulation conditions.

Configurations tested in the vacuum chamber are schematically presented in Figs. 1 and 2. The real flight configuration of Fig. 1 was simulated experimentally by a model 1/20th of the full scale configuration, except for the nozzle exit diameter, which was 1/10th of the exit diameter of the real thruster. The analysis for converting model data to full-scale data allows different scales to be used for the nozzle and the impinged surface, respectively.

Experiments were performed on solar cell cylinders equipped with rough (wooden) and smooth (Kevlar) inner surfaces. In some tests an end ring, composed of 24 elements

Table 1 Full-scale and test conditions

	Flight conditions	Simulation conditions
Stagnation conditions		
Pressure p_o, Pa	6.41×10^5	5×10^5
Temperature T_o, ^oK	2503	1100
Gas properties (at exit plane)	Combustion products ($MMH + N_2O_4$)	Nitrogen
Specific heat ratio γ	1.351	1.40
Molecular weight, g/mole	20.59	28.01
Prandtl number Pr	0.755	0.73
Viscosity law (SI units)	$1.458 \times 10^{-6}\ T^{0.503}$	$\frac{1.374 \times 10^{-6}\ T^{1.5}}{T + 100}$
Nozzle geometry	Contoured	Conical
Throat radius r_c, mm	2.121	0.3
Exit radius r_e, mm	36.733	3.673
Exit 1/2 angle α_e, deg	8.5	8.5
Area ratio ε	300	150
Nozzle flow		
Wall temperature T_w, ^oK	1533 (throat)	450
Throat Reynolds number $R_{e_c} = \frac{\rho_c a_c d_c}{\mu_c}$	2.66×10^4	0.959×10^4
Mass flow rate $\dot{m}$, g/s	5.86	0.169
Exit plane properties		
BL displacement thickness δ^*/r_e	0.300	0.298
Mach number M_e	6.76	6.48
Pressure p_e, Pa	121.2	196.3
Maximum deviation angle θ_∞, deg	54.76	50.90
Density distribution parameter I	0.055	0.054

(Cont. on next page.)

Table 1 (continued)

	Flight conditions	Simulation conditions
Jet properties		
Background pressure p_∞, Pa	≈0	<0.471
Mach disk distance, m	→∞	>1.16
Impingement reference quantities (assuming typical nozzle plate distance equal to:		
h, m	1	0.05
Reference pressure p_N ref, Pa	5.069	30.28
Reference heat flux q_{ref}, kW/m^2	6.151	11.43
Wall temperature, ^oK	300	300
Scale		
Nozzle (except for r_c)	1	1/10
Solar cell cylinder	1	1/20

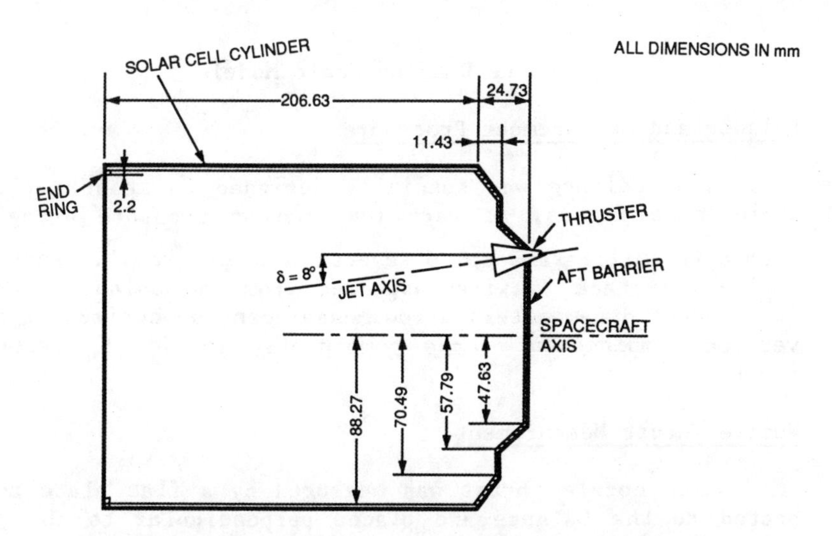

Fig. 1 Axial thruster location and orientation in deployed solar cell cylinder

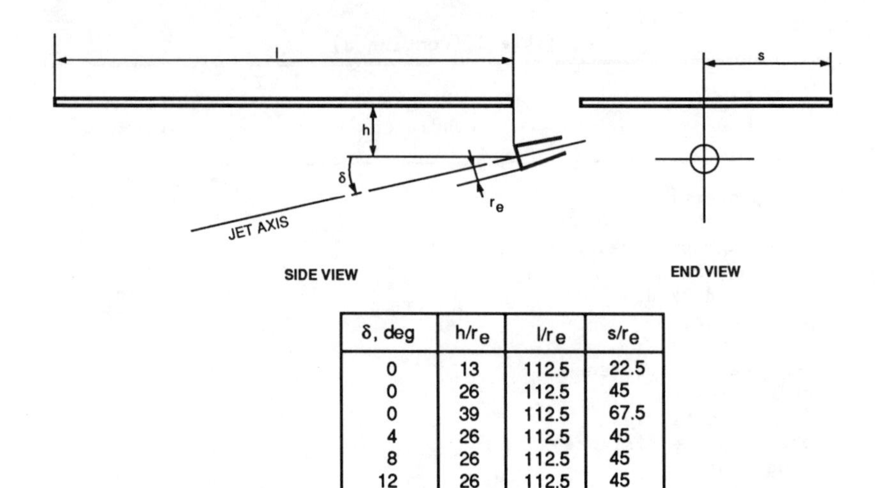

δ, deg	h/r_e	l/r_e	s/r_e
0	13	112.5	22.5
0	26	112.5	45
0	39	112.5	67.5
4	26	112.5	45
8	26	112.5	45
12	26	112.5	45

Fig. 2 Flat plate impingement cases for shear force measurement

2.2 mm thick and about 5 mm long, was inserted around the exit circumference of the solar cell cylinder. For other tests, the aft barrier was removed to investigate the effect of flow confinement.

Besides the flight simulation model, plume impingements were also performed over flat plates (Fig. 2) with rough or smooth surfaces. Flowfield surveys were also made, using electron beam and pitot tube to determine free plume Mach number and density distributions.

Force Data on Scale Model

Balance and Measurement Procedure

The balance was specially designed to measure plume impingement forces, at very low ambient pressure, ranging from a few milligrams up to several tens of grams[2]. When an impinged surface is wire-suspended from the balance, three independent dynamometers allow measurement of horizontal and vertical impingement force components, F_h and F_v respectively.

Nozzle Thrust Measurement

The nozzle thrust was measured by a flat plate connected to the balance and placed perpendicular to the jet axis at a distance of 55 mm downstream from the nozzle exit

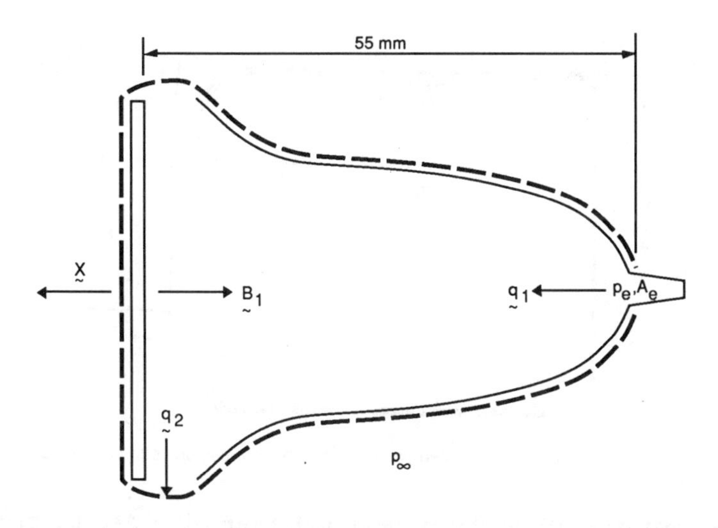

Fig. 3 Analytical model for determining thrust

(Fig. 3). The thrust is defined as $T = |q_1| + p_eA_e$, where q_1 is the rate of momentum through the exit section and p_e and A_e are exit pressure and area, respectively. As defined in the preceding sentence, T is independent of the chamber pressure p_∞. The T must be measured at a chamber pressure sufficiently high so that the jet should be entirely intercepted by the plate and the flow turned perpendicularly to the nozzle axis.

The law of conservation of momentum applied to the dashed contour Σ yields

$$\underset{\sim}{q}_2 - \underset{\sim}{q}_1 = \underset{\sim}{B}_1 + P_eA_e \underset{\sim}{x} + \int_\Sigma P_\infty \underset{\sim}{n}\, d\Sigma$$

where B_1 is the force exerted by the balance on the plate. In the $\underset{\sim}{x}$ direction, we obtained

$$|\underset{\sim}{B}_1| = |q_1| + (p_e - p_\infty)\, A_e = T - p_\infty A_e$$

Force B_1 was measured continuously while the pressure inside the vacuum chamber was progressively increased from 0.5 Pa to approximately 60 Pa. B_1 increased with p_∞, attaining the limiting value $B_1 = 0.203$ N at $p_\infty \approx 30$ Pa. Since the term $p_\infty A_e$ was negligible (1.3×10^{-3} N at $p_\infty = 30$ Pa), the limiting value of B_1 was interpreted as the nozzle thrust $T = 0.203$ N. The thrust calculated from the theoretical boundary-layer data indicated in Table 1 is $T = 0.19$ N, which is in good agreement with the measured value.

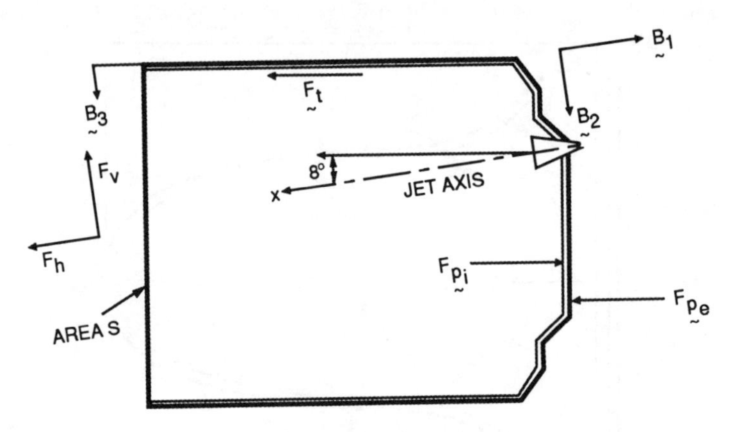

Fig. 4 Analytical model for determining impingement forces on solar cell cylinder

Plume Impingement on Solar Cell Cylinder with Aft Barrier

Considering the solar cell cylinder model connected to the balance, $\tilde{B}_1$, $\tilde{B}_2$, and $\tilde{B}_3$ are the three force components measured by the balance at the suspension points (Fig. 4).

If $\tilde{F}_t$ is the skin friction integrated on the inside of the cylinder, $\tilde{F}_{Pi}$ the wall pressure integrated on the inside of the cylinder, and $\tilde{F}_{Pe}$ the wall pressure integrated on the outside of the cylinder, the following equation holds

$$\tilde{B}_1 + \tilde{B}_2 + \tilde{B}_3 + \tilde{F}_t + \tilde{F}_{Pi} + \tilde{F}_{Pe} = 0$$

Let F_h be the projection of $\tilde{F}_t + \tilde{F}_{Pi} + \tilde{F}_{Pe}$ on the horizontal nozzle axis x oriented downstream and F_v the projection of $\tilde{F}_t + \tilde{F}_{Pi} + \tilde{F}_{Pe}$ on a vertical axis oriented upward. The F_h is directly deduced from the measurement of $\tilde{B}_1$, and F_v is deduced from $\tilde{B}_2 + \tilde{B}_3$.

The variation of F_h/T with p_∞ for the different configurations investigated is plotted in Fig. 5, where T is the thrust measured experimentally (T = 0.203 N). The variation is nearly linear for $p_\infty < 2$ Pa and can thus be easily extrapolated to external pressure $p_\infty = 0$. Let F_{ho} and slope S_h be defined by the relation $F_h = F_{ho} + S_h\, p_\infty$. Values of F_{ho} and S_h were calculated for the different configurations by a least-squares method applied to all data obtained at $p_\infty < 2$ Pa. These values are given in Table 2, with the corresponding correlation parameter r_h. From Table 2 and Fig. 5 it is seen that the influence of the Kevlar sheet covering the wooden surface and the effect of the end ring are negligible within experimental detectability.

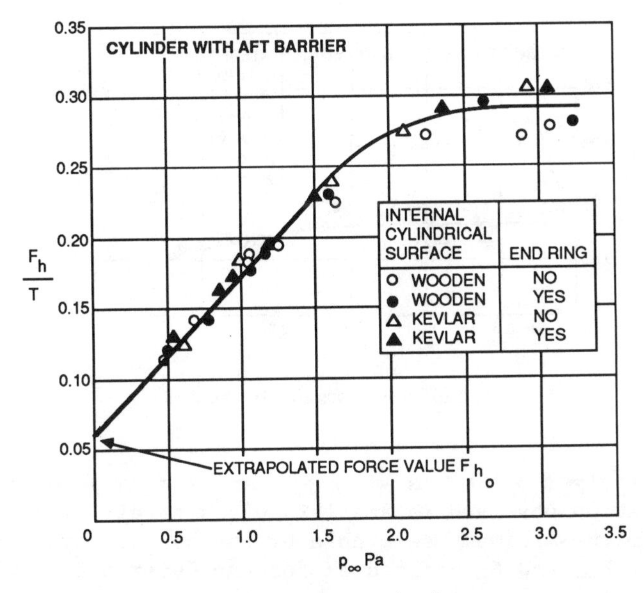

Fig. 5 Horizontal force versus chamber pressure

Table 2 Force data at low external pressure (p_∞ < 2 Pa)

Kevlar sheet	End ring	F_{h_o}, 10^{-2} N	S_h, 10^{-2} m^2	r_h	F_{v_o}, 10^{-2} N	S_v, 10^{-2} m^2	r_v
No	No	1.36	2.14	0.995	0.565	0.104	0.765
No	Yes	1.31	2.17	0.998	0.630	0.121	0.803
Yes	No	1.29	2.26	0.999	0.511	0.864	0.606
Yes	Yes	1.38	2.22	0.999	0.460	0.211	0.95

Assuming a uniform external pressure p_∞ acting on the cylinder, its contributions to F_h is the product of p_∞ by S cosδ, where S is the aft barrier area, minus the area of the hole around the nozzle. From model geometry, S cosδ = 2.41×10^{-2} m^2. This value is close to the value of S_h as indicated in Table 2, which confirms the interpretation of F_h vs p_∞ variation. This indicates that the flowfield inside the cylinder is nearly independent of p_∞ up to p_∞ = 2 Pa.

Force F_v is one order of magnitude smaller than F_h, and the variation of F_v/T with p_∞ is plotted in Fig. 6. The contribution of the external pressure p_∞ to F_v is the product of p_∞ by S sinδ = 3.38×10^{-3} m^2, and F_v should vary by

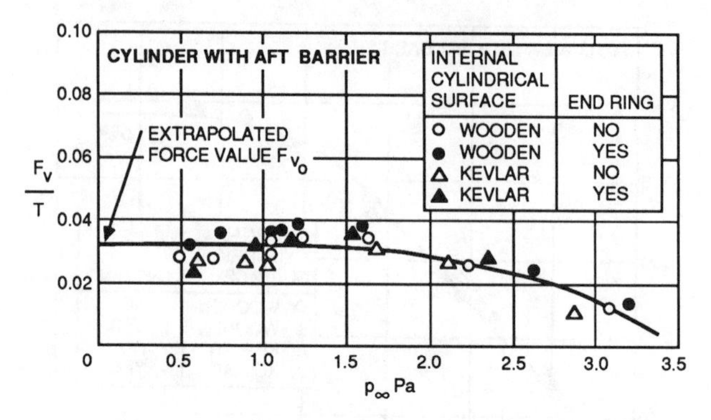

Fig. 6 Vertical forces versus chamber pressure

approximately 5×10^{-3} N when p_∞ varies from 0.5 to 2 Pa. The variation observed is smaller but is hardly distinguishable from the minimum measurable by the balance system. The values of F_{vo} and S_v defined as for the horizontal force by

$$F_v = F_{vo} + S_v p_\infty$$

are given in Table 2. They have been calculated by a least-squares fit based on all data with $p_\infty < 2$ Pa. Note that the corresponding value of the correlation coefficient r_v is smaller than for the horizontal force. The scatter of the experimental points in Figs. 5 and 6 indicates the repeatability of the results.

Plume Impingement on Solar Cell Cylinder Without Aft Barrier

To separate the contributions of the aft barrier and of the side wall of the cylinder, additional measurements were performed with the aft barrier disconnected and removed from the cylindrical side wall, which remained at the same position.

Forces measured by the balance [illegible] ular case, forces exerted only on [illegible] wall. The variation of F_h/T and F_v/T [illegible] Figs. 7 and 8. For $p_\infty < 1$ Pa, both F_h [illegible] F_v a[illegible] stant. The respective contributions of the side wall an[illegible] of the aft barrier can be estimated as indicated in Table 3.

The mean value of the total horizontal force [illegible] vacuum, based on results presented in Table 3, [illegible] 1.34×10^{-2} N, which represents 6.6% of the nozzle [illegible] The same calculation applied to the total vertical force [illegible] vacuum leads to $F_{vo} = 5.4 \times 10^{-3}$ N.

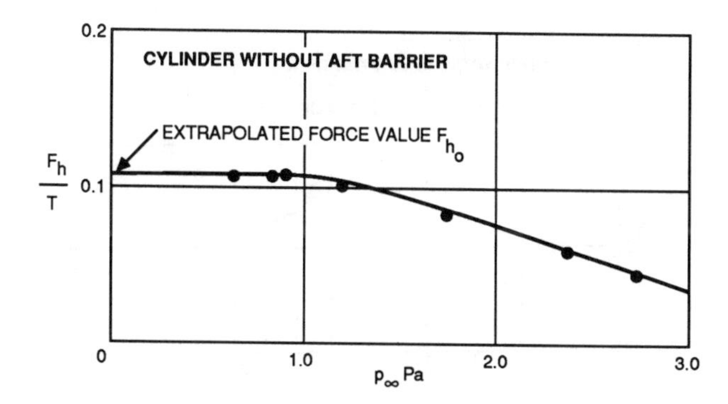

Fig. 7 Horizontal force applied to cylinder without aft barrier

Table 3 Decomposition of forces for model and full-scale conditions

Conditions	F_{h_o}, 10^{-2} N	F_{v_o}, 10^{-2} N	F_t, 10^{-2} N	F_n, 10^{-2} N
Model				
Side wall	2.17	0.45	2.21	0.144
Aft barrier	-0.83	0.09	0.205	-0.809
Total	1.34	0.54	–	–
	F_h, N	F_v, N	F_t, N	F_n, N
Full scale				
Side wall	2.00	0.38	2.03	0.096
Aft barrier	-0.56	0.11	0.189	-0.542
Total	1.44	0.49	-	-

The tangential and normal forces F_t and F_n can be deduced from F_h and F_v; they are shown in Table 3, indicating the contribution of the aft barrier and of the side wall.

Plume Impingement Forces on Flat Plates

Force measurements were performed on flat plates using the same procedure as for the cylinder. The variation of F_h/T with p_∞ for a flat plate parallel to the jet axis is plotted in Fig. 9. The presence of the Kevlar sheet on the surface appears to have no influence on the results. For

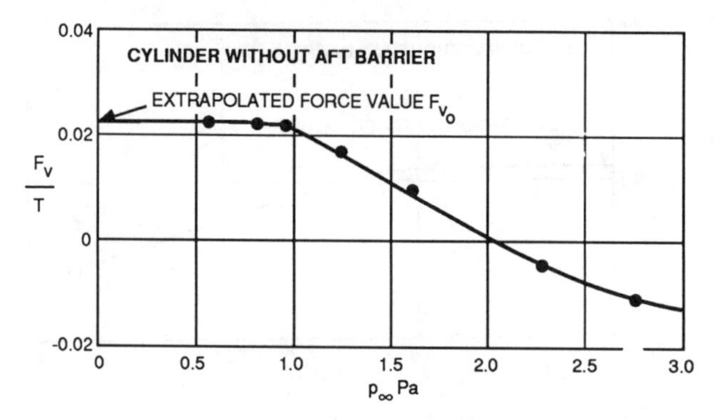

Fig. 8 Vertical force applied to cylinder without aft barrier

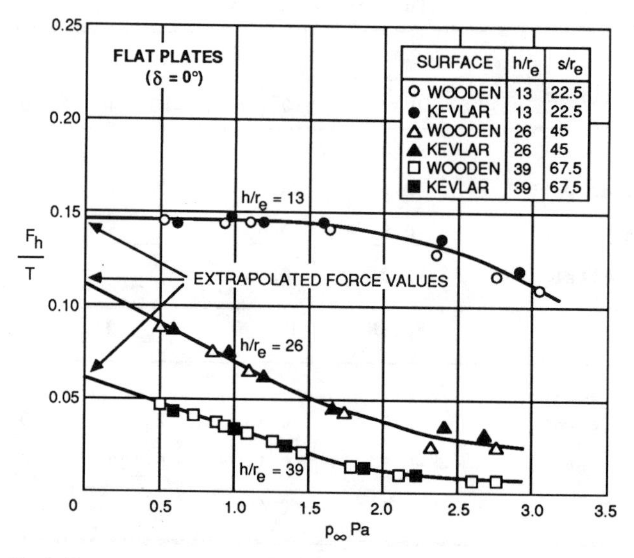

Fig. 9 Horizontal force on flat plates at various distances from plume axis

the smallest distance between nozzle and plate, F_h tends toward a limit as p_∞ decreases, thus indicating that the plate is entirely located inside the plume when p_∞ is lower than 2 Pa. For larger values of h, the pressure dependence of F_h indicates that the flow along the plate is influenced by the external pressure.

Space limitation here does not allow illustration of the variation of F_h/T with p_∞ for $\delta > 0^o$ as tabled in Fig. 2. For the same reason as above, no limiting value of F_h is reached at the lowest value of p_∞. As expected, the impingement force decreases when the plate is tilted away from the jet axis.

Conversion to Full-Scale Forces on INTELSAT VI Configuration

The interaction model explained in Ref. 3 is used as a basis for estimating full-scale quantities from measurements under simulation conditions.

Tangential Forces

The skin friction distribution along the surfaces impinged on by the plume is deduced from the heat-transfer distribution by using the Reynolds analogy

$$\frac{q}{\rho_s U_s C_p (T_{aw} - T_w)} \Big/ \frac{\tau_w}{\rho_s U_s^2} = Pr^{-0.6} \qquad (1)$$

or

$$\tau_w = q \frac{U_s Pr^{0.6}}{C_p (T_{aw} - T_w)} = \frac{q}{q_{ref}} \cdot q_{ref} \frac{Pr^{0.6}}{C_p} \cdot \frac{U_s}{T_{aw} - T_w}$$

The simulation conditions were chosen in such a way that the plume density distribution parameter I in Table 1 is nearly identical under real flight and simulation conditions. Therefore, the reduced heat-transfer distribution q/q_{ref} is also nearly identical. Thus, the ratio of local shear stress under real flight and simulation conditions is

$$\frac{\tau_w}{\tau'_w} = \frac{q_{ref}}{q'_{ref}} \cdot \left\{\frac{Pr^{0.6} U_s}{C_p (T_{aw} - T_w)}\right\} \Big/ \left\{\frac{Pr'^{0.6} U'_s}{C'_p (T'_{aw} - T'_w)}\right\} \qquad (2)$$

where unprimed quantities denote real flight quantities and primes denote the corresponding simulation quantities. The U_s and T_{aw} (respectively, U'_s and T'_{aw}) vary along the impinged surface. Let M (or M') be the Mach number at the boundary layer edge; then,

$$U_s = M (\gamma R T_s)^{1/2} = M \left\{\gamma R T_o / (1 + \frac{\gamma - 1}{2} M^2)\right\}^{1/2} \qquad (3)$$

Recovery temperature T_{aw} is equal to $T_s + Pr^{0.5}U_s{}^2/2$. The factors

$$\frac{Pr^{0.6}U_s}{C_p\,(T_{aw} - T_w)} \quad \text{and} \quad \frac{Pr'^{0.6}U'_s}{C_p'\,(T'_{aw} - T'_w)}$$

in Eq. (2) were calculated for different Mach numbers, and their ratio was found to be nearly independent of Mach number (0.427 - 0.429 when the Mach number ranged from 0.1 to 10); hence, τ_w/τ_w' may be estimated to be 0.428 x q_{ref}/q'_{ref}. This estimation is based on the hypothesis that the Mach number distributions along the wall is the same in full-scale and simulation conditions, but a violation of this hypothesis is expected to have only small consequences because the factor 0.428 is essentially the ratio of temperature differences under simulation and full scale conditions, respectively, and should not be greatly influenced by other quantities. The ratio q_{ref}/q'_{ref} can be estimated from the values of q_{ref} indicated in Table 1 for an arbitrary distance between nozzle and surface:

$$q_{ref}/q'_{ref} = 6.151/11.43 = 0.538$$

Thus, the wall shear stress ratio is $\tau_w/\tau_w' = 0.428 \times 0.538 = 0.230$.

Normal Forces

The reduced pressure distribution p/p_{Nref} is nearly identical in real flight and simulation conditions for the same reasons as mentioned previously. The ratio of wall pressure can thus be estimated as

$$p/p' = p_{Nref}/p'_{Nref} = 5.069/30.28 = 0.167$$

Total force

The total force exerted by the jet on the cylinder is a combination of local tangential and normal forces integrated along the different surfaces of the cylinder. Since τ_w/τ_w' and p/p' are different, the ratio of forces under real flight and simulation conditions depends on the relative contribution of friction and pressure to the total force. There is no reason why that ratio should be equal to the nozzle thrust ratio. The scale of the model is 1/20, the ratio of pressure forces is $(p/p') \times 20^2 = 67$, and the ratio of friction forces is $(\tau_w/\tau_w') \times 20^2 = 92$. These ratios can be applied to the normal and tangential forces exerted on the side wall and the aft barrier, respectively. The values of F_h and F_v deduced separately for the side wall and the aft barrier are presented in Table 3.

Present experimental results indicate that the thrust loss due to plume impingement of the full-scale rocket engine is $F_h = 2 - 0.56 = 1.44$ N; i.e., 9.3% of the thrust and the transverse perturbation force is $F_v = 0.38 + 0.11 = 0.49$ N.

Conclusion

Experiments have been performed in a vacuum chamber to investigate heated nitrogen plume impingement on flat plates and over more realistic configurations as encountered during attitude control maneuvers of the INTELSAT VI satellite.

Experimental force data are presented for all configurations. A conversion model is then applied to calculate full scale force values from the forces measured on the scale model. Under flight conditions, impingement forces resulting from both skin friction and pressure distributions significantly reduce the thruster efficiency. As an example, during axial thruster firing, a thrust loss of about 9% may be expected.

Acknowledgment

This paper is based upon work performed under an INTELSAT VI contract. The views expressed herein are not necessarily those of the International Telecommunications Satellite Organization (INTELSAT). INTELSAT is the nonprofit cooperative of 115 countries that owns and operates the global commercial communications satellite system used by countries around the world for international communications and many countries for domestic communications.

References

[1]Rogers, A. W., Allègre, J., Raffin, M., and Lengrand, J.-C., Plume impingement heat transfer measurements on model simulated INTELSAT VI flight configurations and conversion to real flight conditions," AIAA 22nd Thermophysics Conference, AIAA Paper 87-1604, Honolulu, HI, June 1987.

[2]Allègre, J., Raffin, M., and Lengrand, J.-C., "Aerodynamic balance for high altitude simulation chamber," International Congress on Instrumentation in Aerospace Simulation Facilities, Sept. 1983. Paper published in ICIASF 1983 Record, pp. 141-144 © 1983 IEEE.

[3]Allègre, J., Raffin, M., and Lengrand, J.-C., Forces induced by a simulated rocket exhaust plume impinging upon a flat plate. Proceedings of the 14th International Symposium on Rarefied Gas Dynamics, University of Tokyo Press, Tsukuba, Japan, July 1984, pp. 287-294.

Neutralization of a 50-MeV H^- Beam Using the Ring Nozzle

N. S. Youssef* and J. W. Brook†
Grumman Corporation, Bethpage, New York

Abstract

Measurements are presented of the radial distribution of the neutral particles produced during the passage of a 50-MeV H^- ion beam through the gas flowfield created by pulsed operation of a ring nozzle. Both argon and xenon were used as the neutralizing medium. The experiment was optimized so that the peak neutral fraction was obtained on the ring centerline. The peak neutral fractions observed are in agreement with theoretical predictions. The distributions of the neutral fractions for argon and xenon were uniform to within 2.5% over 40 and 45% of the ring radius, respectively. Optimum conditions required about 30% of the mass flow using xenon as was required using argon. The line integrals inferred from the neutral fraction distributions are in qualitative agreement with the gasdynamic measurements of Farnham and Muntz ("Transient and Steady Inertially Tethered Clouds of Gas in a Vacuum," presented at the 16th International Rarefied Gas Dynamics Symposium, Pasadena, CA, July 1988).

Introduction

The creation of high-density gas targets in an evacuated environment has many practical applications. Muntz et al.[1] previously investigated a method of creating "inertially tethered clouds of gas" under steady flow conditions using the ring nozzle. The ring nozzle consists of a series of small holes in a circular tube, as

---

* Engineering Specialist, Space Systems Division.
† Senior Laboratory Head, Gas Dynamics, Corporate Research Center.

shown by the schematic in Fig. 1. Each hole produces a sonic freejet that issues at an arbitrary angle ω to the ring axis. The flowfield set up by the interaction of these jets sets up the requisite density field. In the experiment to be described, ω = 60 deg.

The objective of this device, which was first proposed as a contaminant purge for an infrared telescope,[2] is to create a collision-dominated flowfield that is not constrained by walls. Pham-Van-Diep and Muntz[3] have proposed using the ring nozzle to produce simulated low-Earth-orbit species at the proper energy for satellite interaction studies. Brook et al.[4] have evaluated the use of the ring nozzle flowfield to neutralize high-energy negative ion beams and found it to be quite suitable. However, considerations of backflow into the beamline and lack of adequate pumping in many facilities suggested that pulsed operation of the ring nozzle would be required. Farnham and Muntz[5] have investigated the gasdynamics of a pulsed ring nozzle to compare with the steady-state results.

In this paper, we present results of an experiment that used the flowfield of a pulsed ring nozzle to neutralize a beam of H^- ions at 50 MeV. The objectives of the experiment were to determine and compare the level and uniformity of the neutral beam using argon and xenon as the ring nozzle gas, and to compared with the gasdynamics results the line integral inferred from the neutral

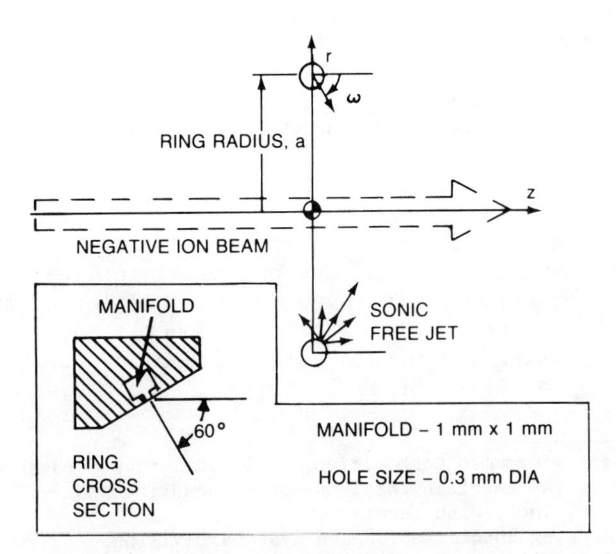

Fig. 1 Schematic of ring nozzle showing cross-sectional shape.

fraction. Monatomic gases were selected because they have a lower expansion angle (90 deg) from sonic conditions to a vacuum than polyatomic gases. The ring nozzle utilized during these experiments was the same as employed by Farnham and Muntz.[5] In what follows, we will describe the connection to gasdynamics, the experimental setup, and the experimental procedure, and we will present and interpret the measurements.

Relation to Gasdynamics

The passage of a negative ion beam through a collection of molecules (gas or solid) results in the stripping of electrons from the beam, producing neutral and, subsequently, positive ions. The connection between the gas dynamic properties of the ring nozzle and its neutralization properties is through the line integral $I_n(r;t)= \int n(r,z;t)\, dz$ where n is the local number density, r and z are cylindrical coordinates (Fig. 1), and t is time. For a negative ion beam passing through any stripping medium, the equations that govern the balance between the fractions of negative (-), neutral (0), and positive (+) ions are[6]:

$$N_- = \exp\left[-(\sigma_{-10}+ \sigma_{-11})I_n\right] \tag{1}$$

$$N_0 = \left\{\frac{\sigma_{-10}}{\sigma_{-10}+ \sigma_{-11} - \sigma_{01}}\right\} \cdot \left\{\exp\left[-\sigma_{01}I_n\right] - \exp\left[-(\sigma_{-10} + \sigma_{-11})\, I_n\right]\right\} \tag{2}$$

$$N_+ = 1 - N_- - N_0 \tag{3}$$

where the cross sections σ_{-10}, σ_{-11}, and σ_{01} refer to the conversion of negative ions to neutral ions, negative to positive, and neutral to positive, respectively. These cross sections depend on the constituents and energy of the negative ion beam and the stripping medium. Equations (1-3) are plotted in Fig. 2 for argon (solid lines for all ion fractions) and for xenon (dashed line, neutral fraction only).

As can be seen, the peak neutral fraction for xenon is very close to that for argon, but it occurs at a smaller value of the line integral. From measurements of the neutral fraction vs radial position in the ring nozzle

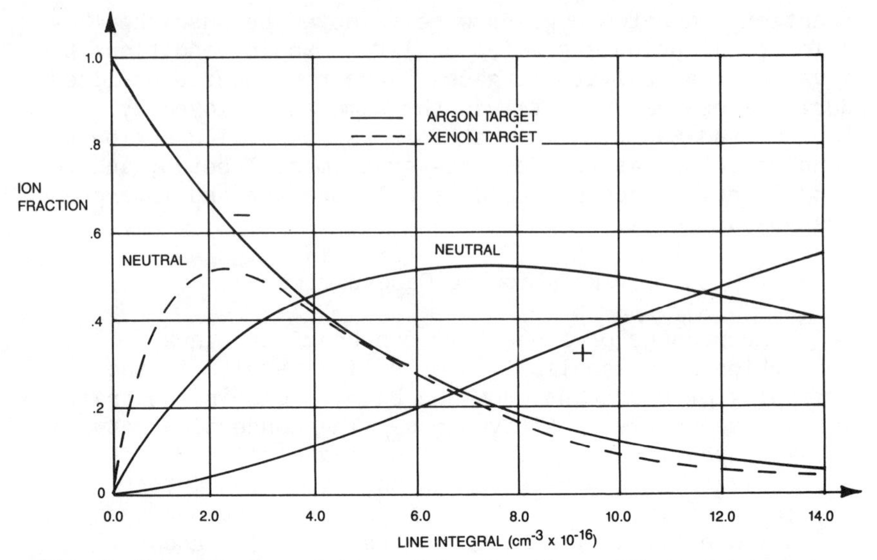

Fig. 2 Theoretical variation of ion fractions with line integral for 50-MeV H^- beam.

flow field, it should be possible to back out the corresponding distribution of the line integral, which can then be compared with the corresponding values computed from the density profiles measured by Farnham and Muntz[5] according to the expression for the line integral presented above.

Experimental Setup

The experiments were conducted at the Argonne National Laboratory (ANL) using a high-energy ion beam consisting of H^- at 50 MeV. The ring (ω = 60 deg) and the fast-acting gas supply valves were mounted in a vacuum chamber 1.6 m long and 1.2 m in diameter, with an internal volume of 1500 liters. The chamber was installed in the beamline between two isolation valves. Four turbomolecular pumps, providing 6000 l/s pumping speed, were attached to the chamber and the pipe sections leading into and out of the chamber.

The ring nozzle had a radius of 7.87 cm and was constructed in two halves. Each half was supplied with gas through a fast-acting injection valve and had a total of 20 orifices 0.3 mm in diameter, supplied by a 1 mm x 1 mm ring manifold (see Fig. 1). Pressures in the manifold were measured by Kulite pressure transducers.

The ring was mounted on a motor-driven, computer-controlled stage to allow motion transverse to the beam so that radial scans of the ring nozzle flowfield could be made. A rotation mechanism allowed 90-deg rotation of the entire ring, so that radial scans in the perpendicular direction could be made. The entire assembly was vacuum leak-checked and the instrumentation calibrated at Grumman before shipment to ANL. A photograph of the vacuum chamber, ring nozzle, and the systems controls is shown in Fig. 3. A closeup of the ring assembly installed in the vacuum chamber is shown in Fig. 4. Mechanical and optical final alignment between the ring nozzle and the beam location was carried out at ANL.

Two H^- beam radii were employed, 2 cm and 3 mm. The former was used for optimization surveys and the latter for finer radial scans of the ring nozzle flowfield. The length of the beam pulse was 40 μs for the 2-cm beam and 20 μs for the 3-mm beam. The positive, negative, and neutral ion currents were measured by means of segmented faraday cups after the test chamber and at a beam stop. The locations of these measurement stations and the general arrangement of the experiment are shown in Fig. 5. The total beam current was measured before entering

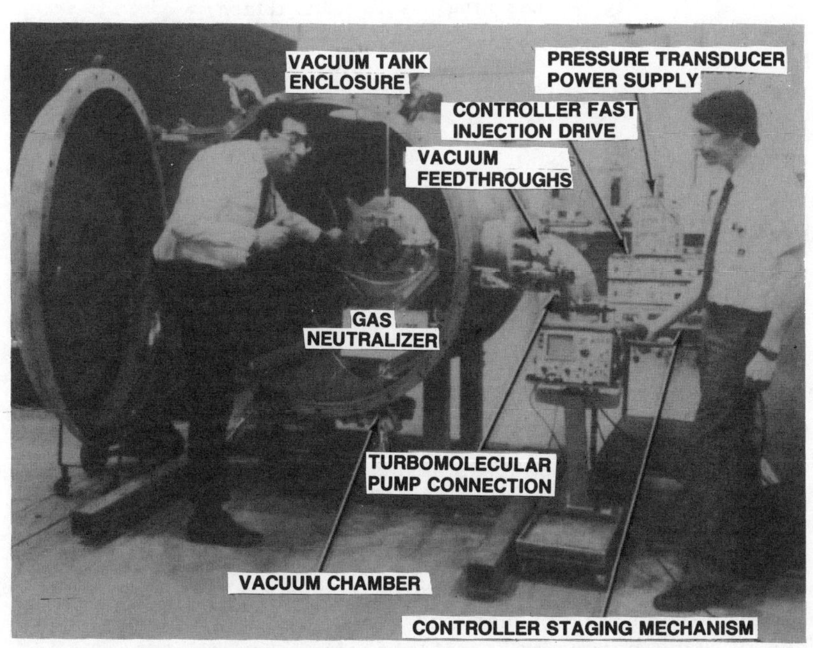

Fig. 3 Experimental setup and checkout at Grumman.

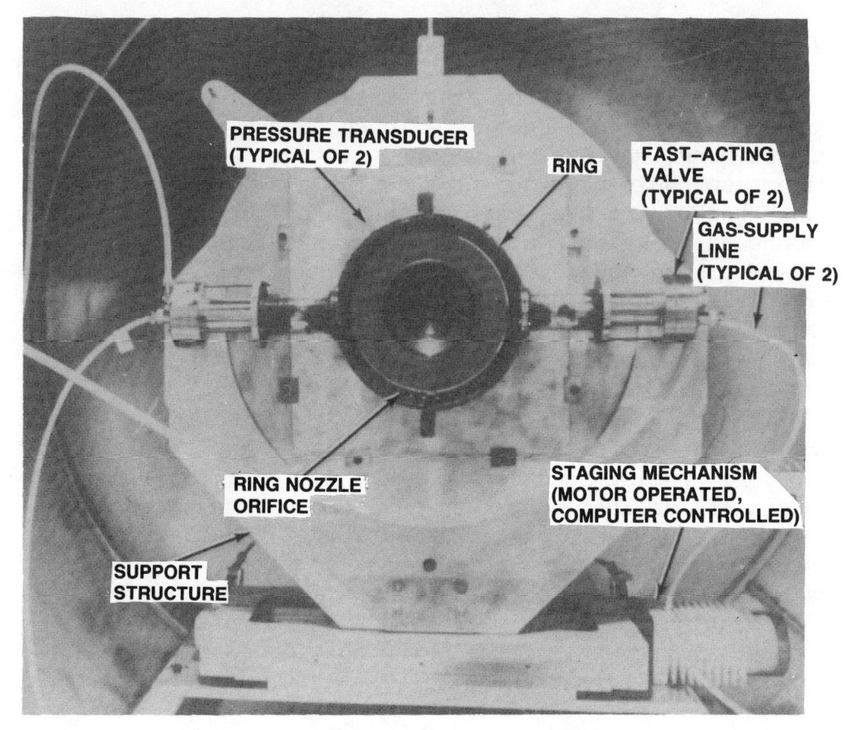

Fig. 4 Gas ring nozzle neutralizer.

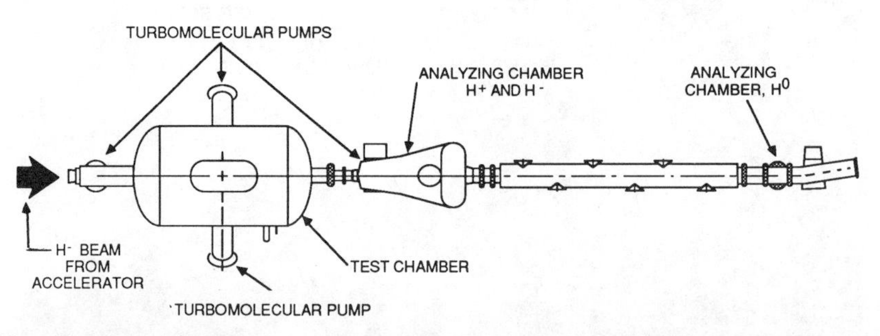

Fig. 5 Setup for ring nozzle experiments on ANL 50-MeV H^- beamline.

the vacuum chamber, but was found not to be accurate and was not used in the data reduction. All data were recorded on-line and analyzed on the ANL computer system.

Assuming room temperature argon to calculate the appropriate flow speeds and using the dimensions of the chamber, we estimated that the test time would be on the order of 4 ms before wall reflections began to interfere with the flowfield. Test times for xenon were expected to be longer due to the lower velocities associated with the higher molecular weight. The initial pressure in the system was assumed to be 10^{-6} Torr (only 10^{-5} Torr was actually achieved during the experiment). After the ring nozzle gas pulse, we estimated that the pressure in the test chamber would rise to about 4×10^{-4} Torr. If we assume that this pressure is uniform in the test chamber and neglect pumping, then the line integral associated with this pressure would be less than 1% of the anticipated line integral on the ring centerline and would not introduce any significant measurement error. With the pumping available, the time to pump back down to 10^{-6} Torr was estimated at slightly more than 1 s.

Experimental Procedure

Since these experiments were conducted before those of Farnham and Muntz,[5] we had to conduct preliminary surveys to establish the parameter space in which we operated. These parameters included mass flow rate, synchronization between the gas and the ion beam pulses, and alignment with the beam as well as rate of rise and duration of the gas pulse valve opening. In this section, we discuss the most pertinent of these surveys, all of which were conducted using the 2-cm H^-beam.

The ion fractions produced by the ring nozzle were calculated from the currents H^i measured in the segmented faraday cups by means of the formula

$$N_i = H^i / (H^- + H^0 + H^+) \quad (4)$$

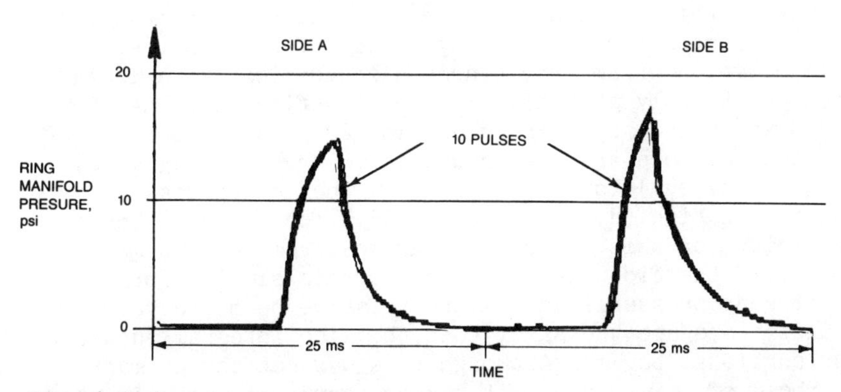

Fig. 6 Repeatability of ring nozzle manifold pressure from pulse to pulse.

where the subscript i denotes -, 0, or +. Assuming that the instantaneous mass flow is proportional to the manifold pressure, the total mass flow per pulse is then proportional to the integral of the pressure with time. This integral is referred to below as the mass/pulse, Δm.

The voltage signals produced by the Kulite pressure transducers in the ring manifold were sampled 50 times per pulse and summed to provide Δm for each of the ring halves. The supply pressures to the gas pulse valves were adjusted independently so that the Δm from each ring half was identical. For each ring nozzle operating condition, ion current measurements were repeated for 5 or 10 beam pulses. Figure 6 shows typical results for the repeatability of the manifold pressures sampled during 10 pulses. It should be noted that the pressures in each manifold are not identical because there are small differences in the individual ring halves, in the supply manifolds, and in the valves themselves. There were no significant effects of the nonuniformity in the flowfield due to transient differences between the manifold pressures in the ring halves.

The averages, as well as the standard deviations, of the ion currents were computed for each operating condition. If the standard deviation for any of the ion currents was greater than 10%, the entire collection of measurements for 5 or 10 beam pulses was rejected. This was seldom observed. The deviations generally were less than 1%. In what follows, the ion fractions are calculated from the average currents.

It was desired to operate the ring so that the line integral on the ring centerline provided peak neutralization (see Fig. 2). To establish the value of Δm to which this corresponded, we measured the ion fractions on the centerline for varying values of Δm. In the optimization of Δm, the synchronization lead time τ was set arbitrarily at 4 ms. The ion fractions obtained for both xenon and argon are shown in Fig. 7 as a function of Δm. Also shown are the theoretical curves from Fig. 2, scaled (by eye) to the value of Δm that gives peak neutralization (4.75 for xenon, 15.5 for argon). The validity for this correlation arises from the assumption that, if the flow field time history is similar for different pressures and if the relative beam and valve pulse times are the same, then the line integral should be proportional to Δm. Although this was not the primary purpose of this survey, it can be seen that there is good agreement between experiment and theory. Remarkably, the

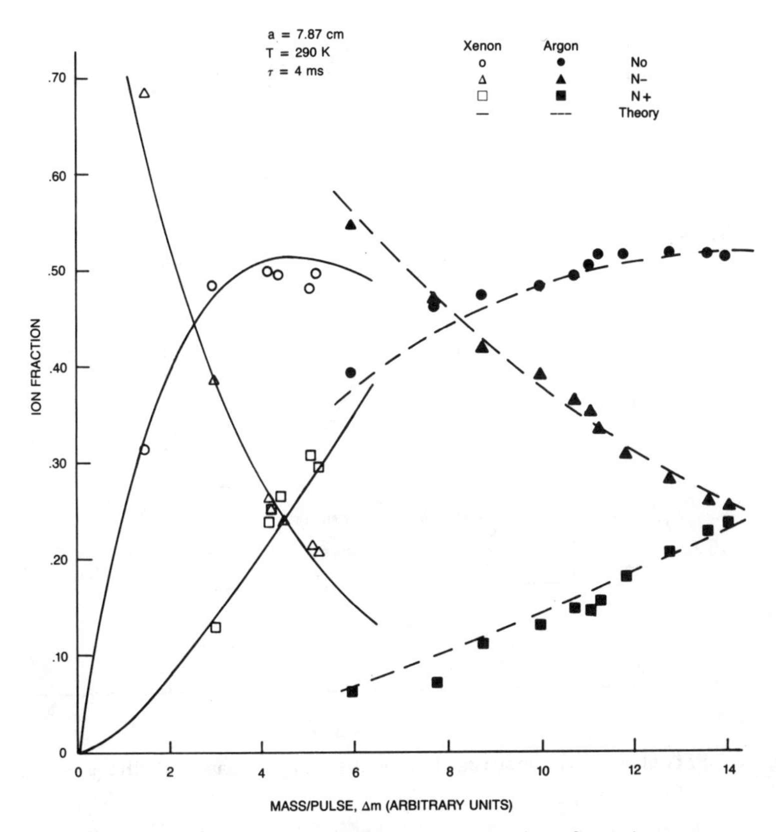

Fig. 7 Effect of gas mass pulse on ion fractions.

ratio ($\Delta m_{Ar}/\Delta m_{Xe}$) turns out to be equal to the ratio of the corresponding theoretical values of the line integrals for peak neutralization (~3.26). Therefore xenon requires approximately 30% of the mass/pulse needed with argon to produce almost the same neutralization.

The H^- beam firing was carried out at regular intervals, controlled by the beamline clock. Therefore, the ring nozzle gas pulses had to be synchronized to the beam firing, with τ approximately equal to the test time (the beam pulse was considered instantaneous). To determine the optimum value of this parameter, we measured N_0 for xenon and argon as a function of τ with Δm fixed.

As can be seen in Fig. 8, N_0 changes significantly outside a range of τ=2-4 ms for argon and τ=2.5-4 ms for xenon. In the data to be discussed below, τ=3.75 and 4 ms were used for xenon and argon, respectively.

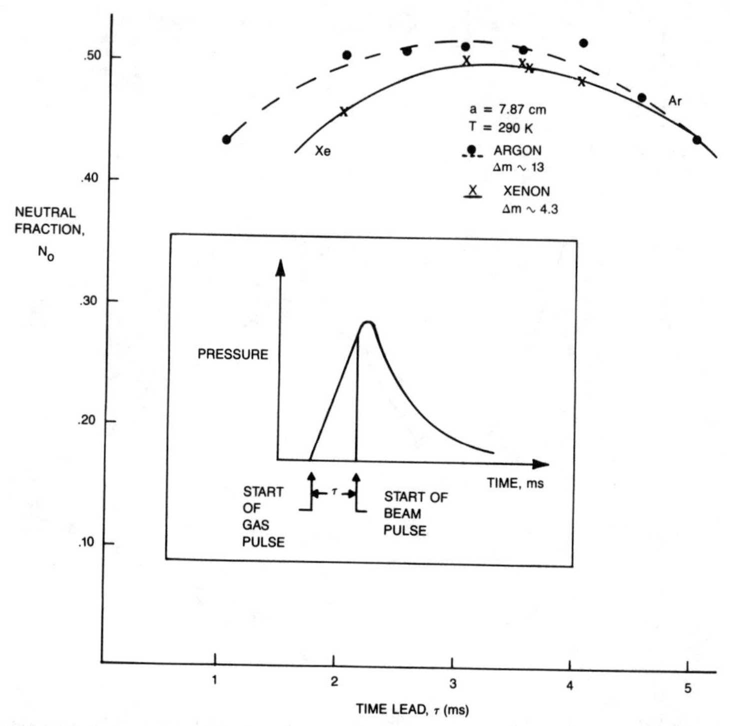

Fig. 8 Effect of synchronization between the beam and the gas pulses.

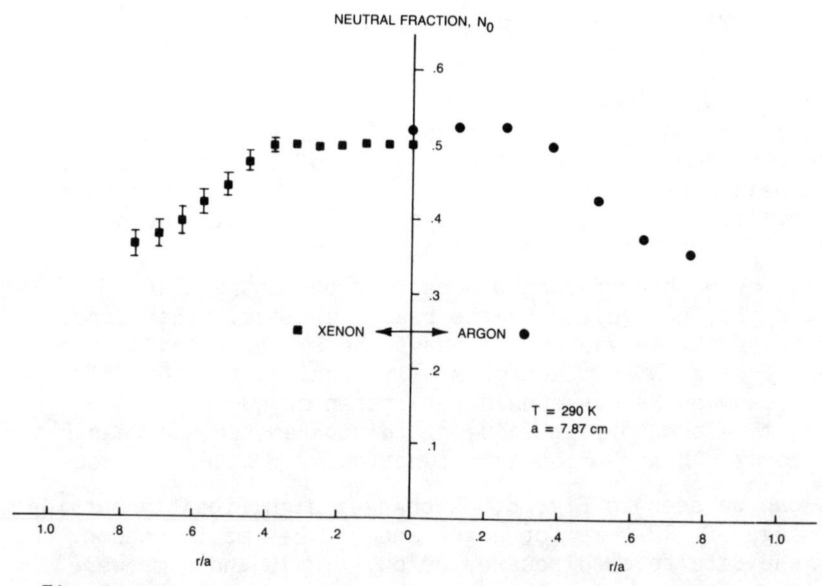

Fig. 9 Distribution of neutral fraction with radial position.

By rotating the ring through 90 deg and varying the radial position of the beam, we determined that the ring nozzle flowfield was symmetric with respect to the beam. We also determined that the rate of opening of the gas pulse valves was not important, nor was the duration of the opening, provided of course that the beam pulse occurred before the valve was closed.

Results and Discussion

Once we had completed the preliminary surveys and had established the optimum operating conditions, we set about making detailed radial surveys (with respect to r) of the ion fractions produced by the ring nozzle flowfield. These surveys were made using the 3-mm H^- beam. These measurements were generally more difficult than for the 2-cm H^- beam because of the lower total H^- beam current. One manifestation of this was that the negative faraday cup had a fairly high background signal level with the 3-mm beam, and the data had to be corrected. This was done in two ways, using the measurements from the 2-cm beam as a guide. First, we assumed that the background signal was proportional to the total current and evaluated the constant of proportionality from the peak values of N_0 obtained from both size beams. The small-beam data were then corrected at all points using this constant (the total current varied by a maximum of about 10% over the entire range of r surveyed). Alternatively, the neutralization fraction for the small beam was adjusted by the ratio of the peak neutralization between the large and small beams. Both methods produced the same results.

These results are plotted in Fig. 9 for both argon (right side) and xenon (left side). The abscissa is the absolute value of the radial position in the ring nozzle flowfield divided by the ring radius a. The ordinate is the neutral fraction N_0. In Fig. 9, N_0 is plotted for both positive and negative values of r/a; the symbols represent the average value. The ends of the bars represent the actual values for positive and negative normalized radial position, r/a, for those cases where the difference is larger than the size of the symbol. As can be seen, the peak neutralization for each species is close to the theoretical values, with N_0 for argon slightly larger than for xenon, as expected. In addition, the neutral fraction is fairly uniform over the central portion of the ring. In fact, N_0 is constant to within 2.5% over approximately 40% of the radius for argon and approximately 45% for xenon.

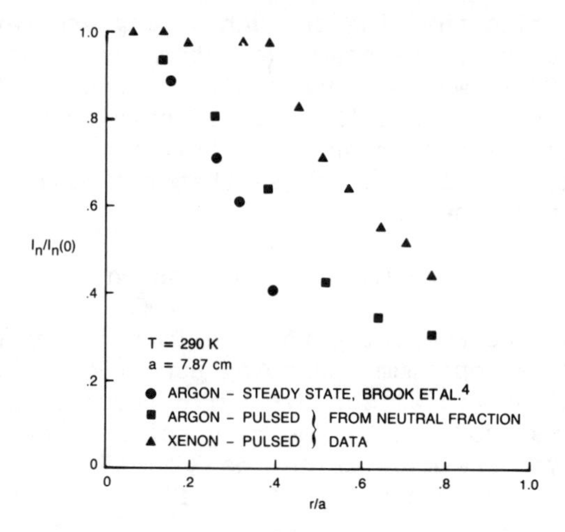

Fig. 10 Normalized distributions of line integrals in ring nozzle flowfield.

Using the results in Fig. 9 in conjunction with Fig. 2, the radial distribution of the line integrals can be obtained. These are plotted vs ring radius in Fig. 10, normalized to the value on the ring centerline (argon: squares; xenon: triangles) to allow a comparison with the gasdynamic results.[5] The results are normalized because of the possibility of slightly different operating conditions being used in each case. At the time this paper was written, only the line integrals obtained by integrating the steady-state density field were available. These are shown by the circles in Fig. 10. It should be noted from Fig. 2 that some errors are involved in extracting the line integral from N_0 near the peak of the curve. The nominally obtained values are plotted in Fig. 10. As can be seen, the width of the distribution of the line integrals is greater for pulsed operation than for steady operation. This observation is consistent with the findings of Farnham and Muntz,[5] which were presented at this symposium. Furthermore, we observe that the width of the distribution of line integrals for xenon is greater than for argon. The reasons for these results are not currently known. One could speculate that the time to establish steady flow is more than just a few typical times based on the thermal speed and the ring radius and that the lower thermal speed of xenon relative to argon is responsible, but, without a detailed flow model, these are just conjecture.

Conclusions

We have found that the level of neutralization of a high-energy, negatively ionized hydrogen beam produced by the flowfield of a ring nozzle during pulsed operation is in agreement with theoretical predictions using both argon and xenon as the stripping medium. Furthermore, the neutral beam produced is uniform to within 2.5% over 40 and 45% of the central region of the flow for argon and xenon, respectively. To produce these results, the total mass injection required for xenon is about 30% of that required for argon. The width of the density field that produces the neutralization is inferred to be greater for pulsed operation than was calculated from steady-state gasdynamics experimental results. This is in qualitative agreement with the results of pulsed gasdynamics experiments. A theoretical model is needed in order to provide a detailed explanation of these results.

Acknowledgments

This work was supported by the Grumman Independent Research and Development program. Time on the Argonne National Laboratory beamline was made available by the U.S. Army Strategic Defense Command. The authors would particularly like to acknowledge the assistance of the ANL staff and Mr. Ron Heuer of the Grumman Corporate Research Center.

References

[1]Muntz, E. P., Kingsbury, D., De Vries, C., Brook, J. W., and Calia, V. S., "Inertially Tethered Clouds of Gas in a Vacuum," Rarefied Gas Dynamics, edited by V. Boffi and C. Cercignani, Teubner, Stuttgart, FRG, 1986, Vol. II, pp. 474-486.

[2]Muntz, E. P. and Hanson, M., "Purging Flow Protection of Infrared Telescopes," AIAA Journal Vol. 22, No. 5, May 1984, pp. 696-704.

[3]Pham-Van-Diep, G. C. and Muntz, E. P., "True Energy Atmospheric Simulator for Low Earth Orbit Species," AIAA Paper 88-0727, AIAA 26th Aerospace Sciences Meeting, Reno, NV, Jan. 11-14, 1988.

[4]Brook, J. W., Calia, V. S., and Muntz, E. P., "Gas Dynamic Properties of a Ring Nozzle Neutralizer," Grumman Corporate Research Center, Bethpage, NY, Rept. RE-718, July 1986.

[5]Farnham, T. L. and Muntz, E. P., "Transient and Steady Inertially Tethered Clouds of Gas in a Vacuum," presented at the 16th International Rarefied Gas Dynamics Symposium, Pasadena, CA, July 1988.

[6]Smythe, R. and Toevs, J. W., "Collisional Electron Detachment from Hydrogen Atoms and Negative Hydrogen Ions Between 4 and 18 MeV," Physical Review, Vol. 139, No. 1A, July 1965, pp. A15-A18.

Chapter 5. Tube Flow

Rarefied Gas Flow Through Rectangular Tubes: Experimental and Numerical Investigation

A. K. Sreekanth* and Antonio Davis†
Indian Institute of Technology, Madras, India

Abstract

The test particle Monte Carlo technique has been utilized for the numerical study of free molecular flows through tubes of rectangular crosssection. A large amount of data have been obtained on direct and total transmission probabilities, density distribution, and molecular efflux distributions as a function of speed ratio, tube geometric parameters (width-to-height ratio and length-to-height ratio), angle of attack, wall temperature-to-gas temperature ratio, reflection and energy accommodation coefficients. Comparison of the data is made with some of the available analytical results. Using a continuous-flow rarefied gasdynamics facility, limited experiments were conducted to measure the mass flow of nitrogen gas through a two-dimensional slit and a rectangular tube in the transition and near free molecular flow regimes at various pressure ratio conditions across the tube geometry. The range of Knudsen numbers, based on the upstream mean free path and tube height, varied from 1 to 0.06 and the pressure ratios from 1 to 20. Comparisons of the measured mass flow data are made with the results of similar flow in a circular cross-sectional tube.

Monte Carlo Solution of Free Molecular Flow Through Rectangular Slits and Tubes

The test particle Monte Carlo method outlined by Fan,[1] which is ideally suited for the study of

---

* Professor, Department of Aerospace Engineering.
† Research Scholar, Department of Aerospace Engineering.

internal flow problems in the free molecular flow regime, was adopted with modification to take into account the rectangular tube geometry of arbitrary shape. The developed program allows calculation of mass flow rates in terms of transmission probabilities, the density distribution (both inside and outside the tube) along the axis and across it, and the angular distributions of molecular efflux as a function of the following parameters: 1) speed ratio, 2) angle of attack, 3) wall-to-gas temperature ratio, 4) tube width-to-height ratio, 5) tube length-to-height ratio, 6) energy accommodation, and 7) Maxwell's wall reflection coefficient.

For calculation purposes, rarefied nitrogen gas was chosen as the test gas. In almost all of the calculations presented, the number of sample molecules considered was 20,000. However, 100,000 molecules were considered for density distribution data. In all of the results presented, the Knudsen number is defined as the ratio of mean free path of the gas molecules in the upstream infinite chamber to the height h of the channel, viz., $Kn = \lambda/h$.

Results of Monte Carlo Calculation

Transmission Probability

As a check for the presently developed program, the total transmission probabilities for

Table 1 Total transmission probabilities for two dimensional channels

L/h	P_t (Monte Carlo)	P_t (Clausing[2])	P_t (Demarcus[3])
0.1	0.96715	0.9525	0.95245
0.5	0.81735	0.8048	0.80472
1.0	0.69175	0.6848	0.68442
2.0	0.55245	0.5417	0.54206
3.0	0.46250	0.4570	0.45716
4.0	0.40230	0.3999	0.39919
5.0	0.35350	0.3582	0.35648
6.0	0.31985	0.3260	0.32339
7.0	0.29460	0.3001	0.29684
8.0	0.27575	0.2789	0.27496
9.0	0.25645	0.2610	0.25655
10.0	0.23940	0.2457	0.24080

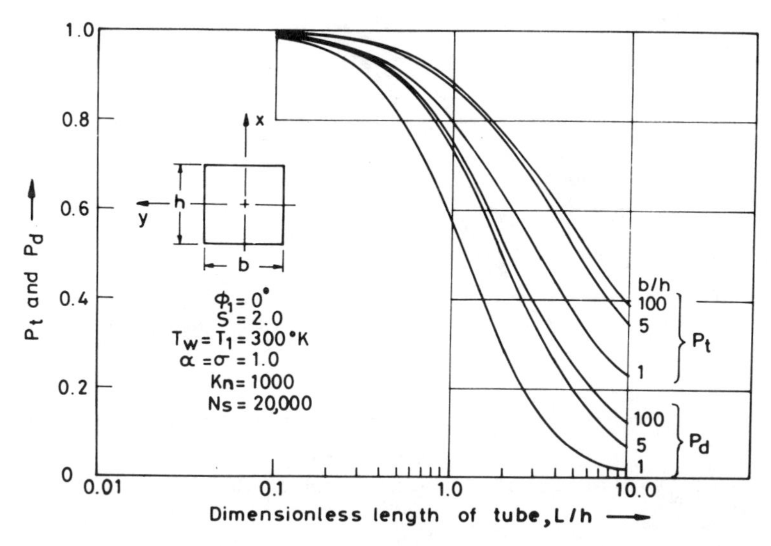

Fig.1 Total and Direct Transmission Probabilities Vs Tube Length.

the flow through a two- dimensional channel having a width-to-height ratio (b/h) of 100 at zero speed ratio, for a Knudsen number of 1000, have been calculated as a function of length-to-height (L/h) ratio from 0 to 10. The results are presented in Table 1 along with the theoretical values of Clausing[2] and Demarcus[3] for comparison purposes. Very close agreement is seen between the Monte Carlo results and the analytical values, thereby validating the present method of studying internal rarefied gas flows.

A considerable amount of data have been obtained[9] on the transmission probabilities (both direct, P_d, i.e. probability of a molecule passing from upstream chamber to downstream chamber without colliding with the wall anywhere inbetween and total, P_t, i.e., the probability of a molecule going from upstream to downstream with or without collisions with the tube) for a wide range of tube geometric parameters (height, width, and length) and flow prameters such as speed ratio S, temperature ratio T_W/T_∞, and angle of attack ϕ.

Some of the representative results are presented in Figs.1 - 3. The flow and geometric parameters are all listed in these figures. Figure 1 shows that, as the tube length approaches

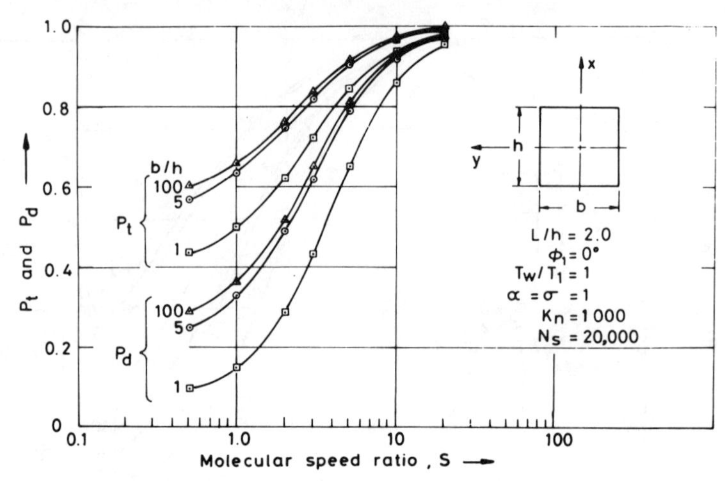

Fig.2 Variation of Transmission Probability with Speed Ratio.

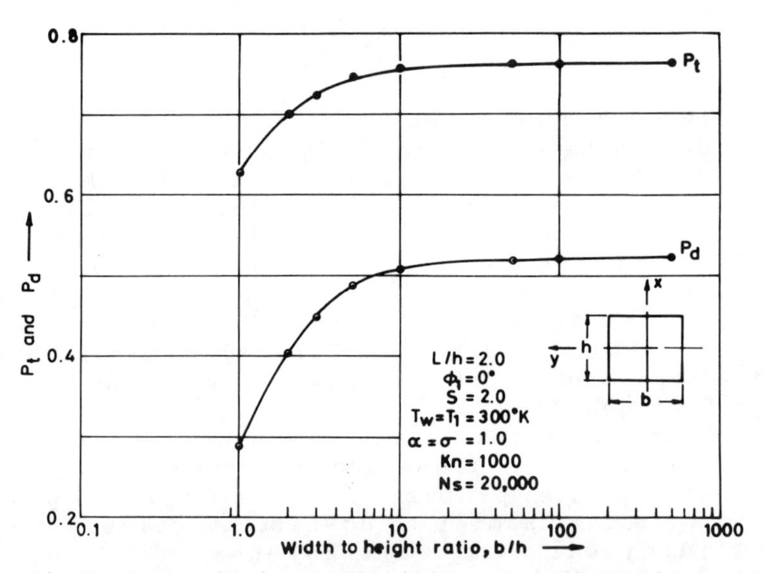

Fig.3 Transmission Probability Vs b/h Ratio.

infinity, both P_t and P_d approach zero and P_d and P_t approach 1 as $L/h \rightarrow 0$. Figure 3 brings out clearly the variation in transmission probability as a function of b/h. It can be seen that a width-to-height ratio of more than 100 is necessary before one can consider the tube geometry to be two-dimensional. In all of the cases considered, for a fixed speed ratio and tube length-to-height ratio, the total and direct transmission probabilities increase as the geometry changes from a square cross section (b/h = 1) to a two-dimensional one (b>>h).

Free Molecular Pressure Probe with Finite-length Slot Orifice

Koppenwallner[4] derived simple expressions for the transmission probabilities for free molecular pressure probes with slot orifices. The assumptions made in the analysis are that the external flow is hypersonic and the length of the slot is small compared to the slot width. The approximate analytical treatment gives "ideal orifice criteria" in terms of S, slot height-to-width ratio, and the orifice inclination α. The same problem is analyzed with the present Monte Carlo approach. For the infinite speed ratio case, for calculation purposes, a value of S = 20 was taken and a slot orifice having a L/h ratio of 0.15 was considered with b/h = 100. The total transmission probability was evaluated for various values of α, the angle of attack. The results are

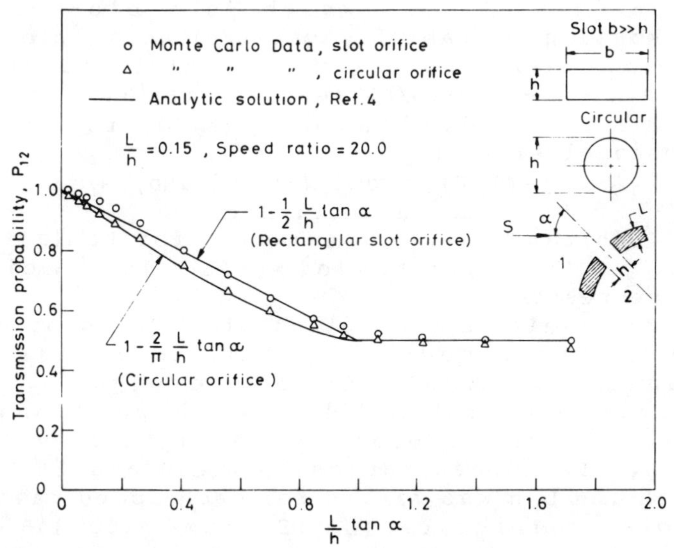

Fig.4 Transmission Probability for a Slot Orifice $P_{12} = P_{12}(S, L/h, \alpha)$.

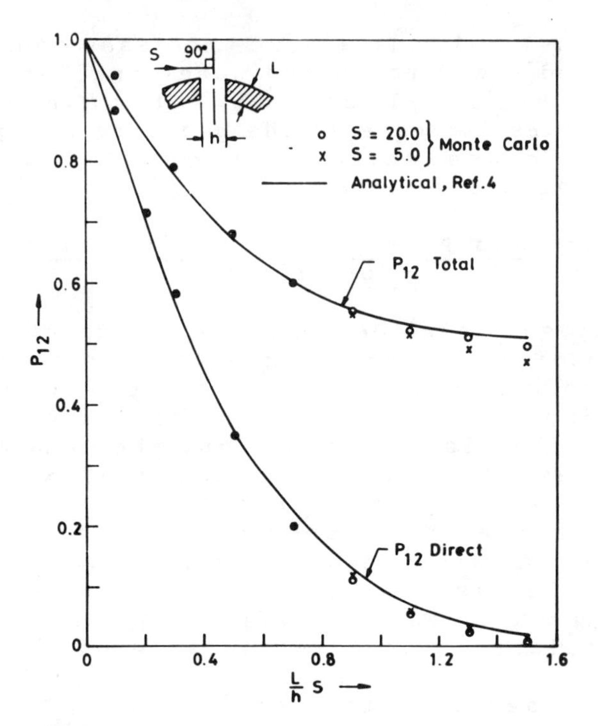

Fig.5 Transmission Probability for Slot Orifice at $\alpha = 90°$.

presented in Fig. 4 which is a plot of total transmission probability vs $(L/h)\tan\alpha$ along with Koppenwallner's approximate analytical relation with $P_{12} = 1-(1/2)(L/h)\tan|\alpha|$ for $\tan|\alpha| < (h/L)$ and $P_{12} = 1/2$ for $\tan|\alpha| > (h/L)$ for the two-dimensional slot and

$$P_{12} = 1-(1/\pi)[\arcsin((L/h)\tan\alpha) + (L/h)\tan\alpha\sqrt{1-((L/h)\tan\alpha)^2}]$$

for $\tan|\alpha| \leq (h/L)$ and $P_{12} = (1/2)$ for $\tan|\alpha| > (h/L)$ for a circular orifice respectively.

For finite-speed ratios when the slot orifice probe is inclined at 90 deg. to the incoming stream, the Monte Carlo results of total transmission probability is shown along with Koppenwallner's approximate analytical values in Fig. 5. Two speed ratios, S = 5.0 and 20.0, were chosen and L/h was varied for each speed ratio case to cover a range of (L/h)S from 0 to 1.5. Very close agreement between the P_{12} values for two different speed ratios for the same (L/h)S values brings out clearly the significance of the parameter (L/h)S and also confirms the results of an analytical analysis of of Koppenwallner[4] by an independent approach.

Density Variation

The density fields for rarefied flow through a square and a rectangular orifice have been theoretically investigated by Weston[5] for a pure effusive flow (S = 0). The same problem was investigated by the present Monte Carlo method, and the results of the axial density variation beyond the exit plane are shown in Fig. 6. Comparisons are also made between the theoretical predictions and Monte Carlo results. The density field is found to be most affected by orifice geometry in the near region. For two-dimensional slits this effect is felt farther from the orifice than for a square cross-section. In the figure the distance from the orifice or slit is nondimensionalized by the square root of orifice area A.

Figures 7 and 8 illustrate the axial and cross-sectional density variation at three axial

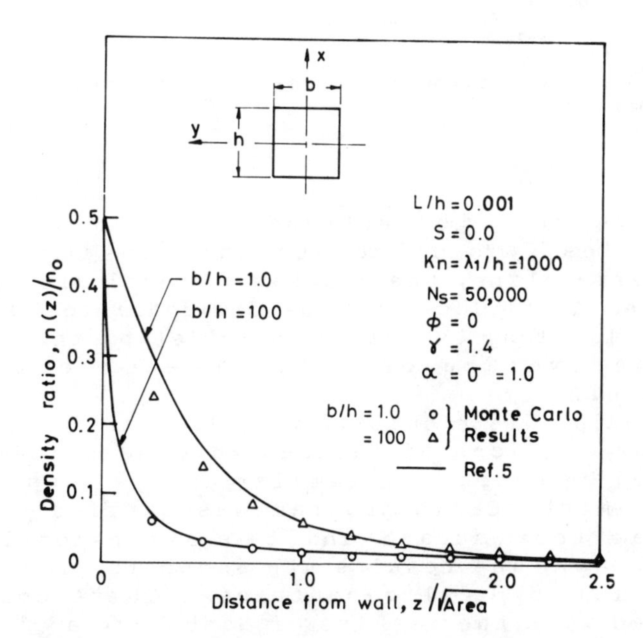

Fig.6 Axial Density Variation Downstream of the Slot Orifice.

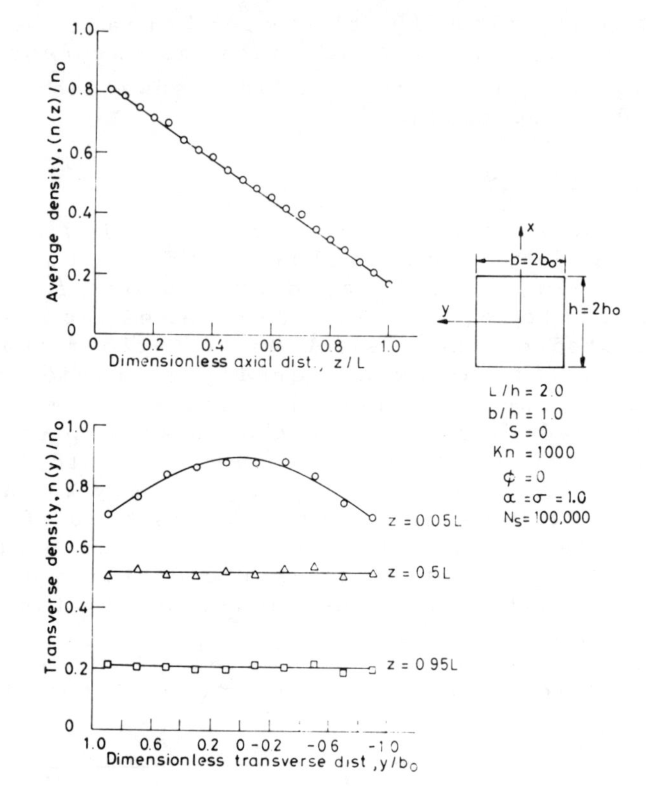

Fig.7 Axial and Transverse Density Distribution inside the Square Tube.

locations from the entrance for two b/h ratio geometries. To calculate the density in the transverse plane, the cross-sectional area at the specified axial location was divided into 10 equal area cells along the width (parallel to the height) and the average density in each cell was calculated.

Angular distributions of molecular efflux from circular orifices of various thickness have been reported by Nanbu.[6] A similar analysis using the present Monte Carlo program was carried out for effluxes from square and two dimensional tube geometries. The results are shown in Figs. 9 and 10 for two typical geometries. These could be compared with the results of Nanbu for a circular geometry to bring out the influence of tube geometry on efflux distribution.

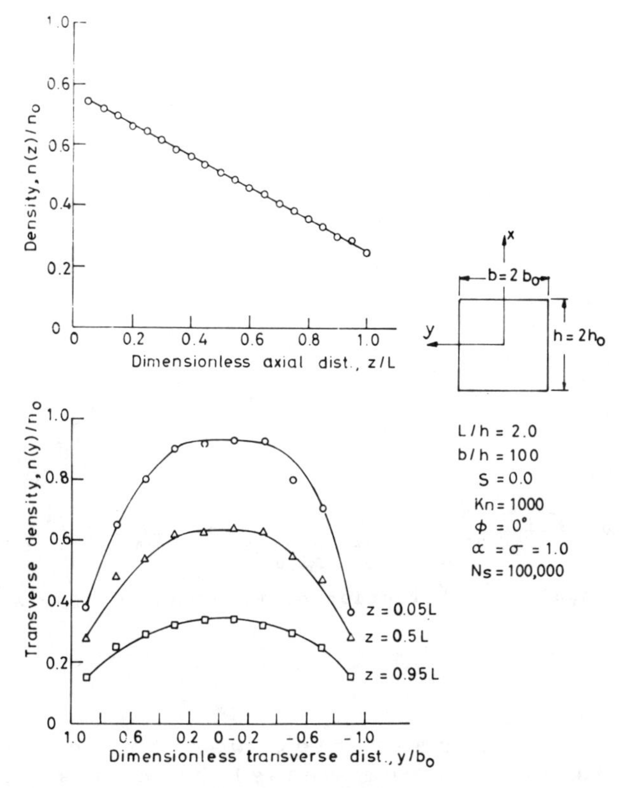

Fig.8 Axial and Transverse Density Distribution inside a Rectangular Tube.

Experimental Investigation of Mass Flow Through Rectangular Tube and Slit

Using a continuous-flow rarefied gasdynamics facility, the mass flow through a rectangular slit and a rectagular tube in the near free molecular and transitional flows were measured. One of the purposes of the experimental study was to make a comparison between the above and the flow through a circular tube for which some mass flow results are already available in the literature.[7,8]

Nitrogen gas at room temperature was allowed to pass through two different types of mass flow measuring devices connected in series and then admitted to the upstream vacuum chamber by means of a throttle valve to reduce the pressure to the desired value. The rectangular slit or the tube

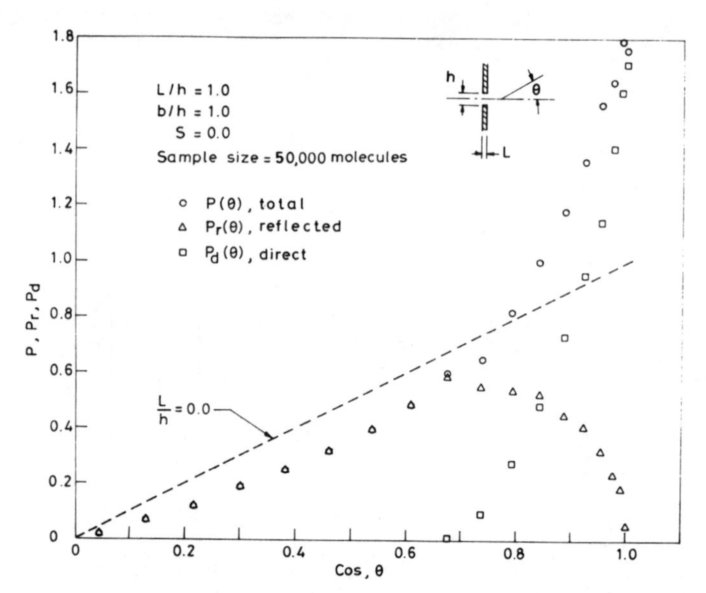

Fig.9 Angular Distribution of P, P_r and P_d for a Square Tube.

through which mass flow was to be measured separated the vacuum vessel into upstream and downstream chambers. The upstream chamber acted as the stagnation chamber for effusive flow studies.

As mentioned earlier, two different types of mass flow measuring instruments were used in series. One was of a thermal type in which the differential cooling of a heated tube caused by the gas flow through it gave a measure of the mass flow (Hastings Raydist model). The other was a laminar flow element in which the pressure drop over a predetermined length due to a flow in it is related to the flow rate by the Poiseulle flow (Furness Control model). Different ranges of instruments that depended on the magnitude of the mass flow to be measured were used. In all of the experiments conducted, the mass flow readings from both types of instruments agreed within $\pm 1\%$. The upstream and downstream pressures were measured by capacitance manometers (Barocell of Datametrics model), McLeod gages, and Pirani-Penning gages. Desired pressure ratios across the tube geometry were maintained by leaking in nitrogen gas through a throttle valve connected to the downstream side.

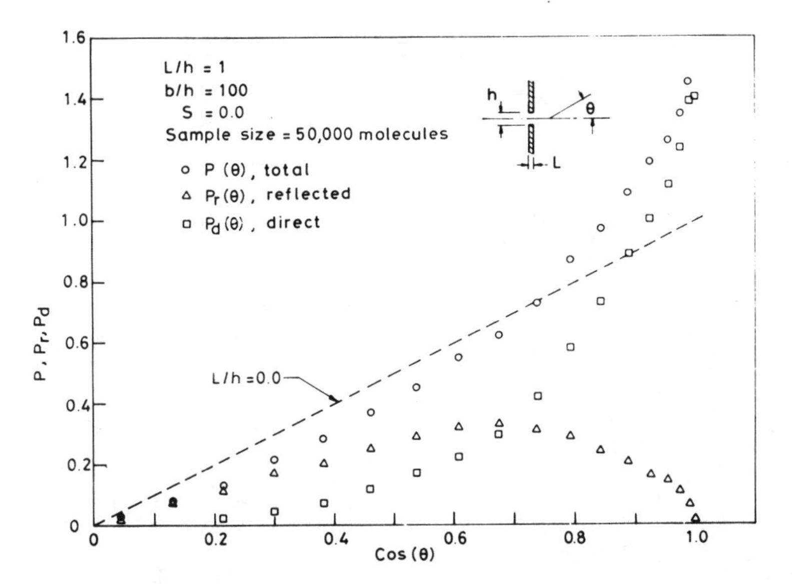

Fig.10 Angular Distribution of P, P_r and P_d for a two-dimensional Tube.

The slit consisted of stretched aluminium foil of 0.05 mm thickness in which an area of 5 mm height and 100 mm width was cut out. The rectangular tube investigated had dimensions of 8 mm x 80 mm crosssection and was 80 mm long, giving a b/h ratio of 10 and a L/h ratio of 10. Different values of upstream pressures were chosen, and for each one, a set of experiments was performed by varying the downstream pressure to get desired pressure ratios. Commercially available 99.9% pure nitrogen gas was used in the experiments.

In both numerical and experimental work the rectangular tube that connected the two large upstream and down-stream chambers had square-edged entry and exit as shown

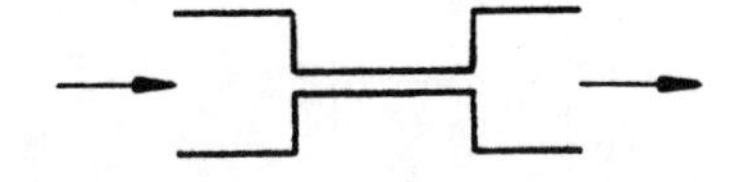

Results and Discussions

For the presentation of experimental results, the measured mass flow rates are nondimensionalized

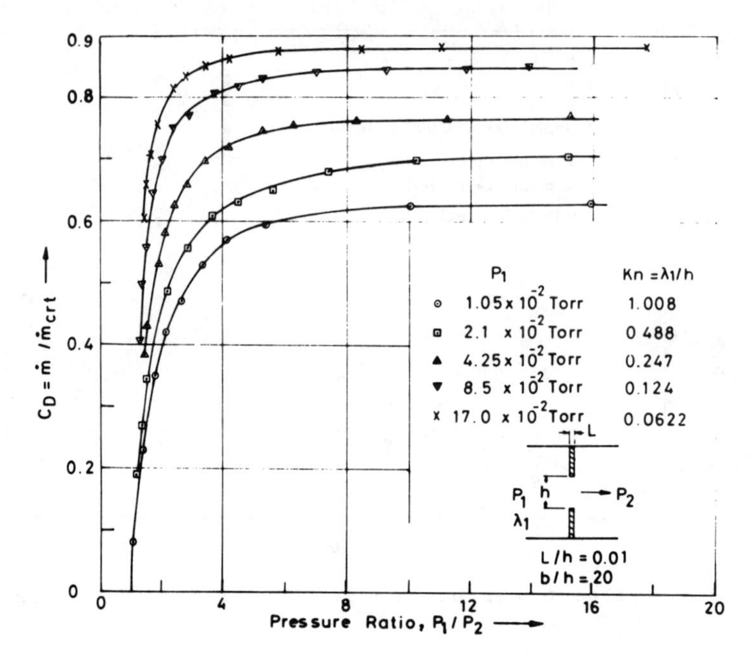

Fig.11 Discharge Coefficient at Various Pressure Ratios for a Slot Orifice.

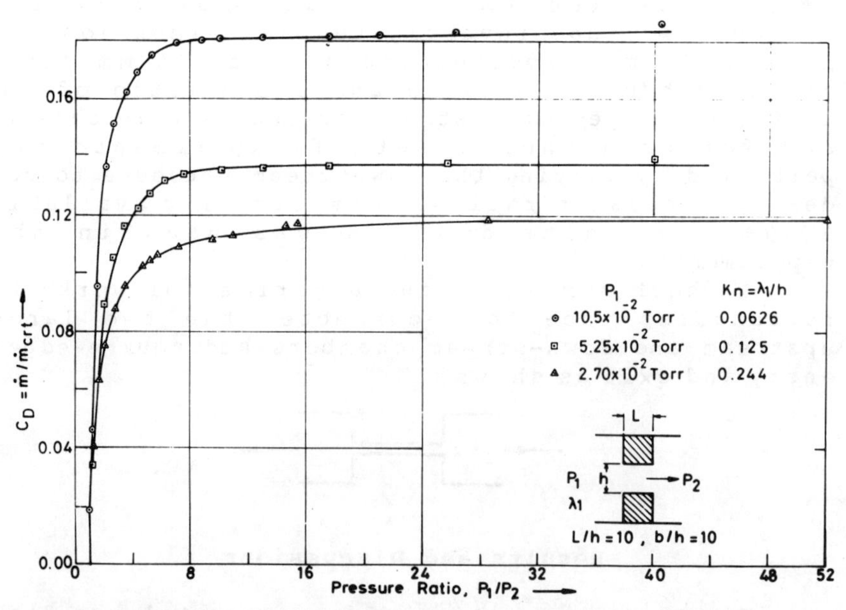

Fig.12 Coefficient of Discharge Vs Pressure Ratio for a Rectagular Tube.

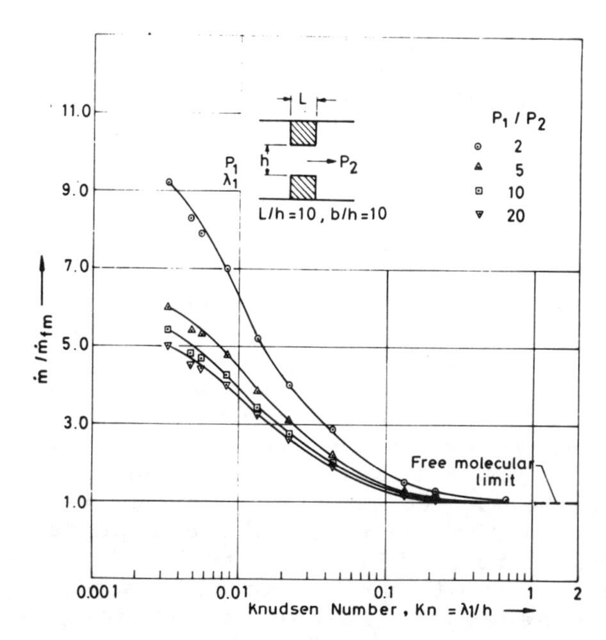

Fig.13 Nondimensionalised Mass Flow Vs Knudsen Number at various Pressure Ratios for a Slot Orifice.

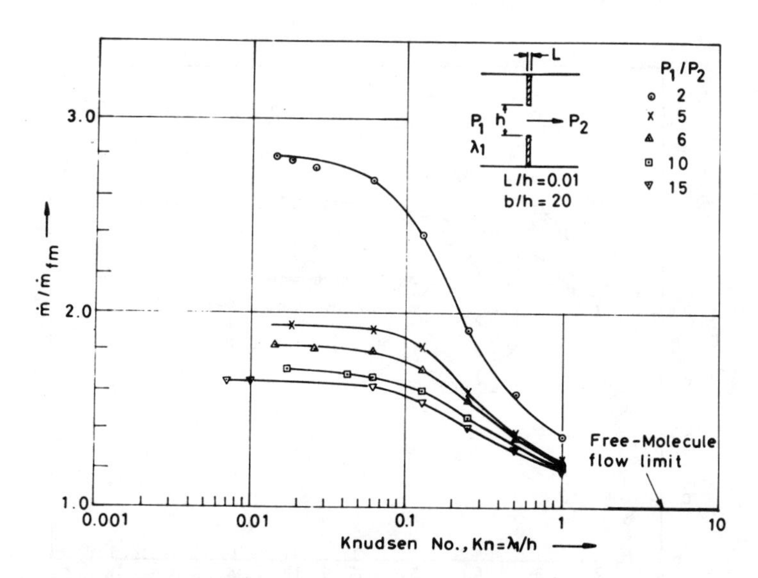

Fig.14 Nondimensionalised Mass Flow Rate Vs Knudsen Number at Various Pressure Ratios for a Rectangular Tube.

either by the theoretical free molecular flow m_{fm} or the theoretical continuum choked mass flow $m_{crt}(=\rho^* A^* V^*)$. The Knudsen number is based on the upstream mean free path and the tube or orifice height:

$$Kn = \lambda_1/h$$

and

$$\lambda_1 = 16\,\mu_1/5P_1\,(RT_1/2\pi)^{1/2}$$

The Reynolds number is based on the tube height h and the measured mass flow rate as follows:

$$Re = mh/(b.h)\mu$$

Figures 11 and 12 are a plots of the discharge coefficient vs pressure ratio for a slot orifice and a rectangular tube, respectively, at various Knudsen numbers. It is seen that, unlike those in the continuum flow regime, where a pressure ratio of around 2 is sufficient to choke the flow, pressure ratios of one or two orders of magnitude higher are necessary in the transition, near free molecular and free molecular flow regimes.

Figures 13 and 14 are plots of the nondimensionalized mass flow vs Knudsen number at various pressure ratios for a rectangular slot orifice and tube respectively. It is observed that

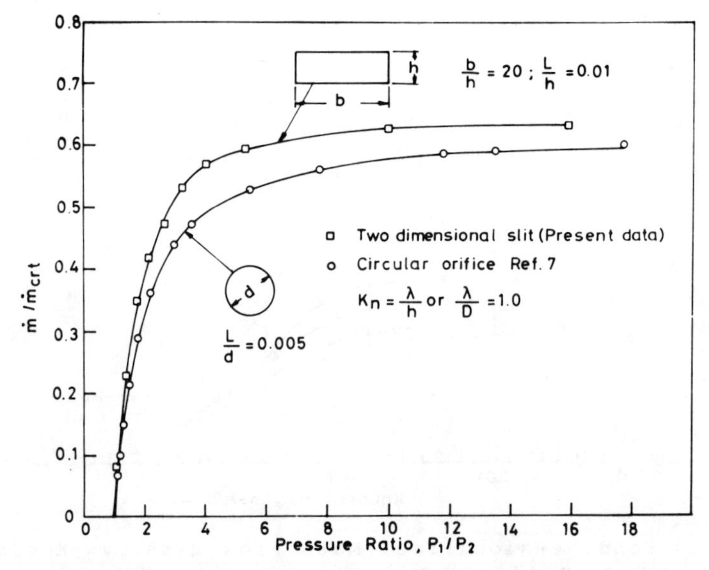

Fig.15 Comparison of Mass Flow for a Rectangular Tube and a Circular Orifice.

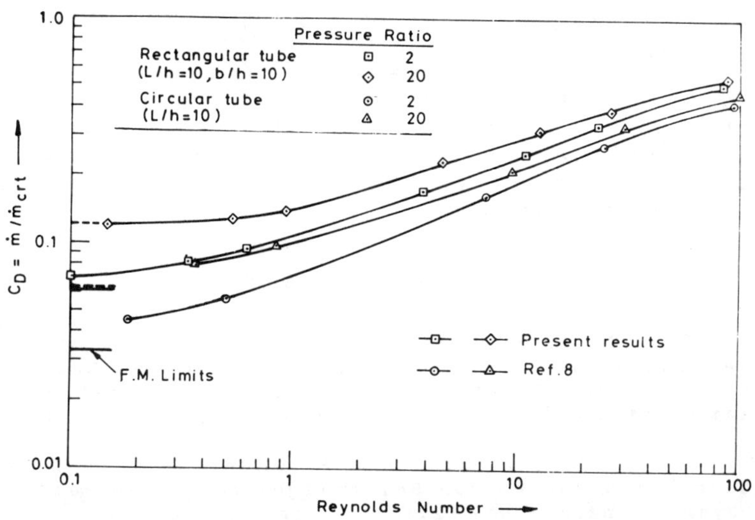

Fig.16 Comparison Between Rectangular and Cicular Tube Mass Flow Rates.

approach to free molecular limit occurs at lower Knudsen numbers for the case of a tube when compared with the orifice.

Figure 15 is a comparison of the measured mass flow for a rectangular slit with that of a circular orifice[7] at Kn = 1.0. It is seen that the measured mass flow rate is always higher at any pressure ratio for a two-dimensional geometry when compared with the circular one. The theoretical free molecular conductance for a two-dimensional slit/tube is always higher than that of a circular one, and the experimental results corroborate this. Comparisons of the flow rates in a rectangular and a circular tube[8] as a function of Reynolds number are shown in Fig. 16 at two pressure ratio conditions, viz., 2 and 20. It is seen that the differences are more pronounced in the free molecular and near free molecular regimes.

Conclusions

A wide range of numerical results based on the test particle Monte Carlo method have been obtained on the free molecular flow of rarefied gases through tubes and slits of rectangular crosssection. Transmission probabilities, density distribution both along the axis and at various crosssections and the efflux pattern have been analyzed.

Limited experimental work involving the measurement of mass flow rate through rectangular slits/tubes was conducted with the main purpose of comparing the obtained results with the already available data of rarefied flow through circular tubes and orifices.

References

[1]Fan, C., "Monte Carlo Solutions of Mass, Momentum and Energy Transfer for Free Molecule and Near Free Molecule Flow Through Circular Tubes." Lockheed Missile and Space Company TR-LMSC/HREC A 791015, Dec. 1986.

[2]Clausing, P., "Uber die Stromung Sehr Verdunnter Gase Durch Rohren von Beliebieger Lange," Annalen der Physik(Leipzig), Vol.12, 1932, pp.961-989.

[3]Demarcus, W. C., "The Problem of Knudsen Flow, Solutions for One Dimensional Systems." Oak Ridge Gaseous Diffusion Plant Rep.K 1302, P.III, March 1957.

[4]Koppenwallner, G., "The Free Molecule Pressure Probes with Finite Length Slot Orifice," Deutsche Forschungs-und Versuchanstalt Fur Luft-und Raumfahrt, AVA Rept.IB 222-84A 15, May 1984.

[5]Weston, M. H., "The Density Field for Rarefied Flow Through an Orifice." NASA Ames Research Center. (Internal Memo)

[6]Nanbu, K., "Angular Distributions of Molecular Flux from Orifices of Various Thickness. "Vacuum, Vol. 35, No. 12, pp.573-576, 1985.

[7]Sreekanth, A. K., "An Experimental Investigation of Mass Flow Through Short Circular Tubes in the Transition Flow Regime." Boeing Scientific Research Lab. Rep.D1-82-0427, 1965.

[8]Marchman, J. F., "The Effects of Length to Diameter Ratio on Flow Through a Short Tube with Square Edged Entrance in the Transition Flow Regime," Ph.D. Dissertation, Dept. of Mechanical and Aerospace Enginering, North Carolina State University, Rayleigh, NC, 1968.

[9]Davis, A., "An Experimental and Numerical Investigation of Rarefied Gas Flow through Rectangular Slits and Tubes," M.S. Dissertation, Dept. of Aerospace Engineering, Indian Institute of Technology, Madras, India, Dec. 1987.

Experimental Investigation of Rarefied Flow Through Tubes of Various Surface Properties

J. Curtis*
University of Sydney, Sydney, New South Wales, Australia

Abstract

Measurements have been made of the helium conductance of tubes of stainless steel, tantalum, and molybdenum both before and after heating to high temperature. The measurements were made in an all-metal vacuum system capable of pressures in the range 10^{-9} to 10^{-10} Torr. An ionization gage was used to measure the pressure increase produced in a known volume by flow through the tube under test. The conductance was calculated from the time rate of rise in pressure and a constant based on the volumes of system components. Conductance values of the tubes used were calculated with the assistance of tabulated values of transmission probability for diffuse interaction. Measured conductance values obtained both before and after heating are presented for comparison with the diffuse values. The measured values after heating are substantially larger than the diffuse values, with those obtained for the molybdenum tube showing the larger increase.

Introduction

Knudsen,[1] in 1908, published his "long tube" equation for the steady molecular flow of a gas through a tube of circular cross section subject to a pressure difference ΔP. This equation,

$$Q = \frac{8R}{3L} \cdot \pi R^2 \cdot \left(\frac{kT}{2\pi m}\right)^{\frac{1}{2}} \cdot \Delta P$$

where Q is the flow rate, R the tube radius, L the tube length, k the Boltzmann constant, T the absolute temperature

---

*Professional Officer, Department of Aeronautical Engineering.

and m the molecular weight, was based on a number of assumptions, including large L/R and a diffuse interaction between gas molecules and the tube wall, with a cosine distribution of scattered molecules. The diffuse interaction assumption was subsequently verified experimentally by Gaede,[2] Clausing,[3] and other workers.

Maxwell[4] and Smoluchowski[5] had recognized the possibility that a proportion of the molecules in rarefied flow may be specularly reflected, and Smoluchowski deduced that for molecular flow of this type Knudsen's expression should be multiplied by the factor $(2-\alpha)/\alpha$, where α is the probability that a molecule will undergo diffuse interaction.

Clausing[6] adopted a different approach to the problem in an effort to develop a theoretical basis for flow in tubes of finite length. His introduction of the concept of transmission probability led to a general statement of the flow rate equation as

$$Q = W \cdot A \cdot \left(\frac{kT}{2\pi m}\right)^{\frac{1}{2}} \cdot \Delta P$$

where W is the transmission probability of the duct and A the duct area, and his subsequent attempts to calculate W for a tube of circular cross section led to an integral equation for which an analytical solution could not be found. An approximation, based on the assumption that $W = 8R/3L$ for long tubes, allowed numerical values to be obtained for tubes of any length.

Clausing also briefly considered the possible effect of a proportion of specular reflection, ε, but concluded that the effect on W was easily determined only for the case in which the probability of specular reflection is the same for all angles of incidence, for which case he noted

$$W_\varepsilon = \frac{1+\varepsilon}{1-\varepsilon} W$$

where W is the fully diffuse transmission probability and W_ε the transmission probability for a probability of specular reflection ε.

Since $\varepsilon = 1-\alpha$,

$$W_\varepsilon = \frac{2-\alpha}{\alpha} W$$

which is, in effect, the Smoluchowski result.

De Marcus[7] adopted a generalized approach to molecular flow that resulted in integral equations of the Clausing type. Various methods of treatment were applied to different cases. In the case of a circular tube a variational method was used which provided more accurate transmission probabilities than those of Clausing, while at large L/R the solution was found to be asymptotic to the Knudsen long tube value 8R/3L. The effect of specular reflection was also considered, and it was concluded that the Smoluchowski result held exactly only for infinitely long tubes. An interpolatory formula was presented that gave a reasonable representation of the numerical results:

$$W_{\varepsilon} = \frac{1 + \varepsilon + W(1-\varepsilon)}{1 - \varepsilon + W(1+\varepsilon)} W$$

Although theoreticians recognized that the presence of a specular component was a possibility in molecular flow, early experiments indicated that, in real flows, diffuse interaction was the norm. More recently, however, a number of small but significant experimental departures from fully diffuse transmission probability values have been reported. In particular, Davis et al.[8] measured flow rates 5% above diffuse values for smooth-walled tubes. They also investigated the effects of surface roughness and showed that large-scale irregularities (up to 15% of radius) could reduce flow rates by as much as 20%, with smaller-scale irregularities showing reduced effects.

Porodnov and co-workers[9] found flow rates to be up to 15% higher than diffuse values in smooth-walled glass capillaries for both slip and molecular flow. They attributed the effect to the presence of a specular component.

Lord[10] measured tangential momentum accommodation (TMA) coefficients for the noble gases on a range of polycrystalline metal surfaces. He used an apparatus in which the metal surface was cleaned by heating to high temperature and maintained in that state by a misch metal getter. The apparatus was filled with the test gas, which was initially of high purity, and was further cleansed as it passed into the equipment. Thc gettering technique used had been developed by Thomas during his measurements of energy accommodation coefficients. Lord's TMA measurements produced coefficients as low as 0.2 for helium on molybdenum.

Steinheil et al.[11] measured normal and tangential momentum transfer between a molecular beam of helium and a metallic surface. Both crystalline and polycrystalline surfaces were used and various surface treatments were employed. They also measured the angular distribution of the scattered beam. Their results indicate that large reductions in TMA can be achieved by removal of surface contaminants and that these reductions are accompanied by an increase in reflection near the specular angle.

Experimental results thus suggest that rates for partially "specular" molecular flow in tubes exceed those observed when the interaction is fully diffuse. The application of clean surface technique to refractory metals is seen to result in surfaces that exhibit a significant proportion of non-diffuse interaction with the lighter inert gases. The application of the cleaning technique to refractory metal tubes should therefore lead to substantial reductions in their helium conductance.

Method

The tube under test was connected between an upstream reservoir, of volume V_1, and a downstream reservoir, of volume V_2. A quick-opening valve was placed between the tube inlet and V_1. With V_1 filled with pure helium to approximately 0.01 Torr and a vacuum of 10^{-9} Torr in V_2, the valve was opened in approximately 1/100 s. The time rate of pressure rise in V_2 was measured by an ionization gage and recorded.

V_2 and V_1 were measured by the expansion method and found to be 57.0 and 3.97 l, respectively. Under typical experimental conditions the pressure in V_2 reached 10^{-5} Torr in about 10 s and 10^{-4} Torr in about 120 s. The data points were taken in this interval, and the pressure was then equalized by a second valve between the volumes, allowing measurement of the final equilibrium pressure P_f.

It can be shown that

$$\ln\left(1 - \frac{P}{P_f}\right) = -C\left(\frac{1}{V_1} + \frac{1}{V_2}\right)t$$

where P is the pressure in V_2 at time t, P_f the pressure in V_2 at time ∞, C the tube conductance and V_1 and V_2 the reservoir volumes.

The conductance may thus be obtained directly from the slope of $\ln(1 - P/P_f)$ vs t and the known volumes.

Since pressure ratio rather than pressure is used, gage linearity is more important to accuracy than correct indication of pressure. The volumes and time are measured to much higher accuracy than pressure and make little contribution to error.

Three other possible sources of error must be considered: 1) ion gage pumping of the test gas in V_2 during the flow; 2) loss of test gas to the walls of V_2 during the flow; and 3) outgassing into V_2 during the flow.

Ion gage pumping has been examined by filling V_2 with helium to various pressures within the range encountered during flow experiments. Measurements with the ion gage would indicate a steady fall in pressure if pumping were taking place. Since in each case examined there was no detectable change in gage indication in a 10-min interval, this possible error source can be ignored.

Adsorption of inert gases by stainless steel surfaces has been the subject of theoretical and experimental examination by vacuum metrologists during development of the expansion method of gage calibration. This work demonstrates that errors arising from adsorption are negligible.

Each conductance measurement was preceded by an outgassing measurement in V_2 conducted over a time equal to that expected for the following flow measurement. In every case the total contribution by outgassing to the final pressure was shown to be less than 1%, and in the majority was less than 0.5%.

Apparatus

The equipment used for the measurement of tube conductance consisted of a Varian FC12E ultra high vacuum system fitted with 80 $l \cdot s^{-1}$ of getter ion pumping capacity and a titanium sublimation pump. The system had been previously used for energy accommodation coefficient measurements and in that capacity had demonstrated the effectiveness of titanium gettering as an alternative to the more frequently used misch metal.

In the apparatus schematic, figure 1, all volumetric capacity to the left of valves A and B constitutes V_2, and all to the right V_1. V_2 thus consists of the FC12E and an auxiliary cylindrical chamber containing the tube under test whereas V_1 consists of a much smaller input chamber, containing a sublimation element, and associated pipework. The tube (200 - 250 mm long) is suspended axially in the chamber from a copper disc, through which it passes. The clearance gap between tube and disc was filled by brazing with a high temperature alloy so that the disc could act both as an electrical contact to the tube and a sealing component between the flanges of the valve and chamber. It was necessary to suspend the tube in this manner to accommodate its thermal expansion during heating.

Electrical contact to the lower end of the tube was made by a clamping block and was carried to the exterior of the system via copper braids E and insulated current feedthroughs F. Heating current was provided by a 6V, 300A

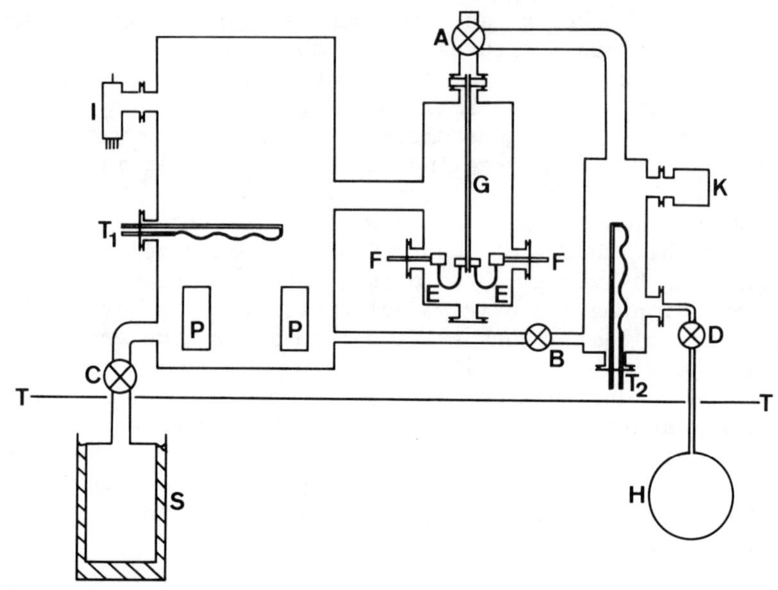

Fig. 1 Apparatus schematic. A: quick-opening valve; B: equalization valve; C: bakeout valve; D: helium admittance valve; E, E: copper braids; F, F: current feedthroughs; G: tube under test; H: helium flask; I: ionization gage; K: capacitance manometer; P, P: getter-ion pumps; T-T: table surface; T_1, T_2, titanium sublimation elements; S: sorption pump.

transformer via connections to the feedthroughs and the copper disk.

The system was pumped as a whole by opening all valves, A, B, C, and D. Initial pumpdown to 10^{-2} Torr was provided by sorption pump S, while the whole system above the table TT was baked to 250°C for 4 h. After bakeout valve C was closed and the getter ion pumps P used to reduce system pressure to 10^{-9} Torr. Pumping continued until acceptable outgassing rates were obtained, at which time gas admittance valve D was closed and the seal on helium flask H broken to admit the test gas to the input of valve D. Thereafter, closure of valves A and B separated V_1 and V_2 and allowed helium admission to V_1. The tube could then be heated at an ambient pressure of 10^{-8} Torr or less, while the getter ion pumps and, if necessary, the sublimation pump T_1 were used to remove gas evolved.

After the completion of tube heating about 45 min was required to allow the tube to cool to room temperature. The test was then carried out by switching off the pumps and opening valve A. The ionization gage output was recorded as a function of time, indicating the rate of pressure rise in the downstream volume.

Results

A summary of results obtained is provided in Table 1, which also lists the diffuse conductance, C, for each tube, calculated from its dimensions and a transmission probability obtained from Cole[12]. A correction was made in each case for Knudsen number calculated on the basis of the mean free path in the inlet reservoir and the tube diameter. An estimate of this correction (of maximum value 8% at Kn=1) was obtained from Porodnov et al.[9]

The results showed a large change in the helium conductance of the molybdenum tube and more modest increases for the stainless steel and tantalum tubes.

It was originally intended to use the stainless steel tube only as a test piece to confirm that the system was functioning correctly. The tube was not expected to show significant changes in conductance because of the complexity of its surface and its relatively low melting point. The results indicate that such assumptions are unwise.

The tantalum results have probably been limited by the temperature attained, which was the highest available with a tube of the size used at maximum available heating current. Lord's TMA coefficient of 0.46 for helium on tantalum was obtained after heating to near the melting point (2996°C) and Gebhardt and Preisendanz[13] found it necessary to outgas tantalum at 2600 - 2800°C in a vacuum of less than 5×10^{-6} Torr to obtain clean conditions.

Lord[10] found molybdenum to have a TMA coefficient of 0.2 for helium. The change in conductance observed for this combination indicates that surface cleanup has been more effective than for the other tubes, as may have been expected given that the temperature reached of 2250°C was much nearer the melting point of 2620°C.

Table 1 Result summary

Tube material	L/R	Diffuse conductance C, $ml \cdot s^{-1}$	Least measured conductance Cmin, $ml \cdot s^{-1}$	Largest measured conductance Cmax, $ml \cdot s^{-1}$	Tmax, °C	Cmax/C
Stainless steel	160	36.0(K_n=1)	39.5	71.3	800	1.98
Tantalum	300	4.87(K_n=10)	5.48	8.11	1650	1.66
Molybdenum	285	4.07(K_n=10)	3.41	10.5	2250	2.58
		4.24(K_n=100)		11.7	2250	2.75

It should be pointed out that the heavy copper current attachments caused appreciable cooling at the tube ends. At least 30 mm in total would be expected not to clean up to any significant extent, and a further proportion would be partially affected. The measured conductance values thus underestimate the increase that would be observed in a uniformly heated tube. This effect would have been more pronounced for the molybdenum tube because its length was 200 mm, whereas the others were 250 mm long.

The maximum cleaning temperature for each tube was reached after a number of conductance measurements with progressively increasing temperatures. Several attempts were made during this sequence to return to the initial surface condition by exposing the tube to the atmosphere. The values of conductance obtained after exposure were always higher than the initial values and tended to increase as cleanup temperature increased.

This effect is attributed to desorption during the bakeout and pumping process that was necessary to return the system to the 10^{-8} to 10^{-9} Torr pressure range after atmospheric exposure.

Lord[10] observed a similar effect in his system, as he measured TMA coefficients of about 0.9 immediately after bakeout. It has also been noted during energy accommodation coefficient measurements in clean systems that it is difficult to obtain values of unity on supposedly dirty surfaces, particularly with the lighter inert gases.

The initial measured conductance of the molybdenum tube was 16% below the calculated diffuse value. When measured 24 h later, after repumping the system to 10^{-9} Torr, it was found to be 9% below that value. An examination of the tube interior showed the presence of numerous projections in the form of attached shavings or "whiskers". It is believed that this roughness was largely responsible for the low measured conductance. Because average internal diameters of the tubes were unobtainable the diameter measured at the ends was used in conductance calculations. Errors from this source are magnified by the dependence of conductance on D^3 and may have contributed to the difference between measured and calculated values.

Conclusion

The conductance for helium of a molybdenum tube of length 200 mm and internal diameter 1.4 mm has been found to increase by a factor of approximately 3 after surface cleaning by heating to 2250°C in a clean environment. Smaller increases have been obtained for stainless steel and tantalum tubes of about the same size, but even these increases are much larger than any previously reported, of which the largest appears to be 15%.

For Maxwell model flow at large L/R, the De Marcus equation indicates that a decrease in the diffuse component from 100% to 52% would be required to cause a conductance increase equal to that observed for the molybdenum tube after heating to 2250°C.

The Maxwell model is used here only because it provides a convenient means of indicating the extent of the change in the gas-surface interaction caused by surface cleanup. Although the present experiments provide no direct indication of the nature of the interaction they suggest that it has been profoundly changed by the removal of surface contaminants.

References

[1]Knudsen, M., "The Laws of Molecular Flow and the Internal Viscous Streaming of Gases Through Tubes," Annalen der Physik, (4), 28, (1), 1908, pp. 75-130.

[2]Gaede, W., "The Outer Viscosity of Gases," Annalen der Physik, (4), 41, 1913, pp. 289-336.

[3]Clausing, P., "On a Measurement of Molecular Velocity and a Test of the Cosine Law," Annalen der Physik, (5), 7, 1930, pp. 569-578.

[4]Maxwell, J.C., "Scientific Papers," edited by W.D. Niven, Cambridge University Press, Cambridge, 1890.

[5]Smoluchowski, M.v., "The Kinetic Theory of Transpiration and Diffusion in Low Density Gases," Annalen der Physik, 33, 1910, pp. 1559-1570.

[6]Clausing, P., "The Flow of Highly Rarefied Gases through Tubes of Arbitary Length," Annalen der Physik, (5), 12, 1932, pp. 961-989.

[7]De Marcus, W.C., "The Influence of Specular Reflection on the Knudsen Conductance of Circular Capillaries," Advances in Applied Mechanics: Rarefied Gas Dynamics, edited by L. Talbot, Academic Press, New York, 1961, pp. 161-168.

[8]Davis, D.H., Levenson, L.L., and Milleron, N., "Effect of "Rougher-than-Rough" Surfaces on Molecular Flow through Short Ducts," Journal of Applied Physics, (3), 35, pt. 1, 1964, pp. 529-532.

[9]Porodnov, B.T., Suetin, P.E., Borisov, S.F., and Akinshin, V.D., "Experimental Investigation of Rarefied Gas Flow in Different Channels," Journal of Fluid Mechanics, 64, pt. 3, 1974, pp. 417-437.

[10]Lord, R.G., "Tangential Momentum Accommodation Coefficients of Rare Gases on Polycrystalline Surfaces," Progress in Astronautics and Aeronautics Vol. 51, Pt. 1: Rarefied Gas Dynamics, edited by J. Leith Potter, AIAA, New York, 1977, pp. 531-538.

[11]Steinheil, E., Scherber, W., Seidl, M., and Rieger, H., "Investigations on the Interaction of Gases and Well-Defined Solid Surfaces with respect to Possibilities for Reduction of Aerodynamic Friction and Aerothermal Heating," Progress in Astronautics and Aeronautics Vol. 51, Pt. 1: Rarefied Gas Dynamics, edited by J. Leith Potter, AIAA, New York, 1977, pp. 589-602.

[12]Cole, R.J., "Transmission Probability of Free Molecular Flow Through a Tube," Progress in Astronautics and Aeronautics Vol. 51, Pt. 1: Rarefied Gas Dynamics, edited by J. Leith Potter, AIAA, New York, 1977, pp. 261-272.

[13]Gebhardt, E., and Preisendanz, H., "On the Solubility of Oxygen in Tantalum and the Accompanying Property Changes," Zeitschrift fuer Metallkunde, 46, 1955, pp. 560-568.

Monte Carlo Simulation on Mass Flow Reduction due to Roughness of a Slit Surface

M. Usami,* T. Fujimoto,† and S. Kato‡
Mie University, Kamihama-cho, Tsu-shi, Japan

Abstract

Mass flow reduction of rarefied gas through a two-dimensional slit due to roughness of the slit surface is investigated numerically. Though the direct simulation Monte Carlo (DSMC) method is very effective in an analysis of rarefied gas flow, it costs a lot of CPU time, especially for the near-continuum regime. Recently, some supercomputers with vector processors are available that allow high-speed calculations. To use them in the most effective way, a computer program must be adequately vectorized. By frequent use of a programming technique called "data collection," a simulation program eight times faster than a conventional one is obtained. Applying this program to flow through a slit, the effect of surface roughness on rarefied gas flow in the transition regime has been clarified. The numerical results are compared with those of experiment.

I. Introduction

Currently there is an interest in applying rarefied gasdynamics not only to low-density flows but to flows at any density through minute flowfields with very small characteristic dimensions. As examples, we consider flows between a magnetic head and a recording media whose separation is extremely small and leakage flows past seals such as the flow past rotor or stator tip clearances in turbomachinery.[1] In an analysis of the flow through these minute flowfields, roughness of the boundary surface cannot be neglect-

*Assistant Professor, Mechanical Engineering.
†Professor, Electronic-Mechanical Engineering. Presently at Nagoya University, Furo-cho, Chikusa-ku, Nagoya, Japan.
‡Professor, Mechanical Engineering.

ed. Several studies of surface roughness on flows through various type of ducts have been conducted in the free molecular flow regime. Davis et al., using the test particle Monte Carlo calculation, have shown that the roughness reduces flow rates about 6% below the prediction with diffuse reflection.[2] Berman and Maegley have studied an internal rarefied gas flow with the backscattering boundary condition,[3] showing that the surface roughness causes a mass flow reduction. However, their approaches are confined to the free molecular flow regime. The effect of surface roughness is not yet sufficiently investigated over all regimes from transition to continuum flow. For example, there is no available information regarding in what part of regimes the mass flow reduction due to the roughness disappears.

The authors have applied the direct simulation Monte Carlo (DSMC) method to rarefied gas flows through a two-dimensional slit or a circular duct and have obtained reliable results.[4,5] In the present study, the mass flow reduction due to surface roughness of a two-dimensional slit is investigated with the DSMC method in the transition regime.

As described in Bird's book and in various papers involving the calculation of the DSMC method, the difficulty in applying this method to any practical flows arises from requirements of large computer time and large computer memory.[6] Fortunately, the rapid progress of computer technology is extending the area to which the DSMC method is applicable. Recently, some supercomputers with vector processors have become available that allow high-speed calculations. To use them in the optimum condition, it is essential that a computer program be adequately vectorized. A program with insufficient vectorization may require a rather higher calculation cost. In this study the practicability of the vector processing of the DSMC computer program improved suitable for the supercomputer is examined, with special emphasis placed on the near-continuum flow regime accompanied by large computer time.

II. Vectorization for the DSMC Method

To realize the vectorization it is convenient to divide the computer program for the DSMC method into four routines: 1) the molecular motion (including interaction with boundary), 2) the identification of the cell in which a particular molecule is located, 3) the molecular indexing, and 4) the intermolecular collision.

For the vectorization of the present computer program, a technique called data collection is frequently used. The data collection that collects the data satisfying a requirement is a preparatory procedure for a DO loop that is expected to be vectorized. A DO loop with an IF statement for the data collection, as shown in Fig. 1, can be vectorized commonly. Note that a statement within the IF block must be a simple instruction since the vectorization of the DO loop is processed with a pattern recognition by the compiler. The numbers N's in Fig. 1 that satisfy the condition IRYO(N)$\leqq$100 are collected in the DO 10 loop, and then the DO 20 loop is executed for only the numbers collected.

Pseudorandom numbers are required to be generated in parallel by the computer. In the present calculation the initial random numbers, as many as the number of cells, are selected out of one row of particular random numbers, ensuring that the initial random numbers are sufficiently far from each other in the row. The FORTRAN statement for the generation of random numbers has to be coded every time in the vectorizable DO loop as shown in Fig. 2, since a statement of SUBROUTINE CALL is prohibited.

A. Vectorization for Molecular Motion

In a computer program for the DSMC method, the routine of the molecular motion has to be designed anew for each flow problem. The routine becomes complicated as the flow boundary becomes complex. Although the motion of each molecule is independent of the others, the vectorization for such a routine is prohibitively difficult due to the prohi-

```
C-------Collection of data-------
C   M and KK( ) are working
C   variables.
C   IRYO(N) contains a certain
C   condition.
C
      M=0
      DO 10 N=1,NM
      IF(IRYO(N).LE.100) THEN
      M=M+1
      KK(M)=N
      ENDIF
   10 CONTINUE
C
      DO 20 J=1,M
      N=KK(J)
          -
          -
          -
   20 CONTINUE
C-------------------------------
```

Fig. 1 Example of data collection.

bition of a backing GOTO statement and a too complicated IF statement in a DO loop. A program in the present study is intended to apply to the transition or the near-continuum regime where the DSMC method costs a lot of CPU time. Since the distance of movement of a molecule during a time step dt_m is not very long in this case, where dt_m is about one-fifth of the mean free time, the number of molecules that interact with a flow boundary within the time step is not very large. The speedup of this routine is realized as follows. First, every molecule moves along a straight path in the physical space during dt_m without regard to the boundaries. After that, the molecules that interact with one of the boundaries or pass out the region are collected. (The procedure is called the data collection.) These two steps of calculation are vectorized. Since the number of the collected molecules is not very large, especially in the near-continuum regime, the recalculation for the correct motions of these molecules does not consume very much computation time, even in the scalar processing.

B. Vectorization for Cell Identification

The cell in which a particular molecule is located is identified rapidly using the method of the coordinate transformations.[1] The method transforms an irregular (unequal cell volumes) physical domain to a regular computational domain in which all cells are equally spaced. At the downstream of the slit in the present simulation, a dimension of a rectangular cell i is varied as a function of the cell position:

$$x_i = ab^{i-1} \tag{1}$$

where x_i is a cell dimension in x direction, i is a cell number assigned in the order of its location, and a and b are constants. The sum of the dimensions up to cell n is

$$\sum_{i=1}^{n} x_i = \frac{a(b^n-1)}{b-1} \tag{2}$$

Therefore, the cell number in which a molecule located at the position of x is involved becomes

$$n = \frac{\log[\frac{x(b-1)}{a} + 1]}{\log b} + 1 \quad \text{(round down)} \tag{3}$$

```
c--------Selection of collision pair-----------
c  IRA(N) is random number (integer).
c  RAND is random fraction ( 0<RAND<1 ).
c  IC1(N) is the number of molecules in cell N.
c  IC2(N) is "Starting address - 1".
c  LCR( ) is cross reference array.
c  L and M are two molecular numbers.
c
   32 IRA(N)=IAND(IRA(N)*48828125,2147483647)
      RAND=IRA(N)/2147483648.
      K=RAND*IC1(N)+IC2(N)+0.9999999
      IF (K.EQ.IC2(N)) K=K+1
      L=LCR(K)
c
   33 IRA(N)=IAND(IRA(N)*48828125,2147483647)
      RAND=IRA(N)/2147483648.
      K2=RAND*(IC1(N)-1)+IC2(N)+0.9999999
      IF (K2.EQ.IC2(N)) K2=K2+1
      IF (K2.EQ.K) K2=IC1(N)+IC2(N)
      M=LCR(K2)
c----------------------------------------------
```

Fig. 2 Selection of collision partners with random numbers.

```
c-------- Molecular Indexing ----------
c   NC is total number of calls.
c   NM is total number of molecules.
c   IP(M) is the cell number in which
c   molecule M is located.
c   IC1(N) is the number of molecules
c   in cell N.
c   IC2(N) is "starting address - 1".
c   IDAM is a dummy variable.
c   IW( ) and KK( ) are working arrays.
c
      DO 4 N=1,NC
    4 IC1(N)=0
c
*VOPTION VEC
      DO 5 M=1,NM
      N=IP(M)
    5 IC1(N)=IC1(N)+1
c
      DO 15 N=1,NC
      IF(IC1(N).LE.10) THEN
      IC1(N)=20
      ELSE
      IC1(N)=IC1(N)*2
      ENDIF
   15 CONTINUE
c
  100 L=0
      DO 6 N=1,NC
      IC2(N)=L
      L=L+IC1(N)
    6 IC1(N)=0
c
      NM2=L
      DO 26 K=1,NM2
   26 LCR(K)=0
c
*VOPTION VEC
      DO 27 M=1,NM
      IW(M)=1
      N=IP(M)
      IDAM=IC1(N)+1
      K=IDAM+IC2(N)
      IC1(N)=IDAM
   27 LCR(K)=M
c
      DO 28 K=1,NM2
      M=LCR(K)
      IF(M.EQ.0) GO TO 28
      IW(M)=0
   28 CONTINUE
c
      L=0
      DO 29 M=1,NM
      IF(IW(M).EQ.0) GO TO 29
      L=L+1
      KK(L)=M
   29 CONTINUE
c
      DO 30 N99=1,L
      M=KK(N99)
      N=IP(M)
      IC1(N)=IC1(N)+1
      K=IC1(N)+IC2(N)
   30 LCR(K)=M
c----------------------------
```

Fig. 3 Routine of molecular indexing using forced vectorization with appropriate postprocessing.

The routine of the cell identification is vectorized easily by this coordinate transformation.

C. Vectorization for Molecular Indexing

The routine of the molecular indexing includes two DO loops over all molecules that cannot be vectorized because of the mutual dependency of list vector data (indirectly accessed data). The statement IC1(N)=IC1(N)+1 within the DO 5 loop in Fig. 3 counts the molecules in each cell. In the vector processing there is a strong possibility that each cyclic run of the DO loop use the same value of IC1(N), though each run of the DO loop in the scalar processing has different value of IC1(N). For this reason, if the DO 5 loop is forced to be calculated in the vector processing, errors in counting occur and the number of molecules in a cell becomes smaller than the correct value. Figure 4 shows the dependence of the ratio of the number of the errors in counting to the total number of molecules on how to number the molecules, where the number of cells is 2,500 and the total number of molecules is 50,000. When the molecules are numbered in the order of the cell in which each molecule is located, the error amounts to 95%. However, the error decreases as the randomness on the numbering of these molecules increases. If the molecules are numbered at random, the error ratio is reduced to only 5%. Figure 4 also shows that the random numbering makes it possible to speedup the calculation.

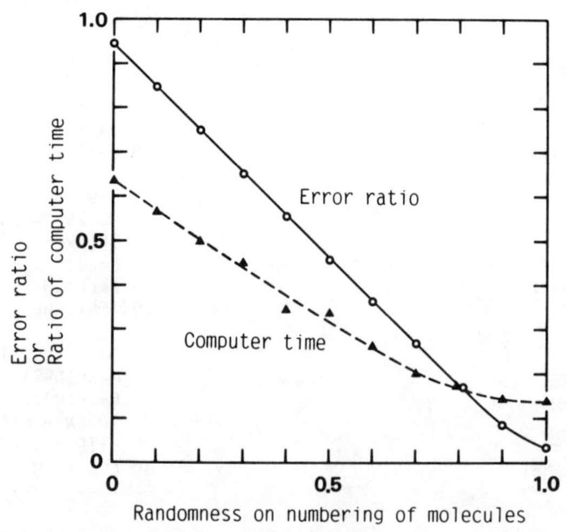

Fig. 4 Ratio of computer time for DO 5 with forced vectorization and error ratio in counting molecules vary with how to number molecules.

The error ratio in the random numbering can be estimated approximately. Consider that the calculation is executed with r parallel processors. The number of cells is n and the total number of molecules is mn, where m is the average number of molecules in a cell. To simplify the discussion, let mn be a multiple of r. Then, a parallel calculation is executed mn/r times. The probability that k molecules in a particular cell are included among the molecules processed by one parallel calculation by r parallel processors is written by

$$ {}_rC_k\left(\frac{1}{n}\right)^k\left(\frac{n-1}{n}\right)^{r-k}, \quad 0 \leqq k \leqq r $$

Since the error in counting occurs k-1 times for K>1, the total of the errors taken over all cells sums up to

$$ n\left\{\sum_{k=2}^{r}(k-1)\,{}_rC_k\left(\frac{1}{n}\right)^k\left(\frac{n-1}{n}\right)^{r-k}\right\} $$

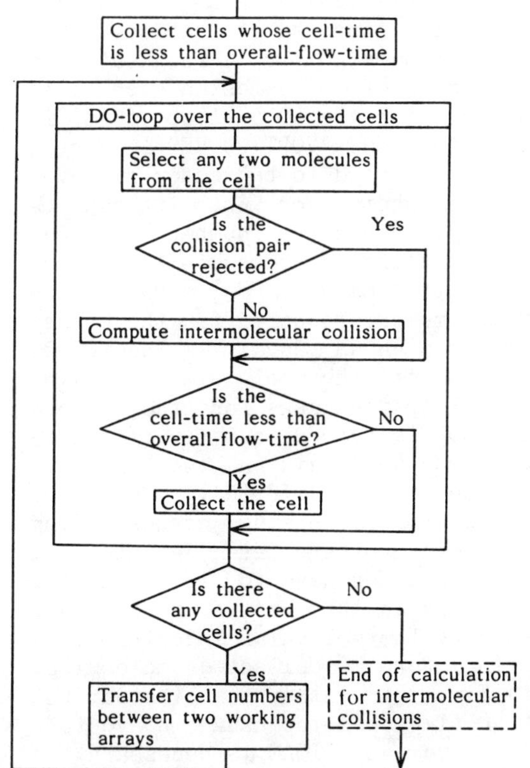

Fig. 5 Vectorization for intermolecular collision.

Each calculation by r parallel processors is independent of one another in the random numbering of molecules. Therefore, the rate of the errors p is given by

$$p=\frac{n}{r}\left\{\sum_{k=2}^{r}(k-1)_r C_k\left(\frac{1}{n}\right)^k\left(\frac{n-1}{n}\right)^{r-k}\right\} \quad (4)$$

Equation (4) is compared with the error actually incurred by the HITAC S810-10 computer in which 256 calculations are executed in a parallel manner. The value predicted by Eq. (4) gives a good approximation of the error that actually occurred. Furthermore, since it is common that n is much larger than r, or n>>r>>1, Eq. (4) is simplified to

$$p = r/2n \quad (5)$$

In practice, if every field molecule is numbered at random initially and is independent of the cell in which the molecule is located, and if molecules flowing into the domain during the calculation are also numbered at random, then every molecule has a nearly random number throughout the calculation. First, the number of molecules in each cell is calculated approximately using the DO 5 loop that is forced to be vectorized (see the code *VOPTION VEC in Fig. 3 which means that the DO loop is forced to be vectorized even if the compiler cannot judge whether or not it can be vectorized without an error). Since the number is then doubled in the DO 15 loop, the starting address - 1 of each cell IC2(N) is determined with a margin for error, even though the amount of computer memory required for the cross-reference array is twice as much as usual. The DO 27 (Fig. 3) for the calculation of the molecular indexing information is the other DO loop that asks for the forced vectorization. If the DO loop is forced to be vectorized, the undercounting of IC1(N) occurs again and two or more molecules have the same value of IC1(N)+IC2(N). Only one molecule calculated last is stored in the corresponding cross-reference array, and the others are eliminated. Therefore, the eliminated molecules have to be collected using the DO 28 and DO 29 loops in the vector processing. The calculation is completed by the molecular indexing routine DO 30 (scalar processing) for the collected molecules. The whole program shown in Fig. 3 is rather complicated, but it requires less calculation time than the usual nonvectorized one. Note that a dummy variable IDAM is necessary.

D. Vectorization for Intermolecular Collision

Although the calculation for intermolecular collision is executed within each cell independently of molecules in other cells, the procedure in which intermolecular collision is repeated until the cell time exceeds the overall flow time requires a backing GOTO statement and cannot be easily vectorized. In the present program, therefore, only one collision is computed in each cell whose cell time is less than the overall flow time. The calculation for selection of two molecules participating in collision is executed according to the computer code shown in Fig. 2, which can avoid the selection of same molecules. (Bird's original code requires a backing GOTO statement.[6]) Since the decision regarding whether or not the molecules selected temporarily collide with each other depends on the relative velocity between the molecules, the temporary pair of molecules may be or may not be rejected. In the case of rejection, the intermolecular collision in the cell is not calculated in the cyclic run of the DO loop about the cell. After one collision per cell is computed except a cell with the rejection, the cells that have still enough time for intermolecular collision are collected. These vectorized procedures are continued until all cell times exceed the overall flow time (see Fig. 5).

III. Simulation Domain and Network of Cells

The simulated domain for flow through a two-dimensional slit is similar to that used previously in the authors' earlier papers.[4,5] It has been proved in those papers that the introduction of the stream velocity normal to the upstream boundary to the equilibrium velocity distribution of inward molecules is effective in correcting the error due to the finiteness of the computational domain. Figure 6 shows the three regions of the domain in the condition 1/Kn=70 with Knudsen number Kn defined by

Kn = upstream mean free path /(slit width x 2)

For large Knudsen numbers the upstream region is enlarged. The shape of cells in the upstream region and in the inner region is a square, and that in the downstream is rectangular, as described earlier. A perfect vacuum is assumed outside of the downstream boundary. The ratio of the slit length t to the slit width s is t/s=1.5. The area of rough surface on the slit wall is restricted to the inner area facing the inner region. The V-shaped grooves (triangular grooves) are substituted for the surface

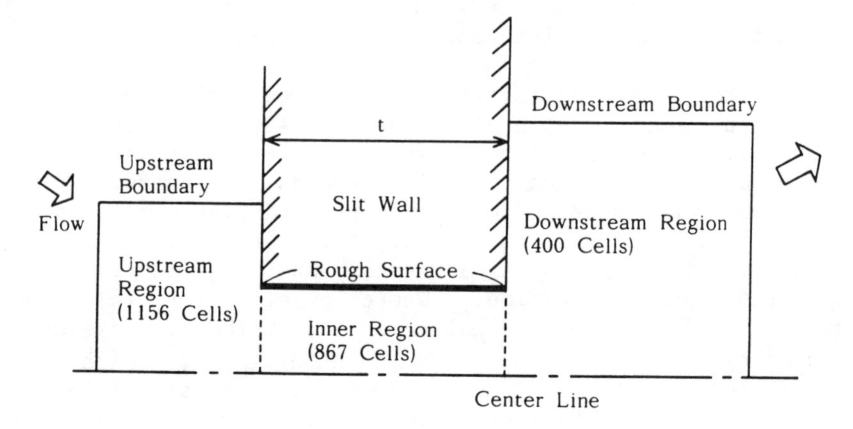

Fig. 6 Computational domain with three regions. (1/Kn = 70)

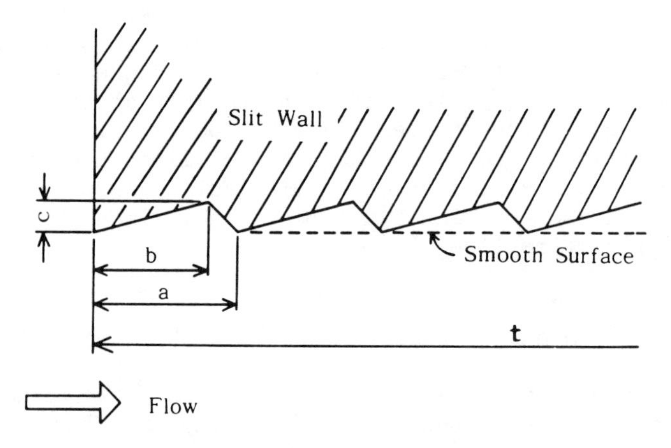

Fig. 7 Triangular grooves substituted for surface roughness.

roughness as shown in Fig. 7. The shape of the triangular groove is varied according to the sizes of a, b, and c.

IV. Results and Discussion

Table 1 shows the comparison between the CPU time for the simulation (1/Kn=70) using the vectorized program and that using the conventional one. The HITAC S810-10 computer is used. Speedup is achieved in every routine, and the speed of calculation is about eight times faster in total. Though the CPU time of full calculation for 1/Kn=70 is less than 5 min, the calculation for 1/Kn=300 requires about 60 min even in the vector processing. Since the calculation for near-continuum regime requires a prohibitively large CPU time in the scalar processing, the vectorization of the DSMC method is essential.

A. Mass Flow Reduction due to the Surface Roughness

1. Dependence on the Upstream Pressure. Figures 8 and 9 indicate that the mass flow reduction due to the surface roughness depends on the pressure of the upstream boundary. Here, three types of the triangular groove are investigated: 1) c/a=0 (smooth surface); 2) a/t=1/15, b/a=0.1, c/a=0.1; and 3) a/t=1/30, b/a=0.5, c/a=0.2. Figure 8 presents the conductance F normalized by the theoretical free molecular conductance of a sharp-edged orifice (t=0), F_{fmo}. The figure also presents the experimental result of the machined

Table 1 Comparison between CPU time of vectorized program and that of conventional one (1/Kn=70)

	Conventional, s	Vectorized, s	Ratio
Molecular motion	272.3	42.7	6.38
Cell identification	503.4	54.2	9.29
Molecular indexing	154.7	36.8	4.20
Intermolecular collision	521.1	40.8	12.8
Sum total	1451.5	174.5	8.32

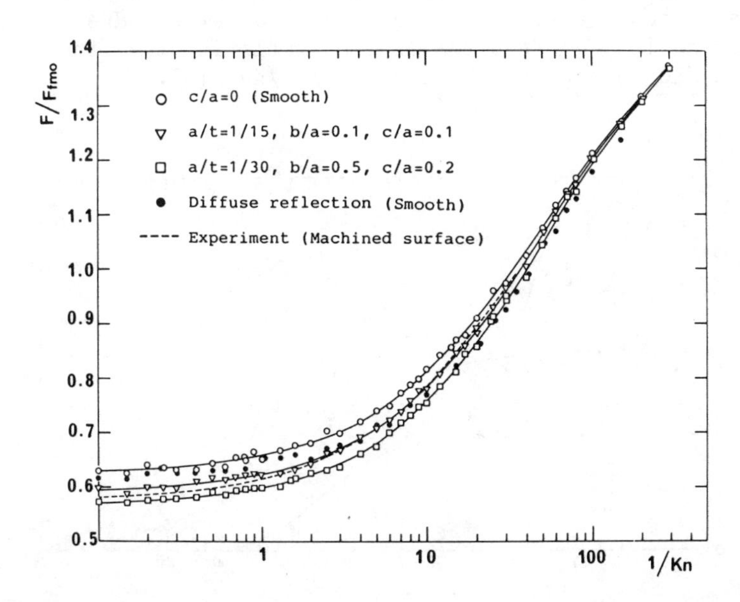

Fig. 8 Influence of surface roughness on conductance. (Pressure dependence)

surface. However, its detailed discussion is postponed until the Sect. IV B. Figure 9 presents the ratio (percentage) of the mass flow reduction to the mass flow through the slit with smooth surface. The fluctuation of the ratio is very large, since the result of the conductance itself includes statistical fluctuations of about 1% due to the limited amount of sampling in this calculation. Nevertheless, it can be seen that the mass flow reduction due to the surface roughness obviously depends on the upstream Knudsen number. These two figures indicate that the mass flow reduction is remarkable in the free molecular regime and decreases gradually as the upstream pressure increases ($1/Kn=5 \sim 50$) and that the surface roughness can hardly affect the mass flow when $1/Kn > 100$. The mass flow reduction of the type 2 surface roughness is less than that of the type 3 roughness in these figures, reflecting the difference of c/a. The effect of the configuration of triangular grooves is described in the next paragraph. The accommodation coefficient of the surface is assumed to be 0.93 throughout the slit wall, which has been chosen in order to ensure a good agreement between the experiment and the calculation as shown by the authors in previous papers.[4,5] Figure 8 also presents the result for the smooth surface with the accommodation coefficient 1.0 (diffuse reflection). The effect of the accommodation coefficient on mass flow is quite different from that of the surface roughness. The mass flow variation due to the accommodation coefficient is remarkable in the transition regime and is appreciable even when $1/Kn > 100$.

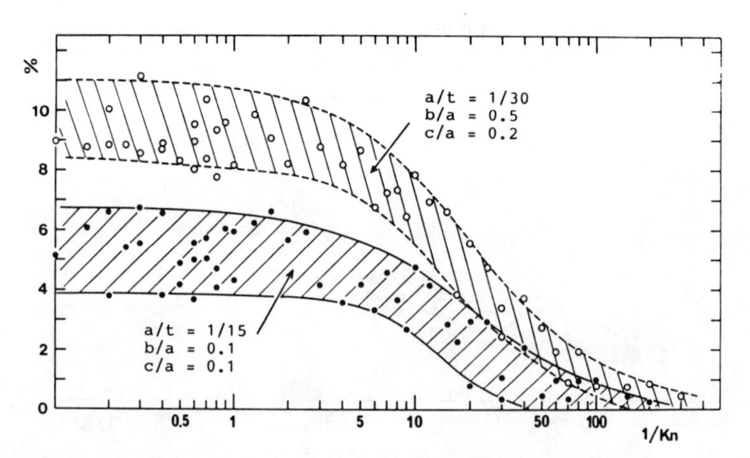

Fig. 9 Ratio of mass flow reduction to mass flow through slit with smooth surface.

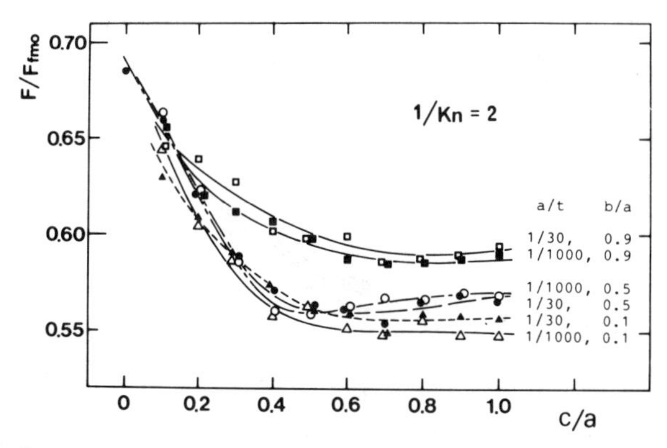

Fig. 10 Influence of configuration of surface roughness on conductance. (1/Kn = 2)

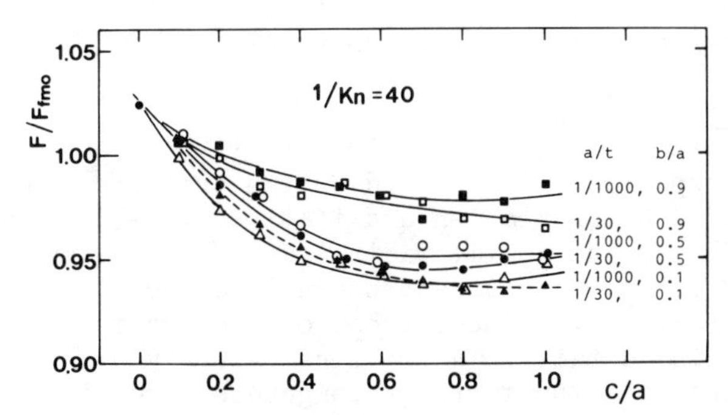

Fig. 11 Influence of configuration of surface roughness on conductance. (1/Kn = 40)

2. Effect of the Configuration of Triangular Grooves. Figures 10 and 11 present the dependence of the normalized conductance on the configuration of the triangular grooves. The numerical values of a/t are 30 and 1000, and those of b/a are 0.1, 0.5, and 0.9. On the other hand, the value of c/a varies at intervals of 0.1 between 0.1 and 1.0. Figures 10 and 11 correspond to 1/Kn=2 and 1/Kn=40, respectively. Though the variation of conductance is greater when 1/Kn=2, both results show a similar tendency. As c/a increases, the conductance decreases, but the rate of mass flow reduction becomes low. The conductance seems to converge to a constant value at the large value of c/a. Regarding the asymmetry of a triangular groove, the mass flow reduction for b/a=0.1 is greater than that for b/a=0.9, though the difference of the mass flow reduction between b/a=0.1 and 0.5 is not as large. Regarding the number of

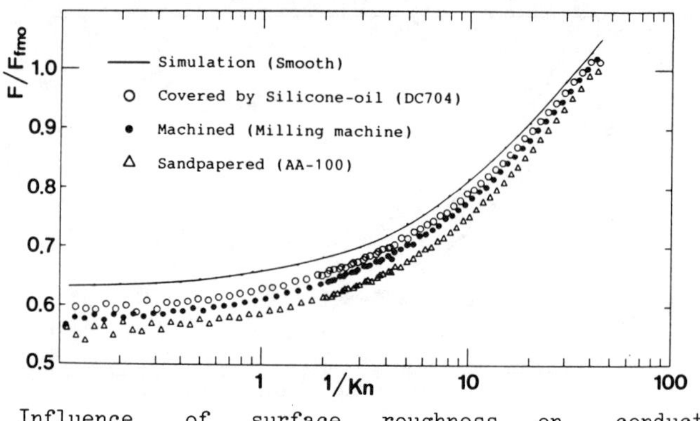

Fig. 12 Influence of surface roughness on conductance. (Experiment)

grooves (fineness of grooves, a/t), the distinct effect cannot be detected in the present calculation. It seems the fineness cannot affect the mass flow.

B. Experimental Results

The experiment is carried out in order to confirm the results of the simulation qualitatively. The test slit made of brass has a slit width (gap) of 1.04 mm, a slit length of 1.5 mm, and an aspect ratio of 29.5. Argon is used as test gas. The methods of measurement has been described in detail in a previous paper by the authors.[7] Three types of the surface roughness are investigated: 1) surface machined by a milling machine, 2) surface striped by sandpaper AA100 at a right angle to flow on the type 1 surface, and 3) surface covered by silicone-oil DC704 over the type 2 surface.

Figure 12 shows that the mass flow through the striped slit is less than that through the machined slit. The mass flow through the slit covered by silicone-oil is greater than that through the machined slit, since the oil covers the striped surface and makes it smooth. The figure also includes the results of the simulation for smooth surface. Although each experimental result for three types of surface roughness is less than the simulated result for smooth surface in Fig. 12, the substitution of triangular grooves for surface roughness in simulation permits the qualitative prediction of the mass flow reduction as shown in Fig. 8.

V. Concluding Remarks

Mass flow reduction of rarefied gas through the two-dimensional slit due to roughness of the slit surface is investigated numerically using the DSMC method with vectorization. The following conclusions are obtained:

1) The mass flow reduction is remarkable in the free molecular regime and decreases gradually as the upstream pressure increases. The surface roughness can hardly affect the mass flow when 1/Kn >100.

2) When the V-shaped grooves (triangular grooves) are substituted for the surface roughness as shown in Fig. 7, the conductance decreases as c/a increases and seems to converge to a constant value at the large value of c/a. Regarding the asymmetry of a triangular groove, the mass flow reduction for b/a=0.1 is greater than that for b/a=0.9. It seems that the fineness of the grooves does not affect the mass flow.

3) Speedup due to the vectorization is achieved in every routine of the DSMC computer program, and the speed of calculation is eight times faster in total.

Acknowledgments

This research was partly supported by the Science Funds of the Ministry of Education (No. 63550134).

References

[1]Merkle, C. L., "New Possibilities and Applications of Monte-Carlo Methods," Proceedings of the 13th International Symposium on Rarefied Gas Dynamics, Vol. 1, edited by O. M. Belotserkovskii, M. N. Kogan, S. S. Kutatecandze, and A. K. Rebrov, Plenum, New York, 1985, pp. 333-348.

[2]Davis, D. H., Levenson, L. L., and Milleron, N., "Effect of Rougher-than-Rough Surfaces on Molecular Flow through Short Ducts," Journal of Applied Physics, Vol. 35, No. 3, Mar. 1964, pp. 529-532.

[3]Berman, A. S. and Maegley, W. J., "Internal Rarefied Gas Flows with Backscattering," Physics of Fluids, Vol. 15, May 1972, pp. 772-779.

[4]Usami, M., Fujimoto, T., and Kato, S., "Direct Monte-Carlo Simulation on Rarefied Gas Flow," Research Reports of the Faculty of Engineering, Mie University, Vol. 9, Dec. 1984, pp. 1-8.

[5]Usami, M., Fujimoto, T., and Kato, S., "Thermal Effect of Orifice Plate on Mass-Flow of Rarefied Gas," Research Reports of the Faculty of Engineering, Mie University, Vol. 11, Dec. 1986, pp. 13-21.

[6]Bird, G. A., Molecular Gas Dynamics, Clarendon, Oxford, UK, 1976.

[7]Fujimoto, T. and Usami, M., "Rarefied Gas Flow Through a Circular Orifice and Short Tubes," Transactions of the ASME, Journal of Fluids Engineering, Vol. 106, Dec. 1984, pp. 367-373.

Chapter 6. Expansion Flowfields

Translational Nonequilibrium Effects in Expansion Flows of Argon

David H. Campbell*
University of Dayton Research Institute, Air Force Astronautics Laboratory, Edwards Air Force Base, California

Abstract

The structure of the flow of argon through a 2-cm-long, 4-mm-diam. tube, around the tube lip and into vacuum, has been investigated using the direct simulation Monte Carlo technique. Two lip geometries have been compared and the shape and thickness of the lip were found to have a significant influence on the flowfield. The number density of the gas has a significant effect on the backflow due to changes in the collisional processes in the expansion flow, which affects the freezing out of the parallel component of the random motion of the gas. Consequently, the backflow flux does not scale directly with the starting number density.

Introduction

The expansion of gases out of a rocket nozzle into a low-density background is an important problem with application to spacecraft contamination, heat transfer, and infrared sensor interference. The expansion of gases around the nozzle lip and into the region upstream of the nozzle exit plane, usually referred to as the backflow region, is a particularly complex gas dynamic problem. It has been recognized for many years[1-6] that the expansion of the subsonic and lower Mach number supersonic regions of the boundary layer around a nozzle lip can produce significant flux into the high angle backflow region not predicted by the conventional Prandtl-Meyer uniform supersonic flow analysis. It has recently been recognized that nonequilibrium effects,[7] due to the rapid rarefaction of the flow out of a nozzle exiting to vacuum or near vacuum, can affect the structure of the flow around a nozzle lip. Consequently, accurate modeling of the flowfield around a nozzle lip and into the high angle backflow region cannot be obtained using standard equilibrium gasdynamic models. The direct simulation Monte Carlo (DSMC) modeling technique,[8] which models a gas flow by following some representative number of molecules

---

*Research Physicist, AFAL On-Site Group.

through simulated collisions, does account for nonequilibrium effects, and is essentially the only technique presently available for accurate prediction of these flows.

Some of the results of a detailed DSMC theoretical analysis of the effects of lip shape and lip thickness on the flowfield around the lip of a nozzle with zero half-angle (tube) will be presented in this paper. In particular, the effects of upstream gas density will be described in detail. As will be seen, the degree of translational mode nonequilibrium in the gas expansion, which is dependent on the gas density, can significantly affect the structure and degree of backflow. In addition, the nonequilibrium processes interact with the different lip shapes and thicknesses in a nonsimple manner. Other results not discussed in this paper have been presented in previous publications[9,10] and will be discussed further in a forthcoming article.[11]

Model Description

To investigate the effects of nozzle lip shape and thickness, the DSMC technique was used to map the flow of argon into vacuum through a 2-cm-long tube of 2 mm radius for two lip shapes (Fig. 1) and for wall thicknesses from 0 to 2.0 mm. Because of the extremely large change in gas number density from the internal flow to the far backflow region, the calculations are conducted in four steps. The first uses a finite-difference Navier-Stokes solving code (VNAP2)[12] to calculate the internal flow to a position 2 mm upstream of the tube exit plane. The Monte Carlo code is then used to calculate the flow from that position to a position 0.15 mm upstream of the tube exit plane. The influence of the tube wall thickness on the flow at this start line was checked and found to be negligible. The third

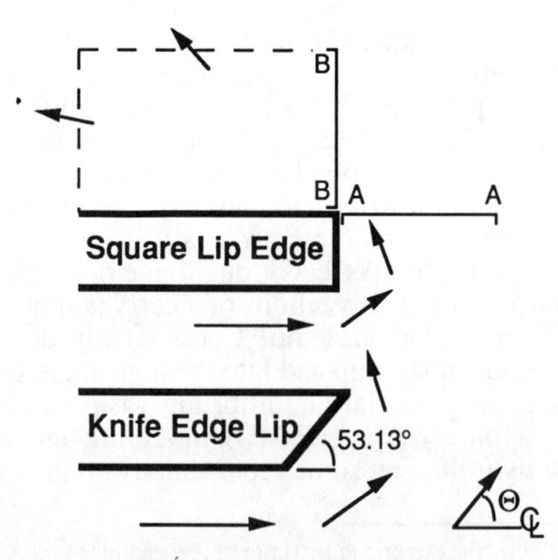

Fig. 1 Lip shapes. Flow angle convention is also shown. A-A and B-B are longitudinal and radial profiles presented in figures to follow. Dotted lines indicate boundary cells used for backflow flux calculation.

step is a full Monte Carlo calculation from the start line 0.15 mm inside the tube out into the forward flow and around the lip into the backflow region. The run is allowed to proceed until good statistics are obtained in the region directly in front of the lip, but not long enough to produce good statistics in the backflow region. The final step is a calculation of the flow from a horizontal line along the inner edge of the lip out into the backflow region, using the results of step three for the input start conditions at that line. Again, the influence of the lip thickness on the start line conditions was checked and was found to be negligible. Consequently, the same start line was used for all lip thicknesses.

The cell structure used in the calculations was the final configuration reached after a number of iterations. The mean free path must be at least two to three times greater than the cell size to avoid flow smearing, and the mean collision time must be two to three times greater than the time step for the calculation to track changes in the flow parameters accurately. The cells were adjusted so that the flowfield and cell structure met or exceeded these criteria in all cells except for those cells far from the lip in the forward flow region, where the mean free path was slightly less than the cell diameter. The flowfield is slowly varying in those regions, so that no flow smearing is expected due to these low ratios.

The effects of random number seed value, cell number and size (within the previously mentioned limits), and number of simulated atoms were investigated and were found to be small in all cases. With respect to the effects of the number of simulated atoms, that number was kept approximately the same for all runs (about 7000 for the last DSMC step as described above).

A stagnation condition of T=300° K and P=1 kPa (7.5 torr) was the standard condition used in the calculations. Diffuse reflection (full accommodation) of gas particles from the tube walls was used for all gas/wall interactions. A strong boundary layer develops for these conditions (Re ~500 at tube end) and fills a large portion of the tube at the exit plane.

Results and Discussion

One measure of the effect of lip shape and thickness is the total amount of flux into the backflow region. In this paper the total backflow flux will be defined as the angular flux (number density x velocity x subtended angle) calculated for each outer boundary cell summed for those cells with flow angles of 90 deg or greater. In Fig. 2 the relative flux of argon into the backflow region is shown. Both tube lip thickness and tube lip shape have pronounced effects on the backflow flux. From these results it can be concluded that the scattering off of the front face of the lip has a significant effect on the flux of gas into the backflow region in this flow regime. Thicker lips produce more forward scattering, i.e., flow angles directed more toward the forward flow direction, which reduces the number of atoms scattered into the backflow. The knife-edge lip shape scatters more atoms into the forward flow direction than the square-edge lip, and thus produces less backflow flux. The ratio of the backflow flux

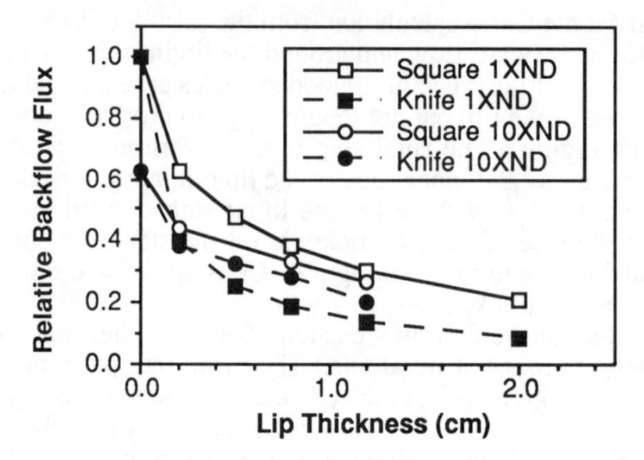

Fig. 2 Normalized flux into backflow region. Dependence on lip shape, lip thickness, and starting gas number density. Flux is normalized by number density multiplier.

between the square and knife-edge lips is shown in Fig. 3. The ratio increases with increasing lip thickness and appears to be leveling off at a value of around 2.5 for this baseline condition.

To investigate the effects of the starting number density of the gas on the backflow structure, the number density at the startline 0.15 mm upstream of the exit plane of the tube was increased by a factor of 10, and the calculations were repeated for various lip thicknesses for the knife-edge and square-edge lips. The boundary layer profile was kept the same as before, with a multiplier used to change the numeric value of the number density, keeping all of the other gas flow parameters the same. A full region run was conducted in each case to get a new start line at the inner edge of the lip, and a lip region/backflow calculation was then run. The total flux into the backflow for the baseline case and for a number density a factor of 10 higher than the baseline case (10XND) is compared in Figs. 2 and 4. The calculated number density of the flow for the higher number density cases have been normalized by the number density multiplication factor (10).

It is apparent that the backflow flux does not scale directly with the starting number density. Specifically, as shown in Figs. 2 and 4, the normalized flux drops about 40% for the higher number density for the 0.0-mm lip thickness. For the square-shaped lip, the normalized flux for the higher-number-density case approaches the values for the lower number density as the lip thickness increases. For the knife-edge lip, the normalized flux for the higher number density is very close to that for the lower density case for a lip thickness of 0.2 mm, and then is higher than that for the lower density case for larger lip thicknesses. Additionally, the difference in the normalized flux for the knife-edge lip compared to the square-edge lip is smaller for the higher number density case than for the lower number density case (Fig. 3).

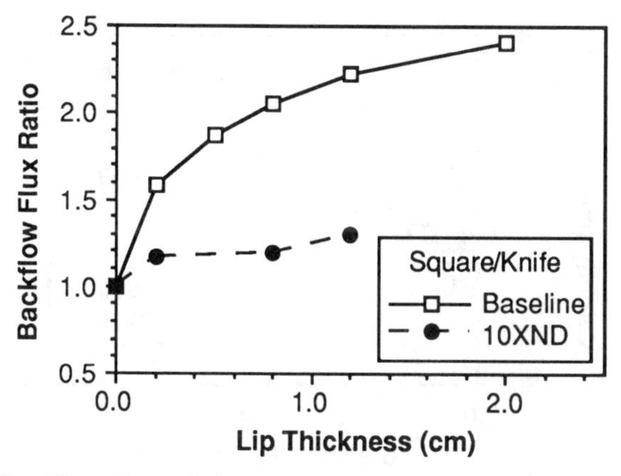

Fig. 3 Backflow flux ratio between square-edge and knife-edge lip shapes.

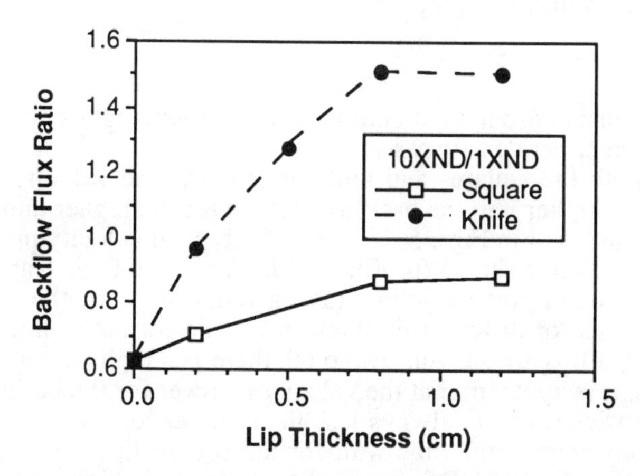

Fig. 4 Backflow flux ratio between 10XND starting condition and baseline condition. Flux is normalized by number density multiplier.

There will be an increase in the total number of collisions when there are more gas particles in the flow, and this change in the collisional process can alter the velocity and angle profiles into the backflow region and, in addition, can change the position in the flow where the breakdown of translational equilibrium occurs.[13] A delay in the decoupling of the parallel translational mode from the perpendicular mode would be expected to increase the directed velocity of the flow, since this delay would allow more energy to be available for the expansion of the gas before energy is frozen into the parallel mode and is no longer available for conversion to directed motion of the gas. Bird's Breakdown Parameter[7] exceeds the value of 0.05 (usually assumed to define the onset of translational nonequilibrium) inside the tube upstream of the tube exit for the lower

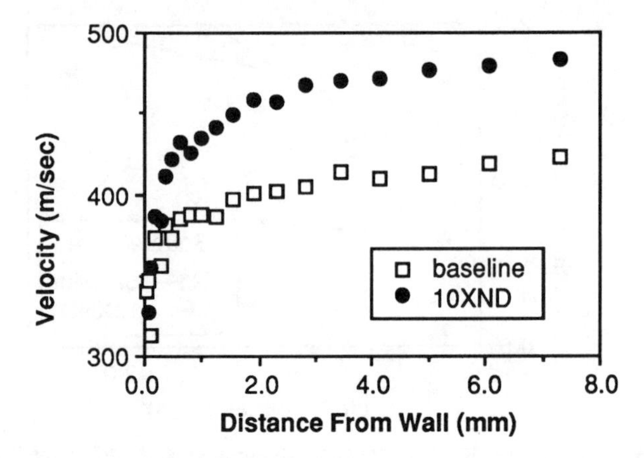

Fig. 5 Velocity profile along B-B for 0.8-mm-thick square lip. Comparison of baseline and 10XND starting conditions.

number density case but not until well into the external expansion flow for the higher number density case.

In both the square- and knife-edge cases the velocity profile was found to be higher into the backflow region for the higher number density case, as illustrated in Figs. 5-7. In Fig. 5 a typical velocity profile into the backflow region is shown (profile B-B in Fig. 1). In Figs. 6 and 7 the ratio of the point-to-point velocities (along B-B) between the baseline and 10XND cases for different lip thicknesses and for the square- and knife-edge-shaped lips are shown. Although there is a high degree of scatter in the ratios, it is apparent that the velocity is lower for the baseline case for all geometries (ratio is always less than one except for one set of cells directly adjacent to the tube wall for the square lip). Additionally, the velocities for the 10XND cases are proportionately higher than the corresponding baseline velocities as the lip thickness increases, with the knife-edge lip showing a stronger propensity for this trend.

The normalized number density profiles into the backflow region (along B-B) show a more complex behavior than the velocity profiles, as shown in Figs. 8-10. For the square-edge lip, the normalized number densities are always higher for the baseline case, and the ratio between the normalized number densities for the baseline and 10XND cases increases with decreasing lip thickness. Consequently, the increased velocities (with increasing starting number density) are not enough to change the trend set by the number density profile, which is to reduce the normalized flux with increased starting number density. For the knife-edge lip, the normalized number densities for the baseline case are higher than those for the 10XND case for the thin lips but are lower for the thicker lips. As a result, the normalized flux for the knife-edge lip is smaller for the higher starting number density case for thin lips, is about the same as the baseline condition for a lip thickness of 0.2 mm, and then is higher than the baseline

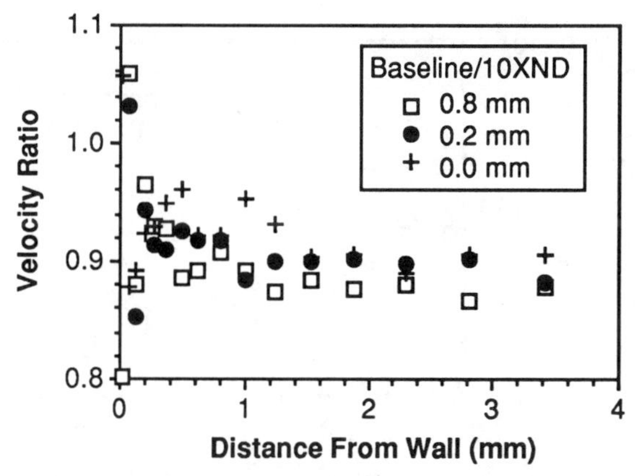

Fig. 6 Velocity ratio profile along B-B for 0.8-mm-thick square-edge lip. Ratio between baseline and 10XND conditions.

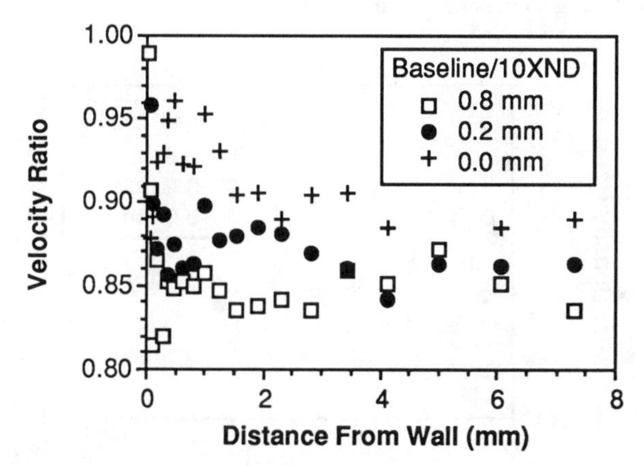

Fig. 7 Velocity ratio profile along B-B for 0.8-mm-thick knife-edge lip. Ratio between baseline and 10XND conditions.

condition for the larger lip thicknesses (Fig. 4). The normalized flux for the higher starting number density appears to level out at a value 50% higher than the flux for the baseline condition for the thickest lips used in this analysis. This is in contrast to the square lip results, where the normalized fluxes for the 10XND cases are all less than the baseline case, and where the difference in normalized flux decreases with increasing lip thickness (Fig. 4).

For both high starting number density cases (square and knife-edge) the increase in velocity is a direct indication that the translational mode is in equilibrium longer than for the higher starting number density cases. The pattern of change in the normalized number density is another more complicated matter. The angle profiles in the backflow region are similar

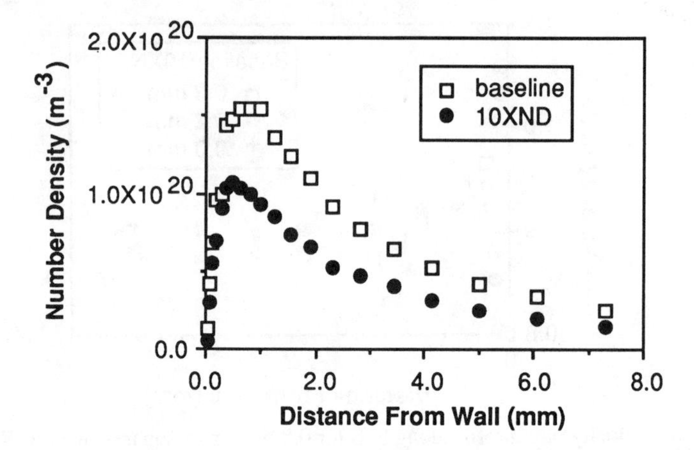

Fig. 8 Number density profile along B-B for 0.8-mm-thick square-edge lip. Comparison of baseline and 10XND starting conditions. Number density is normalized by starting number density multiplier.

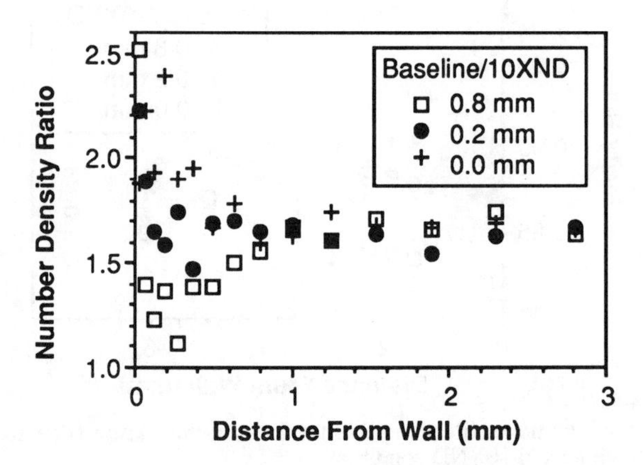

Fig. 9 Number density ratio profile along B-B for 0.8-mm-thick square-edge lip. Ratio between baseline and 10XND conditions. Number density is normalized by starting number density multiplier.

for the two lip shapes (slightly higher angle for higher number density), thus the flow angle in this region does not yield any information about the difference in number density behavior. The horizontal profiles of flow angle in the top row of cells in front of the lip (A-A in Fig. 1), do not change with number density for the square lip and change only a few degrees (and only in a region near the 1 mm position) for the knife-edge lip, hence once again the angle profile does not give a significant clue concerning the backflow behavior. For the higher starting number density case, the temperature profiles in front of the lip (A-A) show a slightly

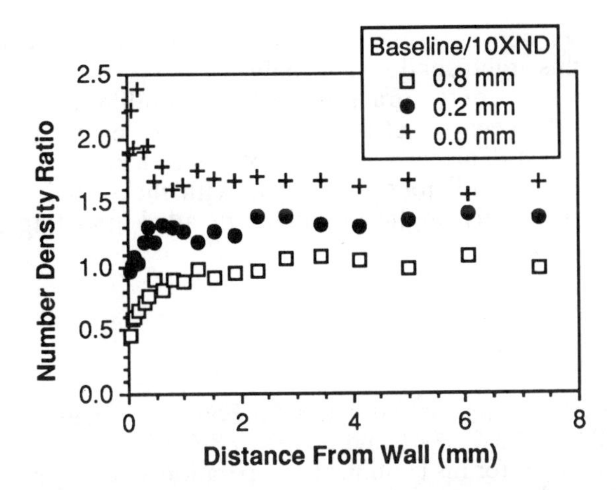

Fig. 10 Number density ratio profile along B-B for 0.8-mm-thick knife-edge lip. Ratio between baseline and 10XND conditions. Number density is normalized by starting number density multiplier.

increased temperature near the face of the lip for the 0.8-mm knife-edge and a similar but weaker trend for the 0.8 mm square lip. For the 0.2-mm lip the temperature was lower for the square-edge lip and approximately the same for the the knife-edge lip near the lip face. The gas is accommodating more to the surface for thicker lips and more for the knife-edge than for the square-edge. For the 10XND case, the normalized number density profile in the lip region for the knife-edge lip shows a significantly lower normalized number density within 0.6 mm of the wall compared to the values for the lower starting number density case, whereas the normalized number density is only slightly lower in that region for the square lip (for all lip thicknesses). For the 10XND case, the velocity profiles are essentially unchanged in the lip region for the knife-edge lip, and for the square-edge lip the velocities are slightly higher for the higher-number-density case.

The correlation of the flow parameters in the region in front of the lip to the backflow number density is not clear. Certainly the lower normalized number density in front of the lip for the 0.8-mm knife-edge lip for the high starting number density is somewhat curious, since the normalized number density is higher in the backflow for this case (Fig. 10). The molecules are "piling up" in front of the lip more for the lower starting number density case. This may be related to the degree of non-equilibrium in the translational mode and to the difference in accommodation to the surface. The relationship between the number density profile in front of the lip and the number density profile in the backflow is similar to that found when comparing the square-edge to knife-edge lip results; i.e., the number density is lower for the square-edge lip in front of the lip but larger in the backflow region.[9] In that case, the difference was explained by the larger angle in front of the lip for the square-edge. For the high vs low starting number density comparison, the angles do not change significantly, hence this argument does not apply.

The interaction of the (translationally nonequilibrium) flow with the wall results in the behavior of the number density being different for the different lip shapes. This interaction between the characteristics of the gas expansion and the collisions of the gas with the wall is obviously a complex one not easily predicted without detailed modeling of the type employed here.

Acknowledgment

This work was performed under Air Force Astronautics Laboratory Contracts F04611-83-C-0046 and F04611-88-C-0020. Special thanks go to Dr. Graeme Bird for his continued co-operation and guidance in the use of the DSMC code.

References

[1]Boynton, F. P., "Exhaust Plumes from Nozzles with Wall Boundary Layers," Journal of Spacecraft and Rockets, Vol. 5, No. 10, Oct. 1968, pp. 1143-1147.

[2]Seubold, J. G., "Use of Viscous Nozzle Flow Program and the Method of Characteristics Plume Program to Predict Plume Expansion in the Back Flow Region for Small Thrusters," JANNAF 9th Plume Technology Meeting, CPIA Publ. 277, April 1976, pp. 1-14.

[3]Smith, S. D., "Improvements in Rocket Engine Nozzle and High Altitude Plume Computations," AIAA Paper 83-1547, June 1983.

[4]Cooper, B. P. Jr., "Computational Scheme for Calculating the Plume Backflow Region," Journal of Spacecraft and Rockets, Vol 16, No. 4, July-Aug. 1979, pp. 284-286.

[5]Simons, G. A., "Effect of Nozzle Boundary Layers on Rocket Exhaust Plumes," AIAA Journal, Vol. 10, No. 11, Nov. 1972, pp. 1534-1535.

[6]Greenwood, T., Seymour, D., Prozan, R., and Ratliff, A., "Analysis of Liquid Rocket Engine Exhaust Plumes," Journal of Spacecraft and Rockets, Vol. 8, No. 2, Feb. 1971, pp. 123-128.

[7]Bird, G. A., "Breakdown of Continuum Flow in Freejets and Rocket Plumes," Progress in Astronautics and Aeronautics Volumn 74: Rarefied Gas Dynamics, edited by S. S. Fisher, AIAA, New York, 1981, pp. 681-694.

[8]Bird, G. A., Molecular Gas Dynamics, Clarenden, Oxford, UK, 1976.

[9]Campbell, D. H. and Weaver, D. P., "Nozzle Lip Effects on Gas Expansion into the Plume Backflow Region ," AIAA Paper 88-0748, AIAA 26th Aerospace Sciences Meeting, Reno, NV, Jan. 1988.

[10]Campbell, D. H. and Weaver, D. P., "Investigations of High Altitude Rocket Nozzle Lip Flow," JANNAF 17th Exhaust Plume Technology Meeting Proceedings, CPIA Publ. 487, April 1988, pp. 111-122.

[11]Campbell, D. H., "Nozzle Lip Effects on Argon Expansions into the Plume Backflow," Journal of Spacecraft and Rockets (accepted for publication), 1989.

[12]Cline, M. C., "VNAP2: A Computer Program for Computation of Two-Dimensional, Time-Dependent, Compressible, Turbulent Flow," Los Almos National Laboratory, Los Almos, NM, Rep. LA-8872, August, 1981.

[13]Bird, G. A., "Breakdown of Translational and Rotational Equilibrium in Gaseous Expansions," AIAA Journal, Vol. 8, No. 11, Nov. 1970, pp. 1998-2003.

Three-Dimensional Freejet Flow from a Finite Length Slit

A. Rosengard*
Commissariat à l'Energie Atomique, Centre d'Etudes Nucléaires de Saclay, France

Abstract

One or two-dimensional jets from an infinite slit taken as a sonic source have been yet examined. This article is devoted to practical three-dimensional situations when finite-length slit is considered, at distances comparable to this length, with slit source flow composed by adding elementary effusive flows. Analytical formulas are given for non-collisional flow and collisional flow simulated by a Monte Carlo particle test method. Densities, velocities and temperatures behavior are considered from noncollisional to continuous flow and approximate fitting formulas of code results proposed, which seem to give good agreement for classical cases (cylindrical and source flows).

Introduction

Many works are devoted to freejet expansion from an infinite slit; one-dimensional cylindrical geometry is mainly used with a sonic source, applying either kinetic theory methods[1] or Bird's Monte-Carlo method.[2] Two-dimensional flow has been taken in account,[3] by the method of characteristics (continuous flow) or Bird's method.

We wish to analyze monoatomic jet gas expansion from a slit, with small width D compared to this length L, and examine mean jet behavior evolution when the distance r from the source increases from two-dimensional situation ($r/L \ll 1$), to three-dimensional one ($r/L \sim 1$) and finally up to a point source flow ($r/L \gg 1$). Moreover, because we wish to analyze the evolution from molecular to continuous flow, the source flow is taken as a sum of elementary effusive flows, with uniform distribution on the rectangular slit surface.

* Engineer, Division d'Etudes de Sēparation Isotopique et de Chimie Physique.

Monte Carlo Calculations

Kogan's[4] iterative particle test method is applied to three-dimensional geometry (Fig. 1).

We briefly describe main points of our version of the test particle methods. Jet atoms are mass m and diameter d with one electronic state. We consider hard spheres collisions (elastic collisions and isotropic scattering from collision mass center). Initial slit conditions are effusive jets with uniform distribution on the slit surface, with temperature T_v and density n_v specified just below the slit (Fig. 2). The n_v value is equivalent to the Knudsen number value,

$$Kn = 1/\sqrt{2}\, n_v\, \pi d^2 D \qquad (1)$$

As a consequence of these assumptions, test particles are thrown with the classical elementary flow distribution from $x = 0, -D/2 \leqslant y \leqslant + D/2, - L/2 \leqslant z \leqslant + L/2$ (Fig 2.)

$$dy\, dz\ .\ vf\ (v)\ dv.(n_v/4\pi)\ \sin\xi\ \cos\xi\ d\xi\ d\varphi \qquad (2)$$

where f(v) is the classical Maxwellian velocity modulus distribution,

$$f(v) = (m/2\pi kT_v)^{3/2}\ \exp-\left[mv^2/2kT_v\right]\ x\ 4\pi v^2 \qquad (3)$$

which corresponds to the total effusive flow

$$Q = (1/4)\ x\ D\ x\ L\ x\ n_v\ \bar{v}\ ,\ \bar{v} = \sqrt{8\ kT_v/\pi m} \qquad (4)$$

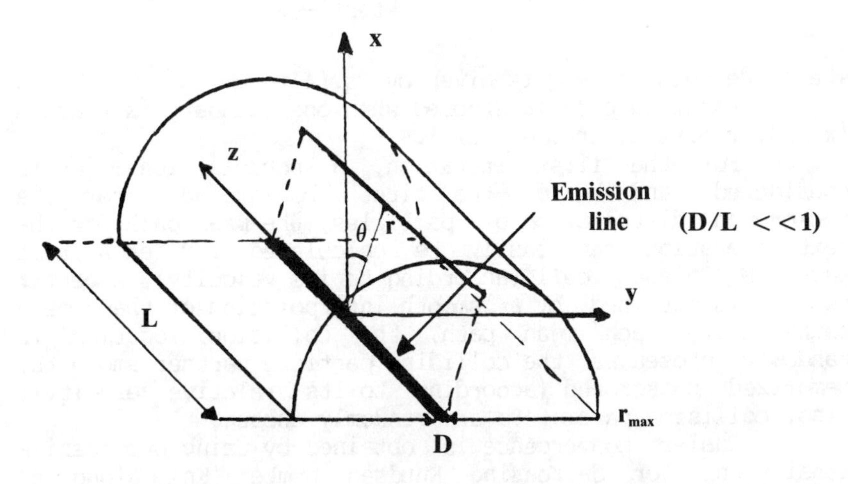

Fig. 1 Three-dimensional geometry.

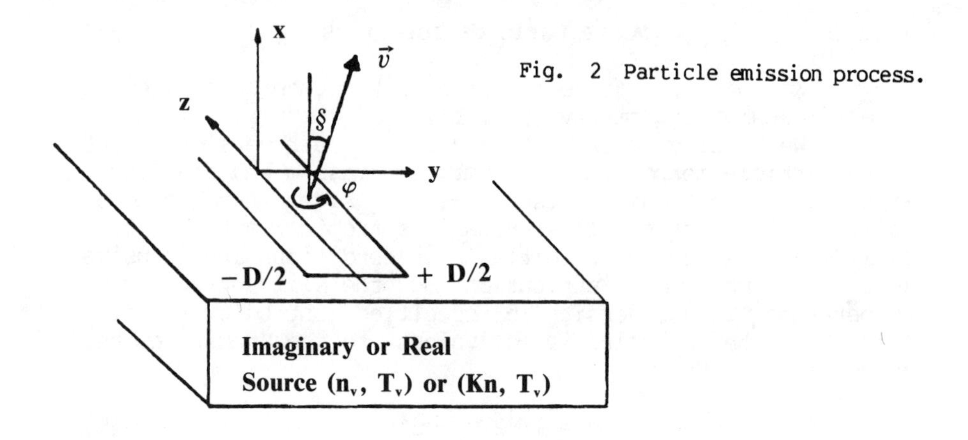

Fig. 2 Particle emission process.

For the mathematical definition of this problem, n_v (or Kn) and T_v are independent specified parameters. For practical problems they can be independent (gases) or related (vapors). N test = 10,000 particles are thrown and individually followed for each iteration. They build an atoms background for the next iteration.

For each particle in state A (velocity for instance) the time path in V cell t_i (A, V) is memorized (Fig. 3). Summation over the particles distribution for each cell V gives the particle distribution functions,

$$N(A,V) \sim \lim_{t \to \infty} \frac{1}{t} \sum_i t_i (A,V) \sim Q \lim_{Ntest \to \infty} \frac{\sum_i t_i (A,V)}{Ntest} \qquad (5)$$

where the total flow Q is given by Eq.(4).

Particle path is stopped when one boundary is reached ($x = 0$, $r = rMAX$, or $z = \pm L/2$).

For the first iteration, a noncollisional jet is considered and build the first background. From the background distribution of particles, the mean path for the next iteration can easily be calculated for each test particle in each cell (according to its velocity). A better result is obtained by a smooth interpolation of these mean paths. With each mean path, the collision location is randomly chosen as the colliding particle partner among the memorized background (according to its relative velocity). Also, collision parameters are randomly chosen.

Easier convergence is obtained by using progressive density n_v (or decreasing Knudsen number Kn) along the iteration process until the desired value. Space (Fig. 3) is uniformly divided in θ into 10 regions ($0<\theta<\pi/2$) and 30 regions in r with approximate geometric progression for small r. A z division is possible but has not been used here, since we are interested in mean jet properties along L.

This mesh is not sufficient for Kn <0. 01, and unrealistic results often occur. Some deviations from energy and momentum conservation that randomly occur for small Kn are corrected by a control procedure during the iteration process. The iteration number varies between 3 and 9 and depends on Kn.

A systematic comparison between this particle test method and the direct simulation Bird's method applied to this problem[5] is going on.

Classification of Regions

Typically four regions appear :

1) $r \sim D$: This is the source flow region, with a prevailing influence of source geometry and the particle emission process. The section $r \ll D$ of this region is the kinetic one-dimensional Knudsen layer.[8,9]

2) $D \ll r \ll L$: This region must exhibit a two-dimensional asymptotic behavior. In this region we can compare our results with published ones.

3) $D \ll r \sim L$: This region typically has three-dimensional behavior.

4) $r \gg L$: Flow tends to appear approximately as a point source flow.

Two-Dimensional Region

Our simulation results are compared for different Kn at $\theta=0$, $r/D = 10$, $L/D=100$ ($r/L = 0.1 \ll 1$) with those of Anderson et al.[3] (Bird's method) and Beylich,[6] (kinetic method) for radial, axial, and azimuthal temperatures T_r, T_z, and T_θ (Fig. 4).

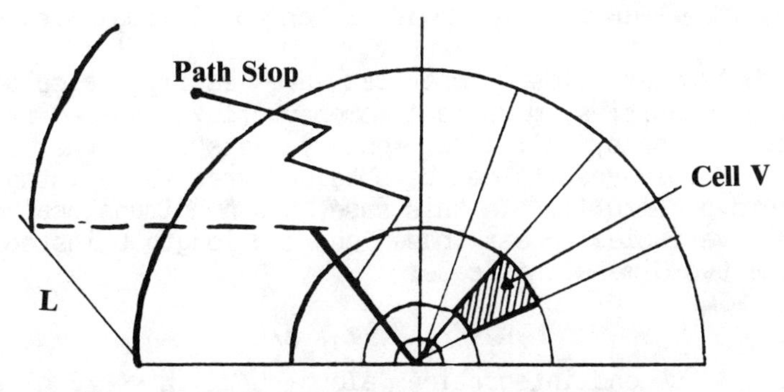

Fig. 3 Monte Carlo particle test simulation.

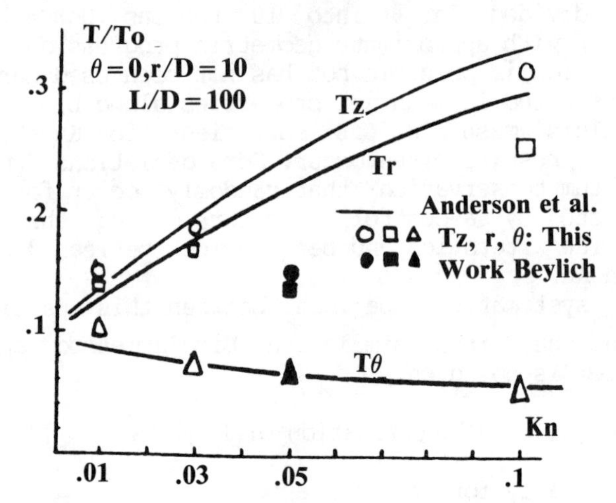

Fig. 4 Temperatures : comparison with Anderson's and Beylich's results.

This comparison is made considering the reference temperature T_v as sonic one, related to stagnation temperature[3,6] by the classical adiabatic relation,

$$T_v = (2T_0)/(\gamma + 1) \tag{6}$$

The agreement is good for T_θ, and a small deviation appears for Kn = 0.1 between Monte Carlo results due to different initial slit conditions (Fig. 4).

Angular density profiles $n(\theta)/n(0)$ for Kn = 0.1 and 0.01 are compared to those of Ref.3 (Fig. 5). Divergence for Kn = 0.1 is also obviously due to different initial slit conditions, whose importance disappears for a near continuous flow (Kn = 0.01).

Three-Dimensional Region : Non Collisional Flow

Molecular flow properties are easily calculated by summing densities, mass and momentum fluxes emited from the emission line by all z'-elementary jets up to the (r,θ,z) point. Simple results can be obtained when the y integration is removed ($r/D \gg 1$). In this case, new nondimensional geometrical variables appear based on the length L instead of D for the two-dimensional case:

$$R = r/L \;,\; Z = z/L \tag{7}$$

A second integration along $-L/2 < z < L/2$ gives the L-length mean flow relations on (R,θ).

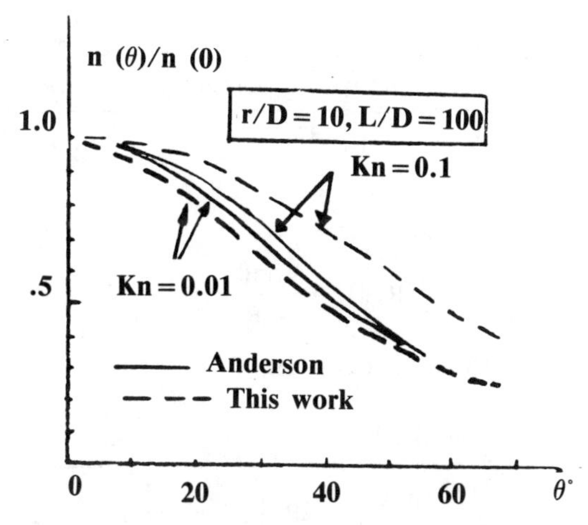

Fig. 5 Density θ profile. Comparison with Anderson's results.

1) L-Mean Density

$$n^*(r,\theta) = \frac{n\ (R,\theta)\ x\ L}{n_v D} = \frac{R\cos\theta}{4\pi}\int_{-\frac{1}{2}}^{+\frac{1}{2}}\int_{-\frac{1}{2}}^{+\frac{1}{2}}\frac{dZ\ dZ'}{[(Z-Z')^2 + R^2]^{3/2}} \qquad (8)$$

therefore

$$n^*(R,\theta) = \frac{1}{2\pi}\left((1 + \frac{1}{R^2})^{1/2} - 1\right)\cos\theta \qquad (9)$$

A new nondimensional variable $n^* = nL/n_v D$ appears that is different from the classical two-dimensional ones n/n_v. This expression is not valid for region 1, but is valid for regions 2, 3, and 4. For region 2, $R \ll 1$ and (9) give the classical two-dimensional noncollisional far-field density

$$n^*(R,\theta) = \frac{\cos\theta}{2\pi R} \text{ or } \frac{n}{n_v} = \frac{D\cos\theta}{2\pi r} \qquad (10)$$

and for $R \gg 1$, the classical effusion source flow

$$n^*\ (R,\theta) = \frac{\cos\theta}{4\pi R^2} \qquad (11)$$

2) L-Mean velocities and Particle Fluxes :
The mean flux along L in the r direction will be

given by

$$\phi_r^*(R,\theta) = \frac{\phi(r,\theta) \times L}{n_v D \bar{v}} = \frac{R^2\cos\theta}{4\pi} \int_{-\frac{1}{2}}^{+\frac{1}{2}}\int_{-\frac{1}{2}}^{+\frac{1}{2}} \frac{dZ\,dZ'}{[(Z-Z')^2+R^2]^2} \quad (12)$$

therefore

$$\phi_r^*(R,\theta) = \frac{\cos\theta}{4\pi R} \, tg^{-1}\frac{1}{R} \quad (13)$$

and the mean radial velocity is

$$v_r^*(R) = \frac{\phi_r^*(R,\theta)}{n^*(R,\theta)} = \frac{tg^{-1}\ 1/R}{2R\left[(1+\frac{1}{R^2})^{1/2}-1\right]} \quad (14)$$

for the two-dimensional region (far field 2) $R \ll 1$, $v_r^* \simeq \pi/4$ and for source flow ($R \gg 1$), $v_r^* \simeq 1$

A similar derivation can be made for particle flux in the z-direction. The L-axial velocity mean value is obviously 0 by symmetry. It is possible to obtain a mean value for the region $0<Z<\frac{1}{2}$:

$$v_z^*(Z>0) = \frac{2\,tg^{-1}\frac{1}{2R} - tg^{-1}\frac{1}{R}}{2\left[(1+\frac{1}{R^2})^{\frac{1}{2}}-1\right]} \quad (15)$$

For the two-dimensional region $R \ll 1$, we find the obvious result $v_z^*(r,z) \sim v_z^*(r) \sim 0$ and for $R \gg 1$ $v_z^*(r,\theta) \sim 1/4R$.

In the θ-direction, $v_\theta \sim 0$ like $(D/r)^2$ as $r \gg D$, from simple geometrical considerations.

3) <u>L- Mean Temperatures and Momentum Fluxes</u> :

The mean momentum fluxes along L in the r direction will be given by

$$q_r = \frac{m n_v \overline{v^2}\, DR^3\cos\theta}{4\pi L} \int_{-\frac{1}{2}}^{+\frac{1}{2}}\int_{-\frac{1}{2}}^{+\frac{1}{2}} \frac{dZ\,dZ'}{[(Z-Z')^2+R^2]^{5/2}} \quad (16)$$

where the mean square velocity v^2 is

$$\overline{v^2} = \frac{3kT_v}{m} \quad (17)$$

therefore since the mean radial temperature is given by

$$T_r^*(R,\theta) = \frac{T_r(R,\theta)}{T_v} = \frac{m}{kT_v}\left(\frac{q_r}{nm} - v_r^2\ (R)\right) \tag{18}$$

We obtain

$$T_r^*(R,\theta) = 2 + (1 + 1/R^2)^{-\frac{1}{2}} - (8/\pi)\ v_r^{*2}(R) \tag{19}$$

For the two-dimensional region (far field 2) $R \ll 1$, $T_r^* \simeq 2 - \pi/2 = 0.429$ and for source flow ($R \gg 1$) $T_r^* \simeq 3 - 8/\pi \simeq 0.453$. Hence, for noncollisional flow, T_r^* variations with R are purely geometric and small.

Momentum flux in the z direction is obtained by a similar derivation. From L-mean of the momentum flux, we obtain the mean T_z^* temperature

$$T_Z^*\ (R,\theta) = \frac{T_Z(R,\theta)}{T_v} = 1 - (1 + 1/R^2)^{-\frac{1}{2}} \tag{20}$$

referenced to zero velocity mean value $\left[v_z^*\ (R) = 0\right]$. If T_z is considered on half-length $0<Z<\frac{1}{2}$, as we prefer, then

$$T_z^*(R) = 1 - (1 + 1/R^2)^{-\frac{1}{2}} - (8/\pi)\ v_z^{*2}(R) \tag{21}$$

For the two-dimensional region ($R \ll 1$),since $v_z^* \sim 0$, $T_z^*(R) \simeq 1$ which means the obvious fact that the z temperature is constant within two dimensional region for noncollisional flows. For $R \gg 1$, for the half-region $0<Z<1/2$ (or symmetrical one)

$$T_z^*\ (R) \simeq (1 - 1/\pi)/2R^2 \tag{22}$$

The azimuthal velocity $T_\theta \sim 0$ as $(D/r)^2$ from simple geometrical consideration : T_θ and T_z behavior are the same as for $R \gg 1$.

General Consideration in the Three-Dimensional Region

Comparison of local (R,θ,Z) values of density, velocities and temperatures with mean values along the L-length $(-1/2 <Z<1/2)$ exhibits very small differences for noncollisional flow. It is thus simpler to consider only the mean values. This consideration and computing times led us to do the same for collisional flow Monte Carlo simulations and fitting formulas. Also, new similarity laws for $r/D \gg 1$, $L/D \gg 1$,

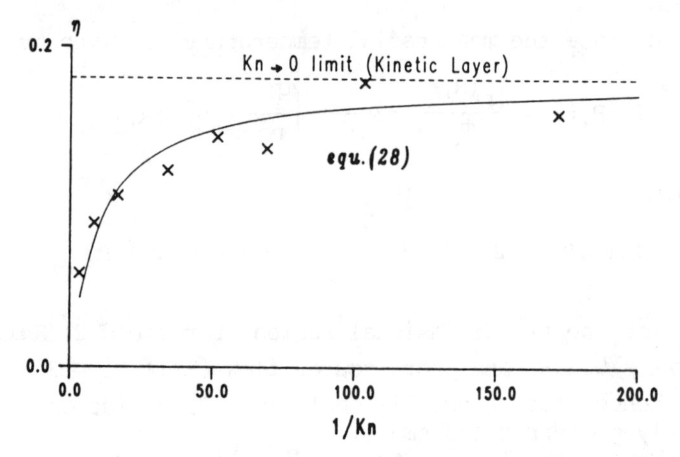

Fig. 6 Backscattered particle proportion.

already demonstrated for non collisional flow, can be verified with the computer simulation; for instance, identical statistical values have been obtained for nL/n_vD, $v_r/\bar{v}$, $v_z/\bar{v}$, T_r/T_v, T_z/T_v, T_θ/T_v for Kn = 0.01 and R > 0.2 ($r/D \gg 1$) when L/D = 20 and L/D = 100 runs are compared. Heuristic continuity considerations lead us to the same result. Classical similarity can be written for L-mean values

$$n/n_v = \varphi_n \;(R,\theta,Kn,D/L) \tag{23}$$

$$T_{r,\theta,z}/T_v = \varphi_T \;(R,\theta,Kn,D/L) \tag{24}$$

$$V_{r,\theta,z}/\bar{v} = \varphi_v \;(R,\theta,Kn,D/L) \tag{25}$$

For fixed Kn and R > 0, as $D/L \to 0$ ($D/L \ll 1$), $n_v \to \infty$ with a finite value for n ; thus $\varphi_n \to 0$, and φ_T, φ_v have finite limits. Existence of a finite derivative for φ_n (which the case for Kn = ∞) induces a finite limit for

$$\frac{n}{n_v} \Big/ \frac{D}{L} = \frac{nL}{n_v\, D} = \varphi_n'(R,\theta,Kn,0) \tag{26}$$

Systematic runs have been made with our Monte Carlo code from Kn = 0.3 to Kn = 0.006. For Kn = ∞, Monte Carlo results gave no statistical bias when compared with results of the preceding formulas. We shall give the main results for collisional flow and propose fitting formulas for the typical three-dimensional region: $0.5 < R < 3$, $0 < \theta < \pi/4$.

We use a three-step fitting method taking for the following analytical form for nondimensional values:

$$\varphi\ (R,\theta,Kn) = \varphi\ (R\ ;\ C_1(Kn),\ C_2(Kn))\ \cos^{q(Kn)}\theta \qquad (27)$$

$C_1(Kn)$, $C_2(Kn)$, $q(Kn)$ are homographic or exponential functions, related to calculated values for $Kn \to \infty$, and with asymptotic convergence for $Kn \to 0$ (continuous flow). For each run, which corresponds to a given Kn, q then C_1 and C_2 are obtained by a least-squares fit. Then homographic or exponential functions of Kn are fitted to q, C_1, C_2 values obtained for the different Kn values.

Back Scattered Particles

The proportion of back scattered particles η is a classical problem, generally resolved by kinetic layer theory [8,9], for $Kn \to 0$: as emission Mach number approaches $M = 1$, $\eta \simeq 0.18$, which agrees quite well with Monte Carlo results (Fig. 6).

A least squares fit for numerical results is

$$\eta = 0.18/[1 - 1/(1 + 0.09/Kn)] \qquad (28)$$

Reduced Density $n^*(R,\theta)$

For $Kn = \infty$, the q_n exponent for $\cos^{q_n}\theta$ variation is 1 [see Equ.(9)] and exhibits a strong variation (Fig.7) as soon as Kn is finite valued, with an asymptotic value for $Kn \to 0$ of about $q_n = 2.7$. This $\cos^{2.7}\theta$ asymptotic law must be

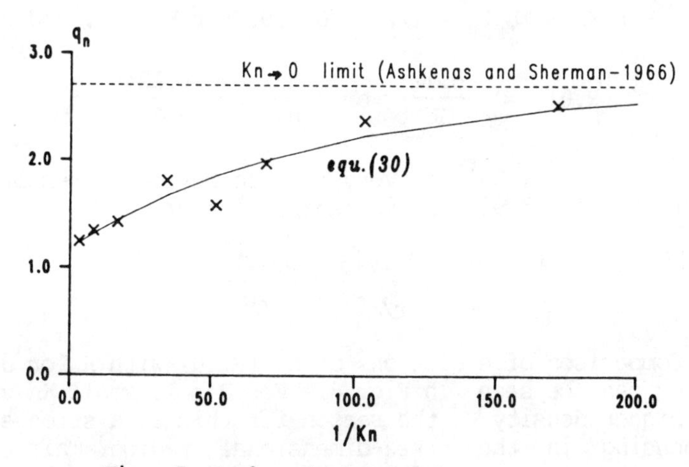

Fig. 7 Cosθ variation density exponent.

compared with the results of Ashkenas and Sherman[7] obtained by the method of characteristics for a continuous source flow, for a monoatomic gas

$$\frac{n\ (r,\theta)}{n\ (r,0)} = \cos^2 \frac{\pi\theta}{2.73} \qquad (29)$$

which is equivalent to $\cos^{2.72} \theta$ for $0 < \theta < \pi/4$. Finally, a least-squares formula for $q_n(Kn)$ (with a small jump from $Kn = \infty$) is given by

$$q_n = 2{,}70 - 1.53 \exp(-0.0115/Kn) \qquad (30)$$

A general relation derived from equ.(9) is :

$$n^*(R,\theta) = \frac{n\ L}{n_v\ D} = C_n \left[(1 + \frac{1}{R^2})^{\gamma\ (R)} - 1 \right] \cos^{q_n}\theta \qquad (31)$$

with $0.5 \leqslant \gamma\ (R) \leqslant 1$, $\gamma\ (R) = 0.5$ for $R \leqslant 0.3$ and for $R > 0.3$, after least squares fits,

$$\gamma\ (R) = 0.5 + 0.295\ (1 - \frac{1}{1 + 0.13/Kn})\ (R - 0.3) \qquad (32)$$

For $Kn \to \infty$, $\gamma\ (R) \to 0.5$ (noncollisional situation).

The coefficient C_n is fitted from simulation results by

$$C_n = 0.120 + \frac{0.039}{1 + 0.25/Kn} \qquad (33)$$

For $Kn \to \infty$, $C_n = 1/2\pi = 0.159$. For $Kn \to 0$, and $R \ll 1$,

$$n^*\ (R,0) \simeq \frac{0.12}{R} \quad \text{or} \quad \frac{n}{n_v} \simeq \frac{0.12}{r/D} \qquad (34)$$

Equation (34) is very near to Beylich's results for cylindrical flow,[6] with sonic radius $r^* = D/2$:

$$\frac{n}{n_v} = \frac{9/16\sqrt{5}}{r/r^*} \simeq \frac{0.126}{r/D} \qquad (35)$$

Comparison of evolutions of n^* (R, 0) with R for different Kn can be seen in Fig. 8. For $R \geqslant 1$, small Kn values give higher density : the reason for this is a stronger jet autofocusing in the three-dimensional region. This effect offsets the lowering effect due to jet accelera-tion and

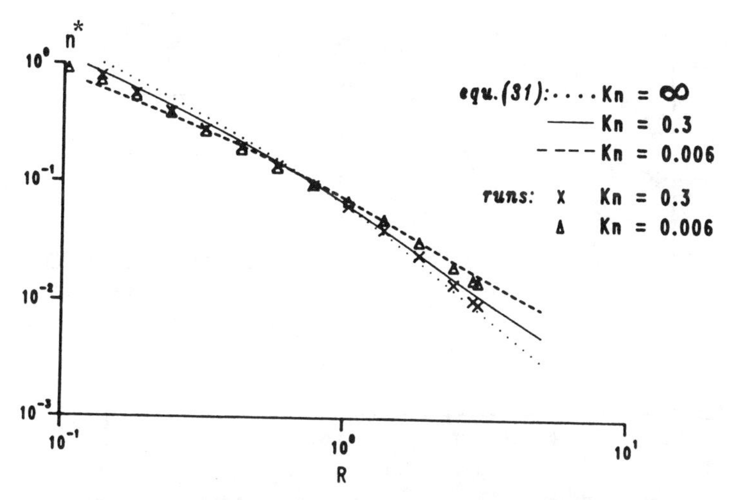

Fig. 8 Reduced density evolution (θ = 0).

backscattered particles, which dominates the two-dimensional region ($R \ll 1$).

Radial Reduced Velocity v_r (R, θ)

Velocity behavior with distance R remains identical for every Kn (Fig. 9) with a limit velocity. For $R > 0.5$, the analytical relation (14) for $Kn = \infty$ can be substitued by the more simple one, within 2% error:

$$v_r^* = 1.02 - (0.0625/R) \tag{36}$$

When Kn decreases, particles are accelerated by collisions, and a small variation with θ appears. The fitting formula generalizes (36) and is

$$v_r^* = \left[\left(1.40 - \frac{0.38}{1 + 0.126/Kn} \right) - \frac{0.0625}{R}\right] \cos^{q_v}\theta \tag{37}$$

with

$$q_v = 0.34 \left(1 - \frac{1}{1 + 0.063/Kn} \right) \tag{38}$$

which is adequate for $R > 0.5$ runs values (Fig. 9).

Radial Reduced Temperature T_r (R,θ)

Temperature behavior with R demonstrates the freezing situation when no more collisions occur ; freezing

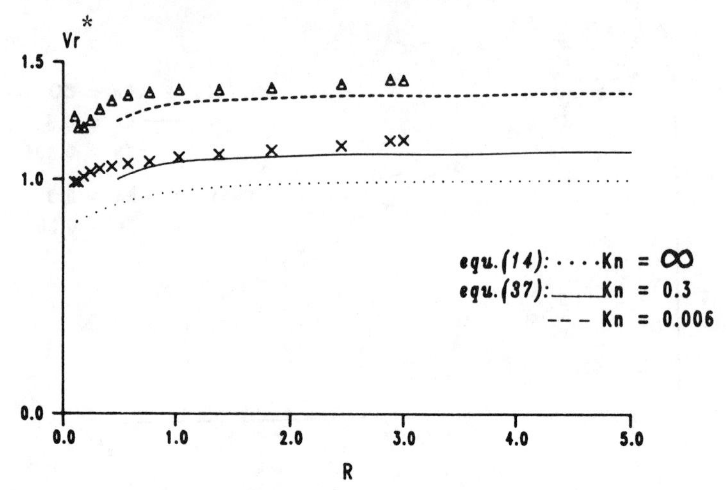

Fig. 9 Reduced radial velocity evolution (θ = 0).

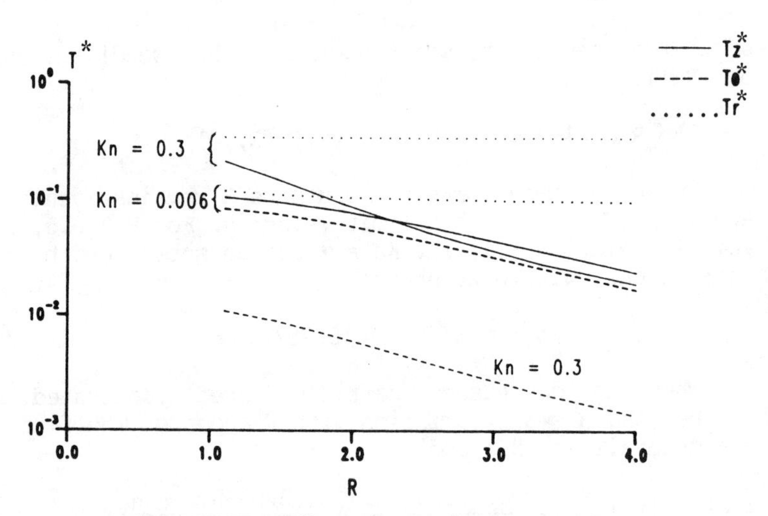

Fig. 10 Comparison of temperature behavior.

appears at increasing distance R, as Kn $\to 0$ (Fig. 10). At fixed distance R, T_r decreases with Kn, until continuous flow is reached. No significant dependence on θ appears.

As v_r, for R > 0.5 the analytical relation (19) can be substituted with 2% error by (Kn = ∞),

$$T_r^* = 0.453 - \frac{0.029}{R} \tag{39}$$

and is generalized by the fitting formula

$$T_r^* = 0.070 + \frac{0.383}{1 + 0.124/Kn} + \left[0.0344 - \frac{0.0634}{1 + 0.30/Kn}\right] x\ 1/R \qquad (40)$$

Axial T_z (R,θ) and Azimuthal $T_\theta(R,\theta)$ Temperatures

For T_z, situations are very different from the two-dimensional region (where $T_z \sim T_r$) and for the three-dimensional one; T_z and T_θ are highly dependent on geometrical considerations; for $R \gg 1$ T_z and T_θ decrease like $1/R^2$ for non-collisional (see Equ. 22) or collisional flow (Fig. 10). For $R \sim 1$, T_z and T_θ exhibit a lower dependence on R as $Kn \rightarrow 0$. Finally, for fixed R, T_z decreases with Kn for $R < 1$ and depends only on geometry for $R \gg 1$, when T_θ is always increasing. We tried to simulate this with the fitting formula

$$T_{\theta,z} = \frac{Tc}{R^{\left|\frac{2\,R^2}{R^2 + Cz}\right|}} \cos^q\theta \qquad (41)$$

with low values of C for T_z, and $Tc \rightarrow 0$ for T_θ, when $Kn \rightarrow \infty$.

For T_z no θ dependence seems to be significant ($q = 0$ for $\theta < 44\pi/2$) and the fitting formulas are

$$Tcz = 0.092 + \frac{0.168}{1 + 0.037/Kn} \qquad (42)$$

$$Cz = 18.6 - \frac{18}{1 + 0.017/Kn} \qquad (43)$$

For T_θ, a θ dependence occurs ($q = 1.2$). It is related to the number of collisions in θ regions. The fitting formulas will be :

$$Tc\theta = 0.098 \left[1 - \frac{1}{1 + 0.039/Kn}\right] \qquad (44)$$

$$C\theta = 13 - \frac{9}{1 + 0.029/Kn} \qquad (45)$$

Comparison of Temperature Components Behavior

Figure 10 compares the behavior of each temperature T_r, T_θ, T_z respectively.

For high Kn, very disparate values occur. Near the two-dimensional region (R < 1), $T_\theta \ll T_z \sim T_r$ whereas for R>1, $T_\theta \ll T_z \ll T_r$.

When Kn → 0, this disparate situation for the temperatures becomes smaller, but never at not too high a distance R ; three-dimensional jet converges to equilibrium, but not uniformly with R : R < 1 or R ~ 1 : $T_\theta \sim T_z \sim T_r$, R ≫ 1 : $T_\theta \ll T_z \ll T_r$.

Acknowledgments

The author would like to thank his co-worker X. Carnet for the particle test method elaboration, F.Doneddu and J.P. Quaegebeur for helpful discussions, and M. Chabert and P. Roblin for computer works.

References

[1]Hamel, B. B., and Willis D. R., "Kinetic Theory of Source Flow Expansion with Application to the Free Jet," Physics of Fluids, Vol.9, May 1966, pp. 829-841.

[2]Bird, G. A., Molecular Gas Dynamics, Clarendon, Oxford, UK, 1979.

[3]Anderson J. B., Foch J. D., Show M. J., Stern, R. C., and Wu B. J. "Monte-Carlo Simulation of Free Jet Flow from a slit", 15th International Symposium on Rarefied Gas Dynamics Vol.I, Teubner, Stuttgart, FRG,1986, pp. 442-451.

[4]Kogan, M. N., Rarefied Gas Dynamics, Plenum, New York, 1969.

[5]Klein, E., and Stephan, Y., "The direct Simulation Method applied to dilute Gas Dynamics,". Monte-Carlo Methods, Lecture Notes in Physics, Springer - Verlag, Berlin, FRG 1985.

[6]Beylich, A. E, "Non Equilibrium Effects in Plane Jets," 12th International Symposium on Rarefied Gas dynamics, AIAA, New York 1980, pp. 710-724.

[7]Ashkenas, H., and Sherman, F. S., "The Structure and Utilization of Supersonic Free Jets in Low Density Wind Tunnels," 4th International Symposium on Rarefied Gas Dynamics, Vol. II, Academic, New York, 1966, pp. 84-105.

[8]Knight, C. J., "Theoretical Modeling of rapid Surface Vaporization with Back Pressure," AIAA Journal, Vol. 17, May 1979, pp. 519-523.

[9]Yttrehus, T., "Theory and Experiments on Gas Kinetics in Evaporation," 10th Symposium on Rarefied gas Dynamics, Vol. II, AIAA, New York, 1976 pp. 1197-1212.

Modification of the Simons Model for Calculation of Nonradial Expansion Plumes

I. D. Boyd* and J. P. W. Stark†
University of Southampton, Southampton, England, United Kingdom

Abstract

The Simons model is a simple method for calculating the expansion plumes of rockets and thrusters and is a widely used engineering tool for the determination of spacecraft impingement effects. The model assumes that the density of the plume decreases radially from the nozzle exit. Although a high degree of success has been achieved in modeling plumes with moderate Mach numbers, the accuracy obtained under certain conditions is unsatisfactory. A modification made to the model that allows effective description of nonradial behavior in plumes is presented, and the conditions under which its use is preferred are prescribed.

Introduction

An analytical model for the prediction of flow properties for rocket plumes expanding into vacuum was developed by Boynton[1] and Simons[2] several years ago. Specifically, this model introduces an expression for the density ρ at a point in the axisymmetric flowfield described for polar coordinates in the form

$$\rho(r, \theta) = \frac{A}{r^2} f(\theta) \tag{1}$$

where A is the plume constant and $f(\theta)$ describes the angular density behavior.

---

* Graduate Student, Department of Aeronautics and Astronautics; currently at NASA Ames Research Center, Moffett Field, California.

† Senior Lecturer, Department of Aeronautics and Astronautics.

The assumptions inherent in formulating Eq.(1) are: 1) The streamlines are straight lines radiating from the point on the plume axis at the nozzle exit. 2) The velocity is everywhere constant and equal to the limiting value. 3) The density decreases radially.

The model divides the flowfield into two distinct regions; thus, the expansion of the boundary layer is treated separately from that of the isentropic core. In this core, the one-dimensional isentropic relationships are employed together with the known stagnation conditions to derive all important flow quantities from Eq.(1). In the boundary layer, the concept of "effective" stagnation conditions is introduced. Under this process, the stagnation conditions are assumed to show an angular dependence. These ideas, together with details of procedures for the calculation of the plume constant A from the gas type and nozzle geometry are fully explained in Ref. 3.

The Simons model has been shown to be a useful, inexpensive engineering tool for the calculation of spacecraft impingement effects in the continuum regime, although clearly many features of the expansion are not treated. At large distances from the nozzle exit, the flow may be regarded as collisionless. In the region lying between the continuum and free molecular limits, the gas is no longer in thermal equilibrium and the continuum equations become invalid. Within the transition flow regime impingement calculations may be performed through the use of relationships bridging the continuum and free molecular results. It is possible to determine the onset of both the transition and collisionless flow regimes from the Simons model. Details of these procedures are also included in Ref. 3.

The use of the one-dimensional isentropic relationships in the transition flow regime should be viewed with caution. Recent calculations of the expansion of the isentropic core of a hydrazine thruster plume using the Direct Simulation Monte Carlo method[4] have shown that for quantities such as drag coefficient and Stanton number, which are important in the determination of impingement effects, the continuum solutions may be up to 100% in error. This is due to the presence of significant areas of thermal nonequilibrium in the flowfield.

The Simons model has been adopted by a number of workers. Calia and Brooks[5] found good agreement for density measurements made at large angles from the plume centerline. Lengrand[6] has investigated the exhaust plumes from a number of small thruster nozzles. In this study, excellent agreement is found between theoretical prediction and experimental data for nozzle exit Mach numbers up to $Ma_E = 5$. The comparisons were made with reference to axial density ratio.

The Simons model has also been used to analyze small hydrazine thrusters[7]. In Ref. 7, measurements of Pitot pressure are compared with Simons model predictions and are found to be in error by a factor of 2. In the hypersonic flowfields of such thrusters the Pitot pressure is approximated by

$$P_{t2} = \frac{\gamma + 3}{\gamma + 1}\frac{1}{2}\rho {U_L}^2 \qquad (2)$$

where U_L is the limiting velocity of the expansion. It is clear that any errors found in the calculation of Pitot pressure must arise from the failure of the Simons model formulation to properly describe the density decay.

Errors of a similar magnitude have also been found by Lengrand[8] for axial density predictions of the two thrusters discussed in Ref. 9. These nozzles, together with those considered in Ref. 7, have exit radii of a few millimetres and thus tend to have large laminar boundary layers, e.g. up to 40% of the nozzle exit plane. Lengrand[8] suggested that the poor results were either due to the presence of a finite chamber pressure acting on very slender expansion plumes, or alternatively due to the failure of the Simons model to adequately deal with flows produced by these thrusters.

The Simons model has been partially improved by the replacement of r by $r\text{-}x_o$ in Eq.(1) and is reported in Ref. 10. The result of such a procedure is to move the effective source of the nozzle to a distance x_o along the plume axis. The experimental derivation of values for x_o of a number of thrusters and test gases is reported in Ref. 10. Rather than resorting to such expensive undertakings, it is clearly more desirable to obtain an improvement in the model from the particular nozzle and gas properties themselves.

An alternative technique for the calculation of plume flowfields is the Method of Characteristics (MOC)[11]. This well-known computational technique provides solutions to the Euler equations and as such should only be applied to the isentropic core expansion. In the present work, this method has been used to obtain solutions for the plumes exhausting into vacuum from a number of nozzles, including those in which the Simons model gave erroneous results. The expansion of the laminar boundary layer was treated by assuming a typical laminar velocity profile for a flat plate[12]. Although such a procedure is rather dubious, variation of this boundary layer velocity profile was found to have no effect on the MOC calculations along the plume axis on which most of the following analysis is based.

In Fig. 1 the solutions obtained with both the Simons model and MOC for the density ratio along the plume axis are shown together with experimental data for the

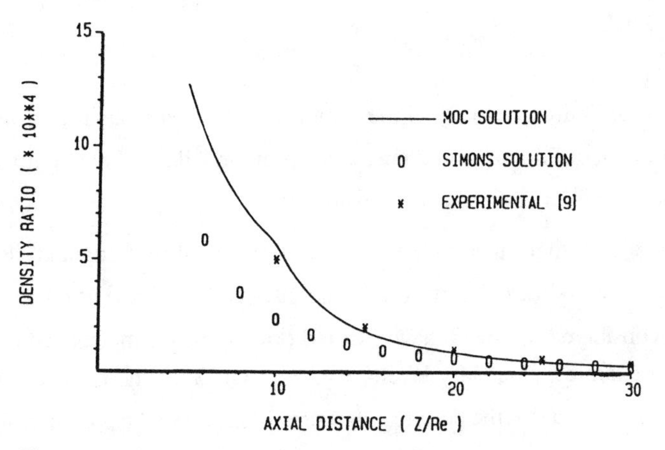

Fig. 1 Density ratio along the plume axis of the radial thruster of Ref. 9.

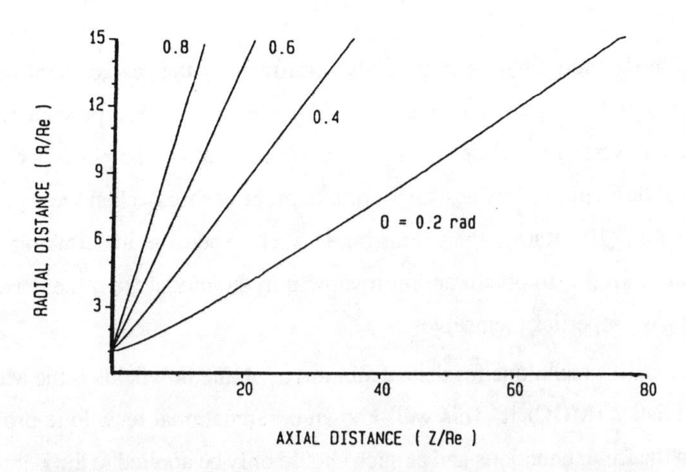

Fig. 2 Contours of constant flow angle for the radial thruster of Ref. 9.

radial thruster described in Ref. 9. For these studies, the working gas is nitrogen. It is clear that the MOC calculations offer a greatly improved correspondence to the experimental results on the Simons model predictions. The latter results are consistently found to be in error by a factor of about 2 along the axis. Similar features were observed for the axial thruster of Ref. 9. On the basis of these calculations, it was concluded that, as the MOC calculations were found to be in good agreement with the experimental data, it must be the Simons model formulation itself that is responsible for the poor results obtained with this method.

Contours of constant flow angle were also derived from the MOC calculations and are shown in Fig. 2. This diagram suggests that the first assumption in the Simons formulation is valid beyond a few exit radii of the nozzle exit. On extrapolation, these contours are found to converge to points very close to the origin. There is therefore no real evidence to justify the translation of the effective source of the nozzle along the plume axis. It was also apparent from the MOC calculations that the assumption that the velocity is everywhere constant is also valid. It is therefore concluded that the assumption of radial density decay is violated and is thus responsible for the poor performance of the analytical model.

A comparison of Simons model and MOC solutions of the flow density along the axis of various nozzles has therefore been undertaken. The intention of such a comparison is to indicate the types of nozzle for which the original Simons model is unsuitable and also to allow the derivation of a modified expression for the density that accounts for the nonradial nature of such plumes.

Analysis

The nonradial nature of a number of flows has been characterized in terms of the exit Mach number Ma_E and the nozzle exit half-angle θ_E. In the present study, the range of exit Mach number investigated was $4.5 < Ma_E < 7$, whereas that for exit angle is $5° < \theta_E < 15°$. The exit Mach number is that calculated from the area ratio reduced by the presence of the boundary-layer thickness.

In each configuration for which calculation is made, both Simons model and MOC solutions for the axial density ratio are generated. On the basis of the previous calculations, the MOC results were assumed to provide an accurate solution. The ratio of the Simons model and MOC solutions suggested that the radial decay law in Eq.(1) should be replaced by

$$\rho(r,\theta) = \frac{A}{r^2 - ar + b} f(\theta) \tag{3}$$

where the constants a and b are to be derived from the nozzle geometry. It has been assumed that these constants are functions of exit Mach number and nozzle exit half-angle only.

The parameters a and b are determined by the best fit of Eq.(3) to the functions defined by the ratio of the Simons model and MOC solutions to all the nozzles investigated. By independent variation of first Ma_E and then θ_E, a correspondence between these constants and the nozzle exit conditions has been obtained.

The large number of calculations made in this study are quite well described by the following simple expressions:

$$a = 3\theta_E^{\frac{1}{2}} Ma_E r_E \tag{4}$$

$$b = 5\theta_E Ma_E^{\,2} r_E^{\,2} \tag{5}$$

where θ_E is expressed in radians and r_E is the nozzle exit radius.

Logarithmic plots of the best-fit constants a, b against nozzle exit half-angle for a number of exit Mach number are shown in Figs. 3 and 4. The straight lines represent Eqs.(4) and (5) for the prescribed exit conditions. Asterisks represent exit conditions for which solutions have been generated. The substitution of Eqs.(4) and (5) into Eq.(3) gives rise to a new flowfield prediction model here termed the Modified Simons model. It is stressed that the single modification made to the procedures described in Ref. 3 is the alteration of the radial density decay behavior. The performance of this model is now discussed with reference to experimental data for a number of real nozzles.

Discussion

The Modified Simons model has been used to calculate a number of thruster flowfields in order to assess the validity of the assumptions made in deriving the expressions given in Eqs.(4) and (5). Figure 5 shows calculations for the axial thruster

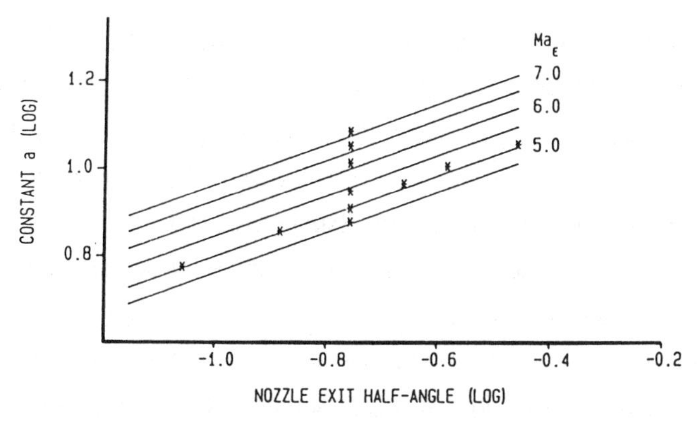

Fig. 3 Best-fit constant a plotted as a function of θ_E and Ma_E.

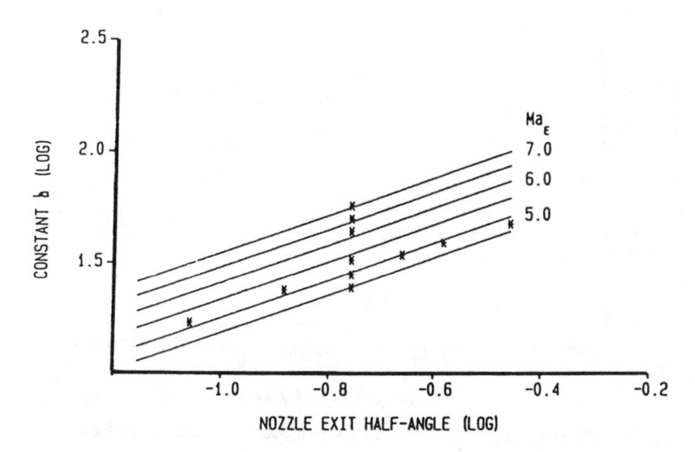

Fig. 4 Best-fit constant b plotted as a function of θ_E and Ma_E.

of Ref. 9 together with experimental data. It may be seen that the Modified model offers considerable improvements on the original calculations. Such improvements were also found for the radial thruster as shown in Fig. 6. In Fig. 7 calculations for the exit conditions $Ma_E = 4.63$ and $\theta_E = 10°$ are shown together with experimental data taken from Ref. 6. Lengrand has previously made successful calculations for this nozzle using the original Simons model. Although it is clear that the introduction of the modifications has not greatly altered the theoretical predictions, it is apparent that the original model offers the better agreement. However, Lengrand[13] believes that small errors made in experimental calibration when rectified should result in the data being pushed more towards the Modified Simons calculations.

In any case, it is clear the procedures for the determination of the preferred model under different conditions is required. Such procedures are considered later.

The calculations shown in Figs. 5-7 illustrate the usefulness of the Modified Simons model and justify the derivation of the parameters a and b. Although these expressions have been derived from calculations made for nitrogen gas, the new model has also been applied to the 0.5N hydrazine thruster discussed in Ref. 7. For this thruster the value of the ratio of specific heats γ is determined to be 1.37 from the matching of transverse Pitot pressure profiles obtained experimentally and with the original Simons model. In Fig. 8 such a transverse distribution is presented

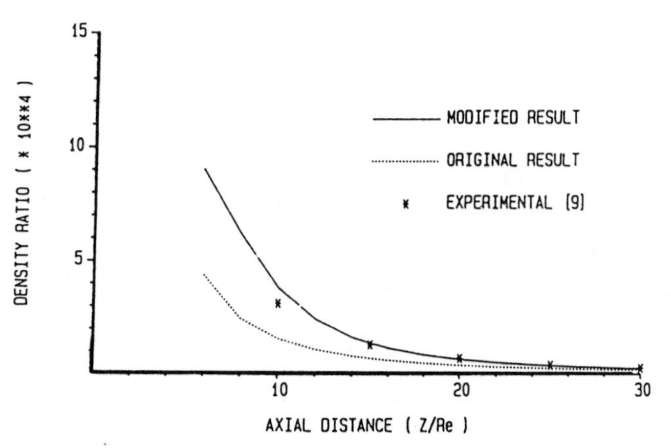

Fig. 5 Density ratio along the plume axis of the axial thruster of Ref. 9.

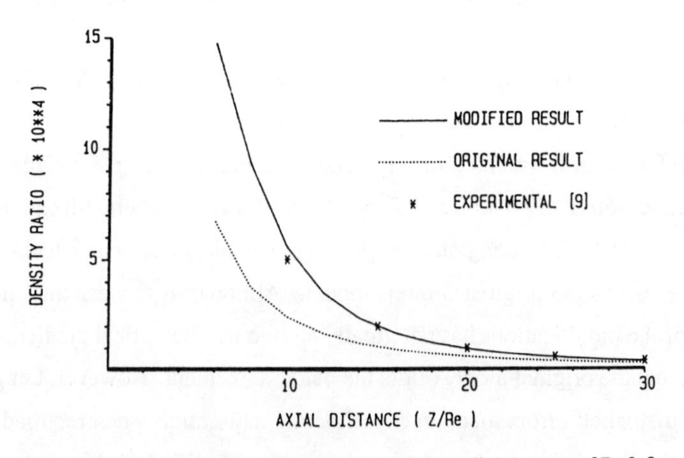

Fig. 6 Density ratio along the plume axis of the radial thruster of Ref. 9.

at an axial distance of 24 exit radii. The compression shock due to the presence of a finite background pressure is clearly distinguishable in the experimental data and cannot be reproduced by the simple analytical models. Calculations using the Modified Simons model are shown for γ = 1.37 and 1.4. It is clear that, for the formulas derived in this study, the analytical predictions near the axis are better described with $\gamma = 1.4$. The calculations made with the new model again show significant improvement when compared to the original results. Figure 9 plots similar calculations made at the axial point 48 exit radii from the nozzle.

The previous calculations have highlighted two important points that depend on the prevailing exit conditions: 1) Under certain conditions the plume expansion

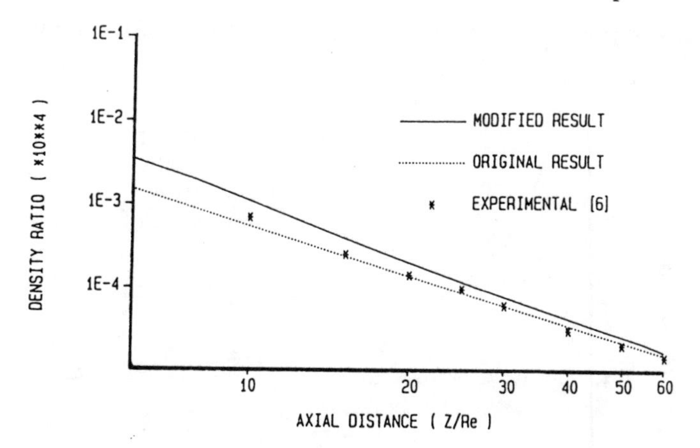

Fig. 7 Density ratio along the plume axis for nozzle number 1 of Ref. 6.

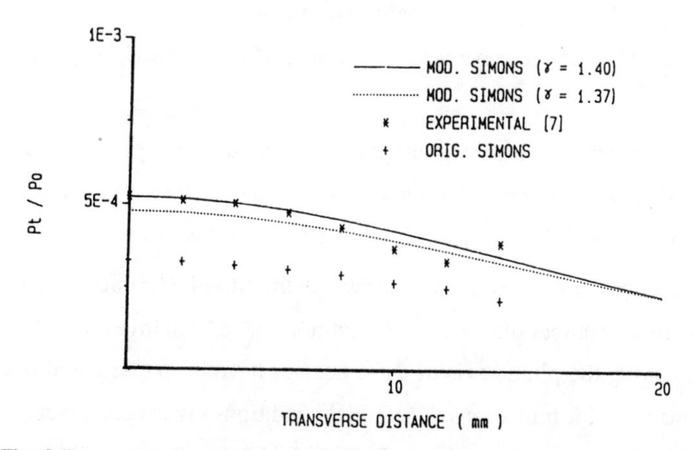

Fig. 8 Transverse pitot pressure profile at $x/r_E = 24$ for 0.5N hydrazine thruster.

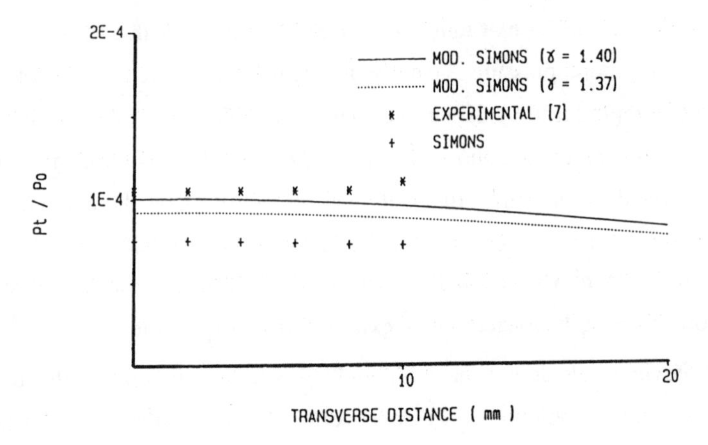

Fig. 9 Transverse pitot pressure profile at $x/r_E = 48$ for 0.5N hydrazine thruster.

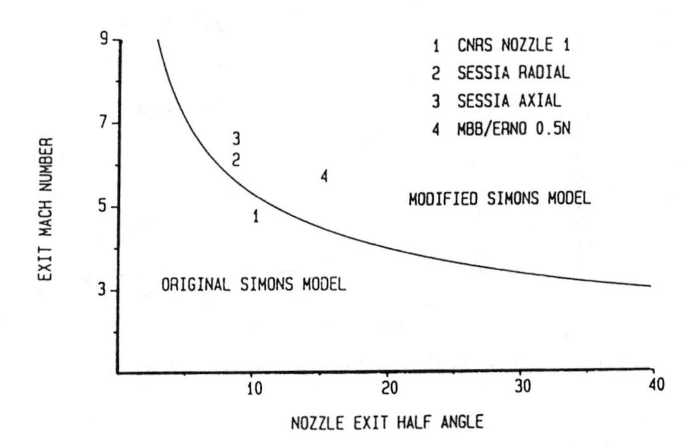

Fig. 10 Critical nozzle exit conditions for onset of nonradial plume expansion.

is readily approximated by the assumption of radial density decay and the Simons model is used. 2) Under other conditions the plume expansion is found to exhibit a degree of nonradial behavior in which case the Modified Simons model is used.

To determine the conditions under which the Modified Simons model is preferred, a further number of nozzle exit conditions have been investigated. The density ratios along the plume axis of these nozzles have been calculated using both the Simons model formulations. The initial conditions investigated were those for the nozzle whose solution is shown in Fig. 7. Under these conditions, the two solutions agree to within 4%. The exit nozzle angle and Mach number were then varied until the solutions were found to diverge by 10%. This is the criterion on which the recommendation for usage of the Modified Simons model is based. Once again, variation of exit Mach number and nozzle exit half-angle was undertaken independently of one another. From this analysis the critical values for the nonradial parameters were found to be given by a=8.7r_E, b=32.2$r_E{}^2$.

In Fig. 10 the conditions under which the nonradial nature of the expansion plume becomes important is shown graphically. The solid line respresents the critical conditions described in the previous discussion. The locations of the four real thrusters investigated in this study are also shown on the figure. Given a set of nozzle exit conditions, it is possible to identify the most appropriate analytical flow model from Fig. 10.

It is interesting to note the form demonstrated by Eqs.(3-5) in terms of the physical parameters that characterize the flowfield. It is clear from our formulation that as the nozzle exit Mach number is reduced the flow approaches the radial approximation. This would appear to be physically sound in that, since Mach number is the ratio of directed velocity to random thermal velocity, then, when random motion becomes predominant, it is to be expected that the flow would approach the effusive, radial condition. Similarly, the dependence on nozzle exit half-angle is reasonable; since effusive flow ideally takes place through a thin orifice, the expansion effectively originates from a parallel-sided aperture. As the aperture takes an increasing angle of divergence, the bulk flow velocity will gain a significant nonaxial component, thus reducing the radial nature of the expansion.

Conclusions

The nonradial nature of plumes exhausting from nozzles with large exit Mach number (Ma_E>5) and/or large exit half-angle has been established. The failure of the Simons model to satisfactorily calculate the flow properties of such expansions is therefore explained. By analyzing MOC solutions of these plumes, a new analytical formulism termed the Modified Simons model has been derived. It has been shown that the modifications result in significant improvement in the prediction of density in nonradial expansion plumes. The critical conditions under which the new model is preferred are presented in graphic format.

Because of the relatively large computational expense of the MOC together with difficulties in effective output of flow properties, the use of the Modified Simons model is recommended for the engineering modeling of the types of nozzle discussed in the present study. The analytical nature of the model allows easy calculation of impingement effects on the complex configurations of modern satellites, so that improvements introduced in this work increase the usefulness of the Simons model in spacecraft design.

Acknowledgments

One of the authors (IDB) recognizes the financial support of a Science and Engineering Research Council/IBM Case award. The contributions made to this work by Dr. J.-C. Lengrand are gratefully acknowledged.

References

[1] Boynton, F.P., "Highly Underexpanded Jet Structure: Exact and Approximate Solutions" AIAA Journal, Vol. 5, No. 9, Sept. 1967, pp. 1703-1704.

[2] Simons, G.A., "Effect of Nozzle Boundary Layers on Rocket Exhaust Plumes" AIAA Journal, Vol. 10, No. 11, Nov. 1972, pp. 1534-1535.

[3] Legge, H. and Boettcher, R.D., "Modelling Control Thruster Plume Flow and Impingement" Rarefied Gas Dynamics, Plenum, New York, 1985, pp. 983-992.

[4] Boyd, I.D. and Stark, J.P.W., "Modelling of Small Hydrazine Thruster Plumes Using Discrete Particle and Continuum Methods", AIAA Paper 88-2631, AIAA Thermophysics, Plasmadynamics and Lasers Conference, San Antonio, TX, June 1988.

[5] Calia, V.S. and Brook, J.W., "Measurements of a Simulated Rocket Exhaust Plume Near the Prandtl-Meyer Limiting Angle" Journal of Spacecraft, Vol. 12, No. 4, July-Aug. 1975, pp. 205-208.

[6] Lengrand, J.-C., "Calculs de Jets Sous-entendus Issus de Tuyeres Supersoniques" Centre National de la Recherche Scientifique, Meudon, France, Rept. 75-4, 1975.

[7] Dettleff, G. and Legge, H., "Hydrazine Thruster Near Field Plume Profile Measurement: Part 2: Pitot Pressure Measurements" Deutsche Forschungs und Versuchsanstalt fur Luft und Raumfahrt, Gottingen, FRG, Rept. IB 222-85 A15, 1985.

[8] Lengrand, J.-C., private communication.

[9] Rogers, A.W., Allegre, J., Raffin, M. and Lengrand, J.-C., "Plume Impingement Heat Transfer Measurements on Mopdel Simulated Intelsat VI Configurations and Conversion to Real Flight Conditions", AIAA Paper 87-1604, AIAA 22nd Thermophysics Conference, Honolulu, HA, June 1987.

[10] Legge, H., Dankert, C. and Dettleff, G., "Experimental Analysis of Plume Flow from Small Thrusters" Proceedings of the 14th International Symposium on Rarefied Gas Dynamics, Vol. 1, Univ. Tokyo Press, Tokyo, Japan, 1984, pp. 279-286.

[11] Vick, A.R., Andrews, E.H., Dennard, J.S. and Craidon, B., "A Comparison of Experimental Free-jet Boundaries with Theoretical Results Obtained with the Method of Characteristics" NASA TN D-2327, 1964.

[12] Schlichting, H., Boundary Layer Theory, McGraw-Hill, New York, 1960.

[13] Lengrand, J.-C., private communication.

Simulation of Multicomponent Nozzle Flows into a Vacuum

D. A. Nelson* and Y. C. Doo*
The Aerospace Corporation, El Segundo, California

Abstract

Several realistic and idealized models of multicomponent nozzle flows into a vacuum have been simulated by Bird's Monte Carlo method. Comparisons with available experimental data show excellent agreement for angular mass flux distributions. Idealized models aid the understanding of basic results and mechanisms leading to flow at large angles from the nozzle axis. The results contradict continuum model expectations and show the necessity of molecular methods for meaningful predictions.

Introduction

The vacuum expansion of a nozzle flow is a recurring problem of spacecraft systems which leads to a number of undesirable effects. These include impingement forces, moments, and heating, as well as the potential for contamination of thermal control devices and instrument components.

It is not possible in a ground base facility to fully characterize a given nozzle flow, let alone the many possible variants for which information may be needed. A few well-planned and executed experiments are, of course, essential, but the real need is for general and accurate analytical methods through which complete predictions of such flows can be made with confidence.

Vacuum expansions from rocket thrusters rapidly transition to rarefied flows, particularly in the region of greatest interest which encompasses expansion turning

*Fluid Dynamics Department.

angles greater than 90 deg in many cases. Since rarefaction is a prominent feature of these flows, it is perilous to base predictions on continuum methods, although this has generally been done in the past due to absence of suitably developed molecular techniques and the lack of familiarity with rarefaction phenomena.

The present work applies Bird's Monte Carlo method to the flow generated by a 22-N bipropellant rocket thruster. The internal flow was calculated initially by a Navier-Stokes solver, but it was found that considerable velocity slip would be expected downstream of the nozzle throat. Consequently, the Monte Carlo method was used from the throat region and out into the external plume. The results for the internal flow have been previously reported[1]; thus, the present paper focuses on the external flow, where rarefaction is most prominent.

The 22-N thruster was chosen because of earlier experimental measurements in a high-vacuum facility.[2] Quartz-crystal microbalances (QCMs) cooled to about 25°K were employed to determine plume angular mass flux distributions for a wide range of parameter variations.

Our calculations explore the effects of nozzle area ratio for comparison with the experimental results and as an extension of the conditions examined there. In addition, the flow of a realistic gas with rotational degrees of freedom is compared to that without rotational freedom but with otherwise identical properties. The reasons for such a comparison will be made clear later in the paper. Calculations were also made for both fully accommodating, diffuse nozzle walls and specular reflectors.

The principal basis of comparison is the angular mass flux distribution for all the analytical results and the experimental data. Other flow features are also shown to illustrate primary rarefaction effects.

Computational Features

Analytical Methods

A one-dimensional expansion is used to estimate the chemical kinetics effects and the nozzle exit plane chemical composition. The major chemical species shown in Table 2 are then used to perform an ideal gas Navier-Stokes calculation of the internal flow with the objective of handing off continuum flow properties as input data for flow continuation by the Monte Carlo method. The handoff surface is determined on the basis of some rarefaction onset criterion that compares the mean molecular scale to some macroscopic gradient scale or, equivalently, the collision rate to a macroscopic rate of change.

For small-scale nozzles one should consider the onset of velocity slip using the criterion

$$Kn_s = \left(\frac{1}{\nu}\frac{du}{dn}\right)_w \geq Kn_s^*$$

where ν is the collision frequency, u is the velocity parallel to the wall, n is the coordinate normal to the wall, and Kn^*_s is a number criterion for onset of slip flow. (Note that Bird's criterion for expansive flows is $P = \nabla \cdot \vec{v}/\nu \geq P^*$, which is very similar.) In the present case our Navier-Stokes solution indicated that $Kn_s > 0.01$ at roughly the nozzle throat location; hence, the Monte Carlo solution was started at the throat plane for all of the work reported herein.

Our Monte Carlo procedures differ from Bird[3] mainly in two respects: an exact cell indexing, is used rather than the fuzzy point indexing, and a net flow input boundary condition is employed in preference to the total inflow condition. The net flow is the difference of the total inflow and total outflow at the input boundary element and is fixed by the continuum solution. Past experience showed that net flow was not generally preserved when the total inflow boundary condition was imposed.

Parameters and Boundary Conditions

A description of the physical parameters for the nominally 22-N thruster is given in Table 1. Table 2 shows the chemical species, composition, and molecular parameters used for the simulation. Molecules were introduced at the throat plane with a Maxwellian velocity distribution, and values of velocity, pressure, and temperature were obtained from the Navier-Stokes solution. The nozzle walls were taken as constant temperature, fully accommodating, diffuse surfaces in most cases. One calculation was made with specular reflections and zero accommodation. The external limits of the spatial domain can be described as perfectly absorbing, since molecular histories were terminated at crossing and no molecules were introduced there. This is the boundary condition applicable to supersonic velocities, a condition that was satisfied in this work.

Table 1 Bipropellant engine characteristics

Thrust, N	22.24
Chamber pressure, N/m^2	6.895×10^5
Chamber temperature, °K	3009.44
Throat area, m^2	1.1676×10^{-5}
Area ratio, nominal	50, 100, 150
Chamber viscosity, $N\text{-}s/m^2$	8.3616×10^{-5}
Viscosity temperature exponent	0.72

Table 2 Various species considered in the Monte Carlo Simulation

Molecular species	Mole fraction	Molecular weight	Reference collison cross section, πd^2, $(m)^2$	Rotational degrees of freedom	Equilibrium ratio of specific heats	Species group[a]
Mixture	1.0	20.79	-	2.27	1.38	
H_2O	0.3183	18	3.505×10^{-19}	3	1.33	1
N_2	0.3026	28	2.660×10^{-19}	2	1.40	2
H_2	0.1689	2	1.333×10^{-19}	2	1.40	3
CO	0.1328	28	2.679×10^{-19}	2	1.40	4
CO_2	0.0358	44	4.681×10^{-19}	2	1.40	5
H	0.0249	1	1.007×10^{-19}	0	1.67	6
OH	0.0123	17	1.439×10^{-19}	2	1.40	7
NO	0.0226	30	2.679×10^{-19}	2	1.40	4
O	0.0012	16	1.439×10^{-19}	0	1.67	7
O_2	0.0009	32	2.679×10^{-19}	2	1.40	4

[a]Species in the same group were modeled as a single species in the Monte Carlo Simulation.

Results and Discussion

Only a small sample of the available results can be presented here. The focus will be on the experimental comparisons, nonequilibrium phenomena, and results that address model weaknesses and aid in the interpretation of our findings. In general our computations achieved a minimum cell average population of 10 molecules. Around the nozzle lip the populations were significantly higher by design.

Table 1 indicates that the primary variable was nozzle area ratio. The nominally 100:1 area ratio (actual value ≈ 86 from contour measurements) is considered our standard case.

Figure 1 shows a comparison of numerical results for the angular mass flux distribution with that measured by a cooled QCM. The excellent agreement is quite remarkable, especially in view of both the experimental and modeling uncertainties. However, it is certainly significant that

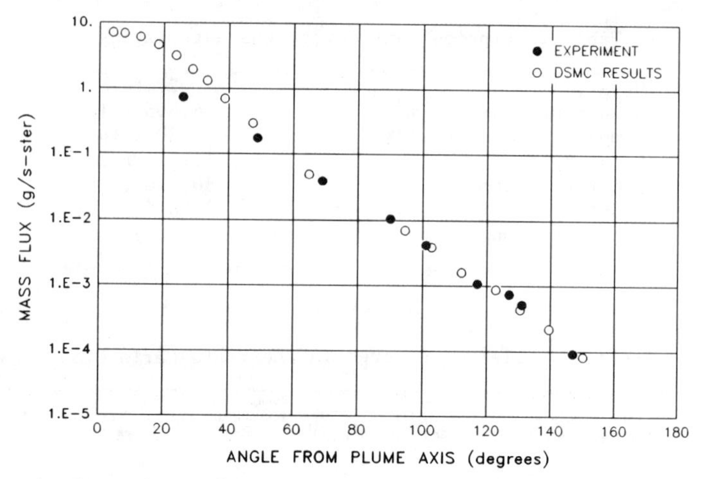

Fig. 1 Comparison of Monte Carlo predictions with measurements of external angular mass flow distribution for the 100:1 nozzle

the slopes agree so well irrespective of those uncertainties. It is believed that the agreement in absolute magnitude is largely due to lack of sensitivity of this flow property to perturbations in the nozzle operating conditions. This is suggested by the results of the experimental program[2] where only minor if any changes were observed for varying chamber pressure, area ratio, and exhaust composition. Only the effect of area ratio has been explored in this work. Figure 2 shows the results of simulation and experiment for a 50:1 nozzle. Once again there is excellent agreement, and, moreover, there is little difference between these and the 100:1 area ratio results, in agreement with the experimental findings. Though it is difficult to determine the cause of this effect unambiguously, it appears likely to be strongly influenced by the highly similar lip flow conditions for these two area ratios[1].

Figure 3 shows results for an area ratio of 150:1 for which experimental data are unavailable. One expects that increasing area ratio will at some point have an influence on the angular mass flux distribution at large angles due to the nozzle exit plane Knudsen number λ/D growth. For the conditions studied here, the 150:1 area-ratio has a relatively small effect, but it does tend to collimate the flow more along the axis and reduce it at larger angles. This is at least partially due to somewhat lower densities and temperatures in the lip region of this nozzle.

It was already mentioned that T-V exchange processes were not modeled in these calculations. As indicated in

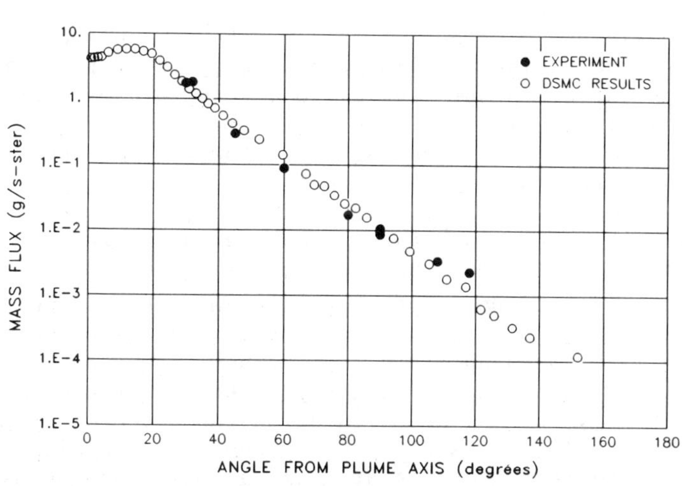

Fig. 2 Comparison of Monte Carlo predictions with measurements of external angular mass flow distribution for the 50:1 nozzle

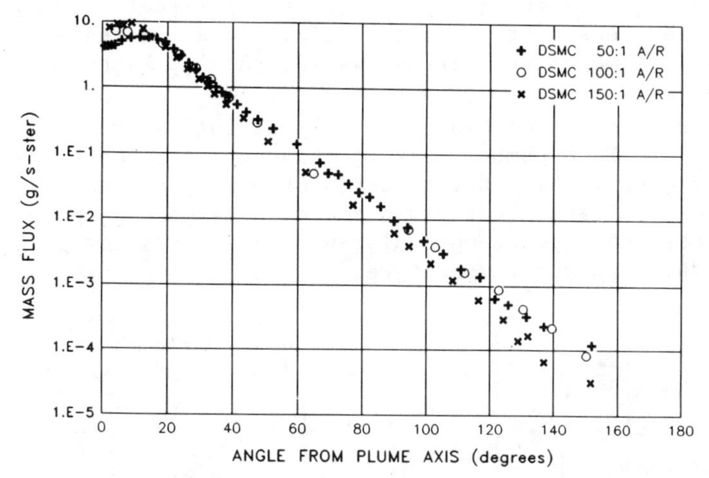

Fig. 3 The effect of nozzle area ratio on the external angular mass flow distribution

Table 2, the effective specific heat ratio was 1.38 when only rotational degrees of freedom were considered. To study the role of internal degrees of freedom, a fictitious gas was created by setting internal degrees of freedom to zero for all molecular species. Figure 4 compares these results with the standard case. Over most of the range shown the changes are not dramatic. The somewhat lower values are again at least partly due to lower densities and temperatures in the nozzle lip region. This comparison suggests that a lack of vibrational degrees of freedom does not influence our results to a large degree. From a one-dimensional expansion with chemical kinetics and T-V equilibrium, it was found that the flow was very nearly chemically frozen with a gas ratio of specific heats that varied from about 1.24 in the chamber to 1.35 at the nozzle exit. This is a rather weak variation in comparison to what was explored in our calculation.

The effect of wall interaction model was examined by considering specular reflection. Figure 5 shows the comparison with the standard case. Here we found no flux of molecules with molecular weight greater than two beyond about 45 deg. Since the calculation can detect a flux at least to the lowest value shown on the figure abcissa, one can conclude that there is at least a four-order of magnitude decrease in the flux beyond about 60 deg from the plume axis for molecules with molecular weight of 18 or more. Figure 6 shows results for the flux when hydrogen is counted. In the specular case the internal flow Mach number exceeds 5 at the nozzle lip. In all cases with fully accommodating, diffuse walls, the lip Mach number is unity.

The presence of hydrogen at large angles even when the exit plane Mach number is well above unity seems to be easily explainable only in terms of molecular scattering and the fact that hydrogen has a much larger velocity variance than higher-molecular-weight species for given mean velocity and temperature.

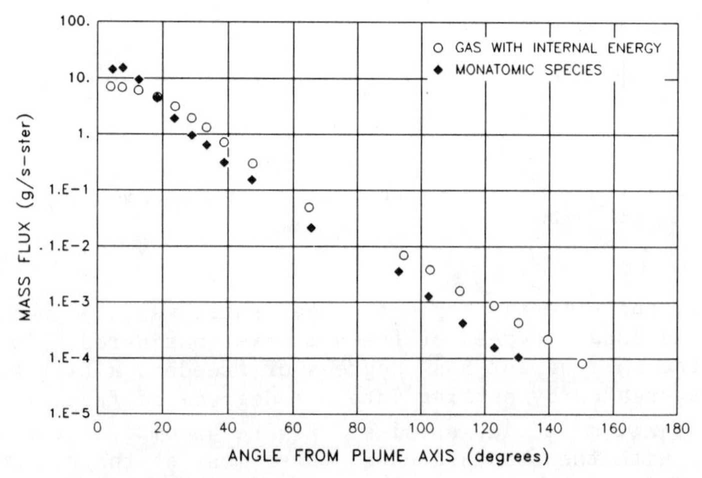

Fig. 4 The effect of internal degrees of freedom on angular mass flow distribution

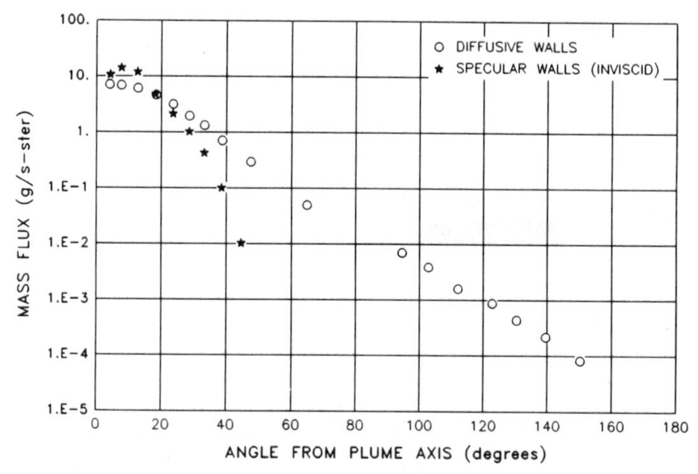

Fig. 5 The effect of an inviscid boundary condition on angular mass flow of molecular species with molecular weights $\geq$ 18

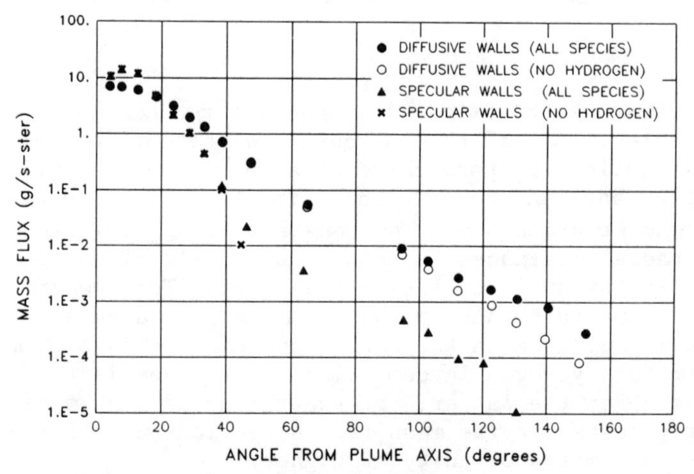

Fig. 6 The effect of an inviscid boundary condition on angular mass flow of all species (molecular weights $\geq$ 1)

The latter two figures make it clear that at large angular coordinates the number density of hydrogen must be quite a bit larger than the other molecular species. Figure 7 illustrates this species separation in terms of mixture molecular weight. For hydrogen (atomic plus molecular) the concentration decreases along the nozzle axis and increases markedly in the region of large flow angle. For heavier species the concentration tends to

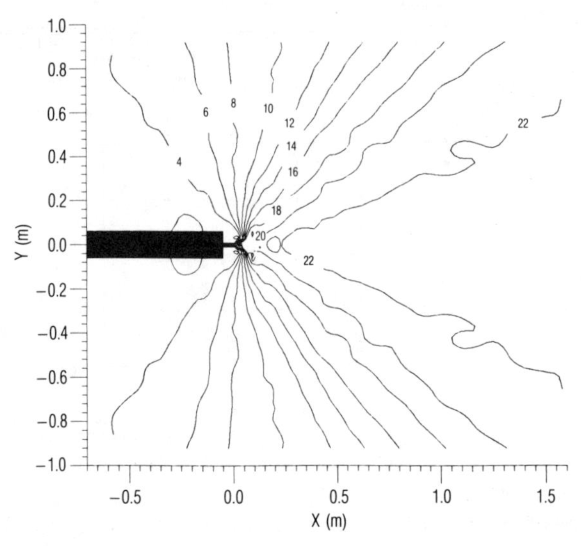

Fig. 7 The mixture molecular weight as an illustration of species separation in a vacuum plume expansion

increase along the axis and is sharply reduced at large flow angles. All of the molecular weight contours seem to be asymptotically approaching straight lines that align with flow angle contours. This fits the intuitive notion that concentration should be constant along flow lines at large radial distances from the nozzle exit plane because of negligibly small collision frequency. This separation effect seems to be due to the fact that the angular scattering diagram in the nozzle coordinates is much more uniform for hydrogen in comparison to the heavier species. For the latter, the scattering is much more strongly peaked in the mean flow direction for all mean flow Mach numbers of unity or higher.

For the internal flow, it was found that T-R processes were essentially in equilibrium. For the external flow, however, the rotational and translational temperatures diverge rapidly. This is shown in Figs. 8 and 9. The rotational temperature freezes rapidly during the lip expansion but continues to decrease along the axis even though it significantly lags the drop in translational temperature. The translational temperature shows clear indications of freezing as well. The entire region to the left of the 200°K contour is essentially at that temperature. The extremities of the W-shaped 100°K contour are freezing, but the centerline values are still

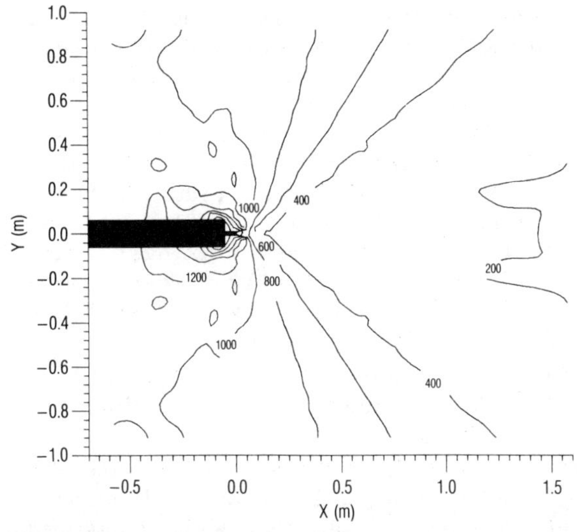

Fig. 8 The freezing of rotational temperature in a vacuum plume expansion

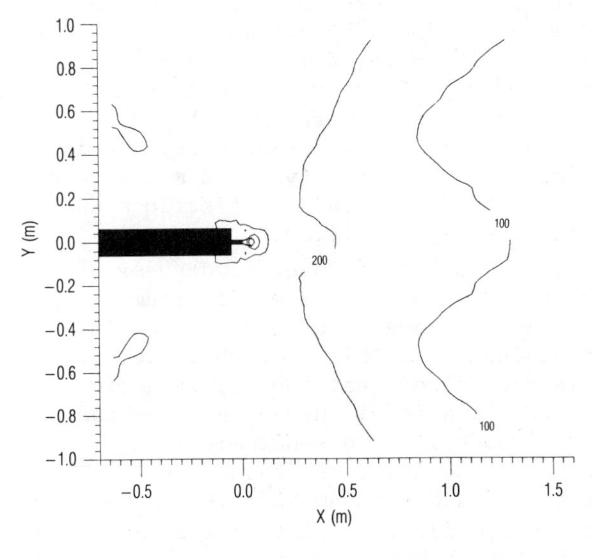

Fig. 9 The freezing of translational temperature in a vacuum plume expansion

dropping. It is quite clear that continuum methods would not apply to the present external flow.

Summary and Conclusions

Direct simulation of several realistic and idealized multicomponent nozzle flows into a vacuum have been carried out for a 22-N rocket thruster.

The realistic simulations were compared to an existing data base, and excellent agreement was obtained. The quantitative agreement is at least partially due to an insensitivity of the angular mass flux to shortcomings in the model; however, it is quite significant that the Monte Carlo method produced the same slope as the data. Furthermore, the measurements showed that reducing the nozzle area ratio from nominally 100:1 (actual 86:1) to 50:1 did not significantly affect the mass flux distribution, and this was also found computationally. These comparisons give strong support to the validity of Bird's Monte Carlo model for rarefied plume expansions.

It was found that increasing the area ratio to 150:1 did not significantly reduce the angular mass flux for this particular system.

Idealized flows were studied to help interpret the results and in particular the mechanism leading to mass flow at large angles. It has been widely assumed for many years that plume expansions can be reasonably predicted by continuum methods if one only allows for a boundary layer in the initial conditions. If there is no boundary layer, then large angle flow is precluded by the mythical Prandtl-Meyer limiting angle. Furthermore, a monatomic substance cannot expand beyond a 90-deg angle in the continuum model even with a boundary layer. These were part of the motivation for studying fictitious flows; one being a gas with zero internal degrees of freedom (viz., a monoatomic substance) and the other having specular reflections and therefore an inviscid boundary condition.

It was found that the number of internal degrees of freedom or, what is the same thing, the equilibrium specific heat ratio, did not have a large effect on the angular mass flux distribution. This is perhaps not so surprising when it is recognized that the lip expansion leads to freezing of the rotational energy, and the gas behaves thereafter as though it were monatomic in any case. It is difficult to see how these facts can be reconciled with the use of a continuum model.

The inviscid boundary condition led to a lip Mach number in excess of 5, yet this did not preclude mass flow at large angles for the hydrogen species though the flux was reduced by about an order of magnitude. This completely contradicts continuum expectations.

The only apparent explanation consistent with all of the previously mentioned findings is that the angular mass flux at large angles is established by molecular scattering processes in a rapidly rarefying flow generated in the lip expansion. If one defines the forward direction by the mean velocity vector, then it is easy to show that, in a mixture of widely different molecular weights, there is considerable backscatter of light species (i.e., hydrogen) even when the mixture speed ratio exceeds unity by a significant margin (i.e., up to five or slightly more). The heavier species, however, though showing significant backscatter at speed ratios close to unity, are barely deflected from the forward direction for larger speed ratios. These considerations provide a qualitative explanation of most, if not all, of the predicted variations in angular mass flux.

When the preceding points are considered along with the freezing of rotational and translational temperatures, it is clear that molecular methods are required for a meaningful prediction of this and many other vacuum plume expansions.

Acknowledgment

This research was supported by the Air Force Rocket Propulsion Laboratory (AFRPL). The Project Manager for AFRPL was Lt. Mark Price.

References

[1]Doo, Y. C. and Nelson, D. A., "Analysis of Small Bipropellant Engine Internal Flows by the Direct Simulation Monte Carlo Method," AIAA Paper 87-1548, June 1987.

[2]Alt, R. E., "Bipropellant Engine Plume Contamination Program," AEDC-TR-79-28, Arnold Engineering Development Center, AFSC, USAF.

[3]Bird, G. A., "Monte-Carlo Simulation in an Engineering Context," Progress in Astronautics and Aeronautics: Rarefied Gas Dynamics, edited by S. S. Fisher, AIAA, New York, 1981, Vol. 74, Pt. I, pp. 239-255.

Kinetic Theory Model for the Flow of a Simple Gas from a Two-Dimensional Nozzle

B. R. Riley*
University of Evansville, Evansville, Indiana
and
K. W. Scheller†
University of Notre Dame, Notre Dame, Indiana

Abstract

A system of nonlinear integral equations equivalent to the Krook kinetic equation for the steady state is the mathematical basis used to develop a computer code to model the flowfields for low-thrust two-dimensional nozzles. The method of characteristics was used to solve numerically by an iteration process the approximated Boltzmann equation for the number density, temperature, and velocity profiles of a simple gas as it exhausts into a vacuum. Results predict backscatter and show the effect of the inside wall boundary layer on the flowfields external to the nozzle.

Nomenclature

A = coefficient in expression for collision frequency
f = Boltzmann distribution function
f_B = Boltzmann distribution function at boundary
k = Boltzmann constant
m = mass of molecules
n = number density
r = plane polar coordinate
$\vec{r}$ = position vector
$\vec{S}$ = boundary position
$\vec{S}'$ = position along characteristic
T = local kinetic temperature
$\vec{u}$ = local mean velocity
u_x = x component of $\vec{u}$

* Chair and Professor, Department of Physics.
† Graduate Student, Department of Physics.

u_y = y component of $\vec{u}$
$\vec{v}$ = velocity vector
η = line integration variable
α = angle for velocity integration
η_0 = distance from grid point to boundary
ϕ = plane polar coordinate

Introduction

Self-induced contamination around a spacecraft can limit its usefulness because the contaminants may adversely affect spacecraft components or experiments being performed on or near the spacecraft.[1] One source of molecular contaminants is the propellant from the propulsion system.[2] Consideration is being given to continuously or semicontinuously firing small, low-thrust (~ 0.1 N) resistojets to partially or totally overcome on-orbit aerodynamic drag on the proposed U. S. Space Station.[3] Thus, there is considerable interest in the gas flow external to the jet's nozzle. It would be useful, for design purposes especially, to know how the external flowfield is influenced by the flowfield inside the nozzle, particularly near the exit plane.

This paper describes a mathematical model and computer code being developed to study the effect of nozzle design on the induced molecular environment around low-thrust nozzles. The model allows one to calculate the flowfield of a simple gas as the gas expands into a vacuum from a two-dimensional nozzle. The model is based on the molecular view of a gas and uses statistical kinetic theory to describe the behavior of the gas flow.

The code has been written in FORTRAN and is being run on the CRAY-XMP computer at the NASA Lewis Research Center in Cleveland, Ohio. Eventually, results are expected to provide insight into the gas flowfield near the nozzle lip, where significant backscattering may occur.

For this paper, three model-dependent external flowfields were generated and compared with each other and with the flowfield that was used as the initial starting approximation. These results are preliminary because the computer program is still being refined.

The numerical techniques used in this two-dimensional model are also to be incorporated into a code to describe gas flow through and from an axisymmetric three-dimensional nozzle.

Background

The mathematical framework for the description of the gas flow is the time independent Boltzmann equation for binary collisions.[4] The complete scattering term was replaced by a Krook-type (collision relaxation) approximation[5-7] to simplify the collision calculations. Starting with a zeroth iteration set of flowfield parameters for temperature, number density, and mean velocity (T_0, n_0, and $\vec{u}_0$), which were generated for the "far field" assuming flow from a point source as characterized by Boynton,[8] the method of characteristics was used to find the first iteration solution to the approximated Boltzmann equation from which first iteration values for T_1, n_1, and $\vec{u}_0$ were calculated. These first iteration values replaced the zeroth iteration values, and second iteration values were calculated. This procedure was repeated until a reasonable convergence was obtained. The Krook approximation in the Boltzmann equation allows a solution in the continuum, transition, and free-molecular regions of the flowfield. All of these regions are present in the flowfield from a low-thrust nozzle.

To the authors' knowledge, the only work that applies the Krook kinetic equation to the gas flow from a two-dimensional nozzle is by Peracchio.[9] The present paper draws substantially from Peracchio's procedures in his analytical study. There are two major differences between this paper and Peracchio's paper. First, Peracchio assumed a uniform line source at the nozzle exit plane (no boundary layer along the inside nozzle wall was assumed), and the velocity used in the exit plane boundary condition was only in one direction (perpendicular to the exit plane line). Second, the flowfield studied by Peracchio was restricted to a region closer to the nozzle exit plane and did not include the region outside the nozzle behind the exit plane (backscatter region).

Mathematical Formulation

For steady state, the Krook kinetic equation[7] or iterate equation[4] in differential form is:

$$\vec{v} \cdot \vec{\nabla}_r f(\vec{r}, \vec{v}) = A n_e(\vec{r})(f_e - f) \tag{1}$$

where f_e is written as

$$f_e = \frac{n_e}{(2\pi kT/m)^{3/2}} e^{-\left[\frac{m}{2kT}(\vec{v}-\vec{u})^2\right]} \tag{2}$$

and is an equilibrium Maxwellian velocity distribution multiplied by the number density.

Using the method of characteristics,[4,7] Eq. (1) can be written along a given characteristic direction in integral form as follows:

$$f(\vec{r}, \vec{v}) = f_B e^{-\left[\int_{\vec{s}}^{\vec{r}} A n_e \frac{ds}{v}\right]} + \int_{\vec{s}}^{\vec{r}} \frac{A n_e}{v} \left[f_e e^{+\int_{\vec{s}}^{\vec{s}} \frac{A n_e}{v} ds''} \right] ds' \tag{3}$$

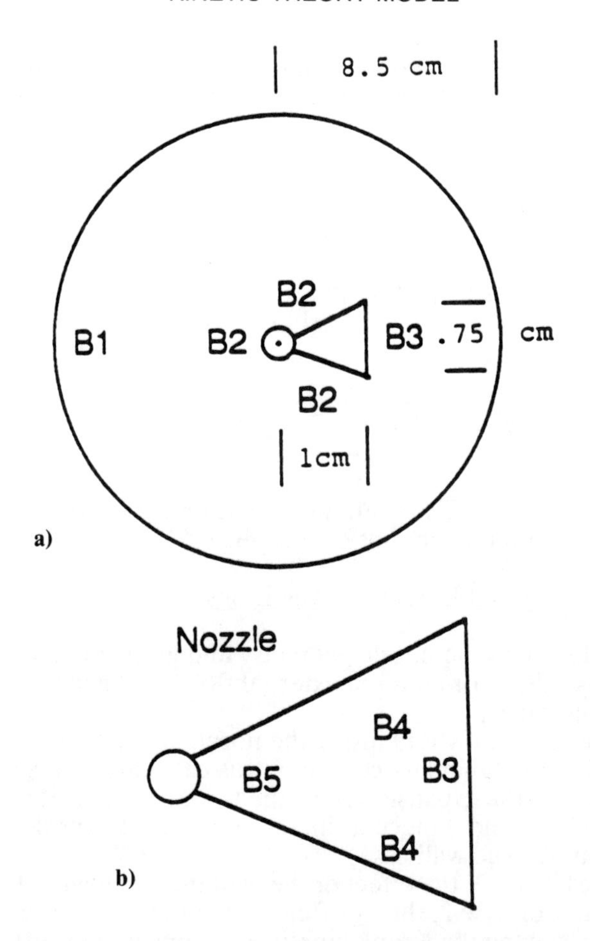

Fig. 1 In the top illustration, two-dimensional nozzle geometry and boundary conditions for three external flowfields are shown. The boundary conditions are $f_{B1} = 0$ for inward component of the velocity, $f_{B2} = 0$ and

$$f_{B3} = \frac{n_{B3}}{(2\pi kT_{B3}/m)^{3/2}} e^{-\frac{m}{2kT_{B3}}(\vec{v}-\vec{u})^2}$$

At the bottom, the two-dimensional nozzle geometry and boundary conditions are shown for the Krook model. The boundary conditions are $f_{B3} = 0$ for inward component of the velocity, based on the

$$f_{B4} = \frac{n_{B4}}{(2\pi kT_{B4}/m)^{3/2}} e^{-\frac{m}{2kT_{B4}}(\vec{v})^2}$$

and

$$f_{B4} = \frac{n_{B5}}{(2\pi kT_{B5}/m)^{3/2}} e^{-\frac{m}{2kT_{B5}}(\vec{v}-\vec{u})^2}$$

With the appropriate boundary conditions for f_B, the first-order solution to Eq. (3) can be obtained. Then the following moments can be obtained:

$$n(\vec{r}) = \int_{-\infty}^{\infty}\int_{-\infty}^{\infty}\int_{-\infty}^{\infty} f(\vec{r},\vec{v})d^3v \tag{4}$$

$$\vec{u} = \frac{1}{n}\int\int\int \vec{v}f(\vec{r},\vec{v})d^3v \tag{5}$$

$$\frac{3}{2}nkT = \frac{m}{2}\int\int\int (\vec{v}-\vec{u})^2 f(\vec{r},\vec{v})d^3v \tag{6}$$

The values of n,$\vec{u}$, and T are put into equation (3) for the second iteration, etc., until one obtains convergence.

Nozzle Geometry and Boundary Conditions

Figure 1a shows the nozzle geometry and gives boundary condition equations for all three external flowfields generated by the model for this paper.

For case 1, the flowfield inside the nozzle (from which the external exit plane boundary condition was calculated) was obtained by assuming that the expansion from the throat to the exit plane was isentropic. Hence, no molecular interaction with the inside nozzle wall was assumed (no wall case).

For cases 2 and 3, the effect of the wall on the flowfield is taken into account. For case 2, the flowfield inside the nozzle was generated by solving the Krook kinetic equation with diffuse scattering assumed for the inside wall boundary condition (see Fig. 1b) The temperature of the wall (T_{B4}) was fixed at 298 K, and the number density at the wall was calculated by $n_{B4} = nT/T_{B4}$, where n and T are the number density and temperature at the centerline at the same distance from the nozzle throat as the point on boundary B4. The macroscopic velocity $\vec{u}$ at the wall was assumed to be zero. A solution was assumed when the mass flow rate at the exit plane was equal to the mass flow rate at the throat. Fifteen iterations were required. The zeroth (initial) flowfield was assumed to be isentropic. The flowfield inside the nozzle for case 3 was generated by Penko[10] with a computer program (SVNAP2) at the NASA Lewis Research Center. The program used by Penko predicts the flowfield inside a nozzle for a compressible fluid described by the Navier-Stokes equations. It was constructed by Cline[11] and is described in detail in the reference manual for the code.

A summary of the exit plane boundary conditions at the boundary grid points for all three cases is given in Table 1.

TABLE 1
Boundary conditions at exit plane grid points

Angle, Deg	number density (molecules/m^3)	u_x m/sec	u_y m/sec	T, K
		No wall		
0	9.75×10^{23}	566.5	0.0	108.4
5	9.75×10^{23}	566.5	19.8	108.4
10	9.75×10^{23}	565.1	39.5	108.4
15	9.75×10^{23}	563.4	59.2	108.4
20	9.75×10^{23}	560.1	78.8	108.4
		Wall		
0	1.04×10^{24}	556.8	0.0	109.0
2	1.05×10^{24}	557.7	19.5	109.0
4	1.05×10^{24}	556.6	38.9	109.0
6	1.05×10^{24}	554.6	58.3	109.0
8	1.05×10^{24}	552.6	77.7	109.0
10	1.05×10^{24}	549.5	96.9	109.0
12	1.05×10^{24}	545.8	116.0	109.0
14	1.05×10^{24}	541.4	134.9	109.0
16	1.06×10^{24}	535.8	152.2	109.2
18	1.16×10^{24}	495.3	150.7	115.5
20	3.80×10^{23}	0.0	0.0	298.3
		Penko		
0	1.10×10^{24}	562.9	0.0	111.2
5	1.10×10^{24}	558.8	62.1	111.2
10	1.10×10^{24}	550.7	105.0	111.3
15	1.10×10^{24}	539.1	155.9	113.1
20	4.50×10^{23}	0.0	0.0	280.1

Grid System

Figure 2 depicts the polar coordinate system of the plane and the corresponding grid points around the nozzle where n, T, and $\vec{u}$ were calculated. It also shows one characteristic line along the direction $\vec{v}$.

Results

The convergence to a solution for the external flowfields for the iteration process for Eq. (1) was assumed when the mass flow rate across the first circle of grid points beyond the end of the nozzle (called R = 6) (see Fig. 2) approached the mass flow rate at the nozzle exit plane. At the same time, the change in the quantities n, T, u_x, and u_y for R = 6 grid points from the previous iteration approached zero. A visual presentation for the entire flowfield as shown in Fig. 2 for the no wall case, is given in Fig. 3 for the number density. This three-dimensional plot is presented to show the

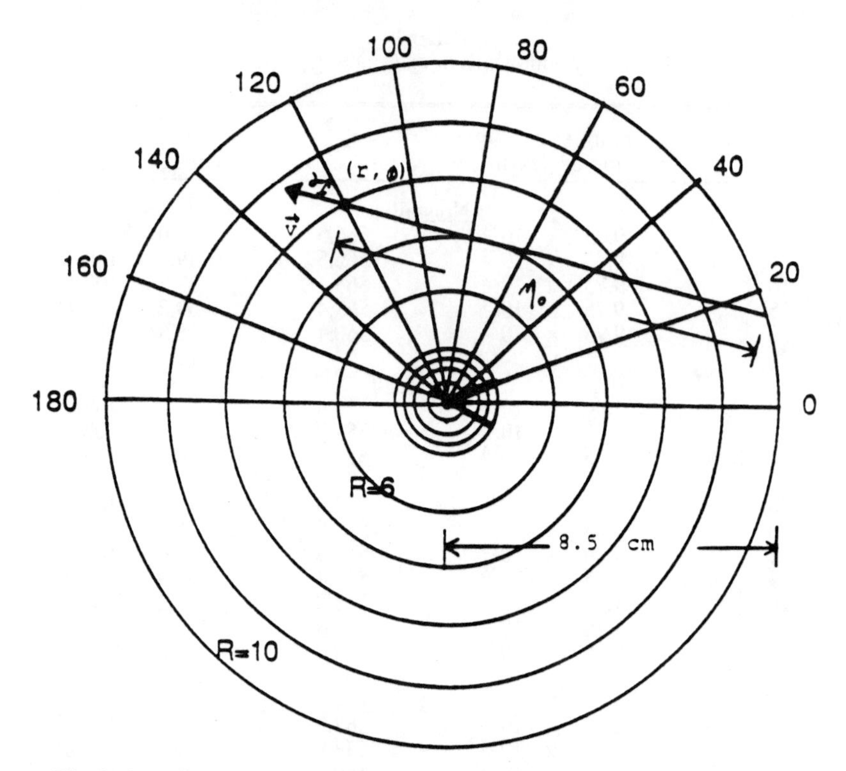

Fig. 2 Coordinate and grid system. The dark "V" at the center represents the nozzle. The nozzle and small circles are enlarged by a factor of 2 as compared to the remainder of the grid. The radii R = 5 and 6 are equal to 11.1 and 25.9 mm, respectively. Only the grid point values in the upper half-plane were calculated since the points of the lower half-plane are symmetrical. A positive component of v_x is along the 0-deg line.

smoothness in the flowfield as a function of position and has the number density plotted along the z axis.

Figure 4 compares more directly the Boynton[8] starting number density as a function of angle with the number densities at R = 6 for the three different internal nozzle flowfields.

Figures 5-7 compare the external flowfield values of T, u_x, and u_y as a function of angle at R = 6 for the three different internal flowfields.

The Boynton[8] external starting solution, of course, produced no back flow beyond the Prandlt-Meyer angle. The effect of the internal nozzle wall on the external flowfield is clear as the number densities beyond 40 deg. for the two cases with the wall effect boundary condition increased over the no wall (inviscid) flow. However, the differences, which are more pronounced for the first

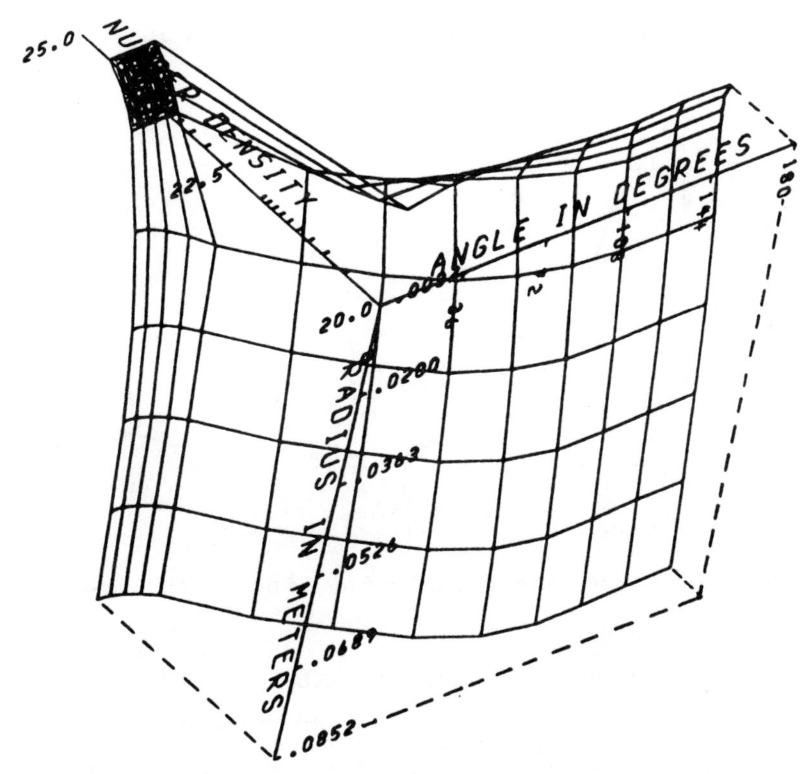

Fig. 3 Three-dimensional number density plot for case 1 (no wall). The z-axis numbers are in powers of 10. The shaded area is inside the nozzle.

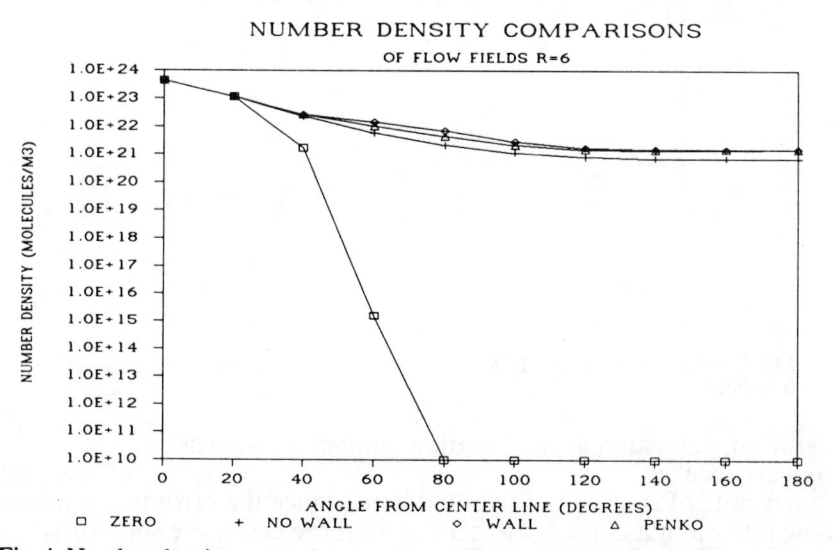

Fig. 4 Number density comparisons at R = 6 for the three external flowfields.

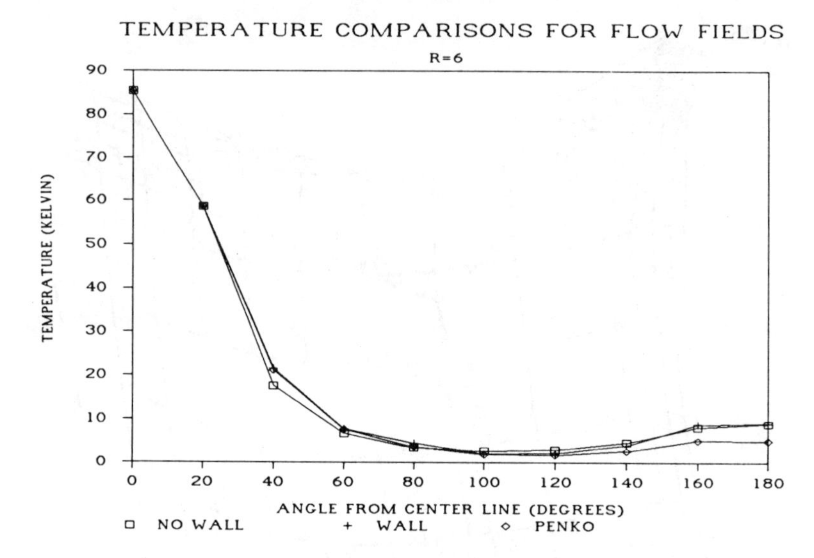

Fig. 5 Temperature comparisons at R = 6 for the three external flowfields.

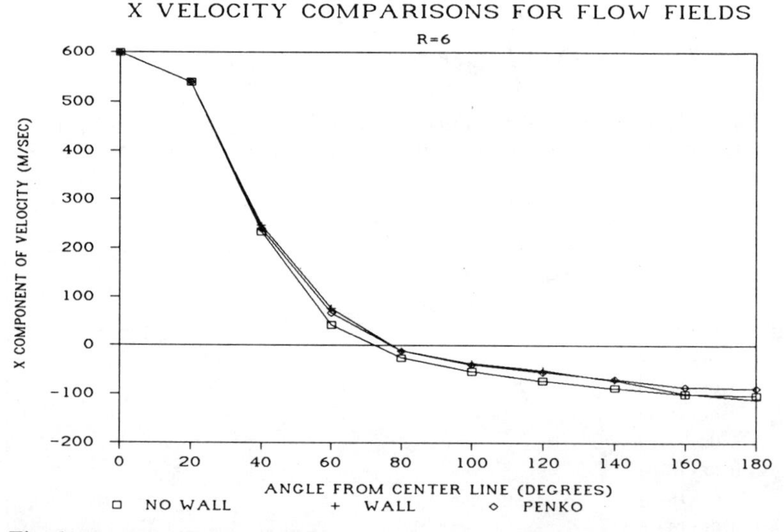

Fig. 6 x-component velocity comparisons at R = 6 for the three external flowfields.

iteration, decreased as the iteration number approached convergence.

A note of caution is warranted here. Since the error in the mass flow rate did not converge to zero in all three cases, the solutions should not be taken as final. The discrepancies may be due to the

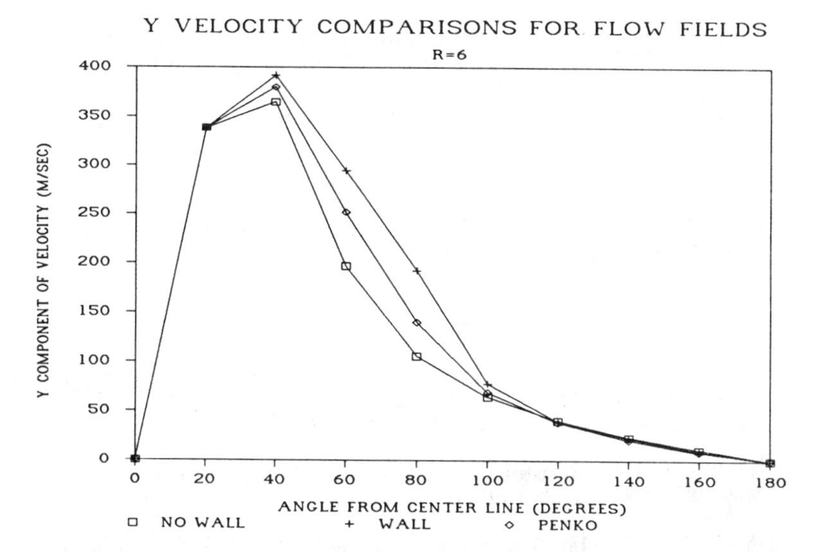

Fig. 7 y-component velocity comparisons at R = 6 for the three external flowfields.

large grid dimensions. In particular, the large difference in values of the quantities (n, T, u_x, and u_y) at the grid point at the nozzle lip and the grid points near it may gloss over details that influence the backscatter. One other assumption that may have a significant effect on the model results is the external wall boundary condition, which was assumed to be f = 0 for the preceding results.

Concluding Remarks

A model-dependent computer code using kinetic theory is being developed to study the flowfields external to two-dimensional low-thrust nozzles. Preliminary external flowfield results have been compared for three different internal flowfields. Although the number density in the backflow region for the no-inside-wall boundary-layer case was less than the two cases with an inside wall boundary layer, the differences are less than might have been expected. The most significant factor that contributes to the number density in the backscatter region seems to be the number density near the nozzle lip. More work on the model needs to be done. The effect of changing the outside wall boundary conditions needs to be studied; in particular, a diffuse boundary needs to be incorporated into the code. The effect of changing the nozzle wall temperature needs to be investigated.

Smaller grid spacings are needed in the nozzle region to determine if the flowfield results presented in this paper are altered substantially, especially at distances a few exit plane diameters from the exit plane. Future work should include a single simultaneous flowfield solution for both inside and outside the nozzle. Thus, the set of exit plane boundary conditions would be removed and the flowfield would extend from the nozzle throat to the exit plane and beyond without the intermediate exit plane boundary conditions.

Acknowledgment

This work was performed with the support of NASA at the Lewis Research Center under Grant NAG 3-746.

References

[1]Jemiola, J. M., "Spacecraft Contamination: A Review," Space Systems and Their Interaction with Earth's Space Environment, edited by H. B. Garnett and C. P. Pike Vol. 71, pp. 680-706, New York University, NY., 1980.

[2]Hoffman, D. J., "Resistojet Plume and Induced Environment Analysis," May 1987. (NASA TM 88957).

[3]Morren, W. E., Hay, S. S., Haag, T. W., and Sovey, J. S., "Preliminary Performance Characterizations of an Engineering Model Multipropellant Resistojet for Space Station Application," AIAA Paper 87-2120, June 1987. (NASA TM 100113).

[4]Patterson, G. N., Introduction to the Kinetic Theory of Gas Flows, Toronto, Toronto, Canada, 1971.

[5]Bhatnagar, P. L., Gross, E. P., and Krook, M., "A Model for Collision Processes in Gases. I. Small Amplitude Processes in Charged and Neutral One-Component Systems," Physical Review, , Vol. 94, May 1954, pp. 511-525.

[6]Krook, M., "Continuum Equations in Dynamics of Rarefied Gases," Journal of Fluid Mechanics, Vol. 6, Nov. 1959, p. 523-541.

[7]Anderson, D. G., "On the Steady Krook Kinetic Equation: Part 1," Journal of Fluid Mechanics, Vol. 26, Pt. 1, 1966, pp. 17-35.

[8]Boynton, F. P., "Exhaust Plumes from Nozzles with Wall Boundary Layers," Journal of Spacecraft and Rockets, Vol. 5, Oct. 1968, pp. 1143-1147.

[9]Peracchio, A. A., "Kinetic Theory Analysis for the Flowfield of a Two-Dimensional Nozzle Exhausting to Vacuum," AIAA Journal, Vol. 8, Nov. 1970, pp. 1965-1972.

[10]Penko, P., private communications, NASA Lewis Research Center, Cleveland, OH, Nov. 1987.

[11]Cline, M. C., "VNAP2: A Computer Program for Computation of Two-Dimensional, Time-Dependent, Compressible, Turbulent Flow", Rept. LA-8872, Los Alamos National Lab., Aug. 1981.

Transient and Steady Inertially Tethered Clouds of Gas in a Vacuum

Tony L. Farnham* and E. P. Muntz†
University of Southern California, Los Angeles, California

Abstract

The production of relatively dense, containerless gas clouds in a vacuum environment is of interest in many processes. Such clouds can be long-term steady-state clouds, or short-term transient clouds produced by a pulse of gas. They can also be varied to produce collisionless or collision-dominated flows. A series of experiments was done in creating both steady-state and transient collision-dominated clouds. The clouds were produced using large numbers of interacting gas jets from a ring-shaped array of small orifices. They were studied by measuring their density distributions using the electron beam fluorescense technique. The cloud densities are a result of a balance between the mass inflow from the jets and the loss of cloud mass to the surroundings. Profiles of the axial and radial density distributions of the clouds were observed for both the steady-state and the transient cases. A comparison was made of profiles to determine any similarities or differences between the two situations. The results indicate that the steady-state cloud is concentrated into a smaller volume than the transient cloud, which tends to spread out farther radially as well as upstream. These differences seem to indicate that a surprisingly long time is required to attain steady flow, which may be due to a long-term buildup of collision products.

---

*Research Assistant, Department of Aerospace Engineering.

†Professor, Department of Aerospace Engineering.

Introduction

The production of relatively high-density gas targets in an otherwise evacuated environment has been of interest in many applications. Providing charge exchange for charged particle beams,[1] a target for molecular interaction studies, or low-temperature transient targets for laser spectroscopic investigations are some examples of situations in which containerless clouds of gas in a vacuum can be of use. Traditionally, gas clouds have been produced using single steady-state or transient jets. To increase the efficiency and produce clouds that have a higher number density for a given mass flow, it is possible to use an array of inwardly directed orifice jets that produce an inertially tethered cloud of gas at the center of the ring. For experimenters with low mass flow pumping capabilities, the gas through the ring jet can be pulsed to produce a transient cloud that is collisionally dominated yet has a minimal mass flow.

The objectives of this study are to look at the generation, formation, and dissipation of a transient cloud of gas produced by a multiorifice ring jet and to compare the results to a long-term steady-state flow. The transient case is related to the steady-state case through a comparison of their respective number density distributions in the flowfield. The shapes of the clouds are also observed and compared to the shape of the theoretical collisionless cloud.

Inertial Tethering of Gas Clouds

The basic flow element (ring jet) used in this experiment is illustrated in Fig. 1. The ring jet contains an array of N inwardly directed orifices that approximate a continuous slot. As it turns out, this is a very good approximation, which breaks down only in the region near the ring where gaps between the orifices are observable. Orifices are used because they are easier to manufacture than a very narrow slot, and the number of jets can be adjusted easily. Typically, the flow through each orifice is in the continuum regime, but the interaction between the individual jets may be either collisionless or collision-dominated. The latter case is studied here because it produces the most dense and confined clouds.

Generally, the external dimensions of the ring are kept as small as possible in order to approach the ideal situation of an isolated, stand-alone gas cloud. In the pulsed jet case, the internal dimen-

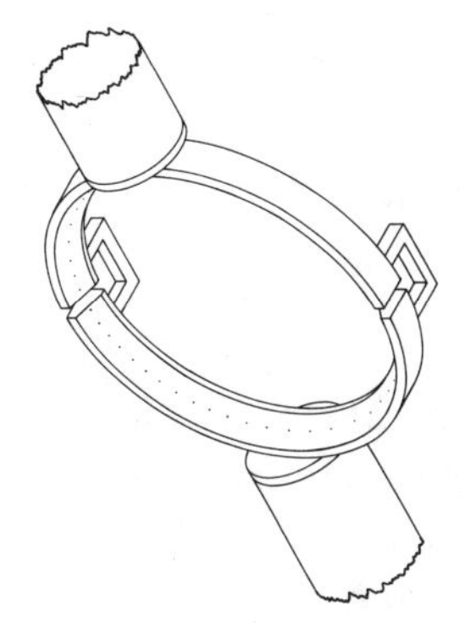

Fig. 1 An array of orifice jets comprising a ring jet – the basic flow element used in this experiment. Includes a cross section through the plenum.

sions of the plenum feeding the orifices are minimized in order to reduce the time required for the gas to fill the plenum.

Ring Nozzle Gasdynamics

1. Steady State

The geometric nomenclature used in this paper is illustrated in Fig. 2. Two types of interaction are of interest here: collisionless flows and collision-dominated flows.[3] In the collisionless case, the flowfield is produced by superposition of the flows from each individual orifice jet. This type of interaction will be used as a baseline flowfield and as a means of comparing the shapes of distribution curves. Using the description of an orifice jet from Ashkenas and Sherman,[2] the number density at any position X along the centerline of a ring with N orifice jets oriented at an angle ω can be found

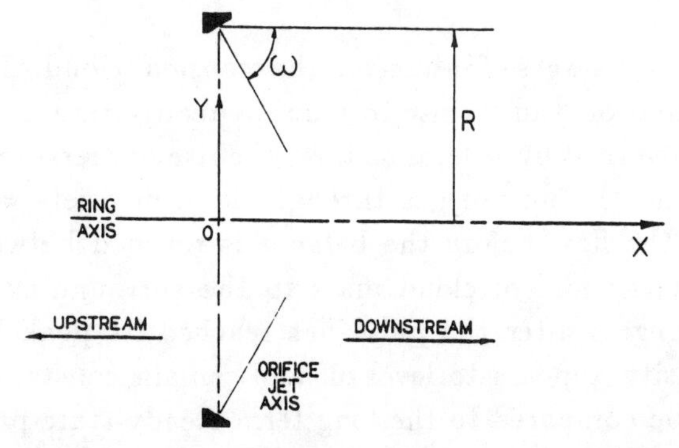

Fig. 2 Geometric nomenclature used in this paper.

analytically by superposition[2] and is

$$\mathcal{N}_{cl} = \frac{4c(\gamma)\dot{M}_R}{\pi f(\gamma) R^2 \sqrt{m\gamma k T_o}} \left[\frac{\frac{X}{R}\cos\omega + \sin\omega}{1 - (\frac{X}{R})^2} \right]^2$$

where R is the ring radius, γ is the ratio of specific heats, $c(\gamma)$ is a streamline scaling term as determined by Ashkenas and Sherman, $\dot{M}_R$ is the total ring mass flow, $f(\gamma) = [2/(\gamma+1)]^{(\gamma+1)/[2(\gamma-1)]}$, m is the gas molecular weight, k is the Boltzmann constant, and T_o is the stagnation temperature. The validity of the superposition prediction has been established in previous experiments with ring jets at very low mass flows.[4]

In the collision-dominated case, the flowfield is a result of the balance of gas influx from the jets and the loss of the cloud mass to the surroundings. The number density in this flowfield must be calculated using the Monte Carlo simulation method[5] or a combination of this and Navier-Stokes codes. These calculations are not presently available and will not be considered here.

2. Transient Flow

The transient case consists of a cloud that exists for only a short period of time (on the order of 10 ms compared to several seconds in the steady-state case). The investigation of this short-term cloud is an extension to the study of the steady-state case. Observing what happens in the formation and dissipation of the transient case provides insight into the dynamics of how the steady-state cloud is produced.

There are three stages of interest in the transient cloud. The first is the flow formation and its rise to a quasi-steady-state condition. This stage is governed by several factors: the rise of pressure within the ring plenum, the flux of gas through the orifice jets, and the interaction of the flow before the balance is reached between the gas influx and the loss of cloud mass to the surroundings. The second stage begins after the pulse has reached its peak level at which the density appears to level off and remain constant. This situation can be compared to the long-term steady-state profile to determine any effects that result from a continuous mass flow. The third stage that is of interest begins after the valves have shut off, allowing the cloud to dissipate to the surroundings.

In this experiment, the maximum duration of the pulse is limited to 40 ms. This limit is determined by the performance of the valves. After approximately 40 ms with the valves open, a flutter appears that interferes with the gas flow. To study the flow between 40 ms and a few seconds, it would be necessary to install new valves that could operate reliably for longer than 40 ms.

Experimental Apparatus, Operation and Results

1. Experimental Setup

The experimental apparatus is the same as in Ref. 3 for the steady-state case. It consists of a vacuum chamber two meters in diameter with a gaseous helium liner that is cooled to approximately 30°K. The ambient pressure with no mass flow into the chamber is in the $3–4 \times 10^{-6}$ torr range. The gas is argon [$\gamma = 1.67, c(\gamma) = 0.15, f(\gamma) = 0.562, m = 6.68 \times 10^{-23} g, T_o = 300°K$]. The ring jet contains 40 orifices and has a radius $R = 7.87$ cm and an angle $\omega = 60$ deg. Each orifice has a diameter $D = 0.0305$ cm. (With these conditions, the approximation of a continuous slot breaks down only within about 1 cm of the orifices.)

The ring is constructed in two halves to allow the electron beam used for diagnostics to penetrate to the center of the flowfield. Two Newport Research Corporation BV-100 high speed valves are used to create the pulse of gas. The two valves are driven open and closed using two BV-100D valve drivers. The valve drivers can be adjusted independently in such a way that both valves are triggered at the same time. In normal operation, these valves produce a pulse of 1–10 ms in duration. With slight modifications, they can produce a pulse of up to 40 ms in duration. In the transient experiments, the valves are generally used with a pulse of 8 ms, however, several trials were observed with a 40-ms pulse to determine any effects that might develop after the 8-ms cutoff.

The plenum that feeds the orifices has a cross section that is a square with 1 mm per side (see Fig. 1 cross section). The valve connections to the two half-plenums (Fig. 1) allow the gas to travel from the gas valve inlets to the end of the plenum in less than 0.5 ms. For the transient case, this time is small on the scale of the fill-up time. The pressure response within the plenum can be calculated by looking at the difference between the mass flows in and out of the plenum and the resulting increase in pressure. The following expression is obtained and shows how pressure changes with time:

$$p(t) = p_v \frac{A_{\text{in}}}{A_{\text{out}}} \left[1 - e^{-t/\tau}\right]$$

where p_v is the pressure in the valves, A_{in} is the effective cross sectional area of the inlet tube to the plenum, A_{out} is the total cross sectional area of the orifice jets, $\tau = \mathcal{V}/V^* A_{\text{out}}[(\gamma+1)/2]^{1/(\gamma-1)}$ is the time constant, $\mathcal{V}$ is the volume of the plenum chamber and $V*$ the choked speed of sound.

To compare the calculated values with what is really going on, a high-speed piezoelectric pressure transducer was installed in the ring plenum. The measured pressure signal of an 8-ms pulse is shown in Fig. 3, along with two calculated curves. The first is a calculation that was done using only the volume of the plenum. The large discrepancy from the experimental result caused some concern until it was discovered that the installation of the transducer also introduced additional volume into the plenum. This added volume

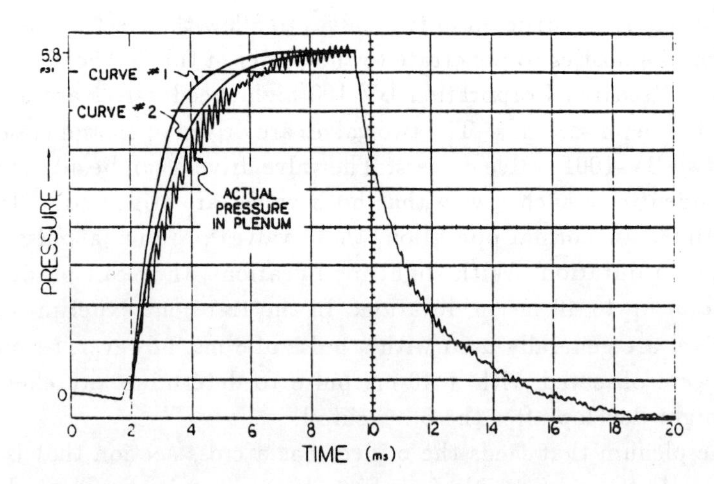

Fig. 3 Pressure in the ring plenum during an 8-ms pulse. Curve 1 shows the calculation using the volume of the plenum only. Curve 2 is the calculation with the extra volume from the pressure transducer taken into account. Curve 1 is used as the representative pressure in this experiment.

increases the time response. The second calculated curve takes this extra volume into account and has a better fit to the measurements.

It should be noted that the data presented here were taken without the pressure transducer installed. For this reason, the calculated values that were obtained without the additional volume will be used to represent the plenum pressure. The mass flow through the ring can then be estimated using the pressure in the plenum and the relation

$$\dot{M}_R = 2.7 \times 10^{19} m \frac{p_o}{14.7} N f(\gamma) \sqrt{\gamma \frac{k}{m} T_o}\ \frac{\pi D^2}{4}$$

where p_o is the pressure in the plenum (in psi). The stagnation pressure within the valves is 74.7 psi. In the pulsed mode at this pressure, the mass flow through the ring during a pulse is equivalent to approximately 0.4 g/s in the steady-state mode.

The electron beam fluorescense technique (see Ref. 6 for a detailed description) is used to measure the number densities at various axial and radial positions in the flowfield. In this process, a beam of high-energy electrons is used to excite the gas molecules in the cloud. The excited molecules then emit light, which can be measured. The intensity of the light emitted is proportional to the number density at that point. Light from the electron beam-induced fluorescense is focused on the photocathode of an EMI photomultiplier tube through a narrow passband filter centered at 4609.6 Å.

The electron beam current and the data from the photomultiplier are recorded using a Hewlett Packard HP-3325A waveform recorder, which stores the data in digitized form. Each run records 2048 points of data for both the electron beam current and the photomultiplier signal. For the 10-ms pulses, a period of 20 ms is recorded, and for the 40-ms pulses, a period of 80 ms is recorded. The waveform recorder is triggered by the same signal that triggers the valves.

Calibration is accomplished by measuring the intensity at a known mass flow through a single calibration orifice at a given distance from the electron beam. Using the known mass flow, the density field can be calculated from the relation for a free-flowing jet:[2]

$$\mathcal{N}_{FF} = \frac{4c(\gamma)\dot{M}_c}{\pi f(\gamma) d^2 \sqrt{m\gamma k T_o}}$$

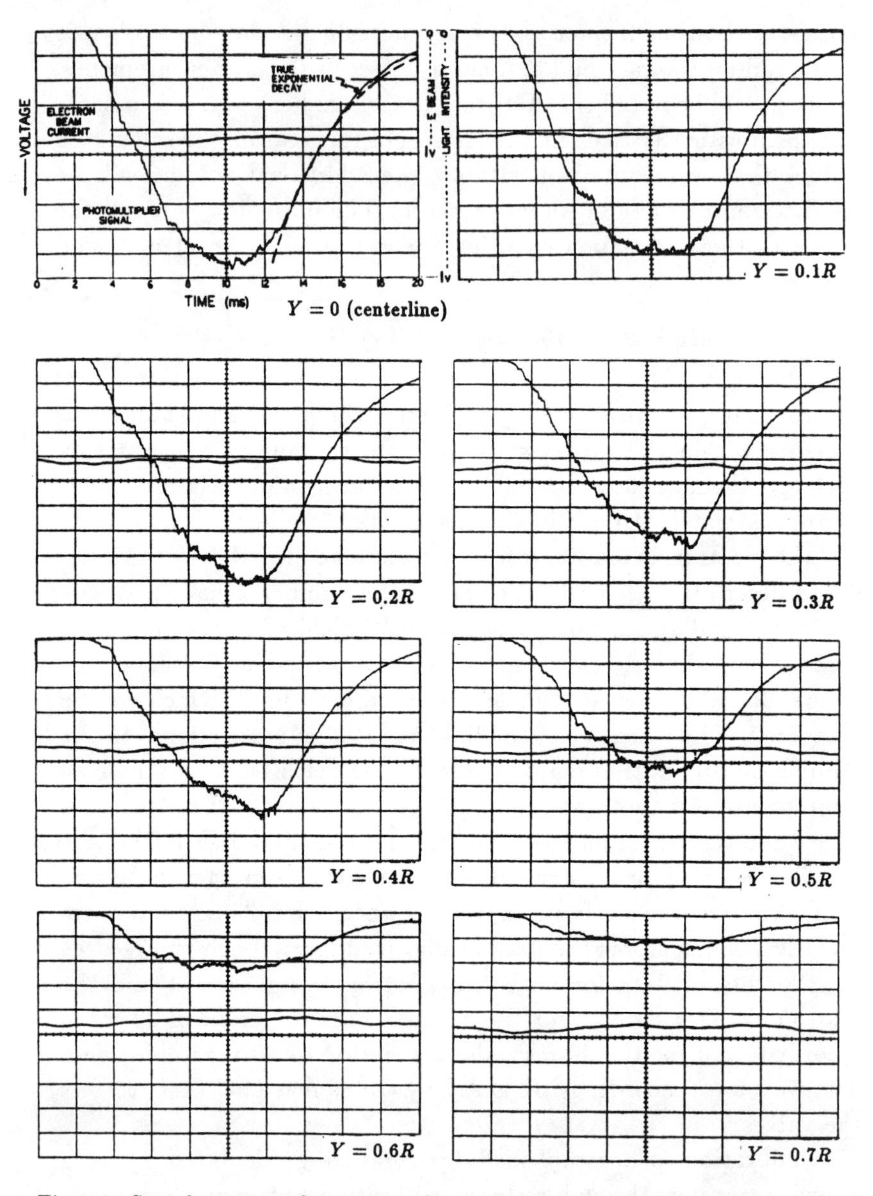

Fig. 4 Sample traces of transient data. Axial position X = 0.25R. The centerline plot illustrates a true exponential decay.

where d is the distance from the calibration orifice to the electron beam, and $\dot{M}_c$ is the mass flow through the orifice. The calculated number density is then related to the voltage of the photomultiplier signal to give a calibration.

2. Operation

In the experiments, the ring jet is traversed along its axis through a 20-keV electron beam that is oriented perpendicular to the axis. The position is determined with the use of a calibrated potentiometer that is connected to the traverse mechanism.

Because of the symmetry of the ring, it is assumed that the flowfield is axially symmetric. With this assumption, a given position along the electron beam is representative of the flowfield at that radial position in any direction. To observe off-axis positions in the flowfield, the photomultiplier is traversed in such a way that the point of interest along the electron beam is focused on the photocathode.

Typical samples of the transient pulse data are shown in Fig. 4. Illustrated are a sequence of radial measurements at a position 0.25 radii downstream of the ring. The horizontal axis is time (in 2-ms divisions), and the vertical axis gives both the photomultiplier output and the electron beam current. The zero position is the top line, the range of the photomultiplier signal is 1 V, and the range of the electron beam signal is 2 V.

The steady-state data are taken using the same general method. The differences are that the high-speed valves are removed and a

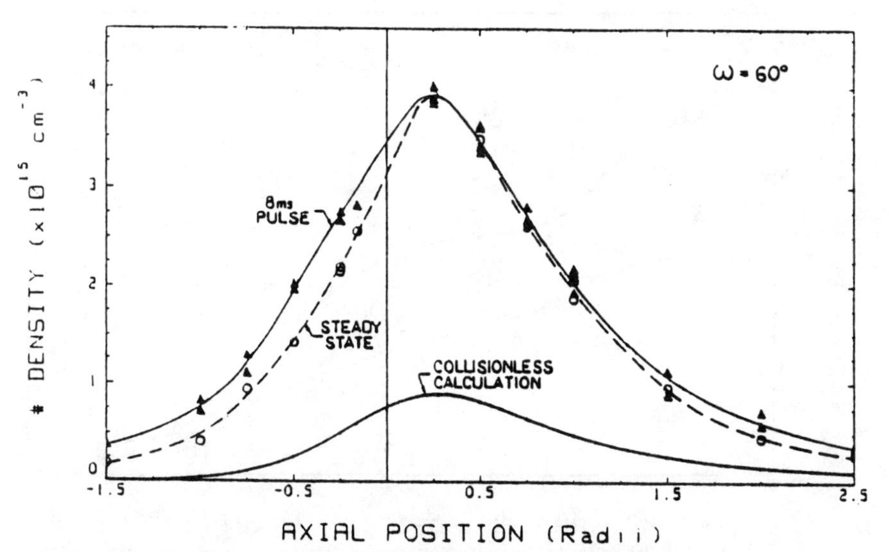

Fig. 5 Axial number densities for steady-sate and pulsed flow. Also shown is the calculation at the same mass flow for the collisionless assumption.

needle valve is installed to control the flow through a meter. A known mass flow is then allowed to run for several seconds, allowing the density field to build up to a constant steady-state condition before the waveform recorder is triggered to take the density reading. A mass flow of 0.4 g/s is used to provide a direct comparison to the transient data. The background pressure in the tunnel rises to a value of 2×10^{-4} torr and remains steady at this level during these steady-state runs.

3. Results

From the plots shown in Fig. 6 as well as other transient data, the formation of the gas cloud can be observed. The density rises to its peak value at an approximately exponential rate. However, the time constant for the cloud density is 4 ms (±10%), and, from the calculated value of the pressure rise in the plenum (Fig. 3), the rise time is 1 ms. This discrepancy is apparently due to some characteristic in the formation of the flowfield; however, the exact details are not understood.

Number densities of both the peak of the transient signal and the steady-state flows were obtained at 13 different axial locations along

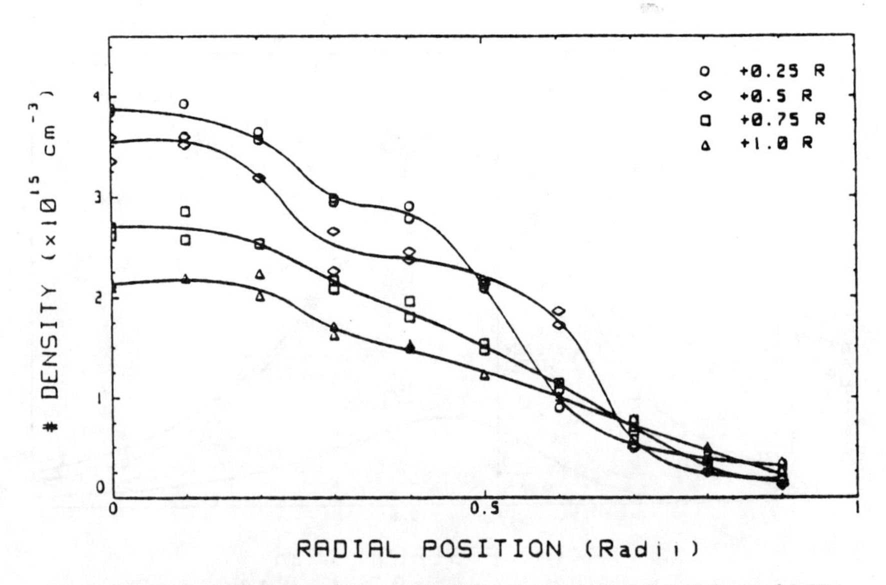

Fig. 6 Radial number densities of transient clouds at axial positions of 0.25, 0.5, 0.75, and 1.0 radii.

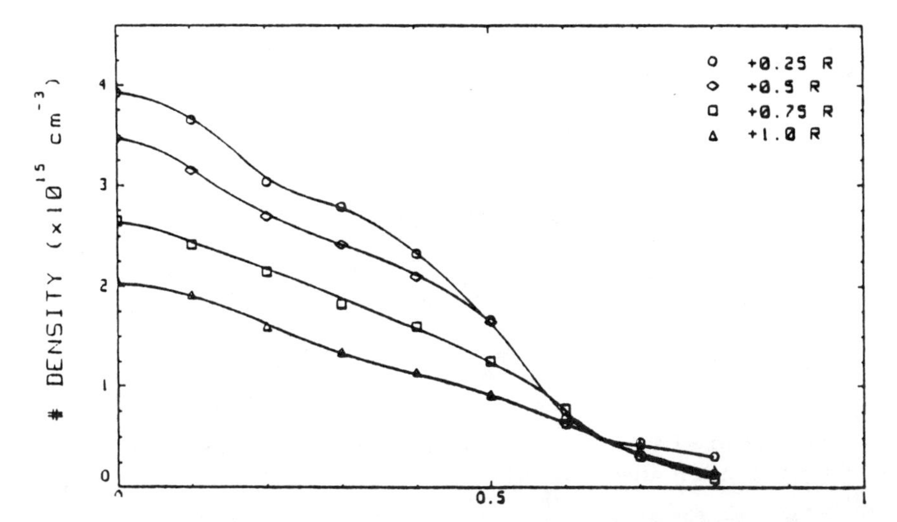

Fig. 7 Radial number densities of steady-state clouds at axial positions of 0.25, 0.5, 0.75, and 1.0 radii.

the centerline from 1.5 radii upstream to 2.5 radii downstream. For off-axis number densities, six axial positions were looked at (X/R = -0.5, -0.25, 0.25, 0.5, 0.75, 1.0), and each was traversed radially at increments of one-tenth of a radius. Several measurements were taken at each point in order to average out variations between pulses. (Under identical conditions, the pulsed gas cloud densities vary by an amount up to 8%.)

The number densities on the centerline for both the pulsed and the steady-state cases are shown in Fig. 5. Also shown are the calculated values using the collisionless model at the same mass flow. Radial densities of the transient cloud are shown in Fig. 6 for the axial positions 0.25, 0.5, 0.75, and 1.0 radii, and the steady-state densities are shown in Fig. 7. Because the densities within the cloud are relatively high, the background pressure (2×10^{-4} torr, due to the high mass flow) should not affect the major portion of the cloud.

Figure 8 illustrates the flowfield densities with respect to position for both the transient and steady clouds. Using this plot, differences in the flowfields between the two cases are easily distinguished. The peak densities on the centerline at the $0.25R$ position are the same for both cases; however, the steady-state case drops off in density faster than the transient case as position moves outward.

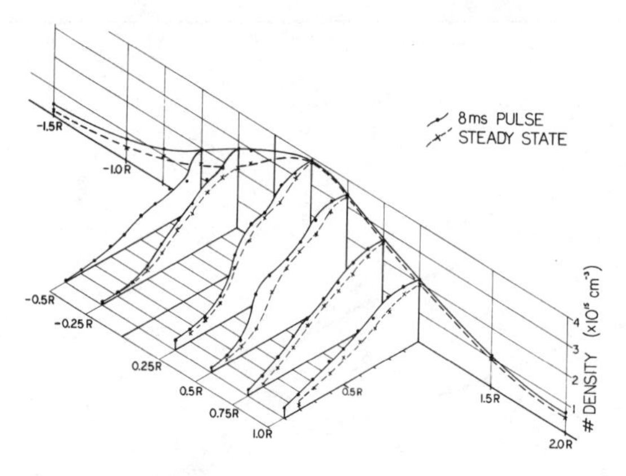

Fig. 8 Transient and steady axial and radial number densities, plotted with respect to their relative positions.

This deviation is observed in both the axial and the radial directions and is at its greatest upstream of the ring and in the region of the shock.

The details concerning what produces this effect are not fully understood, but it is believed to be caused by the shock-formation mechanism in the flowfield. When the cloud reaches a steady-state condition (when the mass flows in and out have reached a balance) the shock establishes itself at a stable position. Fig. 9 indicates that this position is located in the region around $0.5R$.

The 8-ms transient case, however, apparently does not have time to reach a state of complete equilibrium, and Fig. 8 shows that the shock is located at a position outside the steady-state position. Because of the finite length of time for the valves to open and the flowfield to form, the creation of the shock is not fully understood. It appears at the $0.6R$ position at the start of the flow, but after the steady flow has run for a few seconds, it has moved to a position of $0.5R$.

Because of a flutter introduced by the modification of the valves (to increase the pulse length), the peak number densities in the 40-ms transient data are not very uniform, having a variation of up to 20% in some places. However, from the data that were obtained, the location of the shock appears to remain at its outermost position throughout the duration of the 40 ms. This indicates that the shock takes longer than 0.05 s to settle at its steady-state position.

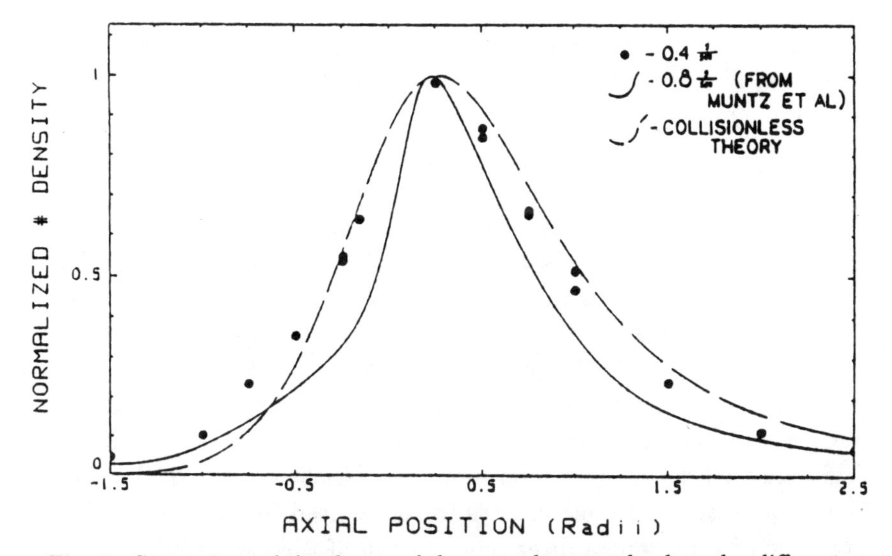

Fig. 9 Comparison of the shapes of three steady-state clouds under different conditions. Experimental results with mass flows of 0.4 and 0.8 g/s and a theoretical curve for the assumption of collisionless flow.

The shock disappears when the flow from the ring stops. Without this confining flow, the cloud simply dissipates in an approximately exponential decay, as the gas diffuses to the surroundings in all directions. The centerline plot in Fig. 4 shows a comparison to a true exponential decay.

The shape of the steady-state cloud (mass flow 0.4 g/s) was compared to the profile of a steady-state cloud with a mass flow of 0.8 g/s that was previously studied by Muntz et al.[3] and to the shape of the cloud in the theory of a collisionless flow. Fig. 9 shows the normalized number densities (along the ring axis) of these clouds. Note that the peak number densities are at the same position in all three cases. Collisionless theory number densities are calculated using the method described in Ref. 7.

Downstream of the -0.5R position, the profile at the lower mass flow fits closely to the theory of the collisionless flow on the ring axis, exhibiting only a small decrease in the width of the cloud. On the other hand, the high mass flow cloud has its high-density region much more concentrated at its center. The radial profiles of the clouds differ significantly, but this is expected due to the shock that forms in the collision-dominated flow. It appears that the higher influx of gas and the formation of the shock make the density in the central region increase more than the density in the outer areas. As the mass flow increases, a greater percentage of the gas is confined to the nucleus by the higher number of collisions.

Conclusions

In this experiment, inertially tethered clouds of gas in a vacuum were studied. Observations were made of both transient and steady-state clouds, and a comparison was made of their peak number densities at various positions. The formation and dissipation of the clouds were observed and appear to be well behaved and repeatable.

Given the dimensions of the ring and the pressure in the valves, the pressure response within the plenum can be reliably determined. This can then be used to determine the mass flow through the ring during a pulse. The same mass flow can be used to produce steady-state clouds, allowing a direct comparison of the two cases.

Although their peak densities are at the same location, the pulsed and the steady-state clouds have variations between their flowfield densities. First, the transient case has a broader distribution than the long-term case, and second, the shock that forms around the ring axis is located at different positions in the two clouds. This is apparently a time dependent issue, because the transient case is merely the beginning of a long-term flow.

It is not understood why the transient cloud tends to have a higher density upstream of the ring, but the higher density in the radial direction is related to the different position of the shock. The shock first appears at the $0.6R$ position. It remains there for at least 50 ms and eventually shrinks to a steady-state position of $0.5R$. The time that it takes for the shock to move was not studied because of the required modifications to the experimental apparatus.

Using the method of superposition, it is possible to calculate the shape of the cloud produced by collisionless interaction of the orifice jets flows. The axial profile of the collisionless flow gives a good estimate of the axial shape (but not the density levels) of the steady-state cloud with small ring mass flows (up to 0.4 g/s). The radial profiles differ significantly, however, due to the formation of the shock in the collision-dominated flow. For higher mass flows, collisions begin to affect the axial shape as well as the density, and the clouds tend to have their high-density regions concentrated in a smaller nucleus.

Acknowledgment

This work was supported in part by Grumman Corporation and by NASA Contract NAGW-1061.

References

[1]Youssef, N. S. and Brook, J. W., "Neutralization of a 50 MeV H^- Beam Using the Ring Nozzle," presented at the Sixteenth International Symposium on Rarefied Gas Dynamics, Pasadena, CA, July 1988.

[2]Ashkenas, H., and Sherman, F., "The Structure and Utilization of Supersonic Free Jets in Low Density Wind Tunnels," *Rarefied Gas Dynamics*, Academic, New York, 1966, Vol. 2, pp. 84-105.

[3]Muntz, E. P., Kingsbury, D., DeVries, C., Brook, J., and Calia, V., "Inertially Tethered Clouds of Gas in a Vacuum," *Rarefied Gas Dynamics*, FRG, 1986, Vol. 2, pp. 474-475.

[4]Muntz, E. P., Hanson, M., *AIAA Journal*, Vol. 22, pp. 696-704.

[5]Bird, G. A., *Molecular Gas Dynamics*, Oxford Univ. Press, London, 1976.

[6]Muntz, E. P., "The Electron Beam Fluorescence Technique," AGARD, Paris, 1968, *AGARDograph*, 132.

[7]Brook, J. W., and Muntz, E. P., Brookhaven National Laboratory, New York, 1986, Rept. RE-718.

Radially Directed Underexpanded Jet from a Ring-Shaped Nozzle

K. Teshima*
Kyoto University of Education, Kyoto, Japan

Abstract

A radially directed underexpanded jet from a ring-shaped nozzle into a quiescent gas is studied experimentally by flow visualization using a laser-induced fluorescence technique and by density measurement using laser Rayleigh scattering. The radially expanded flow makes a core, which becomes a new source of the secondary jet. The structure of the secondary jet is essentially the same in its scale as that of a free jet issued from a circular orifice having an equivalent orifice area. By closing one side of the ring, the secondary jet becomes about 1.3 times larger. Since the gas molecules experience many more collisions before the final expansion than in a free-jet expansion, this type of jet will be useful as a molecular beam source, that is required to be more internal energy relaxed, i.e. in lower vibrational and rotational temperatures.

Introduction

Underexpanded jets issuing from a toroidal or ring-shaped reservoir (ring nozzle) and directed radially toward the center of the ring have been studied associated with applications in an aerodynamic resonator and a technique for producing gas clouds in a vacuum. In the former application,[1] the pressure ratio between the reservoir and the ambient pressure is kept small so that the resonator can perform efficiently. Muntz et al.[2] used an array of small orifices on the inner surface of a ring-shaped reservoir instead of a continuous slit, operated it at high pressure ra-

*Associate Professor.

tios, and obtained inertially tethered gas clouds in a vacuum. They measured the density distribution of the jet at a very large stagnation-to-vacuum-chamber-pressure ratio for flow conditions from a very low stagnation pressure, where the flow is nearly collisonless, to a high stagnation pressure, which produces a collision-dominated interaction. The purpose of the present study is to understand the general feature of the jet issuing from a slit slotted on an inner surface of a ring nozzle into a quiescent gas under a condition of collision-dominated interaction. This was performed by using flow-visualization and density-distribution measurements of the flow at a wide range of pressure ratios. In order to obtain other possible applications of the jet from this type of nozzle, experiments also were made for the case in which one side of the ring was closed so that the jet could expand only into the other direction of the ring axis.

Experiments

Nitrogen gas at room temperature was expanded through a radially slotted sonic orifice on the inner surface of a ring-shaped reservoir (Fig. 1). Two nozzles with different dimensions were used: the smaller one had a slit width (W) of 0.2 mm and a ring diameter (D) of 4.8 mm; in the other, W measured 0.3 mm and D, 7.2 mm. The D/W ratio was 24 for both nozzles and is the same as in Ref. 1. The outer diameter of the ring was about six times that of the inner one, and the gas was supplied through eight holes drilled in the outer peripheral wall of the ring-shaped reservoir. A photograph

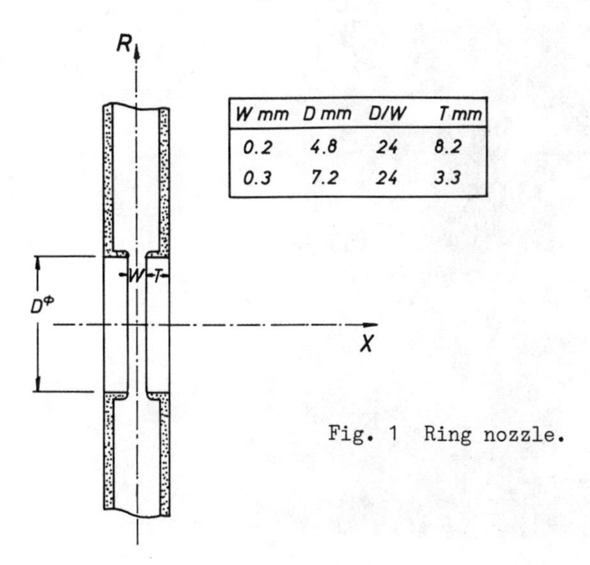

Fig. 1 Ring nozzle.

of the whole assembly of the smaller ring nozzle is shown in Fig. 2. Both nozzles were used in the flow visualization, but the smaller one was used only in the density measurement. The larger nozzle was used mainly to observe the primary gas expansion. The stagnation pressure was kept at 200 or 400 Torr, and the ambient pressure was changed so that the pressure ratio ranged from about 10 to 200. Experiments were made for two cases: 1) the both sides of the ring were open so that the gas expanded in both directions of the ring axis, and 2) the one side of the ring was closed so that the jet was forced to expand only into the other direction of the ring axis.

The flowfield was visualized by a laser-induced fluorescence (LIF) technique. Fluorescence of a small amount of iodine molecules traced in the flow medium and excited by the 514.5-nm argon ion laser was recorded photographically. Details of the present LIF technique are given in Ref. 3. The flow structures were observed in the X-R plane, including the ring axis, and in planes normal to the X axis at different axial distances.

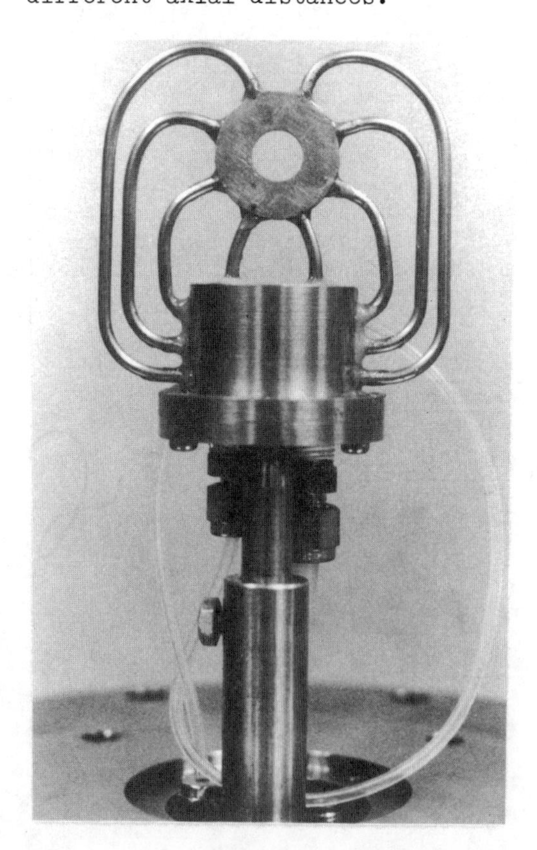

Fig. 2 Photograph of ring nozzle assembly.

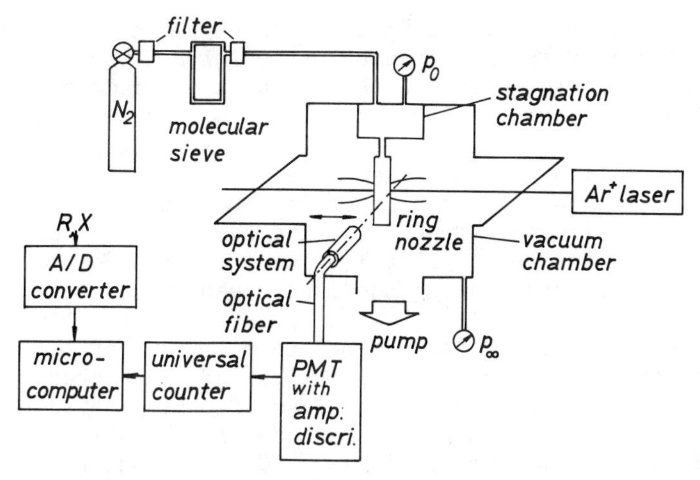

Fig. 3 Schematic diagram of Rayleigh scattering measurement.

Axial and radial molecular number density distributions of the jet were measured using Rayleigh scattering of a laser beam from gas molecules. Since the Rayleigh scattering cross section of nitrogen is almost independent of the gas temperature and the internal state of the molecule, the scattered light intensity is directly proportional to the number density of the molecules in the flow and can be easily converted to an absolute molecular number density by calibrating the sensitivity of the whole optical system with scattered light intensity of a known amount of the test gas. Because the scattered light intensity is very weak, a photon counting system was used. A schematic view of the density measurement system is shown in Fig. 3. The same system had been used for the density measurement of the jet from a circular orifice. It had been shown that the system is reliable in the number density measurement of a flowing gas in which the number density is higher than $10^{17}/cm^3$. The scattered light by a narrow laser beam of less than 0.05 mm was collected by a lens and is guided to the surface of a photomultiplier through an optical fiber. The resolution of the present light collecting system in the direction of the laser beam was less than 0.3 mm. In order to remove small particles and condensable impure molecules from the test gas, the gas was flowed through several filters and a container of molecular sieves. The laser was operated at the multilateral single mode of 514.5 nm at 0.1 W.

The light pulses of the photomultiplier were discriminated after preamplified and amplified pulses were counted using a universal counter. The counted numbers were recorded and processed by a microcomputer together with the digitalized signal of the light detector position. The density distribution measurement on the jet axis was made by moving the detector along the laser beam, which went through the center of the ring. In the radial density distribution measurement, the ring was placed facing the detector, and the measurements along the laser beam were made for several different distances between the ring surface and the laser beam. The scattered light pulses were counted for 10 s at each 0.5-mm interval along the laser beam. Since stray light intensity from bodies in the test chamber was relatively strong and strongly dependent on the positions of the detector and the ring nozzle, distributions of the background signal also were measured without gas flows. They were subtracted from the measured apparent density signals in order to reduce the net density signals. Calibration of the sensitivity of the whole measurement system was made by measuring the signals from a known amount of nitrogen molecules in the test chamber. About 67 pulses/s/10^{18}/cm^3 were obtained. This rather low sensitivity is mainly the result of reducing the effective collection area in order to obtain a better space resolution.

The experiments were made for two cases; both sides of the ring were open so that the gas expanded in both directions of the ring axis (double-sided expansion), and one side of the ring was closed so that the gas could expand only in the other direction of the ring axis (single-sided expansion). The density measurements were made for both cases, but the axial measurement was not made for the single-sided expansion because of the strong scattered light from the plate attached on the side of the ring nozzle.

Results and Discussions

A photograph of the visualized jet for the double-sided expansion in the X-R plane and cross-sectional views at two different axial distances are shown in Fig. 4. Although most of the primary jet was inside the ring and could not be observed, the whole structure of the jet will be illustrated (see Fig. 5). The radially expanded flow interacts and forms a shock wave similar to a hyperboloid of one sheet of revolution or a cusp-shaped cylindrical shock wave, then the confined gas caused by this cylindrical shock wave becomes a new source of the secondary jets, which expand transversely in both directions of the ring axis. The structure of this secondary jet is very similar to that issuing from a circular sonic orifice: the barrel shock, the Mach disk, the reflected shock, and the slip line appear. At smaller pressure

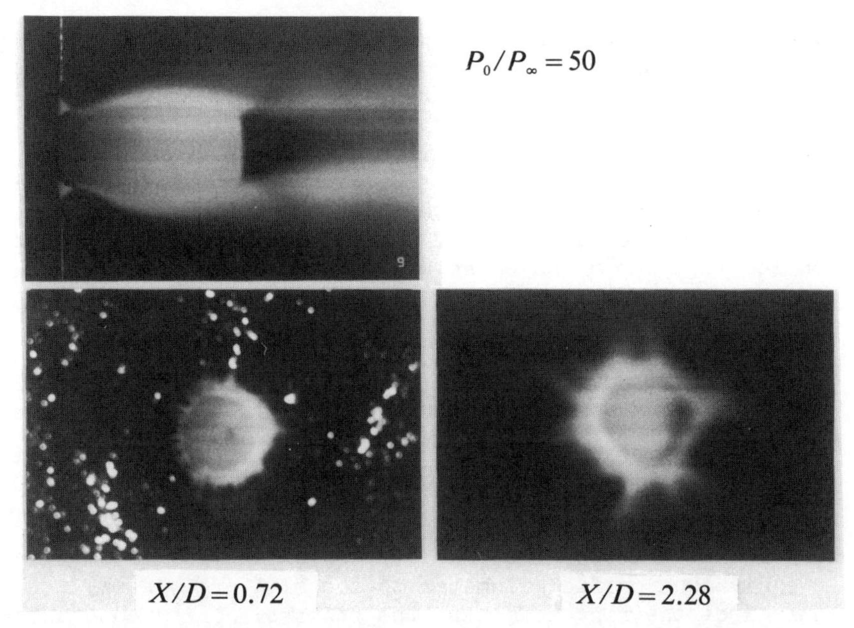

Fig. 4 Visualized jet issuing from a ring-shaped nozzle and expanding into both sides of the ring in a plane, including the jet axis and different cross-sectional views at two different axial positions. The pressure ratio was 50.

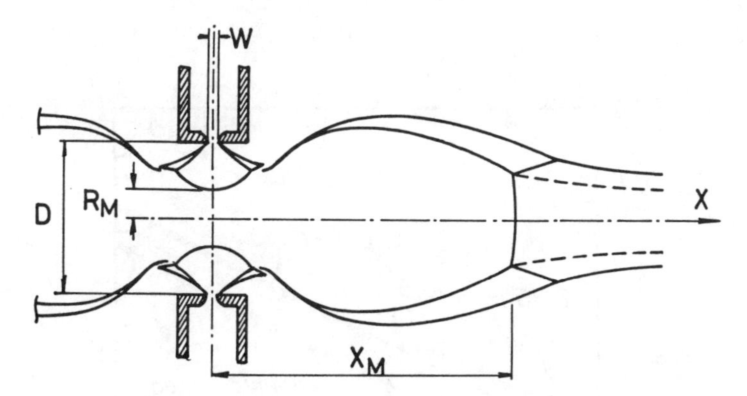

Fig. 5 Structure of a radially underexpanded jet issuing from a ring-shaped nozzle and expanding into both sides of the ring.

ratios of less than 20, the secondary jet showed an apparent repetitive cellular structure. At a still smaller pressure ratio of less than 3, the jet became very unstable and a distinct structure could not be observed. For the single-sided expansion, the structure was nearly the same, except in dimension.

The distance of the cylindrical shock wave from the axis, R_M, was observed to decrease and to reach a certain value by increasing the pressure ratio. For the double-sided expansion, the distance reduced by the ring radius approaches about 0.43 at pressure ratios larger than 100; for the single-sided expansion, the value was somewhat larger (about 0.56 at a pressure ratio of 200 and seemingly still decreasing). Because most of the primary jet was hidden inside the ring and could not be well observed, these values were not measured accurately. However, the value for the double-sided expansion coincided well with the prediction by Muntz et al.[2]

The distance of the Mach disk from the center of the nozzle, X_M, measured with the ring diameter is plotted against the pressure ratio in Fig. 6. In the figure, the open circles show the data when the shock wave did not form a normal shock wave on the axis. The distance of the Mach disk increases with the square root of the pressure ratio for both expansions but the proportionality constant is different; for the single-sided expansion it is about 1.3 times larger than that for the double-sided expansion. If we take an equivalent orifice diameter to the slit area, $2(DW)^{0.5}$, instead of the ring diameter, the proportionality constants become 0.64 and 0.86 for the double- and the single-sided expansions, respectively. The former value is almost the same as the value (about 0.67) for a circular orifice jet. This means that when the gas expands into both sides, the scale of the single jet is almost the same as that of the

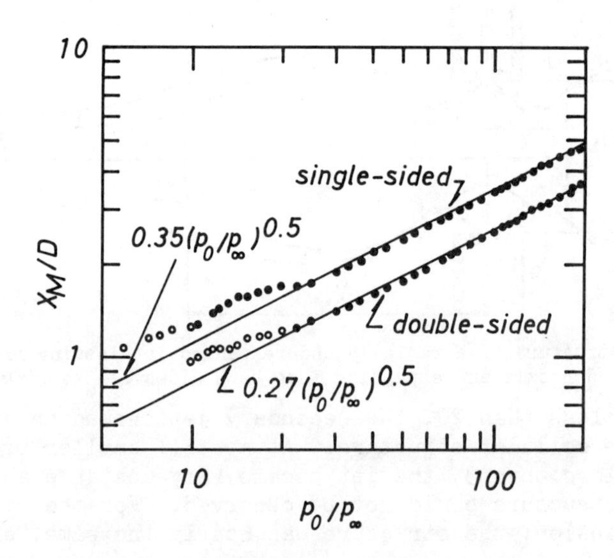

Fig. 6 Distance of Mach disk measured by the ring diameter plotted as a function of the pressure ratio.

free jet issuing from a circular orifice with the same orifice area. Therefore, if the gas is forced to expand in one side of the ring, the dimension of the jet becomes about 1.3 times larger than the equivalent circular orifice jet. This is probably due to the fact that by closing one side of the ring, the pressure balance between the primary jet and the core gas changed the location of the primary cylindrical shock and increased the effective diameter of the source for the secondary jet. Since the prediction of the location of the cylindrical shock by Muntz et al.[2] is independent of the D/W ratio, the relations for the location of the Mach disk will be applied to jets issuing from ring-shaped nozzles with different D/W values.

As another different feature of the jet between the double- and single-sided expansions, we noticed that the flow was more stable for the latter case, although it was rather unstable for both cases in comparison with the jet from a circular orifice.

Axial density distributions of three different pressure ratios for the double-sided expansion are shown in Fig. 7. The stagnation pressure was 400 Torr. Because of the strong scattered light from the nozzle surface, the gas density signals close to the wall could not be measured. The density distributions for pressure ratios of 200 and 100 are almost the same except small bumps. The density decreases with the distance from the nozzle center until a gradual increase caused by the Mach disk. For the jet of a pressure ratio of 50, the density upstream of the Mach disk is lower than those of the jets with larger pressure ratios.

Radial density distributions for three different pressure ratios at various distances from the nozzle center are

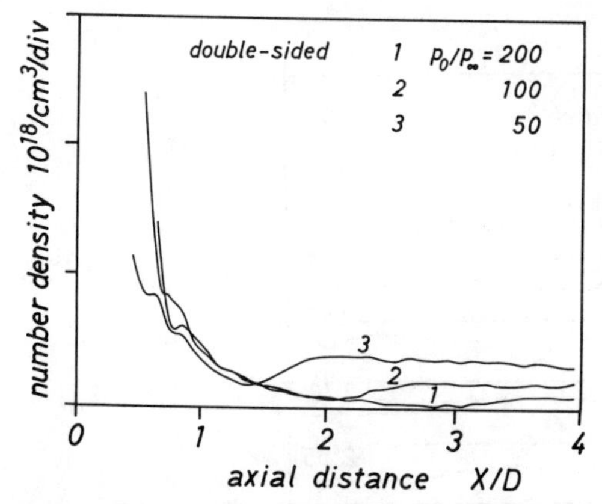

Fig. 7 Number density distributions on the ring axis for different pressure ratios.

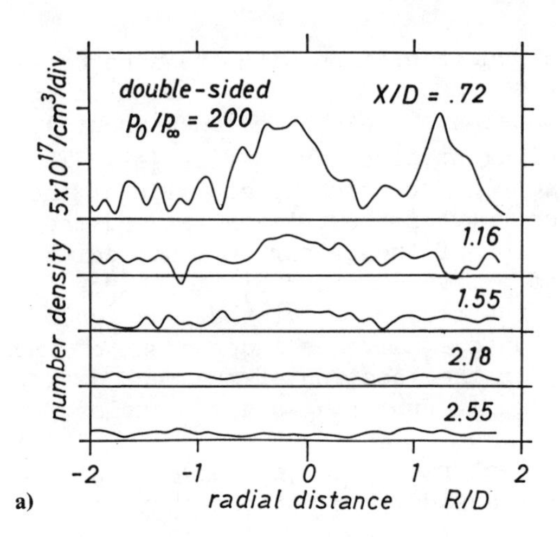

a)

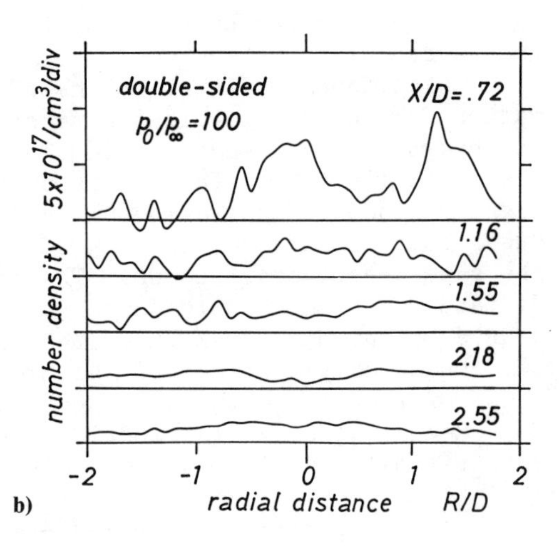

b)

Fig. 8 Radial number density distributions of a jet for double-sided expansion at different distances from the nozzle center:
a) pressure ratio of 200;
b) pressure ratio of 100;
c) pressure ratio of 50.

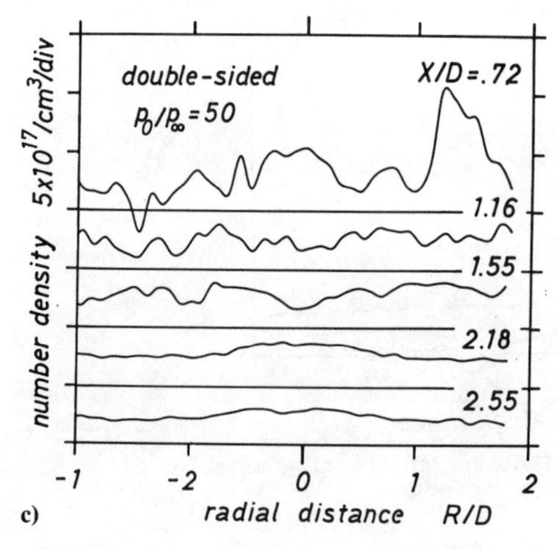

c)

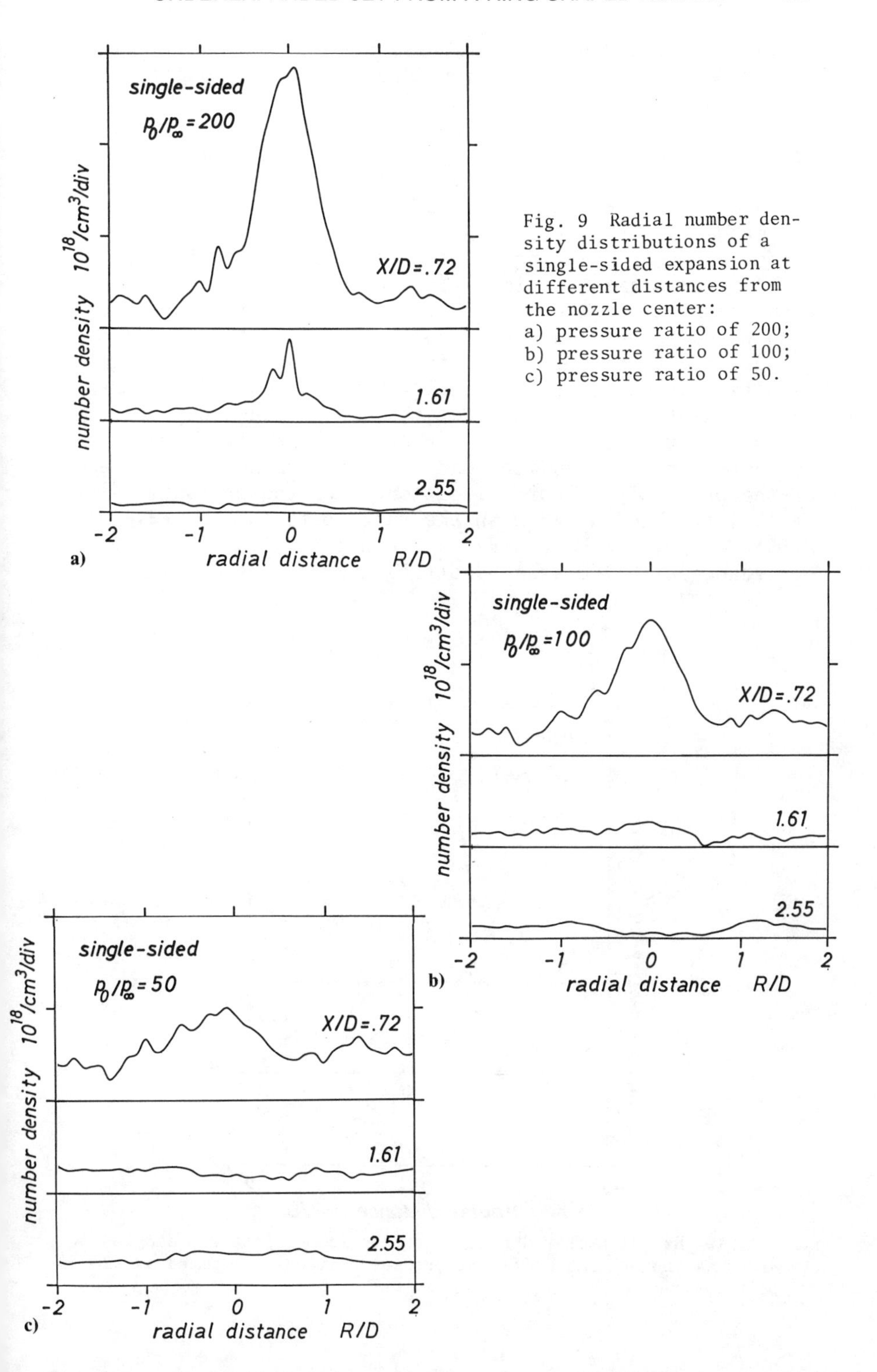

Fig. 9 Radial number density distributions of a single-sided expansion at different distances from the nozzle center:
a) pressure ratio of 200;
b) pressure ratio of 100;
c) pressure ratio of 50.

shown in Fig. 8 for the double-sided expansion and in Fig. 9 for the single-sided expansion. Rather strong density changes occur along the radial direction. As can be seen from the cross-sectional views of the jet in Fig. 4, the jet does not appear to be radially symmetric, especially outside of the jet core. Several traces of the jet boundary were often observed in the flow visualization. Therefore, the jet is rather unstable, but not merely randomly fluctuated in times; probably changing the stable positions of the jet boundary from time to time. The amount of density fluctuation is small for the single-sided expansion in comparison with the double-sided expansion, which means that the single-sided expansion is more stable. The abnormally large density increase near the nozzle exit just outside of the inner diameter of the ring for the double-sided expansion shown in Fig. 8 cannot be explained at present. For the pressure ratio of 200, the Mach disk is further downstream of the presently measured location. For the pressure ratio of 100, the Mach disk is in the range between X/D = 2.18 and 2.55; for 50, the Mach disk exists between 1.55 and 2.18 in accordance with the axial distribution measurements.

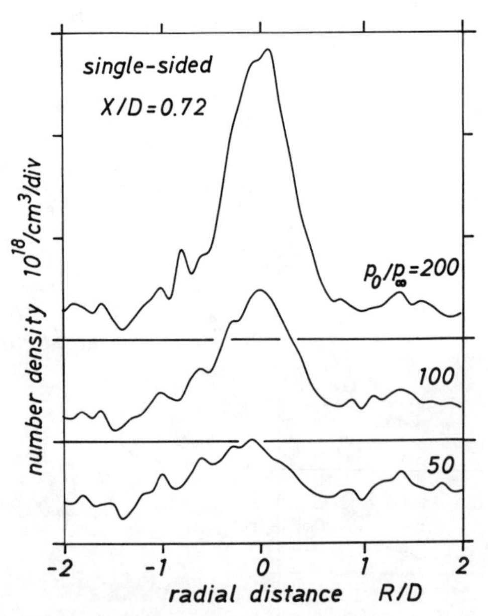

Fig. 10 Radial number density distribution of jets for single-sided expansion with different pressure ratios at a fixed axial position.

When one side of the nozzle was closed, the density on the jet axis more than tripled that of double-sided expansion at X/D = 0.72 and the abnormal density distribution that appeared outside of the ring disappeared; almost radially symmetric density distributions were observed. In this case, the absolute density depends largely on the pressure ratio; that is, for the same stagnation gas density, the measured number density at the same location on the axis differs largely with the pressure ratio (see Fig. 10). This was also observed for the double-sided expansion between the pressure ratios of 50 and 100, although the difference was small. This probably occurs due to a difference in the core gas density, which is confined by the primary cylindrical shock wave; when one side of the ring was closed, the location of the cylindrical shock wave was still approaching the center for a pressure-ratio range up to at least 200. On the other hand, for the double-sided expansion, it has already reached its final value at the pressure ratio of 100. When the effective diameter of the jet source changes, the two-dimensional effects on expansion become significant, especially near the expansion source, and a similar expansion rate, as in the one-dimensional flow, cannot be expected. This two-dimensional effect combined with the different effective source gas density resulted in the density distributions shown in Fig. 10.

The gas molecules in the single-sided expansion experience a slow two-dimensional expansion in the primary jet, compression by the cylindrical shock wave, and then an expansion slower than the free-jet expansion. Therefore, internal energy modes of molecules will be relaxed or cooled down more during the expansion in comparison with those in the simple free-jet expansion.

Conclusions

The structure of the radially expanded jet from a ring-shaped nozzle has the following features in contrast with that issuing from a circular orifice.

1) The primary jet from the slit forms a cylindrical shock wave and compresses the gas into a particular core region. The confined gas expands as the secondary jets into the both direction of the ring axis.

2) The secondary jet is very similar to the free jet from a circular orifice in its structure. The distance of the Mach disk from the center of the nozzle increases with the square root of the pressure ratio and is essentially the same as that for the free jet from a circular orifice having an equivalent area.

3) If we close one side of the nozzle, the main jet

becomes about 30% larger and the density near the ring exit depends largely on the pressure ratio.

4) The flow is rather unstable compared with the jet from a circular orifice especially at a low pressure ratio, but the single-sided expansion is more stable than the double-sided expansion.

5) Since the molecules experience many more collisions before the final expansion than in a free-jet expansion, this type of jet can be used as a molecular beam source, that is required to be more internal energy relaxed, i.e. in lower vibrational and rotational temperatures.

Acknowledgments

The author would like to express his thanks to Mr. M. Sumida for his effort in making the ring nozzles. A part of the present study was supported by the Ministry of the Grant-in-Aids for Scientific Research from the Ministry of Education and Culture under contract 620550046.

References

[1]Wu, J. H. T., Elabdin, M. N., and Neemeh, R. A., "Structure of a Radially Directed Underexpanded Jet," AIAA Journal, Vol. 15, Nov. 1977, pp. 1651-1653.

[2]Muntz, E. P., Kingbury, K., De Vries, C., Brook, J., and Calia, V., "Inertially Tethered Clouds of Gas in a Vacuum," Rarefied Gas Dynamics, Vol. II, edited by V. Boffi and C. Cercignani, Teubner, Stuttgart, 1986, pp. 474-485.

[3]Teshima, K and Nakatsuji, H., "Visualization of Rarefied Gas Flows by a Laser Induced Fluorescence Method," Rarefied Gas Dynamics, Vol. I, edited by H. Oguchi, Tokyo Univ., Japan, 1984, pp. 447-454.

Three-Dimensional Structures of Interacting Freejets

Tetsuo Fujimoto* and Tomohide Ni-Imi†
Nagoya University, Furo-cho, Chikusa-ku, Nagoya, Japan

Abstract

Three-dimensional structures of two interacting identical supersonic freejets of Ar are studied by flow visualization using laser-induced fluorescence of I_2 molecules seeded in Ar. The experiments are carried out for various angles θ between the jet axes from 45 to 180 deg and with various ratios of the source pressure Ps to the pressure in the expansion chamber Pb. The flowfield of interacting freejets shows various patterns depending on θ and Ps/Pb. For θ≦90 deg, the flowfield is symmetrical with respect to the interacting plane that bisects the jet axes, and there a second cell appears surrounded by shock waves, leading to a ϕ-shaped structure in the plane perpendicular to the interacting plane. For θ>90 deg, many types of flowfields appear as Ps/Pb varies, suggesting the complexity of flow stability.

Introduction

Knowledge of the flowfields of interacting freejets is important for high technologies and the space sciences related to species separation, molecular beam epitaxy, control of satellites, etc.

For the species separation, spatial distributions of velocities and species molal fraction in the neighborhood of the interacting zone of opposed two-dimensional freejets of Ar-He mixture were measured quantitatively using the optical method and mass flow probe by McDermott and Hurlbut.[1] Also, the local molecular velocity distribution of opposed two-dimensional jets of $He-C_7F_{14}$ was investigated by Bley and Ehrfeld.[2] These reports gave details on the interacting free-

---

*Professor, Department of Electronic-Mechanical Engineering.

†Research Associate, Department of Electronic-Mechanical Engineering.

jets over a wide range of parameters but did not mention the structures of the flowfields and their shock-wave systems.

The structure of the flowfield of two interacting parallel freejets was studied by Soga et al,[3] and Dankert and Koppenwallner,[4] and Dankert[5] by means of flow visualization using an electron beam and glow discharge, respectively. They also measured density,[3] pressure,[4,5] Mach number,[5] and rotational temperature[3] distribution on the line of symmetry and the cross section perpendicular to that line.

In the present study, three-dimensional structures and shock-wave systems of two interacting identical freejets of Ar are studied by flow visualization. The experiments are carried out for various angles θ between the jet axes from 45 to 180 deg and with various ratios of the source pressure Ps to the pressure in the expansion chamber Pb. For flow visualization, the laser-induced fluorescence (LIF) of I_2 molecules seeded in Ar is employed. The intensity of LIF of I_2 is proportional to the n-th power of local density of Ar when the laser frequency is detuned slightly from the center frequency of an absorption line of I_2 (off-resonant fluorescence). The flow visualization is achieved on the following cross sections: the plane that involves two nozzle centerlines, the interacting plane that bisects the angle between two centerlines of nozzles, and the planes perpendicular to the former two.

The structures of the flowfields of interacting freejets depend on θ and the pressure ratio Ps/Pb. For $\theta \leq 90$ deg, the structures of the flowfield are symmetrical with respect to the interacting plane irrespective of Ps/Pb, and a cell

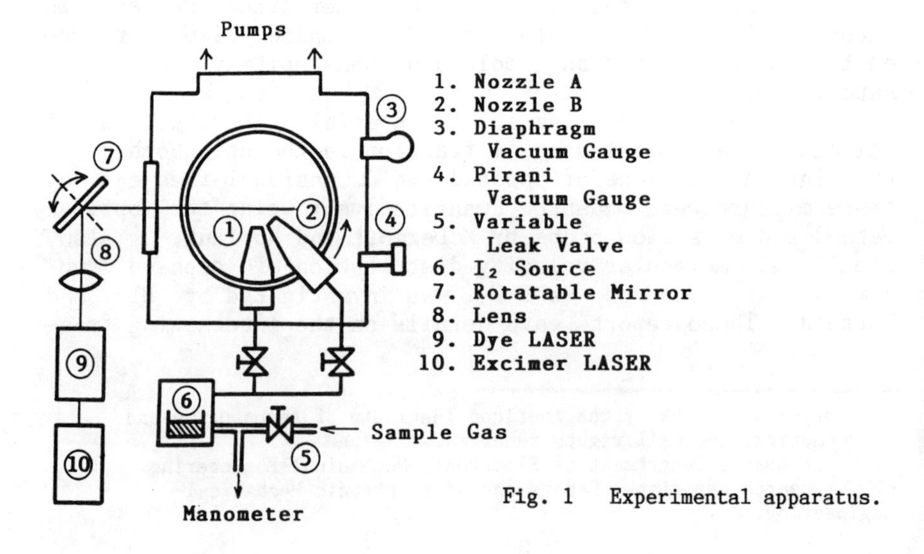

Fig. 1 Experimental apparatus.

surrounded by shock waves is formed near the interacting plane when Ps/Pb is relatively large. The shock-wave system has a ϕ-shaped structure in the plane perpendicular to the interacting plane. It is rather peculiar that the Φ-shaped structure tends to be planar in the interacting plane as the flow proceeds far downstream. For θ>90 deg, many types of flowfields appear, depending on Ps/Pb, suggesting the complexity of flow stability. In particular, either symmetrical or asymmetrical structure of the flowfield with respect to the interacting plane is observed in spite of the same pressure ratios.

Experimental Apparatus and Procedure

A schematic diagram of the experimental apparatus is shown in Fig. 1. Carrier gas (Ar) is supplied through a variable leak valve into the source chamber and mixed with sublimated I_2. The mixture is expanded through two identical sonic nozzles into the expansion chamber which is evacuated by rotary pumps and an oil diffusion pump.

The Ps is varied from 5.32×10^3 to 2.66×10^4 Pa and the Pb from 1.33×10^2 to 1.33×10^{-1} Pa. The source pressure is measured by a mercury U-tube manometer and the expansion chamber by a capacitance manometer (MKS, Baratron, model 220BA). Carrier gas temperature is kept in the range of 20±1 °C, at which the vapor pressure of I_2 is 26.5 Pa.

Two sonic nozzles with exit diameter D=0.513 mm are made of Pyrex glass. Each exit of nozzles A and B is set at an equal distance L from the intersecting point of their centerlines, with L/D from 6.5 to 11.0. The angle between two nozzle centerlines can be varied easily by rotating the ring on which the nozzle B is set. In this study, θ is varied over a range of 45 to 180 deg.

For flow visualization, the LIF of I_2 molecules seeded in Ar is adopted. The total flowfield can be visualized by the fluorescence of I_2 molecules, since I_2 molecules do not disturb the flow as long as the molal fraction of I_2 is low. A laser system consists of a tunable dye laser (Lumonics Hyperdye-300, energy conversion efficiency of about 10%) and an excimer laser (Lumonics TE-431T, operation on XeCl, wavelength 308 nm, power 4.0 W/150 Hz) as a pump source. Coumarin 500 is used as a laser dye with a tuning range of wavelength from 482 to 545 nm. Since I_2 molecules have many absorption lines in the range of the laser wavelength, relatively intense fluorescence is radiated.

Figure 2 shows the coordinate system used in this study. The line that bisects the angle between two centerlines of the nozzles is selected as the X axis, and Y and Z axes are

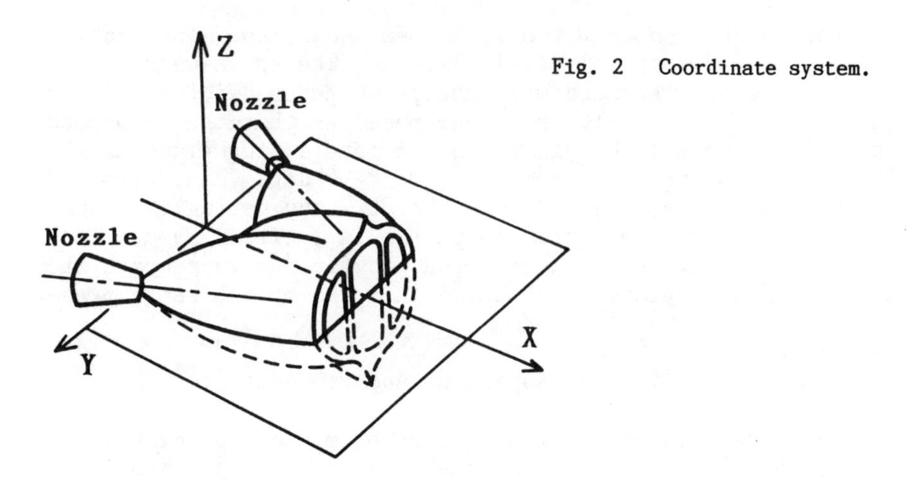

Fig. 2 Coordinate system.

defined as shown in the figure. The flow visualization is achieved on the X-Y and X-Z planes and also on the planes parallel to the Y-Z plane. To photograph the entire flow-field, the laser beam is scanned on these planes by a rotatable mirror. The scanning speed of the laser beam is varied from 0.14 to 0.4 mm/s, depending on the laser power. A camera is set perpendicular to the visualized plane and at about 50 cm from the plane of the visualization. The camera system consists of a telephoto lens (200 mm, F2.5) and a bellows in order to obtain a nearly isometrical image on film. To intercept the original laser beam, a cutoff filter (cutoff wavelength below 560 nm) is employed. Because of very weak fluorescence of I_2, the photographic film (ASA400) is processed with sensitization.

Results and Discussion

Fluorescence Intensity and Local Density

Fluorescence intensity of I_2 depends on many physical quantities, such as density, temperature and pressure in the flow field. In addition to these quantities, the intensity is also influenced by the laser frequency relative to the absorption frequency of I_2. If we define the laser detuning by

$$\Delta\nu = \nu - \nu_L \tag{1}$$

where ν is the absorption frequency of I_2 and ν_L the frequency of the laser, the fluorescence intensity depends strongly on $\Delta\nu$. Fluorescence by the laser with $\Delta\nu=0$ is called "resonant" and with $\Delta\nu\neq0$ "off-resonant." Figure 3 shows photographs of a single jet taken under the two detuning condi-

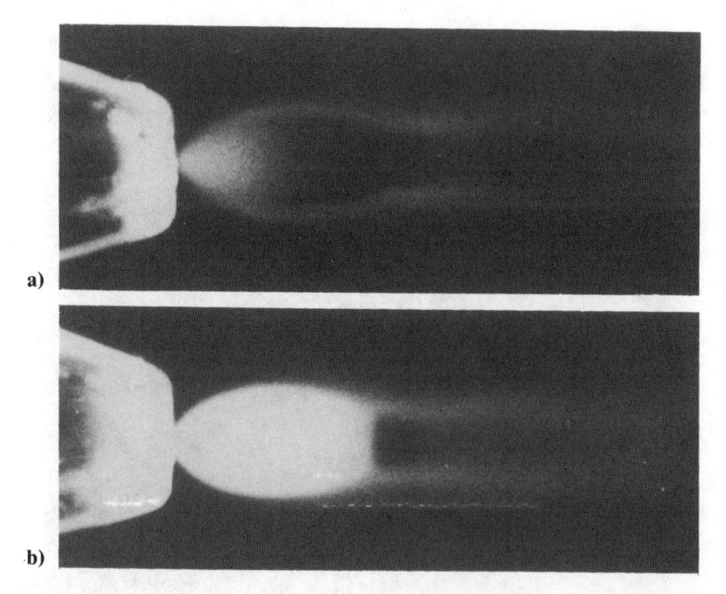

Fig. 3 Photographs of a single jet by off-resonant and resonant fluorescence: a) off-resonant ($\Delta\nu\neq0$); b) resonant ($\Delta\nu=0$).

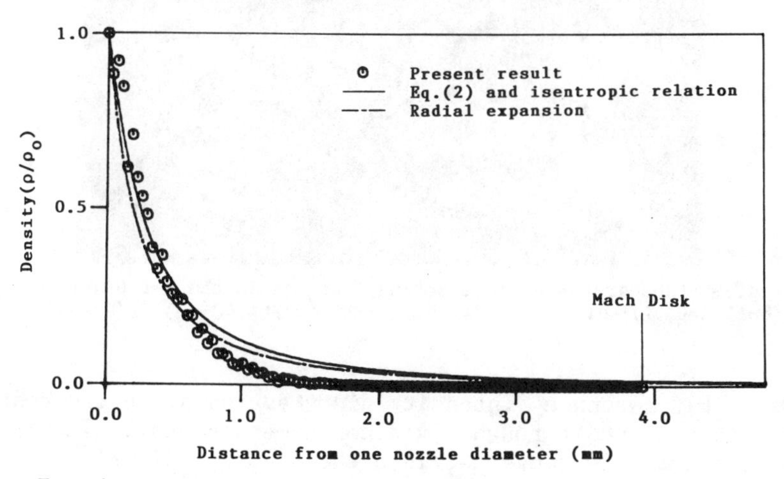

Fig. 4 Density distribution on centerline of a single jet.

tions. The photograph at $\Delta\nu=0$ (Fig. 3b) clearly indicates that the fluorescence intensity is not proportional to the local density of Ar. In off-resonant fluorescence, it can be inferred from McDaniel's[6] experimental results that the fluorescence intensity S_F is roughly proportional to the cube root of density, $S_F \propto \rho^{1/3}$, in that pressure is low and $\Delta\nu=1\sim2$

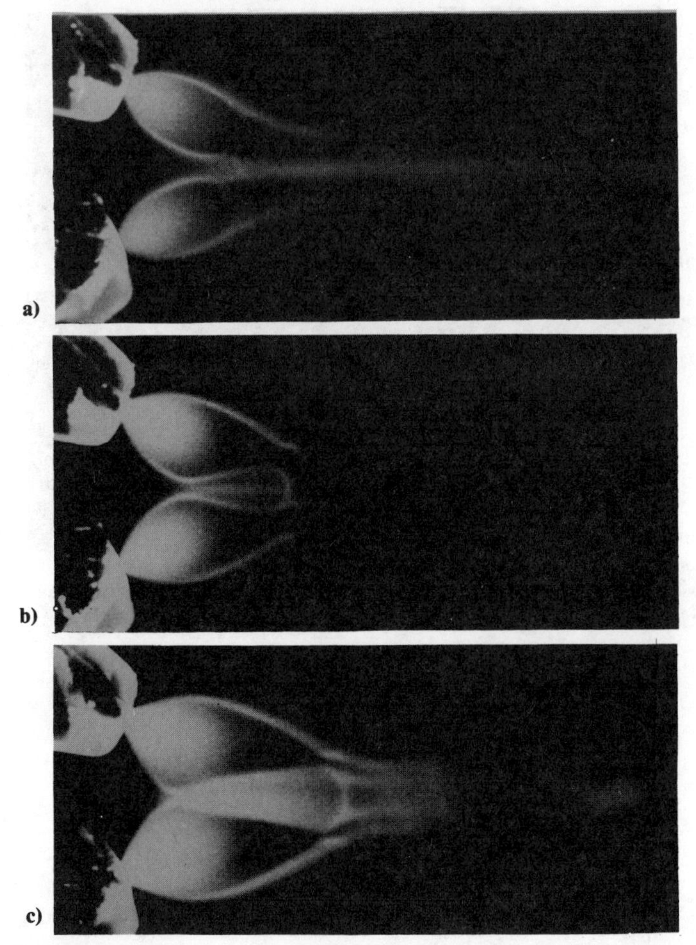

Fig. 5 Photographs of interacting freejets in the X-Y plane (θ=45 deg, L/D=10.7): a) Ps/Pb=129; b) Ps/Pb=240; c) Ps/Pb=342.

GHz. Fluorescence intensity distribution along the centerline is measured through a vidicon camera system. In Fig. 4, this result is converted into the density distribution by means of the cube root relation, where the local density ρ is normalized by the density ρ_o at one nozzle diameter from the nozzle exit plane. This is compared with two theoretical predictions, i.e., one based on the isentropic relation and Mach number given by[7]

$$M = A\left(\frac{x - x_o}{D}\right)^{\gamma-1} - \frac{1}{2}\left(\frac{\gamma + 1}{\gamma - 1}\right) \Big/ A\left(\frac{x - x_o}{D}\right)^{\gamma-1} \tag{2}$$

where x is a distance from the nozzle lip, γ is the specific heat ratio (=1.67), A=3.26, and x_o/D=0.075, and another based on the radial expansion from a point source at the nozzle exit. It can be seen from the figure that the simple relation between the fluorescence intensity S_F and density ρ, such as $S_F \propto \rho^{1/3}$, represents only qualitative tendency. Therefore, it seems necessary to decide appropriate n depending on the local pressure, if one wishes to employ the $S_F \propto \rho^n$ relation. Photographs that appear later in the paper were taken by means of the off-resonant fluorescence.

Flowfield Structures for θ=45 and 90 deg

Figures 5 and 6 provide photographs of interacting free-jets visualized in the X-Y plane for various Ps/Pb and for θ=45 and 90 deg, respectively. Typical photographs for θ=45 deg in the three visualized planes are given in Fig. 7 and illustrated by schematic pictures in Fig. 8.

For $\theta \leq 90$ deg, the structure of the flowfield in the X-Y plane is symmetrical with respect to the interacting plane (X-Z plane), irrespective of Ps/Pb (see Figs. 5 and 6). As Ps/Pb increases, the interacting region grows and oblique shock waves appear, intersecting with normal shock waves. As shown in Figs. 5c, 7a, and 8a, a cell surrounded by the shock waves is formed near the interacting plane when Ps/Pb is relatively large. We call this the second cell, in contrast to the first cell formed just downstream of the nozzle. For θ=90 deg, however, the second cell is formed only when L/D is shorter and Ps/Pb is larger than that for θ=45 deg.

Figure 6c is a photograph in which only one jet is visualized. In this case, two source chambers with equal pressures are used in order to seed I_2 molecules only in the one chamber. The gas of a jet scarcely flows into the region of the other jet beyond the interacting plane.

With the flowfield for θ=45 deg in the X-Z plane, the structure of the second cell is very similar to that of a single jet, consisting of barrel shock waves and a normal shock wave, as shown in Figs. 7b and 8b. Just downstream of the second cell, another cell is formed, and its shock wave is also observed in the X-Y plane (Fig. 7a).

The series of photographs in Fig. 7c is taken at various cross sections parallel to the Y-Z plane, where each number at the bottom corresponds to the visualized cross section shown in Fig. 7a using the same number. As shown in Fig. 7c-1 and 2, the flowfield in the plane relatively close to the nozzle lips has the ϕ-shaped structure that is named after its geometric similarity to the Greek letter ϕ. Thus, the second cell stretches its jet boundary in the X-Z plane.

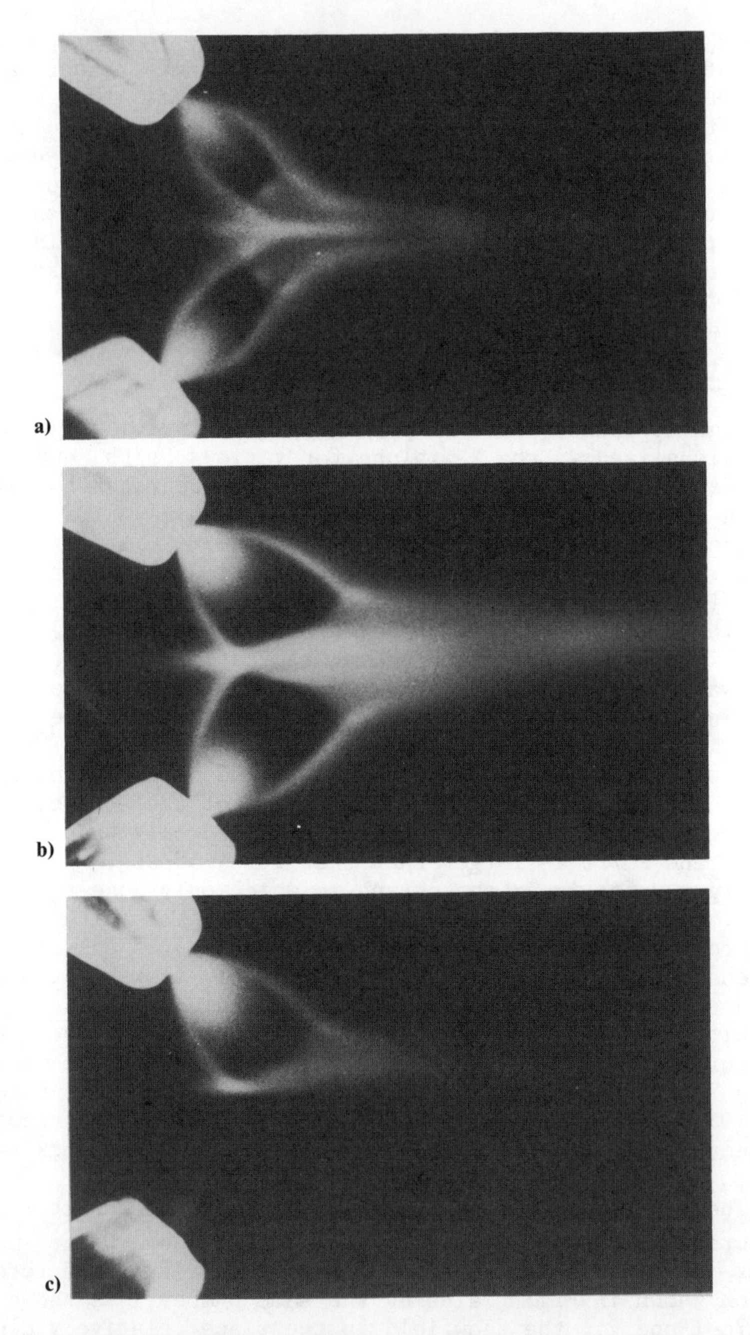

Fig. 6 Photographs of interacting freejets in the X-Y plane (θ=90 deg, L/D=10.4): a) Ps/Pb=115; b) Ps/Pb=330; c) Ps/Pb=264.

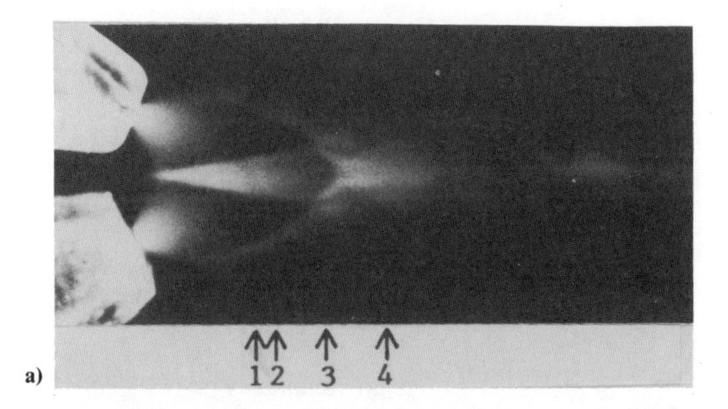

a)

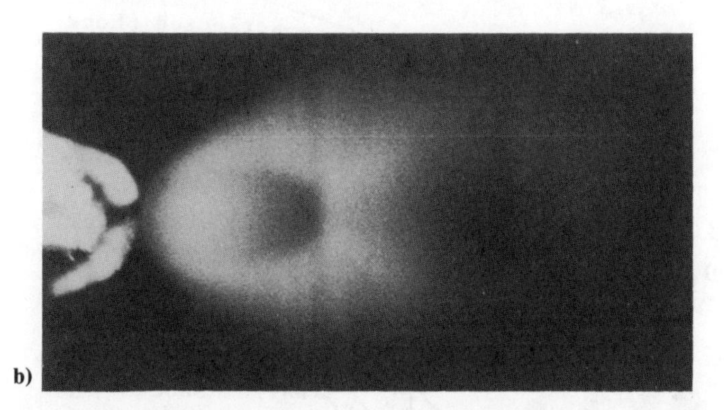
b)

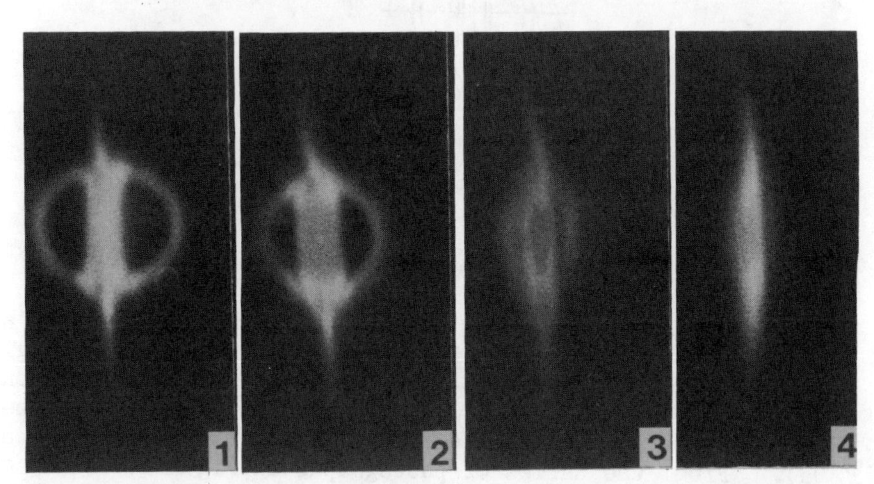

c)

Fig. 7 Photographs of interacting freejets in several planes (θ=45 deg, L/D=8.9, Ps/Pb=266): a) X-Y plane; b) X-Z plane; c) planes parallel to Y-Z plane.

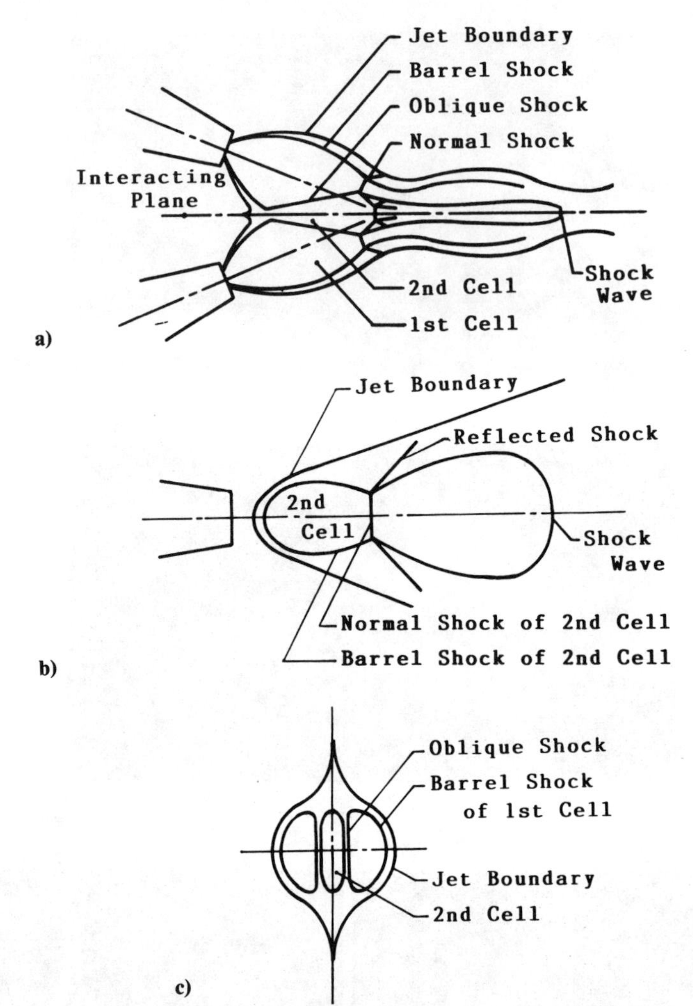

Fig. 8 Structures of interacting freejets for θ=45 deg: a) X-Y plane; b) X-Z plane; c) Y-Z plane.

In these two photographs, the half-circular shock waves are the barrel shock waves of the first cell, and the region between the two chord lines of the half-circles corresponds to the second cell as shown in Fig. 8c. Therefore, these chord lines are the oblique shock waves (see Figs. 7a and 8a). From the series of photographs in Fig. 7c, one can observe that the Φ-shaped structure tends to become planar in the interacting plane as the flow proceeds farther downstream. This becomes more noticeable as θ increases.

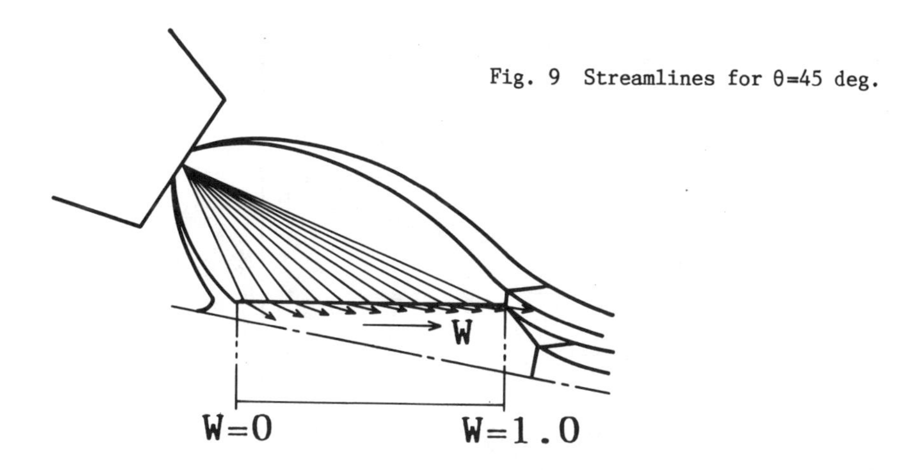

Fig. 9 Streamlines for θ=45 deg.

To analyze the flowfield downstream of the oblique shock waves, we examine the direction of the streamlines, Mach numbers, and the static pressures just behind the shock waves, assuming that 1) the streamlines in the first cell are radial with a center at the nozzle exit, 2) the Mach numbers along these lines are approximated by eq. (2), and 3) the flow in the first cell is isentropic. The directions (deflection angle) of the streamlines and Mach number M_2 just behind the oblique shock wave are manipulated in terms of Mach numbers M_1 just in front of the shock wave and the inclination angles between the streamlines in the first cell and the oblique shock wave, which are measured from the visualized photographs, and the static pressure P is deduced from the Rankine-Hugoniot relation for the oblique shock wave.

Figure 9 indicates the directions of the streamlines by arrows for θ=45 deg and relatively large Ps/Pb. The streamlines just behind the oblique shock wave are not parallel to the interacting plane near the starting point of interaction but shift direction parallel to this plane at locations apart from the point. The distributions of Mach numbers and the static pressures are shown in Figs. 10 and 11, respectively, according to whether or not the flowfield downstream of the oblique shock waves is closed off by the shock wave generated from the points at which the oblique shock waves intersect with normal shock waves of the first cells [i.e., the flowfield with the second cell (θ=45 and 90 deg) or without it (θ=90 deg)]. In these figures, the abscissas W are normalized by the length of the oblique shock wave (see Fig. 9) and the static pressure by the source pressure Ps. As shown in Fig. 10, M_2 do not exceed 1.0 near the starting point of the

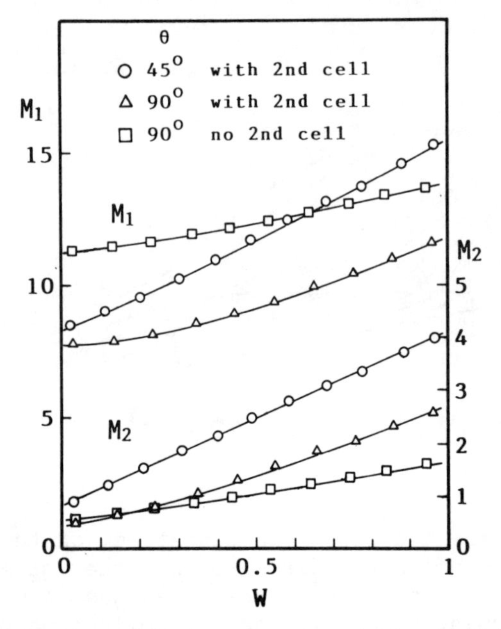

Fig. 10 Distribution of Mach number just in front (M_1) of and just behind (M_2) the oblique shock wave.

interaction under any condition. Therefore, this shock wave changes along itself from a strong shock wave ($M_2<1$) to a weak one ($M_2>1$) in the same way as a detached shock wave. It is also evident from Figs. 10 and 11 that M_2 is larger and the gas behind the oblique shock wave expands more thoroughly when there the second cell appears, compared to the case without it.

Flowfield Structures for θ=135 and 180 deg

For θ>90 deg, many types of flowfield structures are observed depending on Ps/Pb. As an example, photographs in the X-Y plane for θ=135 deg are shown in Fig. 12. When Ps/Pb is relatively small, the flowfield is symmetrical with respect to the interacting plane as shown in Fig. 12a. For Ps/Pb larger than 60, however, either symmetrical or asymmetrical structure with respect to the interacting plane appears (Figs. 12b and 12c), even if the value of Ps/Pb is nearly identical. Whenever we stop supplying gas to the nozzle A temporarily and start supplying gas again, the flowfield turns out to be asymmetrical as shown in Fig. 12c or 12d; i.e., the jet expanding from nozzle B bends, remarkably deviating from the centerline, or collides with the jet from nozzle A. As Ps/Pb increases further, only a symmetrical

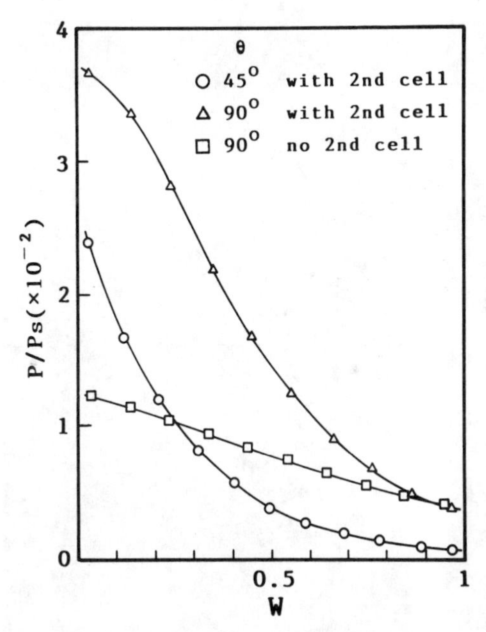

Fig. 11 Pressure distribution just behind the oblique shock wave.

structure is observed again as shown in Fig. 12e. The three-dimensional structures for these flows are so complex that we are not able to infer the details of the interaction. It seems necessary to study the flow stability in order to clarify the flowfield. Figure 12f is a photograph in the free molecular regime [Classification into the free molecular regime is judged by the parameter $\xi = D(Ps/Pb)^{1/2}/T$ (Pa.m/K), where T is source temperature which was introduced by Muntz et al.[8]] One can observe a high density region formed near the interacting plane. The same results are obtained for other θ.

Figure 13 shows the photographs of the flowfields of interacting opposed jets with θ=180 deg in the plane containing the centerline. These structures are symmetrical with respect to the jet axis irrespective of Ps/Pb. When Ps/Pb is relatively large, as shown in Fig. 13a, the flowfield is also symmetrical with respect to the interacting plane that bisects the interval between the two nozzles. For relatively small Ps/Pb, however, the flowfield becomes asymmetrical as shown in Fig. 13b; i.e., the structure of the jet expanding from nozzle A retains the same shape as a single jet, but the jet from nozzle B is pushed back to the nozzle. This phenomenon is attributed to the interaction between the first cell of the jet from nozzle B and one of the cells formed repeatedly behind the first cell of the jet from nozzle A. One can find normal shock waves in the region of

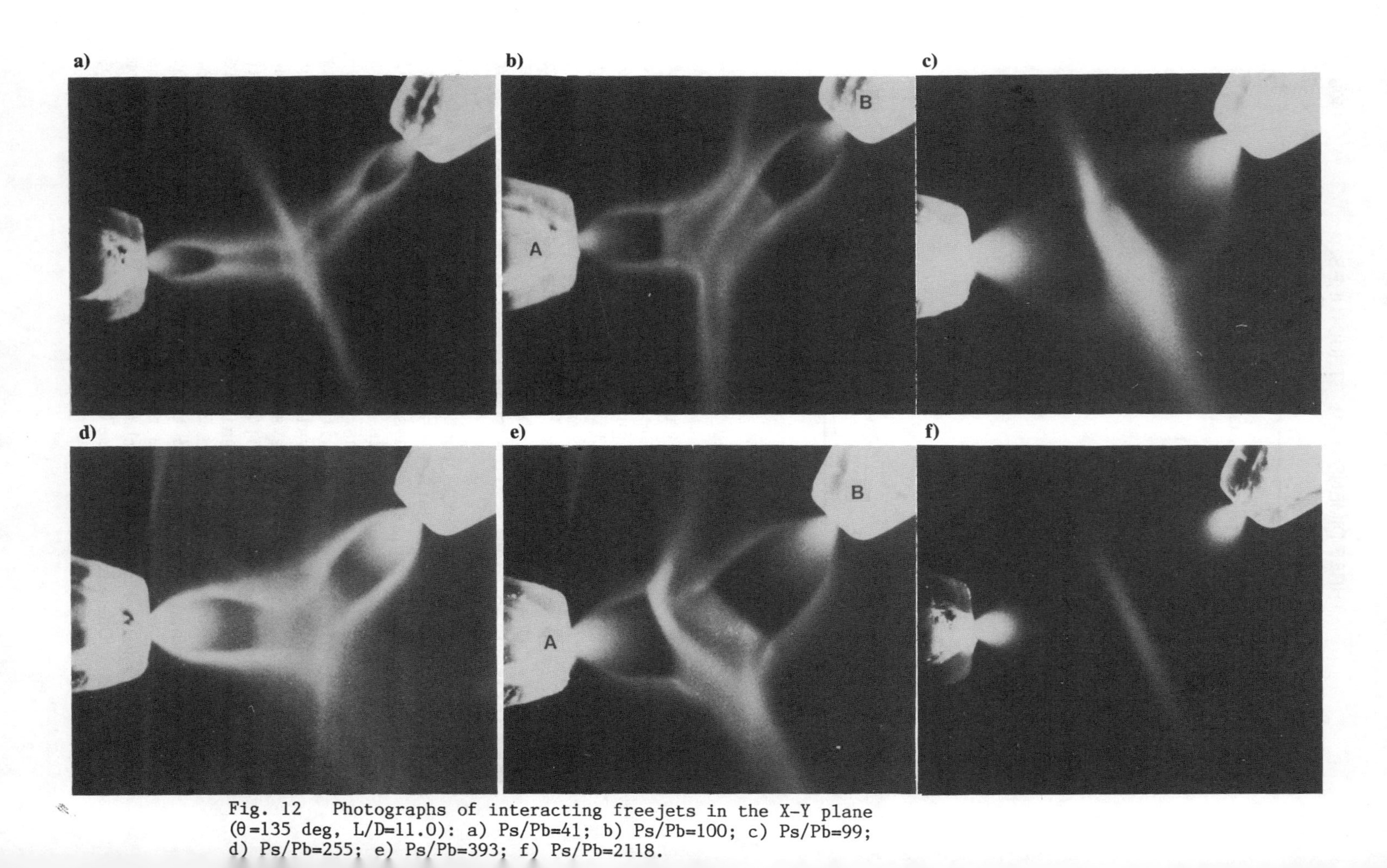

Fig. 12 Photographs of interacting freejets in the X-Y plane (θ=135 deg, L/D=11.0): a) Ps/Pb=41; b) Ps/Pb=100; c) Ps/Pb=99; d) Ps/Pb=255; e) Ps/Pb=393; f) Ps/Pb=2118.

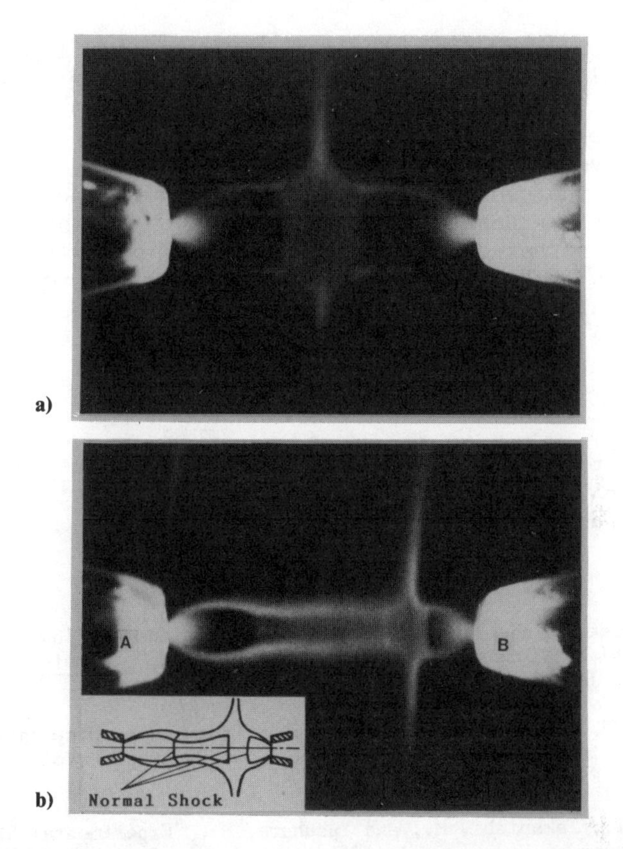

Fig. 13 Photographs of interacting freejets in the X-Y plane (θ=180 deg, L/D=10.0): a) Ps/Pb=263; b) Ps/Pb=90.

interaction by careful observation of this photograph (see schematic picture, Fig. 13b). In a particular range of the pressure ratio (Ps/Pb), either symmetrical or asymmetrical structure can be observed, in spite of the same pressure ratios, as for θ=135 deg. For θ=180 deg, an intense radial expansion normal to the jet axis from a stagnant region is observed, since the pressure is raised high as a consequence of the interaction of the two jets.

Concluding Remarks

In analyzing the structure of the interacting two freejets of Ar by flow visualization using LIF of I_2 molecules seeded in Ar, the three-dimensional structures and the shock-wave systems of its flowfield are investigated.

The flowfields of interacting freejets displayed many types of structure depending on θ and the pressure ratios

Ps/Pb. For $\theta \leq 90$ deg, the structures of the flowfield are symmetrical with respect to the interacting plane irrespective of Ps/Pb, and the second cell surrounded by the shock waves is formed near the interacting plane when Ps/Pb is relatively large. The ϕ-shaped structure is also observed in the cross section perpendicular to the interacting plane near the nozzle lips. For $\theta > 90$ deg, many types of flowfields appear depending on Ps/Pb, suggesting the complexity of flow stability. In particular, either symmetrical or asymmetrical structures of the flowfield with respect to the interacting plane are also observed in spite of the same pressure ratios.

Acknowledgment

The present work was supported by a grant-in-aid for Scientific Research from the Ministry of Education, Science and Culture.

References

[1] McDermott, W. and Hurlbut, F. C., "Observations on Flow Fields Generated by Opposed Free-Jets of Gas Mixtures," Physics of Fluids, Vol. 27, Jan. 1984, pp. 60-71.

[2] Bley, P. and Ehrfeld, W., "Molecular Dynamics of Disparate Mass Mixtures in Opposed Jets," Rarefied Gas Dynamics, Vol. 1, edited by S. S. Fisher, AIAA, New York, 1980, pp. 577-589.

[3] Soga, T., Takanishi, M., and Yasuhara, M., "Experimental Study of Interaction of Underexpanded Free jets," Rarefied Gas Dynamics, Vol. 1, edited by H. Oguchi, University of Tokyo Press, 1984, pp. 485-492.

[4] Dankert, C. and Koppenwallner, G., "Experimental Study of the Interaction between Two Rarefied Free Jets," Rarefied Gas Dynamics, Vol. 1, edited by H. Oguchi, University of Tokyo Press 1984, pp. 477-484.

[5] Dankert, C., "Flow in the Interaction Region of Two Parallel Free Jets," Rarefied Gas Dynamics, Vol. 2, edited by V. Boffi and C. Cercignani, B. G. Teubner Stuttgart, 1986, pp. 486-494.

[6] McDaniel, J. C., "Investigation of Laser-Induced Iodine Fluorescence for the Measurement of Density in Compressible Flows," Ph. D. Dissertation, Dept. of Aeronautics and Astronautics, Stanford University, Stanford, CA., Jan. 1982.

[7] Ashkenas, H. and Sherman, F. S., "Experimental Methods in Rarefied Gas Dynamics," Rarefied Gas Dynamics, Vol.2, edited by J. H. de Leeuw, Academic Press, New York, 1966, pp. 84-105.

[8] Muntz, E. P., Hamel, B. B., and Maguire, B. L., "Some Characteristics of Exhaust Rarefaction," AIAA Journal, Vol. 8, Sept., 1970, pp. 1651-1658.

Flow of a Freejet into a Circular Orifice in a Perpendicular Wall

A. M. Bishaev, E. F. Limar, S. P. Popov,
and E. M. Shakhov
*Computing Center of the USSR Academy of Sciences,
Moscow, USSR*

Abstract

This paper deals with the problem of an axisymmetrical freejet of monatomic gas impinging on a normal flat wall with a circular orifice at the axis of symmetry. The wall separates the gas of the jet from the vacuum behind it. Walls of both zero and finite thickness are considered. For the wall of zero thickness the problem was solved using both the Euler equations and the model kinetic equation. In the continuum limit the computations show that a regime of steady oscillations occurs in the throat when the wall is located at a moderate distance from the nozzle. The mass flux through the orifice as a function of the distance of the wall from the nozzle is shown to be practically independent of flow rarefaction. If all the impinging molecules are absorbed on the surface of the normal wall, an inlet flow into and through the tube is established. This process was studied primarily in the free molecular regime. A deviation from a linear distribution of the macroparameters along the tube is indicated. For a wall of finite thickness and diffuse surface interaction the problem was solved at moderate Knudsen numbers using the model kinetic equation. The influence of boundary conditions was also studied.

Introduction

We consider a supersonic or sonic jet of monatomic gas expanding into a vacuum chamber (1 in Fig. 1) from a circular orifice or a nozzle exit with a radius r_e. At the distance L from the nozzle exit a flat wall is located perpendicular to the jet with a circular orifice in it with radius R (see Fig. 1). This wall has a finite thickness ℓ and the orifice has the form of a tube with circular cross section. Behind the wall the second vacuum chamber (2 in Fig. 1) is located. Gas coming to the partition initially flows along the plate in a radial direction and then flows through the orifice. The object of this study is to find the mass flux through the hole and to study both the inlet flow and the flow through the tube/orifice. For simplicity, we assume

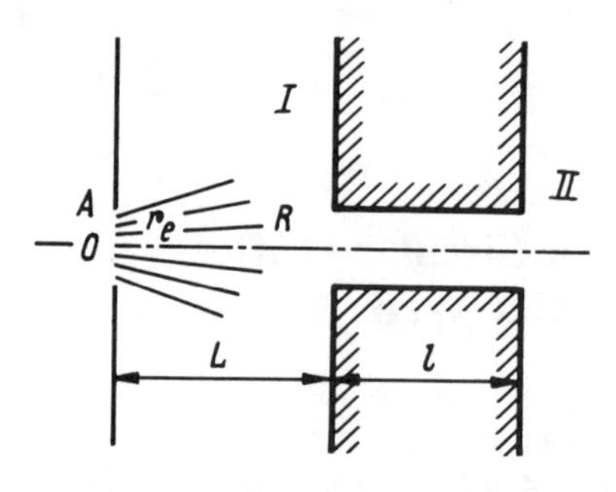

Fig. 1 Scheme of flow.

the flow to be uniform at the initial cross section so that the gas parameters such as density ρ_e, temperature T_e, pressure P_e, and velocity u_e are constants. Subscript e denotes the parameters at the nozzle exit. At infinity the condition of vacuum is assumed. We assume that the orifice radius r_e is equal to the nozzle exit radius $r_e = R$. The flow pattern is essentially defined by the parameters L, ℓ, M, and Kn as well as the gas-surface interaction law. Here, L is the dimensionless distance from the initial jet section to the partition, ℓ is the dimensionless thickness of the wall (and hence, the length of the tube), M is the Mach number of the issuing jet, and $Kn = \lambda_e/r_e$ is the Knudsen number, which is given by the Maxwellian mean free path λ_e divided by the radius r_e. The radius r_e is chosen as the characteristic length for normalizing all length dimensions.

Three typical kinds of flow are considered: 1) inlet flow into the circular orifice of a perpendicular partition; 2) a jet coming into a circular tube and flow through the finite length tube; and 3) jet flow into an orifice in a wall and then through a circular tube of a finite length, which is the combination of 1) and 2). All three problems can be treated as one with the characteristic boundary conditions being specified. We consider these three kinds of flow step by step.

Jet Impinging on a Plane with Orifice-Gas Dynamic Regime

Numerical studies of a jet impinging on a plane partition with a circular orifice at the axis of symmetry were carried out in two directions: 1) based on the Euler equations (gasdynamic limit when $Kn \to O$) and 2) using a kinetic model. Unfortunately, we were not able to cover the whole range of Knudsen numbers, but comparison of the solutions based on such different approaches illuminates rarefaction effects. Moreover, the solutions for the gasdynamic limit and the free molecular regime are found to be rather close when comparing the macroscopic parameters.

The gasdynamic flow regime was studied using the Euler equations that were integrated numerically by use of the SHASTA method.[1] When the Mach number is fixed the solution depends only on the parameter L. It is useful before solving the problem to imagine how this parameter affects the flow pattern. Under condition $L \ll 1$, when the distance from the nozzle exit to the partition is negligible, the jet reaches mainly the orifice and only its small peripheral part flows away from the axis along the wall.

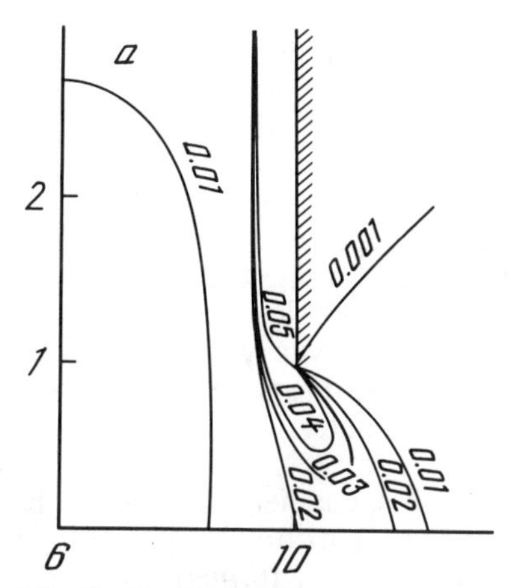

Fig. 2 Oscillating flow through the orifice. Supersonic regime at the axis of symmetry. Density contours.

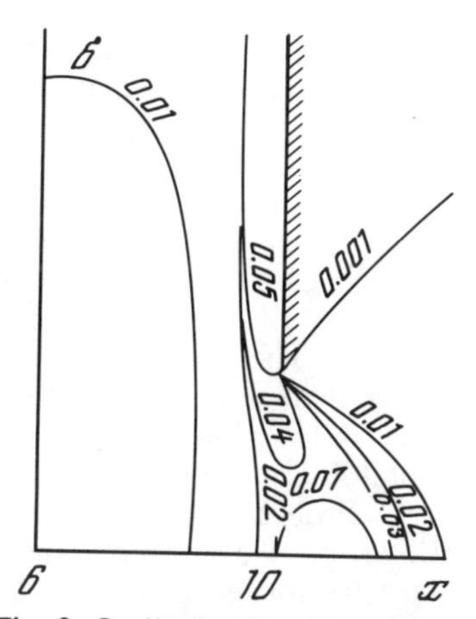

Fig. 3 Oscillating flow through the orifice. Subsonic regime at the axis of sym metry. Density contours.

The influence of that part of the jet on the flow through the orifice is therefore negligible. In the opposite case of a large distance of the wall from the initial section of jet, i.e., L>>1 only a small part of jet mass flux comes to the orifice and the flow is expected to be dominated mainly by the jet-partition interaction. As is well known when a jet impinges on a perpendicular plane without any hole, a shock is formed in front of the partition with subsonic flow behind it and with the flow being then accelerated up to supersonic speed in the radial direction through the hole.

Numerical studies of these kinds of flows were performed in Ref. 2 and a typical flow pattern is given in the Ref. 3. The presence of the orifice transforms the flow structure in the vicinity of the orifice, so that under condition L>>1 a normal shock is expected to cover the entrance into the orifice. When $L \gtrsim 1$ transition from free inlet flow to the regime with a normal shock is expected. It is worth pointing out that there is an accumulation effect that occurs near the axis of symmetry for a sonic jet but the flow is still supersonic.

The computations concerning the sonic jet expansion show that for $L \cong 6$-10 a regime of steady oscillation arises - the principle structure of the arising flow now described. The orifice region can be divided into three subregions. In two of them adjacent to the edge of the partition and to the axis, respectively, the flow is subsonic, but in the third, the intermediate region is characterized by an oblique shock and supersonic flow behind it. However, this configuration is unstable and the solution has a quasi-periodic structure in time. In Figs. 2 and 3 the flow density at different instants of time is depicted for L = 10. In Fig. 2 the location of the shock wave corresponds to the curve of constant density $\rho = 0.02$ (normalized by the density at the nozzle exit). Flow near the axis is supersonic. The maximum value of density is at the axis of symmetry. In contrast, in Fig. 3 the density contours are given for that time when the density at the axis reaches its maximum value. The flow is subsonic with an inertially tethered gas cloud near the axis. After that the cloud is transformed into the form represented in Fig. 2. The solution is then repeated again and again.

Figure 4 presents the variation of density at the orifice center in time. Oscillations with modes of both high and low frequency are observed. The magnitude of the oscillations in density is two to three times greater than the mean value. The corresponding pressure amplitude is about 10-15. Note that the oscillations were calculated very carefully with about 300 time steps per period.

Probably the simplest explanation of the phenomena obtained is the following. Gas flowing from the edge of the partition to the axis of symmetry forms an accumulating jet that results in the formation of a gas cloud with raised pressure in it. At the stage of maximal accumulation (Fig. 3) this region of raised pressure makes the incoming gas turn out from the axis of symmetry and that generates a screening of the cloud from the incoming gas. As a result, the gas cloud of increased pressure moves downstream, enters the region of low pressure, and then begins to freely expand. The flow then comes into the state depicted in Fig. 2. Then, again, gas from the partition edge flows toward the axis and the accumulation process begins anew. The process described characterizes the low frequency mode of the oscillations. The high frequency mode is apparently defined by vibrations in the cloud because this frequency is about equal to a characteristic speed of sound divided by the diameter of the cloud.

To answer further questions that arise concerning the flow behavior, more detailed computations are needed with the oscillations being isolated and investigated as a perturbation to the main flow, which should be computed in advance.

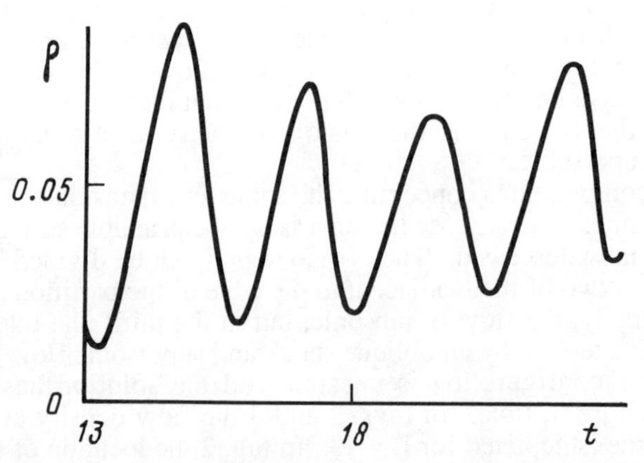

Fig. 4 Oscillations of density at the central point of the orifice.

Jet Impinging on a Plane with Orifice-Kinetic Regime

When gas in the oncoming jet is rarefied enough the gasdynamic description breaks down and it is necessary to describe the flow on a molecular level. The gas motion was studied using the following model kinetic equation (s-model) that is written in cylindrical coordinates (x,z,φ).[4]

$$\xi_x \frac{\partial f}{\partial x} + \xi \cos\omega \frac{\partial f}{\partial z} - \frac{\xi \sin\omega}{z}\frac{\partial f}{\partial \omega} = \upsilon(f^+ - f)$$

$$f^+ = f^o \left[1 + \frac{4}{5}(1-P_2)\, S_\alpha c_\alpha \left(c^2 - \frac{5}{2}\right)\right]$$

$$c_i = \frac{\xi_i - u_i}{\sqrt{2\,RT}}, \quad \xi_1 = \xi_x, \quad \xi_2 = \xi_z, \quad \xi_3 = \xi_\varphi$$

$$\xi_2 = \xi \cos\omega, \qquad \xi_\varphi = \xi \sin\omega$$

$$\upsilon = \frac{4}{5\sqrt{\pi}}\frac{1}{Kn}\frac{nT}{\mu}$$

Here and below all the quantities are normalized by the corresponding values at the nozzle exit.

The following boundary conditions are assigned: at the initial cross section OA (see Fig. 1) a Maxwellian distribution function with given macroparameters is assigned:

$$f = f_e = \frac{1}{\pi^{3/2}}\, e^{-(\underline{\xi} - u_e)^2} \quad \text{for} \quad \xi_x > 0$$

On the wall surface the boundary condition of diffuse scattering with full accommodation is assumed:

$$f = f_w = \frac{n_w}{(\pi T_w)^{3/2}}\, e^{-\xi^2/T_w} \quad \text{for} \quad \xi_n > 0$$

Here ξ_n is the projection of molecular velocity ξ on the outer normal to the reflecting surface. The wall temperature is usually assumed to be constant. Typically, $T_w = 1$. The density of reflected molecules n_w is defined by the condition for mass flux on the surface to be zero:

$$n_w = \frac{2\sqrt{J_1}}{\sqrt{T_w}} \int_{\xi_n<0} \xi_n\, f\, d\underline{\xi}$$

The problem was solved for vacuum condition at infinity. In practice the condition $n_w = 0$ for $\forall(x,z) \in \Gamma$ was assigned at a contour Γ. The latter was chosen in such a way that the result of the calculations does not depend on its location. Generally, the removal was about 5 units (i.e., nozzle exit radius).

For numerical solution of the problem two methods were employed: 1) the finite difference method[4] and 2) a method based on reduction to integral equations. The first method is very efficient and convenient for problems in which the effects of discontinuities of the distribution function on characteristic cones is negligible. These effects will increase with Knudsen number. Phase space should be divided into subspaces of influence in order for discontinuities to be taken into account. In the method of integral equations the discontinuities are automatically considered. However, this method is more expensive in the sense of computations as compared to the method of finite differences.

The results are presented using both approaches with use of the second method for checking of the first one. Keeping the role of discontinuities in mind, the accurate calculation of the flowfield near corner point A at the nozzle exit was made using a combination of the two methods. To be correct, the integral method was employed for positive velocity components $\xi_x > 0$ in space region I and the finite difference method was used for other situations.

One of the quantities of interest is the mass flux through the orifice as a function of distance L from the initial section of the jet and as function of the Knudsen number. In Fig. 5 the mass flux ratio to its free molecular value for a sonic jet is represented. It is interesting to point out that this

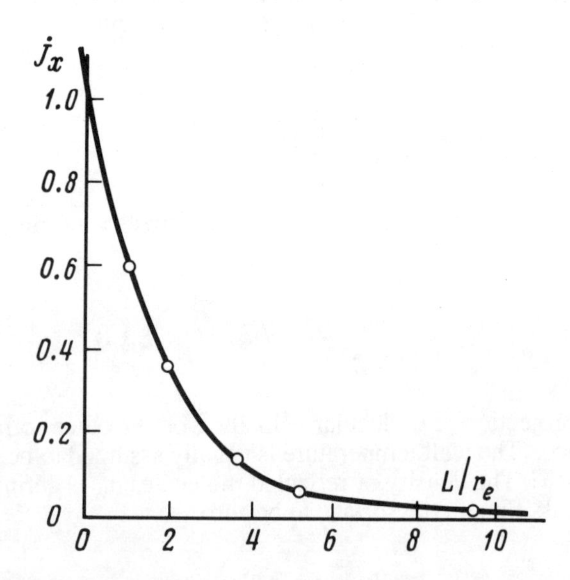

Fig. 5 Normalized mass flux through the orifice: solid line - free molecular solution, circles - gasdynamical regime.

quantity is approximately independent of the degree of rarefaction. The solid line is the free molecular solution ($Kn \to \infty$) and the circles represent the gasdynamic limit ($Kn \to 0$). Points of numerical solutions are placed on the curve with reasonable accuracy.

Computations were performed both for sonic and supersonic jets. Here we consider a supersonic jet for the conditions $M = 3$, $L = 3$. These conditions are slightly different from those given in Ref. 3 for a jet impinging on a wall without a hole. Consider the influence of rarefaction for $T_w = 1$. In the free molecular limit, when $Kn \to \infty$, the molecules reflected from the wall move without collisions, there is no gas pillow in front of the partition.

Increasing the density at the nozzle exit, i.e., decreasing the Knudsen number, results in formation of a higher density region in front of the orifice. Even for $Kn = 0.3$ the flow is similar to that predicted by the analysis based on the Euler equations. In particular, an accumulation zone becomes visible near the axis. With decreasing Kn this property appears more and more distinctly.

The results of our computations carried out to distance $L = 10$ do not give any evidence of the existence of the auto-oscillation regime. Of course, we could not obtain an oscillation in time solution by the method of stationary iterations used. The regime of oscillations may exist at Knudsen numbers when the iterations diverge. By our calculations it may happen for Knudsen numbers smaller than 0.03. Unfortunately this range of Kn was beyond the range of application of our analysis technique.

Flow into a Circular Tube and Through the Tube-Free Molecular Regime

Flow through a tube of finite length is usually studied as one element of the flow from one reservoir to another. If the tube is long enough the stream is generally treated as one-dimensional Poisseille flow. On the other hand, the inlet flow into a tube with a closed end has been investigated in connection with the problem of impact pressure probe.[5] In this section we consider the entrance of a supersonic jet (or uniform stream) into a tube and the flow inside the tube with a vacuum chamber following the tube. Here the flow is assumed to be free molecular.

Formulation of the problem and flow scheme is obtained from the general one under the condition that the density of particles reflected from the plane surface facing to the stream is equal to zero, i.e., $n_w = 0$ for $x = L$. The tube is assumed to have walls of zero thickness. Interaction of molecules with the inner surface is assumed to be diffuse scattering with full accommodation of momentum and energy. The temperature is constant.

The principal peculiar feature of such a formulation is that the molecular flux entering the region of integration is assigned at the circle with radius OA (Fig. 1) rather than at the whole initial plane. Therefore, in the case under consideration the distance L from the nozzle exit is an essential parameter and hence, the divergence of the stream is of primary importance.

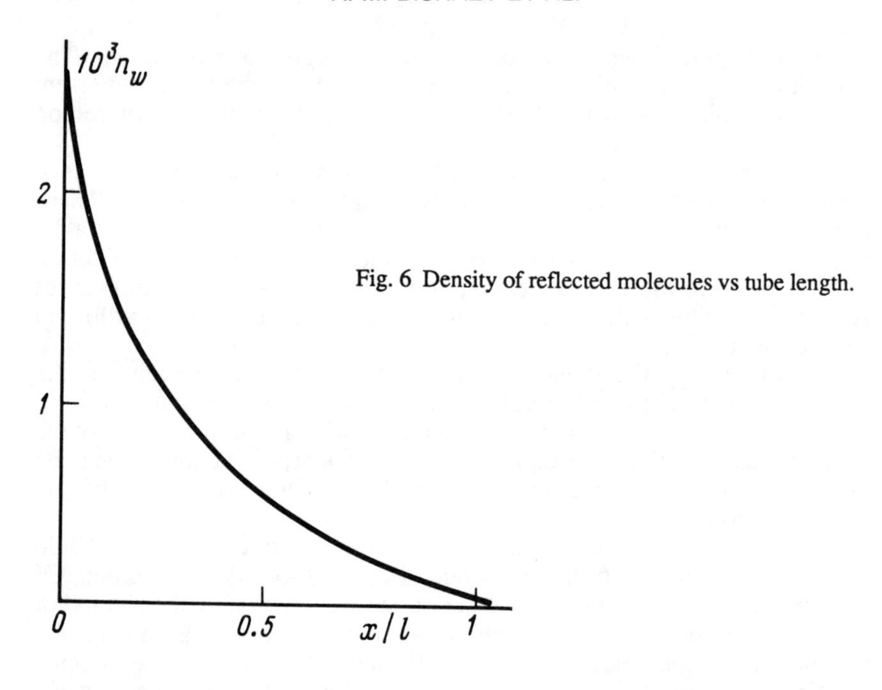

Fig. 6 Density of reflected molecules vs tube length.

The problem was solved by a method of integral equations.[5] The non-homogeneous integral equation for density n_w of reflected particles was constructed. It is of the form

$$n_w = F(x) + \frac{1}{\ell} \int_0^{\ell} K\,(x,x')\; n_w\,(x,x')\; dx'$$

The function F(x) is defined by the conditions at the initial section OA and at the tube exit. This equation was solved by successive iterations with use of the Simpson rule for evaluation of quadratures. Convergence of the iterations becomes slower with increasing tube length ℓ.

Gas flowing into the tube is retarded by the walls because the molecules coming on the solid surface lose their tangential momentum. Accommodation of molecular velocities to the Maxwellian distribution with temperature being equal to the wall temperature takes place. In other words, thermalization of the flow occurs. The gas state at a distance x from the tube entrance depends on the total tube length ℓ as well. In Fig. 6 the density n_w is plotted vs x for $\ell = 33$. The density decrease is not in accordance with linear law as adopted for very long tubes joining two reservoirs, and is even faster than x^{-2}. This fact is mainly caused by the divergence of the jet. For a uniform stream coming into the tube of the same length the density n_w drops approximately linearly.

A density distribution along the tube represents essentially non-linear curves with two corner points corresponding to the entrance and exit of the tube respectively. The density distribution derives from the distribution of the free expanding jet, but the behavior of macroparameters when $x \to +\infty$ follows the same asymptotic law as if free expansion takes place.

The solution for the free molecular regime was obtained by using two methods. The results were found to be rather close, implying that the method of finite differences is efficient enough for our purposes.

Inlet Flow into Orifice of Finite Length-Kinetic Regime

The entrance flow into an orifice in a partition of finite thickness is accompanied by a number of interesting phenomena including the accumulation effect at the axis of symmetry. Recent experiments[6] concerning the accurate investigation of flow structure in a short tube joining a chamber of high pressure with vacuum showed the formation of strong shocks and their reflection from the walls of the tube. We have pointed out that the accumulation effect takes place even for a partition of zero thickness (ideal orifice). The accumulation effect is caused by the partition and the axisymmetrical geometry. In this section a flow is considered for a partition that does not play any role in determining the flow structure. Inlet flow into a tube with walls of zero thickness and flow through the tube is investigated as a result of both molecular-surface and intermolecular interaction. Formulation of the problem includes the assumption that the density n_w of molecules reflected from the partition surface facing the jet is zero.

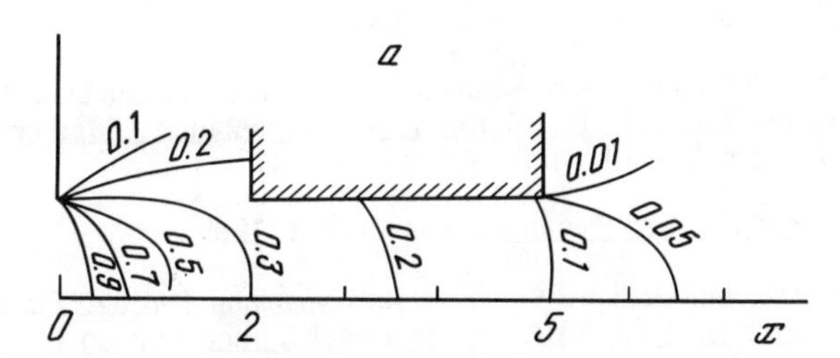

Fig. 7 Density contours for the flow through a tube for cryogenic wall conditions and Kn = 0.5.

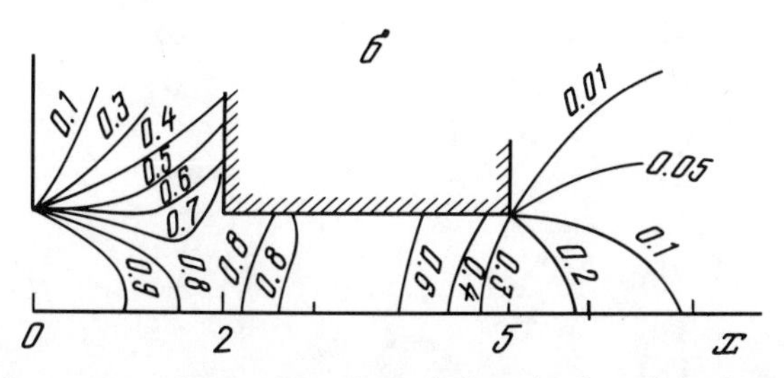

Fig. 8 Density contours for the flow through the tube for adiabatic stagnation wall temperature conditions and Kn = 0.5.

Typical computations were completed for Kn=0.5 under sonic conditions at the nozzle exit located at the distance L=2 from the tube entrance. We studied the effect of boundary conditions prescribed by the wall temperature T_w. The two simplest versions are: 1) $T_w \to O$ {the absorbing cryogenic surface (Fig. 7)} and 2) $T_w = T_o$, where T_o is the adiabatic stagnation temperature (Fig. 8). As expected for moderate Knudsen numbers and cryogenic conditions the flow looks like a free expanding jet. Of course, the flow for the hot surface differs essentially from the flow with an absorbing cryogenic surface. The first qualitative distinction is that the density distribution along the x-axis is non-monotonic, the distribution has a minimum in front of the tube. But this fact holds even for the free molecular regime. The second characteristic feature is that the tendency for accumulation at the axis is observed in spite of the fact that there is no perpendicular partition.

References

[1]Boris, J. P. and Book, D. L., "Flux-corrected Transport I. SHASTA, a Fluid Transport Algorithm that works," Journal of Computational Physics, Vol. II, Nov. 1973, pp. 38-69.

[2]Dubinskaya, N. N. and Ivanov, M. Ya., "On computation on Interaction of Ideal Gas Jet with Perpendicular Plane Partition," Scientific Note TsAGI, Vol. 6, No. 5, 1975, pp. 38-44 (in Russian).

[3]Godunov, S. K., et al., "Numerical Solution of Multi-dimensional Problems of Gasdynamics," Nauka, Moscow, 1976 (in Russian).

[4]Shakhov, E. M., "Solution of Axi-symmetric Problems of Rarefied Gas Dynamics by Finite difference Method," USSR Journal of Computational Mathematics and Mathematical Physics, Vol. 14, April 1974, pp. 970-981.

[5]Kogan, M.N., Rarefied Gas Dynamics, Plenum, NY, 1969.

[6]Koppenwallner,G. and Dankert, C., "Free Jet Expansion Through Finite Length Orifices," Proceedings of the XIIIth International Symposium on Rarefied Gas Dynamics, Book of Abstracts, Vol. 2, Novosibirsk, 1982, pp. 464-467.

Chapter 7. Surface Interactions

Particle Surface Interaction in the Orbital Context: A Survey

F. C. Hurlbut*
University of California at Berkeley, Berkeley, California

Abstract

This work summarizes our present understanding of interactions between gas molecules and solid surfaces under conditions of extremely high relative velocities, and reviews the relevant literature. Within recent years there has been a broadening of anticipated operational regimes for aerospace vehicles so that the mean relative velocity of particles intercepted by the vehicle surface now considered of interest has risen to as much as 45000 ft/sec or about 14 km/s. The translational energy of oxygen atoms at this velocity is about 16.3 eV. We are particulary interested in the distributions of translational velocities of molecules scattered from surfaces and in their rotational and vibrational states. Knowledge of these consequences of scattering bears directly on the prediction of flight behavior of vehicles and also upon aerothermodynamic effects. Throughout the range of energies from those of low Earth orbit to those of planetary entry a great variety of inelastic and reactive scattering process are energetically possible. The majority of studies discussed in this review are ground based using ion or neutralized beam methods and yield scattered flux distributions or, in a few cases, both flux and energy distributions. Several examples are included of close correspondence between experimental mappings and trajectory simulation. It is concluded that earth based experimentation by beam methods is practicable for the relevant energy range and will be able to provide a valuable compliment to flight experiments.

Presented as an Invited Paper.

*Professor of Aeronautical Sciences.

Introduction

This paper is written to summarize our present understanding of interactions between gas molecules and solid surfaces under conditions of extremely high relative velocities, that is, velocities of orbital order. We focus with particular interest on those aspects of the gas/wall interaction that directly influence the dynamical behavior of flight vehicles and bear also upon aerothermodynamic effects. Accordingly, we are interested in the distributions of translational velocities of molecules scattered from the surface and in the related distributions of rotational and vibrational states. From the totality of these distributions, we can infer the transferred momenta and energy. These distributions of velocity and state are also of interest for their importance to the ongoing structure and chemistry of the flowfield.

The words "orbital context" imply that we are interested primarily in the interaction at high approach velocity of major atmospheric constituents with those material surfaces that are commonly employed or considered for employment on aerospace vehicles. But there is a true scarcity of information concerning these matters from direct experiment. Witness to this point is the present program. Historically the RGD Symposium has been the first forum for the reporting of such work. This year, the Symposium does not include a single paper that addresses the results of new measurement made fully within the orbital velocity context.

On the other side of the coin, developments of the past decade have brought a significant broadening of anticipated operational regimes for aerospace vehicles and a corresponding growth in the requirement for new information. To illustrate: in low Earth orbit, the energy of molecular particles is about 0.323 eV/AMU or 5.15 eV for oxygen atoms and about 12.0 eV for argon. The mean relative velocity of particles intercepted by the satellite surface is 7.89 km/s or 25.89×10^3 ft/s. Other velocities, also viewed as relevant, are considerably higher. The nations of the world are preparing for the next steps in the exploration of space and in its economic use. To implement either the space station in stationary orbit or a major planetary exploration to Mars, NASA sees the necessity for a space transportation system based upon the concept of the aeroassisted orbital transfer vehicle (AOTV).[1] As you know, the AOTV is a lifting configuration that enters deeply into the atmosphere during an orbital change maneuver. The entry velocity from a stationary Earth orbit is about 33,500 ft or 10.2 km/s, which is equivalent to an energy of 8.63 eV for oxygen

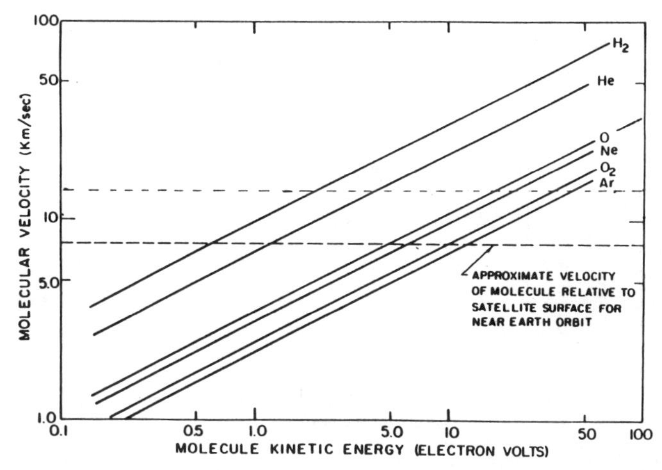

Fig. 1 Velocity vs kinetic energy for six atmospheric species.

atoms. The spacecraft, upon return from a planetary mission, may approach the Earth at perhaps 45,000 ft/s, or about 14 km/s, bringing the energy for oxygen atoms to 16.27 eV. Values of energy vs speed for various gases are shown in Fig. 1. It should be realized that oxygen dissociates at 5.08 eV and N_2 at 9.756 eV. The first ionization potential for O_2 is 12.21 eV and that for N_2 is 15.58 eV. For all atmospheric constituents except H and He, the ionization potential is <u>below</u> the energy associated with surface impact during planetary returns. The point needs no further elaboration; it is clear that complicated and very energetic surface chemical processes will occur.

From the Earth's surface to an elevation of about 90 km, the composition of the Earth's atmosphere changes little. From that altitude on up, increasing compositional changes occur, owing principally to the solar dissociation of oxygen and to the escape of light constituents.[2] At 150 km, the atomic oxygen fraction is nearly as large as that of nitrogen and is a major constituent of the atmosphere at conventional Shuttle altitudes. Farther out from the Earth's surface, helium and then hydrogen become the principal constituents. In the lower atmosphere, the temperature decreases rapidly with altitude, then increases substantially as the chemosphere is entered at about 90 km. Below this level, values of the mean distance between molecular collisions, mean free path (mfp), can be calculated on a straightforward basis, but the determination of mfp becomes somewhat more conjectural above 90 km as a result of both temperature and compositional changes. Total number densities and mean free paths to 250 km are illustrated in

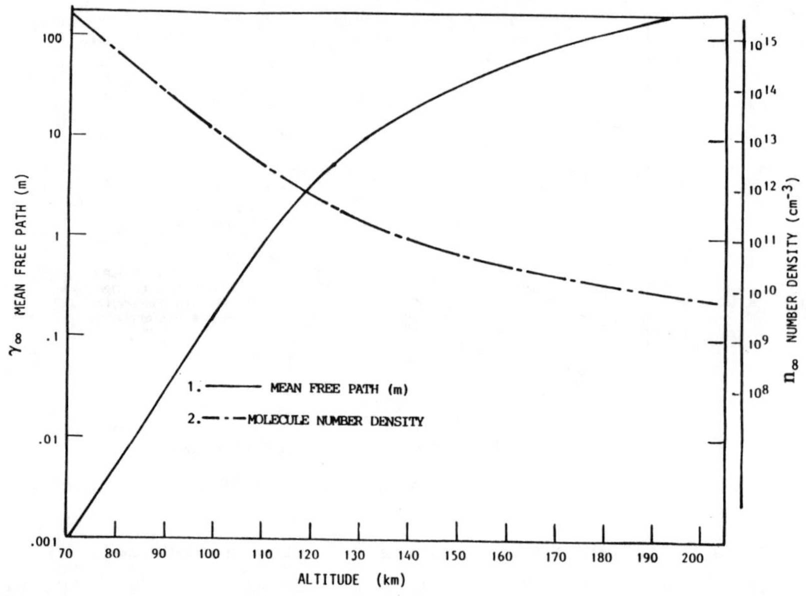

Fig. 2 Mean free path and number density as a function of altitude 70-200 km.

Fig. 2, and the number densities of principal species are shown as a function of altitude in Fig. 3. It is appropriate, in connection with Fig. 3, to call your attention to one obvious point: Earth-based scattering studies using atomic oxygen as the specimen gas must be actively pursued.

With some few exceptions, scattering studies have been conducted using single crystal surfaces as targets, a point that will be amply illustrated in this survey. From the possible options, a great many experimenters have elected to study the (111) face of silver. The use of these and other clean, well-defined surfaces is essential for progress in surface science. However, the surfaces of the Shuttle consist, for the most part, of a glassy envelope over thermal protection tile, whereas certain other surfaces, such as the nose cap, are formed of carbon-carbon. Exposed metallic surfaces on the Shuttle or on satellites may be presumed to be oxide-coated. Just as we must initiate studies using the gases that are most important, so we must study scattering and the attendant chemistry using the appropriate surface materials.

In the next sections, I shall discuss experimental results that, at this date, seem to contribute most effectively to our understanding. A section following the development of experimental work will be devoted to considerations of the representation of experimental data. A general discussion and summary will follow.

Experimental Results: Space-Based Experiments

Atmospheric science has been materially aided by data from sounding rockets and later most particularly from flights of the Shuttle Orbiter. As an example, let me cite the recent paper by Blanchard, Hinson, and Nicholson,[3] in which high-resolution accelerometers revealed details of atmospheric wave structures. Although rather little information concerning the local phenomenology of momentum and heat transfer under free molecule conditions has come from these flights, accelerometer data have, however, permitted much improved measurements of overall Shuttle aerodynamics through determination of aerodynamic coefficients. The lift-to-drag ratio in the free molecule regime, for example, has been calculated by Blanchard to be approximately 0.13, a value that implies a significant departure from diffuse, fully equilibrated re-emission of molecules at the wall.[4] However, the Orbiter presents a complex configuration of

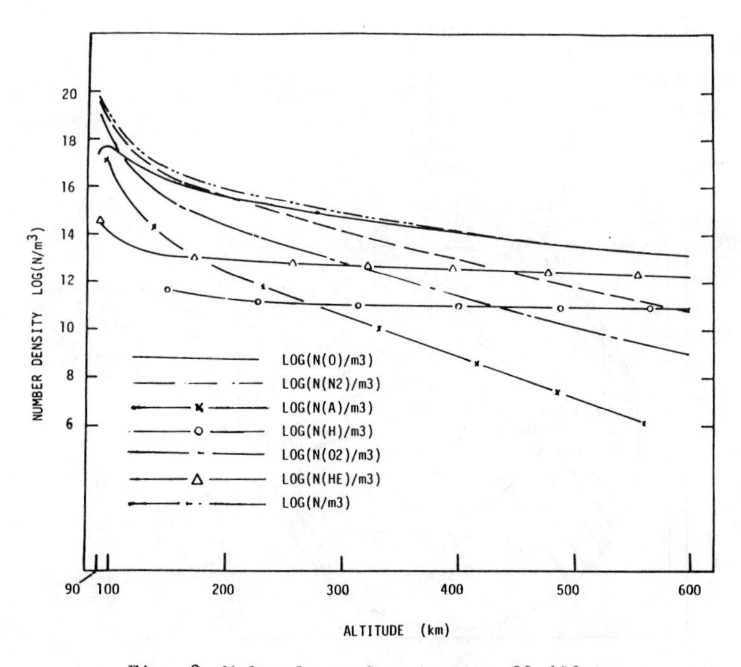

Fig. 3 Molecule number density 90-600 km.

surfaces to the incident flow. The development of localized information concerning momentum transfer from these measurements would be subject to a good deal of uncertainty.

A similar consideration applies to the application of observation of satellite orbital decay, as has been noted by numerous authors over the years. An opportunity of some promise was given by the Explorer VI, an early satellite which was spin-stabilized to fix the orientation of its solar panels. The configuration was that of a four-bladed propeller set in rotation against the incoming stream. About 60% of the satellite surface was coated with silica and the remainder with a black polymer. The rotational rate was precisely monitored and, in consequence, the rate of rotational decay was well known. Moe[5] has calculated that the decay in rotational frequency induced by re-emitted momentum flux was 10 times that which would have resulted if the molecules were diffusely scattered, with velocity determined by the surface temperature (Maxwellian re-emission). This result, like those for the Shuttle Orbiter, does not support a particular scattering model but does rule out Maxwellian re-emission as the predominant scattering mode.

In a much more recent Shuttle-based experiment, flown September 1983, Gregory and Peters[6] measured the angular

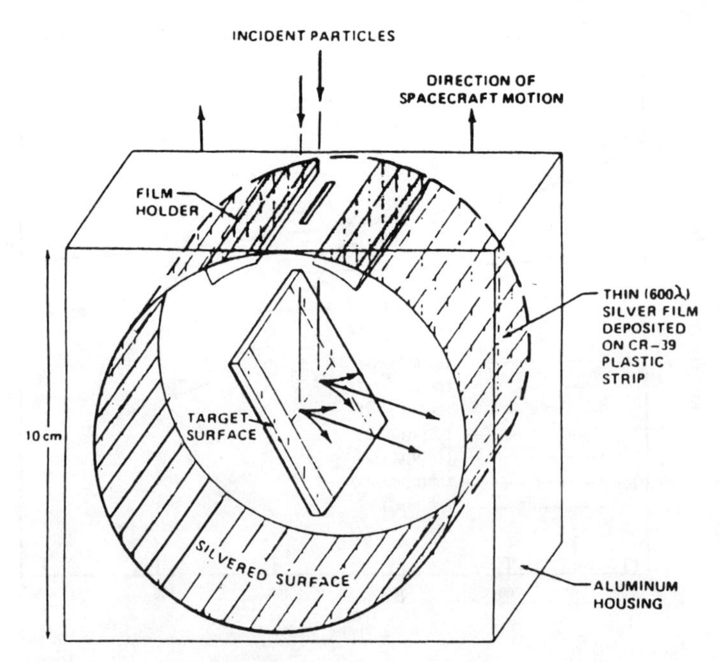

Fig. 4 Oxygen atom reflectometer flown on Shuttle STS-8.[6]

flux distribution of atomic oxygen scattered from polished vitreous carbon. The target surface was oriented with its normal at an angle of 55 deg from the direction of vehicle motion. Flow, primarily atomic oxygen, entered through a narrow slot and either reacted with the carbon or was scattered to the thin silver films serving as a detector (Fig. 4). Recombination efficiency was estimated to be about 20%.

In 1974, the entry of the Viking Lander to the surface of Mars yielded atmospheric and accelerometric information from which good flight dynamic performance data resulted. The Viking Lander was a sphere-nosed, broad (140 deg) cone without projections on the forebody, providing a configuration of relative simplicity. Blanchard and Walberg[7] have analyzed data for the free molecule regimes in terms of the HSN (Hurlbut, Sherman, Nocilla) wall interaction model[8] and have developed a best-fit range of parameters governing scattered velocity distributions and energy transfer. Further discussions of this model follow in another section of the present paper.

Ground-Based Experiments

The domain of energies for particle wall encounters of most immediate interest lies between 5 and 15 eV. Those who are familiar with particle beam experiments are aware that this range presents problems of special difficulty for the production of ion or neutral light particle beams having sufficient flux density. The energies are above those available directly from steady thermal sources (> 50,000 K) but may be obtained for heavy species by seeded beam techniques. Flux densities of particle streams using ion beam technology, with or without charge exchange, are severely limited by beam spreading at such low energies. Production of oxygen or atomic oxygen beams has presented its own set of problems stemming from the chemical nature of oxygen. Recently, new possibilities have appeared on the horizon, such as laser stagnation heating,[9] laser blowoff,[10] and new developments in plasma beam generation.[11] From discussions at the NASA-JPL Workshop on Atomic Oxygen Effects, November 1986,[12] it was estimated that about 25 groups are developing oxygen beam generators. Although these are not for aerothermodynamic studies, the experience will be of value to our area.

For understandable reasons, surface scientists and aerodynamicists have, for the most part, worked at higher or at lower energies than those within our target range, and very few have attempted to observe detailed distributions of velocity. On this account, I shall discuss certain of those studies that appear to contribute most effectively to our understanding of wall/gas processes and to the accuracy of our conjectures concerning behavior in the relevant energy range.

It has become customary to characterize regimes of scattering, following a suggestion by Gerlach-Meyer and Hulpke,[13] on the basis of the product $\omega_D\tau$, in which τ is the duration of the encounter and ω_D is the frequency of lattice vibrations based upon the surface Debye temperature. According to this scheme, for the first regime, the quasielastic regime, $\tau > \omega_D^{-1}$; for the second, the inelastic regime, $\tau \approx \omega_D^{-1}$; and for the third, the impulsive regime, $\tau \ll \omega_D^{-1}$. There is some rough validity to this classification for incident particles that are much lighter than the target species. The second category is sometimes divided to allow a "trapping-hopping" regime at the lower energy range, and other variations in language and concept are occasionally introduced. However, the character of scattering cannot be represented by such a simple scheme; the mass ratio of gas atom to target atom is of critical importance as is the nature of the gas atom or molecule, the wall atoms, and the interaction potential. The third regime is more aptly called the structure scattering regime, as suggested originally by Oman,[14] since the energetic incident gas particle can sense more deeply the target lattice configuration. Structure scattering can be impulsive but, in the cases most relevant to our interests, i.e., where the mass ratio is of the order of one, the dynamic process is one of direct, inelastic scattering. The fast incoming particle slows markedly in a strongly anharmonic collision involving two or more surface atoms as well as the substrate before the repulsive fields, which have become engaged, turn the particle away from the surface. Within the context of our interest, structure scattering is associated with an anharmonic interaction with the surface and possibly to some depth into the solid.[15]

It is useful to look at a series of early observations of structure scattering in which Devienne et al.[16] scattered beams of charge neutralized argon at 100-3000 eV from the surface of copper, silver, and aluminum, and at 1000-3000 eV from glass (Fig. 5a-c). The metal surfaces were microcrystalline and were assumed to possess no preferred orientations. These results exhibit characteristic features, not always simultaneously. As illustrated in Fig. 5c, three lobes predominate, of which those marked T and Y are insensitive to changes in incident direction or in target material, whereas the lobe marked S, directed approximately along the specular ray, moves in accordance with changes in incident direction. We may surmise, as in

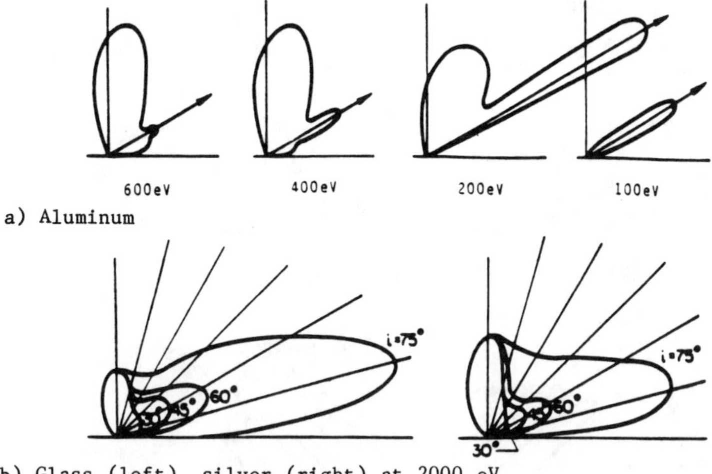

a) Aluminum

b) Glass (left), silver (right) at 2000 eV

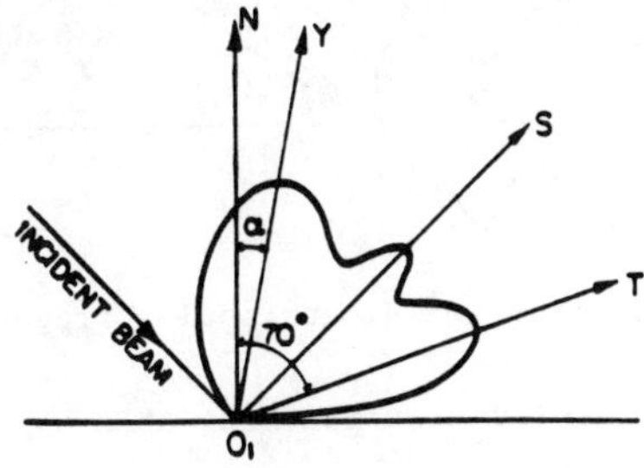

c) Schematic, argon on aluminum

Fig. 5 Scattering of high-energy neutral argon from various surfaces.[16]

Ref. 16, that the surfaces were oxide-coated. Assuredly, the aluminum surface would be oxide-coated, and the others as well, as evidenced by the similarity between patterns of scattering from the glass surface and from the silver. These patterns bear a resemblance to those obtained from argon at 1 eV on epitaxially deposited silver (Fig. 6) after a 5-h delay following deposition.[17] Doubtless, roughness plays some role, as suggested by the work of Smith,[18] but dynamics of the oxide layer must be of equal importance in these patterns. Note, in Fig. 5a, Ref. 16, the remarkable predominance of the single specular lobe at 100 eV. Some uncertainty in the flux representation is created because the method of detection was by direct secondary electron emission from a sensitive surface, for which there is a sharp threshold in sensitivity.

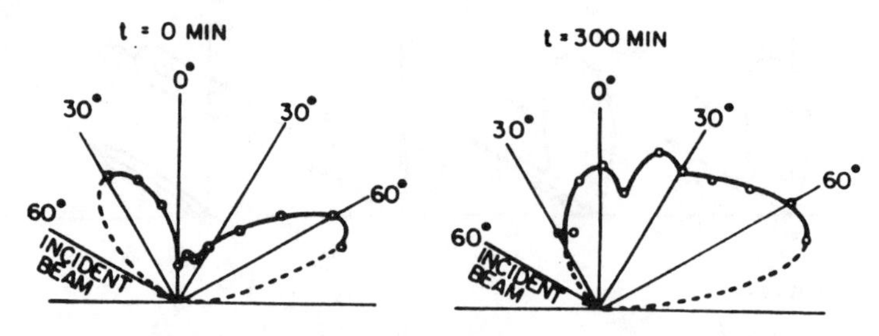

a) Scattered from epitaxially depositied silver[31]

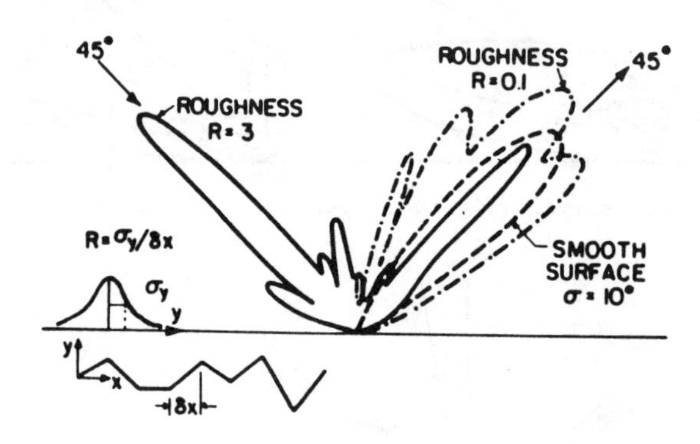

b) Reflections from two-dimensional corrugated wall[18]

Fig. 6 Density distributions of 1 eV argon.

Two separate groups have recently conducted closely allied studies on the scattering of potassium ions (K+) from clean tungsten surfaces at energies lower than 50 eV. The objective in each case was the study of surface configurations and lattice scattering behavior at an impact energy too high for trapping but low enough for the particle to interact with several surface atoms simultaneously. Both groups conducted analyses of test ion trajectories based upon similar assumptions, and both found reasonable agreement with experiment.

The first of these, Hulpke and Mann,[19] publishing in 1983, describe experiments in which the initial energies of the K+ beam were Ei = 6, 11, 16, 21, 36 eV and the incoming polar angle θ_i was varied from 0 to 70 deg. The K+ was scattered from either the W(110) or the W(001) surface. Lithum ions were also used, but the results were not subject to rigorous examination because the neutralization of charge was undetermined. The K+ ion, however, is known to retain its charge throughout the interaction. In trajectory calculations, the incoming particle was allowed to interact with

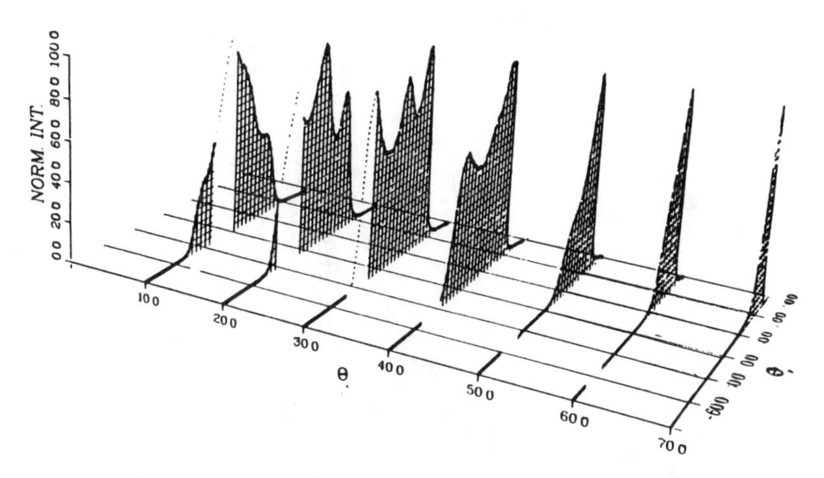

Fig. 7 K^+ on W(110). Measured angular distributions for different impact angles θ_i, fixed beam energy E_i=21.5 eV, and exit angles θ_r.[19]

five surface atoms simultaneously, four in the lattice diamond and one in a second layer at the middle. Born-Mayer exponential repulsive potentials were vectorially summed for the repulsion while a general coulomb potential was used for the attractive portions. Differential equations of motion were numerically solved. Vibrational motion of the lattice and motion of lattice atoms during the collisional impulse were inconsequential. A sequence of measured angular distributions for E_i = 21.5 eV at impact angles from 10 to 70 deg is shown in Fig. 7. Comparisons of two- and three-dimensional calculations with measured results clearly demonstrated that three-dimensional analyses are required. Similarly, it is found that the two-dimensional analysis missed the lower energy tail in distributions of scattered particle energies (Fig. 8).

The second of these groups, with Tenner as first author,[20-22] undertook studies very similar to those just described. Tenner et al., however, elected to direct the K+ beam at normal incidence only and to use a primary beam energy of 35 eV for the bulk of the studies. The group also provided a charge exchanged ion source producing a neutral 35 -eV beam of potassium atoms. Tests of scattering from this source very closely duplicated scattering from the ion source, very neatly supporting the assertion that neutrals of potassium are ionized at close distance of approach to a metal surface. In Fig. 9, we observe typical trajectories of particles scattered from the 35 -eV potential surface. In Fig. 10, we see a measured distribution of scattered

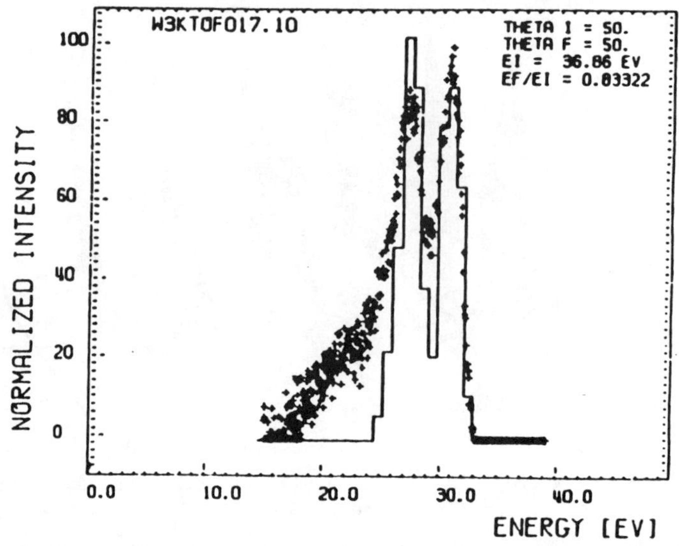

Fig. 8 Measured and calculated (histogram) energy distributions. K+ on W(110), l=0, θ_i=50 deg, θ_f=50 deg, E_i=36.86 eV.[19]

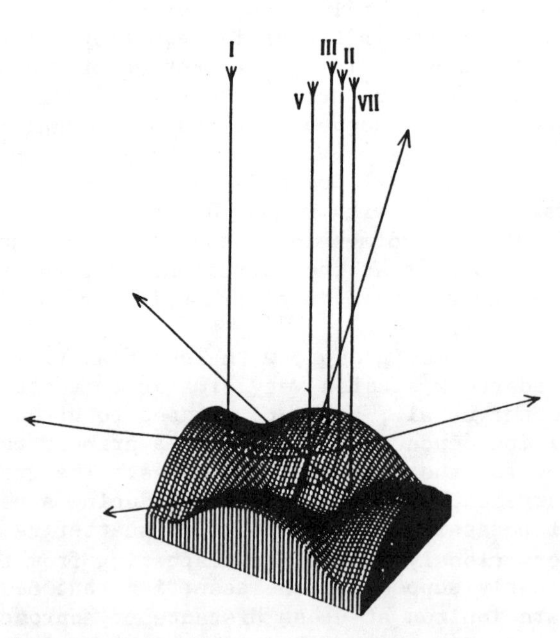

Fig. 9 Gas atom trajectories at normal incidence on 35 eV equipotential surface.[21]

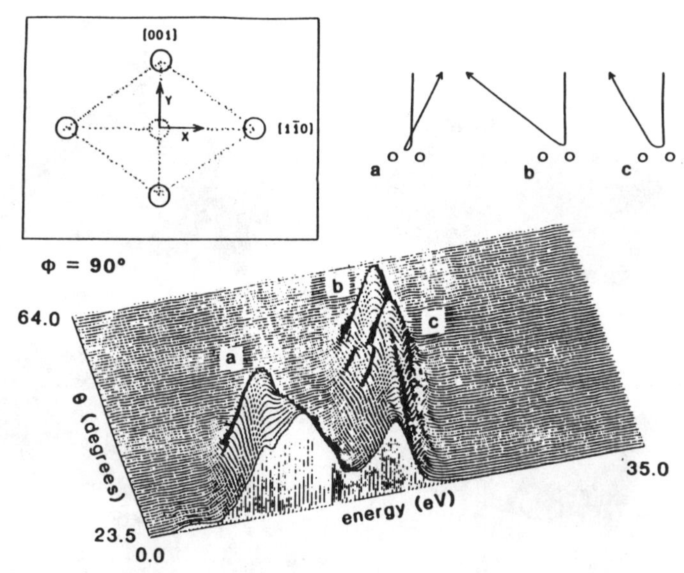

Fig. 10 Measured distribution of the scattered intensity of the K+ ions for an incident energy of 35 eV as a function of the scattering angle θ.[20]

intensity on which trajectory types are identified with peaks and, in Fig. 11, we see comparisons of experimental and simulated spectra. It can be seen that much of the detail of the physical interaction relating to momentum and energy transfers is correctly simulated by the trajectory analysis.

I have devoted some time to the studies of Hulpke and Mann[19] and of Tenner et al.[20-22] because these provide the basis for certain conjectures of some importance. To begin, we are assured that trajectory analysis, with its various assumptions, for example, the assumption that the potentials of interaction are vectorially additive, can yield simulations that represent details of the physical scattering process very accurately. Next, we are given additional evidence that three-dimensional calculations are essential. We are further informed that experiment must accompany analysis. In this instance, in addition to providing overall confirmation, the experimental results were essential to finding the best-fit values for the interaction potential parameters.

Several investigators have used direct balance measurements to determine the momentum transferred from energetic ion or neutral beams to target surfaces. Boring and Humphris[23] reported studies of neutral beams of N_2 at 7-200

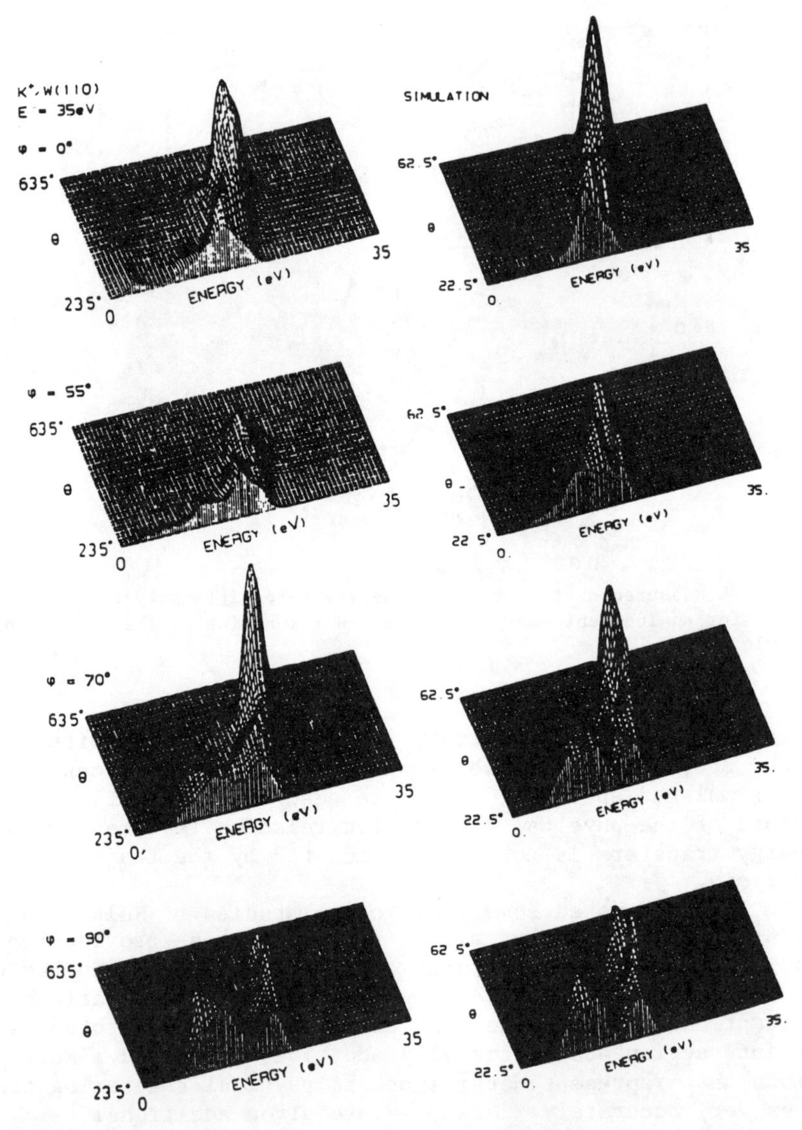

Fig. 11 Experimental spectra (left) and simulated spectra (right), E=35 eV.[21]

eV directed at normal incidence to three target materials. These were a coating of Al on Mylar, of Al coated with amorphous phosphate, and of gold foil, all presumed to be covered with adsorbed N_2 molecules. Momentum flux of the incident beam was calibrated using a momentum trap. Values of the ratio of reflected to incident momentum flux were quite low, dropping from 0.2 at 7 eV to ~ 0.04 at 200 eV. No significant differences in behavior for the three surface materials were observed. Doughty and Schaetzle,[24] in a somewhat more ambitious series of experiments, also used a balance system to observe normal and tangential momentum transferred to targets at neutral beam incident energies from 4 eV to 200 eV. Air, nitrogen, and argon were scattered from technical surfaces of aluminum and fresh varnish at incident angles from 90 to 20 deg, measured from the surface tangent. A momentum trap was used for beam calibration. Normal momentum transfer was found to diminish from values such as 0.8 or 0.9 at normal incidence with energies in the lower range to ~ 0.2 at 150 eV with 30 -deg incident angle. Extrapolation of the measured data suggests very low values at glancing incidence. Similar trends were observed for tangential momentum transfer, but the values of transfer ratio were larger and evidence for backscatter appeared. Normal momentum transfer results of N_2 and A on aluminum are shown in Fig. 12. Knechtel and Pitts[25] performed similar experiments in the energy range 15-50 eV using ions of argon incident on targets of vapor-deposited gold or aluminum. Normal momentum transfer diminished from values of about 0.9 near normal incidence to about 0.3 near 60 deg but, contrary to observations of Knechtel and Pitts, was lower at lower than at higher energies. Angular trends for tangential momentum transfer were less well defined, but values were significantly diminished at the lower energies, again contrary to Ref. 25.

Additional measurements are needed, of course, to clear up these differences, but we can make certain rather well-founded conjectures. It appears that particles of light to moderate mass at velocities of orbital order transfer a fraction of their initial normal momentum and a fraction of their translational momentum to a technical surface, these fractions diminishing in magnitude as the incident angle θ_i moves from the normal toward the glancing direction. The parallel behavior of these momentum transfers implies, at least at incident directions fairly far from normal, a strongly lobular configuration of the scattered momentum flux, with maximum near the specular direction. Finally, it seems quite likely that surface contaminants, probably metal oxides, mask the substrate and reduce its role in the interaction.

Scattering distribution and energy transfer measurements by Kolodney et al.[26] emphasized heavy atom scattering, in this instance, of Hg at 1-9 eV scattered from MgO. Seeded beams of Hg in H_2 were scattered from the clean crystal surface, the final velocity distribution being almost as narrow as the incident. The scattered lobes were slightly subspecular, the directional difference between lobe maximum and specular diminishing to 1.5 deg to 8 -eV incident energy. It must be inferred that the fractions of tangential and normal momentum transferred were about equal. The translational energy transfer rose from 28% at 2.2 eV to 44.7% at 8.7 eV. Trajectory modeling yielded energy exchange and scattered flux distribution in good agreement with experiment (Fig. 13). Kolodney and Amirav[27] have also investigated the scattering of I_2 from polished sapphire crystal at energies of 1 to 10 eV. Specular peaks approximately 20 deg in angular width at half-maximum (Fig. 14) were centered about the specular angle. The iodine dissociation probability, shown in Fig. 15, exhibited an observable threshold at ~ 1.7 eV and rose to ~ 35% at 10 eV. The experimental points were well approximated by the relation $P = 0.514 (E_k - 1.54)^2$ in which P is the dissociation probability and E_k the kinetic energy of the incident iodine molecule. The iodine bond energy is 1.54 eV. Changes in sapphire temperature over the range 100-1100°C produced an

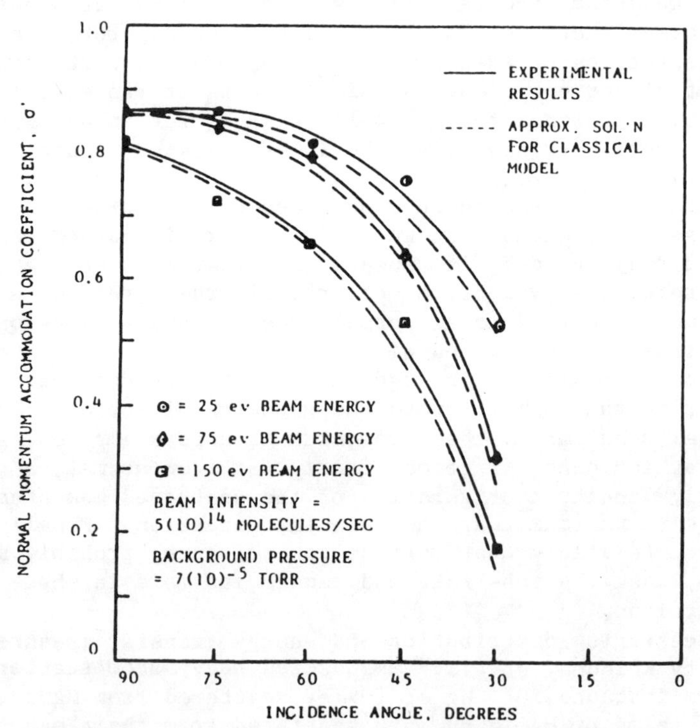

Fig. 12 Normal momentum A.C. as a function of incidence angles for N_2 on aluminum.[24]

extremely small effect on the dissociation probabilities, 1.7% increase at the highest temperature. From these results, the dynamics of dissociation appear to be primarily dependent on molecular kinetic energy, a point corroborated by classical trajectory calculations.

A number of studies have been reported over the years in these symposia and in the archival literature in which beam energies spanned the range of transition from low-energy inelastic scattering with trapping to inelastic scattering with no trapping. Those few studies that included velocity measurement were performed below incident energies of about 1.5 eV. Perhaps three of the experimenters worked with energies up to about 5 eV, and one team, consisting of Hays, Rogers, and Knuth[28] using argon in a helium arcjet, produced beams of an astonishing 20 eV. Certain of these studies of particular importance in the satellite context have been recently reviewed by this author[15] and so will be only briefly mentioned. These include Miller and Subbarao,[29,30] Alcalay and Knuth,[31] Callinan and Knuth,[17] Jih and Hurlbut,[32] Fisher and Pjura,[33] Bishare and Fisher,[34] Romney and Anderson,[35] Sau and Merrill,[36] Smith, O'Keefe, and Palmer,[37] Jakus and Hurlbut,[38] and Jackson and French.[39]

More recently, working also at energies in a range from a few hundred meV to 1. eV, a group based at the University of Chicago has reported a series of experimental results in which velocity distributions were obtained of Ar, Xe, and N_2 scattered from surfaces of W, Ag, and Pt. The group included K.C. Janda, J.E. Hurst, G.A. Becker, J.P. Cowin, L. Wharton, and D.J. Auerbach (see Refs. 40-43). The experimental equipment permitted clean surface conditions. Seeded beam techniques provided beam energies up to 20,000 K, where

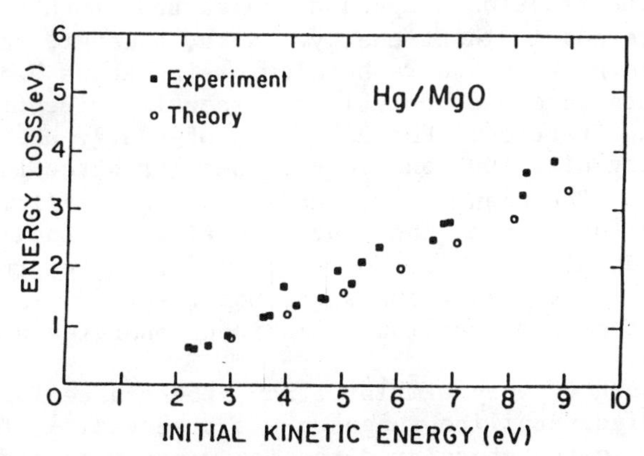

Fig. 13 Mercury energy loss in high kinetic energy scattering from MgO. (■), experimental results; (O), results of the trajectory calculations.[26]

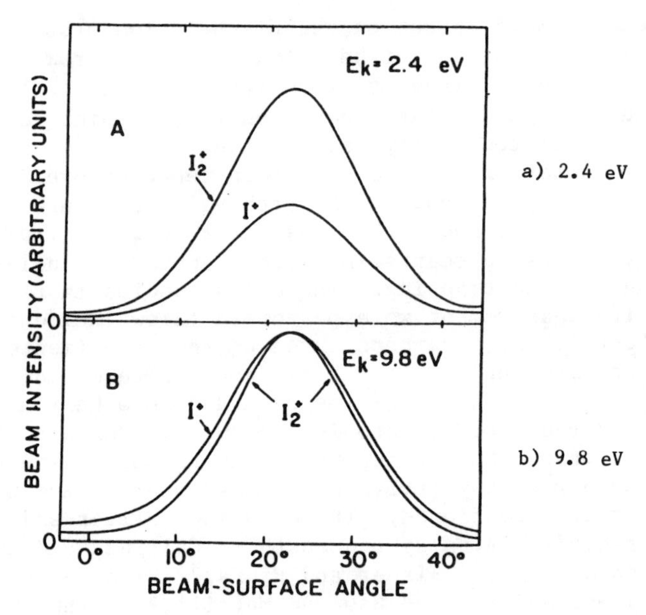

Fig. 14 Molecular and atomic iodine signal vs the surface-beam axis angle. Sapphire single crystal (0001 orientation) target.[27]

$KE_i/2k$ is measured in K, v is the particle velocity, m its mass, and k the Boltzmann constant. In an experiment using $^{15}N_2$ at energies up to 3380 K impinging at 45 deg on polycrystalline tungsten, Janda et al.[40] were able to distinguish between a fraction of the beam trapped, then re-emitted, and a fraction undergoing direct inelastic scattering. The fraction trapped depended upon both surface temperature and incident energy. Note, however, that the highest energy here was rather low, being about 0.3 eV. Measurements were made only in the specular direction and in the normal direction. For all beams of energy above 2400 K, the velocity distributions in the specular direction were supersonic. The mean energy accommodation coefficient was about 0.46 for 45 -deg incidence, considering both normal and specular directions. The fraction of energy transferred because of rotational changes was not observed. Normalized exit energies as a function of incident energies are shown in Fig. 16.

In another, very similar study, the scattering of the Ar from polycrystalline tungsten was observed.[41] The scattered flux in the specular direction was supersonic at all incident energies above 1000 K, and the averaged thermal accommodation coefficient was about 0.26. In another

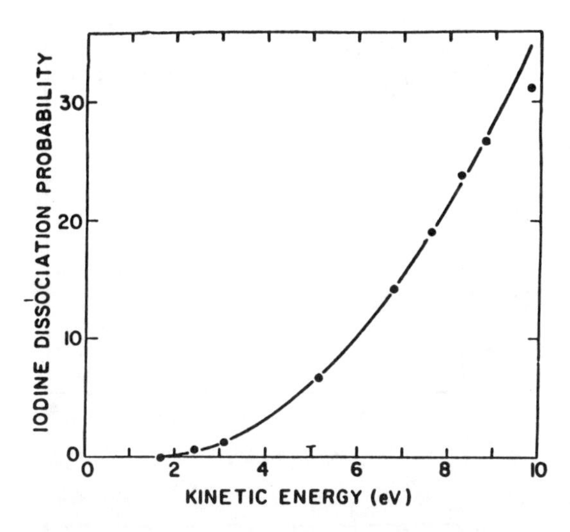

Fig. 15 Molecular iodine dissociation probability after a single collision with a sapphire single crystal vs its kinetic energy.[27]

interesting experiment, the group[42] scattered Xe at a maximum energy of 1685 K from the clean Pt (111) surface. In this experiment, differences in the time of flight spectra permitted the separation of the trapping mode from the direct inelastic scattering mode. Time of flight curves are shown Fig. 17.

In a substantially larger paper Hurst et al.[43] measured the scattering of Ar from the Pt (111) surface at energies to 20,000 K (~ 1.7 eV). The angle of the incident beam was usually equal to 45 deg but, in a few cases, was equal to 75 deg. As one of the significant results, it was found that the speed ratio of scattered atoms ranged from 5 to 10 at incident energies greater than 5000 K. At the highest energy, $E_i = 19760$ K, $\theta_i = 45$ deg, $\theta_f = 48.5$ deg, $U_{iy} = U_{iz} = 2.02 \times 10^5$ cm/s, $U_{fy} = 1.88 \times 10^5$ cm/s, and $V_{fz} = 1.66 \times 10^5$ cm/s, from which we can estimate that the proportional total energy lost (thermal accommodation coefficient in this instance) was 0.23. The coordinate z lies in the normal direction, whereas y lies in the tangential direction.

Velocity and density distribution measurements were made in this same range of energies by Subbaroa and Miller,[29] who studied the scattering of neon and argon at energies to 1.4 eV from the silver (111) surface, and by Jih and Hurlbut,[32] who studied argon scattering at the silver (111) surface at incident energies up to 0.8 eV. A few of the results are represented in Fig. 18a-c. Mean speeds of scattered particles are found by Jih and Hurlbut to be

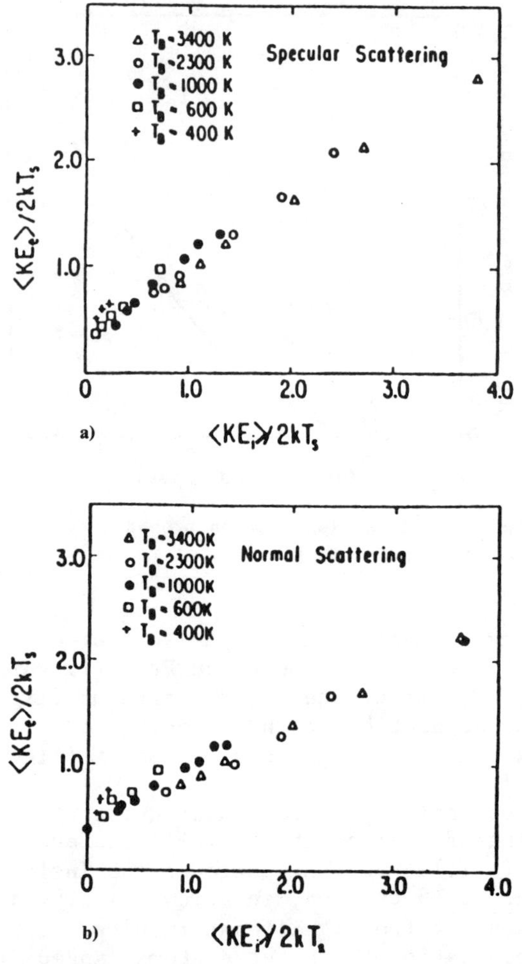

Fig. 16 a) $(KE_e)/2kT_s$ vs $(KE_i)/2kT_s$, specular scattering of N_2-W, 45 deg incidence.[40]

b) $(KE_e)/2kT_s$ vs $(KE_i)/2kT_s$, normal scattering of N_2-W, 45 deg incidence.[40]

lowest near normal incidence and highest as the tangential direction is approached. However, as may be seen at each angle of incidence, the scattered mean speed passes through a dip in the region of the density maximum. Subbarao and Miller, working at 50 -deg incidence, found a similar dip at 0.3 eV but, at an incident energy of 1.3 eV, found the mean speed to be slightly rising. Unfortunately, this excellent work was not extended beyond the limits of the specular peak. Curve G in Fig. 5b was produced by DSMC, Hurlbut[44] using a fully coupled three-dimensional lattice model with

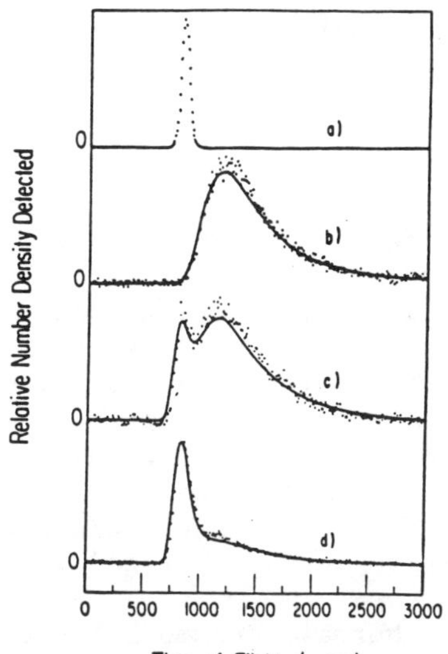

Fig. 17 Experimental time-of-flight spectra for T_s=185 K, angle of incidence 75 deg from normal. Curve A, incident Xe beam with (Ekin)i/k = 1615 K; curve B, Xe scattered at 0 deg (normal); curve C, 45 deg; curve D, 75 deg (specular). Solid lines are predicted TOF spectra from model.[45]

five active atoms in the Ag (111) lattice and several fixed atoms on the surface and also within the lattice. Agreement with experiment in location of the peak and in the general form of the density distribution provides an early example of the success of the classical trajectory approach.

Inelastic processes proceeding from energetic encounters involve many atoms below the surface layer. These must necessarily contribute to the restitutive surface reaction where the mass ratio is of order unity, as in O on SiO_2, or, as in Ref. 27, Hg on MgO. How to model the role of lower level atoms in such cases remains unresolved. An example of current research is found in one approach known as the three-dimensional generalized Langevin model. This model, which rests upon the work of several investigators, is described by Tully.[45,46] Details of the surface/gas interaction are calculated as in a trajectory simulation. Secondary lattice atoms are coupled to the primary atoms through two additional terms in the equations of motion, one accounting for effects of dissipation in the lattice and the second, a fluctuating force, accounting for the effects of lattice vibrations on the surface atoms. The lattice is harmonically coupled. The method works well for low-energy, low-mass-ratio encounters and may suggest directions useful in the orbital context.

Interpretation

Results of fundamental scattering studies such as those discussed here can often be sufficiently well simulated by trajectory analysis to reproduce subtle features of the exit distributions. Ideally, for practical application, observations of momenta and exit velocities should be made over a grid of initial and exit parameters sufficiently dense to support full description by parametric representation. We may suggest three general levels of modeling, the first and most detailed of which is direct simulation. At the next level, results are described through parametric representation and, at the third level, averaged values replace detailed information.

The traditional thermal accommodation coefficient (AC or α) is one such thoroughly averaged quantity, determined at conditions approaching thermodynamic equilibrium in a conductivity cell apparatus. Clearly, results from the direct experimental measurement of α should be subject to corroboration from sources of a more detailed nature. It is useful to set down once again definitions of α, representing the efficiency of thermal energy transfer at the surface and of its analog σ, the tangential momentum transfer coefficient. The third coefficient, σ', is a measure (in this case, not strictly an efficiency) of the normal momentum transferred. Thus,

$$\alpha = \frac{E_i - E_r}{E_i - E_w} \tag{1}$$

$$\sigma = \frac{\tau_i - \tau_r}{\tau_i} \tag{2}$$

$$\sigma' = \frac{P_i + P_r}{P_i + P_w} \tag{3}$$

In Eq. (1), E_i is the total energy, translational and internal, brought up to the surface by the gas, and E_r is the total energy removed by the gas upon scattering. E_w is that total energy that would be removed from the surface by the gas in thermal equilibrium at the surface temperature. In Eq. (2), τ_i and τ_r are the tangential components of momenta brought to, and removed from, the surface, respectively. In Eq. (3), P_i, P_r, and P_w are normal components of momentum. The sign given P_r and P_w reflects the reversal of momentum between incidence and re-emission, the notation reflecting a suggestion by Liu et al.[47] The range of σ' is from 1. to 2.

It is clear that the transfer coefficients as written here apply only to flows of very low speed ratio, that is, at near equilibrium conditions. In high speed ratio, high-enthalpy flows the incident molecular velocities are closely clustered around the velocity of flow, and the kinetic temperature of the incident molecular stream may be three orders of magnitude higher than that of the surface. Re-emission at the wall temperature contributes very little to momentum or energy transfer, and so we may write the so-called differential or partial accommodation coefficients as

$$\alpha(\theta_i) = \frac{E_i(\theta_i) - E_r}{E_i(\theta_i)} \tag{4}$$

$$\sigma(\theta_i) = \frac{\tau_i(\theta_i) - \tau_r}{\tau_i(\theta_i)} \tag{5}$$

$$\sigma'(\theta_i) = \frac{P_i(\theta_i) + P_r}{P_i(\theta_i)} \tag{6}$$

The quantity $\sigma'(\theta_i)$ may also be written as an efficiency by analogy with Eq. (4) in the form

$$\sigma'(\theta_i) = \frac{P_i(\theta_i) - P_r}{P_i(\theta_i)} \tag{7}$$

Since Eqs. (6) and (7) are employed with about equal frequency, care is needed to avoid confusion. The angle θ_i is measured either from the surface or from the surface normal.

Values for these quantities can come, as we have seen, from heat-transfer and momentum-balance experiments. A very useful summary of the data of Refs. 23-25 has been prepared by Knuth,[48] in which $\sigma'(\theta_i)$, Eq. (6), determined by these experiments, are found to lie within about 20% of the line $\sigma'(\theta_i) = (1. + \theta_i/90 \text{ deg})$ (Fig. 19). I have removed points from Knuth's plot for which incident energies were less than 15 eV.

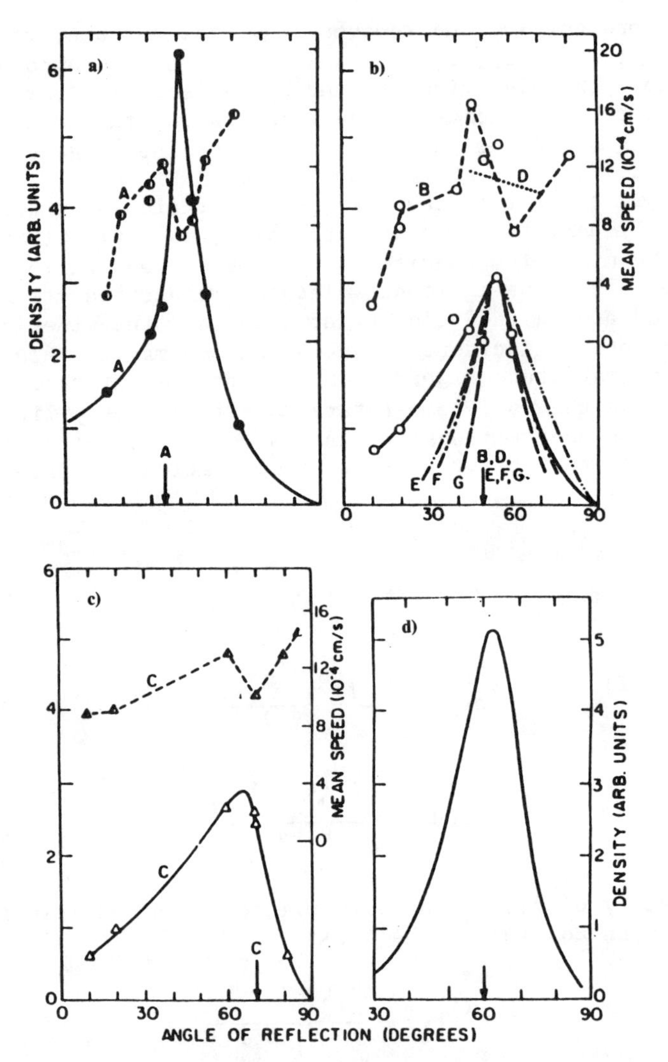

Fig. 18 Scattering of argon on silver (111) surfaces: a–c) 0.8-eV incident argon, solid lines show density and broken lines show mean speed;[32] in b) curve E 1.23 eV argon and curve F 0.3 eV argon show density; curve D 0.3 eV shows mean speed[29] and curve G 0.8 eV DSMC argon;[44] d) Reference 28.

Accommodation coefficients can be very useful in aerodynamic and aerothermal estimation. They do not, however, specify those distributions of velocity or species number density required for DSMC although they are often incorporated into models for these distributions. In the absence of sufficient experimental information, it has been customary to meet this requirement through certain simplified models of historical origin. In the most widely used of these, the Maxwellian re-emission model, all molecules scatter with a Maxwellian distribution of velocities at the surface temperature. For this model $\alpha = 1.$, all incoming particles are fully adjusted to the surface temperature. The flux distribution follows the cosine law, which provides that the outgoing flux at θ, measured from the surface normal, is proportional to cosine (θ). As we all know, the employment of this model is justified for low-velocity flows over technical surfaces but is inaccurate for flows of orbital velocity. A second model, also applying to all incident particles, is like the first except that the Maxwellian distribution is based upon a nonsurface temperature given by an assumed value of the thermal accommodation coefficient; thus, $0. \leq \alpha \leq 1.$ Pure specular reflection is also used occasionally to define scattering behavior. More often, it is used in a composite model in which f, a certain fraction, of incident particles, is assumed to undergo Maxwellian re-emission as in the first case while a fraction (1-f) is assumed to be specularly reflected. The fraction f is often confused with the thermal accommodation coefficient but, in fact, the meaning is quite different. These models are very frequently used in DSMC since they imply well-defined distributions from which to select the next increment of the

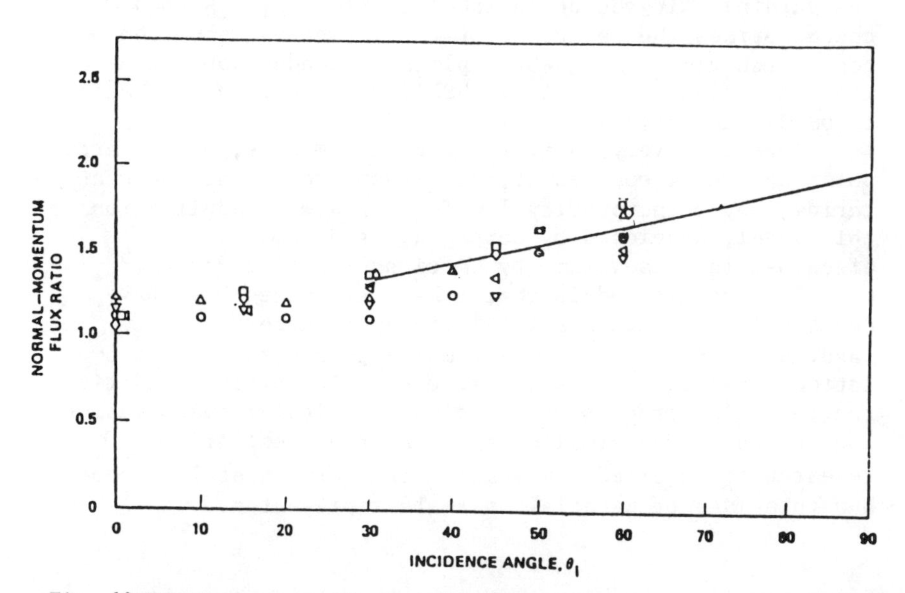

Fig. 19 Normal-momentum flux ratio as function of incidence angle θ_i for several gas-surface conditions.[48]

trajectory. They have a serious disadvantage, however: they do not accurately represent interaction at orbital velocity. Additionally, they do not provide the lobular patterns of particle density distribution, nor do they give correctly the exit velocity distributions.

Various parametric models more closely representing reality have been devised. The Schamberg model is cited as an early example.[49] I shall, however, discuss the HSN (Hurlbut, Sherman, Nocilla) model,[8] which is based on more modern physical observation. The HSN model grew from Nocilla's recognition,[50] now many times confirmed, that along any outgoing ray of the scattered flux of particles, the velocity distribution is well represented as a drifting Maxwellian. Nocilla went on to suggest that the re-emitted lobe could be parametrized in terms of a macroscopic velocity, a thermal spread, and a direction for the macroscopic velocity vector. Nocilla was able to show very close comparisons with existing data. As a model it is incomplete, as has been frequently noted, because in it there is no connection between the incident distribution and the issuing. This essential connection was subsequently added with the construction of the HSN model. Once parameters of the model have been supplied through observation of the behavior of a physical system, the exit distribution is determined for each incident condition. Parameters required are $\alpha(\theta_i)$ and $S_r(\theta_i)$, S_r being the molecular speed ratio. The angle θ_r of the guiding exit vector is taken equal to θ_i. $S_r(\theta_i)$ characterizes the departure of the reflected lobe from cosine behavior. Large S_r implies a slender lobe, $S_r = 0$ implied cosine re-emission. HSN lends itself quite readily to DSMC calculations.[51]

A second, very interesting model employs, as a starting point and major constraint, the requirement that the scattering obey a reciprocity law for surfaces. Application of this model, developed by Cercignani and Lampis,[52] is discussed in this volume by Cercignani and Frizzotti.[53]

The dynamic models that I have discussed here have dealt only with translational energy transfers. Gasdynamicists require more complex parametric representations than these because, at energetic initial conditions, the collision process will include chemical transformation and changes in internal energy. A great deal of new research is required. We must think very carefully about how this mass of material is to be represented.

Summary

A few elements of our modest knowledge of particle surface interaction have been examined in this paper. It is clear that the last decade has seen a very considerable growth in the quantity of experimental information and in the sophistication of experimental methods. Certain views of the scattering process have become reinforced, and others have been rejected. The substantial increase in the number of low-energy ion experiments (below 50 eV) and the corresponding increase in very high-velocity neutral atom experiments in overlapping energy ranges (Refs. 19-27) have shown that Earth-based experimentation in this energy range is feasible. It is interesting that Ar^+ is always neutralized (20) but K^+ is never neutralized at metal surfaces (19). In either case, the experiments can be performed with a stable particle. For these fast particles, momenta are never fully accommodated to the surface but are more accommodated in normal directions than in the specular.[40] Normal momentum transfer decreases as the polar angle θ_i increases. The scattering is always lobular, maximizing near the specular direction for clean surfaces. These observations do not constitute new information but information that has been reinforced by recent observation.

There are now several examples of very close correspondence between thorough experimental mappings of scattered particle distributions and trajectory simulations. This correspondence in subtle exit velocity configurations has led to greatly increased confidence in the validity of trajectory simulations. It is accepted that, in this energy range (our energy range!), classical dynamic descriptions are appropriate. It also appears that pairwise summation of attractive/repulsive potentials between test particle and surface atoms and among surface atoms provides correct model representations in this regime. It is also accepted that we must use a detailed atomic model for the surface. For similar reasons, the simulations must be three-dimensional.

At energies in the orbital range and even substantially below, the trapping-hopping scattering mode is not observed. In the profound interaction of projectile and lattice, a significant fraction of the projectile energy may be transferred. The energy of lattice vibrations is of the order of 10^{-3} smaller than the incident projectile energy and, in consequence, has little effect on the scattering process. We have little direct experimental information on molecular scattering at glass surfaces in the correct energy range. However, trajectory simulation studies suggest that the trends in momentum-transfer behavior resemble those for oxide-coated surfaces.

Directions for future work seem clear. It will be essential to extend Earth-based research with a focus on glasses and oxided metals in interaction with particle beams of atmospheric constituents, particularly atomic oxygen. Experimentation should include the study of internal energy transfers and chemical changes. Development of simulation models and of parametric representations suitable for aerodynamics and thermophysics estimations should be undertaken. It is equally essential that flight experiments be conducted that provide information inaccessible to Earth-based equipment or that provide calibrations and validations of laboratory and computational study.

References

[1]Walberg, G.D., "A Survey of Aeroassisted Orbit Transfer," Journal of Spacecraft and Rockets, Vol. 22, Jan.-Feb. 1985, pp. 3-18.

[2]Jacchia, L.G., Thermospheric Temperature Density, and Composition: New Models, Smithsonian Astrophysical Observatory Special Report 375, March 1977.

[3]Blanchard, R.C., Hinson, E.W., and Nicholson, J.Y., "Shuttle High Resolution Accellerometer Package Experiment Results: Atmospheric Density Measurements Between 60-160 km," AIAA Paper 88-0492, January 1988.

[4]Blanchard, R.C., "Rarefied Flow Lift to Drag Measurements of the Shuttle Orbiter," Paper No. ICAS 86-2.10.1, 15th Congress of International Council of Aeronautical Sciences (ICAS), London, September 1986.

[5]Moe, K., "Absolute Atmospheric Density Determined from the Spin and Orbital Decays of Explorer VI," Planetary and Space Science, Vol. 22, November 1966, pp. 1065-1077.

[6]Gregory, S.J. and Peters, R.N., "A Measurement of the Angular Distribution of 5 eV Atomic Oxygen Scattered off a Solid Surface in Earth Orbit," Proceedings of the 15th Symposium on Rarefied Gas Dynamics, Vol. 1, edited by V. Boffi and C. Cercignani, B.G. Teubner, Stuttgart, FRG, 1986, pp. 644-656.

[7]Blanchard, R.C. and Walberg, G.D., "Determination of the Hypersonic Continuum/Rarefied Flow Drag Coefficient of the Viking Lander Capsule 1 Aeroshell from Flight Data," NASA TP-1793, December 1980.

[8]Hurlbut, F.C. and Sherman, F.S., "Application of the Nocilla Wall Reflection Model to Free-Molecule Kinetic Theory," Physics of Fluids, Vol. 11, March 1968, pp. 486-496.

[9]Cross, J.B., Spangler, L.H., Hoffbauer, M.A., Archuleta, F.A., Leger, L., and Visentino, J., "High Intensity 5 eV Atomic Oxygen Source and Low Earth Orbit Simulation Facility," Proceedings of the NASA Workshop on Atomic Oxygen Effects, Jet Propulsion Laboratories Publication 87-14, Pasadena, CA, June 1987, pp. 105-118.

[10]Brinza, D.E., Coulter, D.R., Liang, R.H., and Gupta, A., "Production of Pulsed Atomic Oxygen Beams via Laser Vaporization

Methods," Proceedings of the NASA Workshop on Atomic Oxygen Effects, Jet Propulsion Laboratories Publication 87-14, Pasadena, CA, June 1987, pp. 143-150.

[11]Mahadevon, P. and Herr, N.C., "Production and Beam Surface Interactions of a Low Energy (2 eV to 10 eV) High Flux Beam of Oxygen Atoms in the Laboratory," AIAA Paper 85-1067, June 1985.

[12]Brinza, D.E. (Ed.), Proceedings of the NASA Workshop on Atomic Oxygen Effects, Jet Propulsion Laboratories Publication 87-14, Pasadena, CA, June 1987.

[13]Gerlach-Meyer, U.G. and Hulpke, E., "Low Energy Ion Scatterings," Topics in Surface Chemistry, edited by E.K. Kay and P.S. Bogus, Plenum, New York, 1978, pp. 195-223.

[14]Oman, R.A., "The Effects of Interaction Energy in Numerical Experiments on Gas-Surface Scattering," Proceedings of the 6th International Symposium on Rarefied Gas Dynamics, Vol. II, edited by L. Trilling and H. Wachman, Academic, New York, 1969, pp. 1331-1343.

[15]Hurlbut, F.C., "Gas/Surface Scattering Models for Satellite Applications," Progress in Astronautics and Aeronautics: Thermophysical Aspects of Re-entry Flows, Vol. 103, edited by J.N. Moss and C.D. Scott, AIAA, New York, 1986, pp. 97-119.

[16]Devienne, F.M., Souquet, J., and Roustan, J.G., "Study of the Scattering of High Energy Molecules by Various Surfaces," Proceedings of the 4th Symposium on Rarefied Gas Dynamics, Vol. II, edited by J.H. de Leeuw, Academic, New York, 1966, pp. 584-594.

[17]Callinan, J.P. and Knuth, E.L., "An Experimental Study of the Particle, Momentum, and Energy Flux Distributions of Products of Collisions of 1 eV Argon Atoms and Surfaces," Proceedings of the 6th International Symposium on Rarefied Gas Dynamics, Vol. II, edited by L. Trilling and H. Wachman, Academic, New York, 1969, pp. 1217-1220.

[18]Smith, M.C., "Computer Study of Gas Molecule Reflection from Rough Surfaces," Proceedings of the 6th International Symposium on Rarefied Gas Dynamics, Vol. II, edited by L. Trilling and H. Wachman, Academic, New York, 1969, pp. 1217-1220.

[19]Hulpke, E. and Mann, K., "Surface Rainbow Scattering of Alkali Ions from Metal Surfaces," Surface Sciences, Vol. 133, July 1983, pp. 171-198.

[20]Tenner, A.D., Gillen, K.T., Horn, T.C.M., Los, J., and Kleyn, A.W., "Rainbows in Energy and Angle-Resolved Ion Scattering from Surfaces," Physical Review Letters, Vol. 52, June 1984, pp. 2183-2186.

[21]Tenner, A.D., Saxon, R.P., Gillen, K.T., Harrison, D.E., Horn, T.C.M., and Kleyn, A.W., "Computer Simulations and Rainbow Patterns of Alkali Ion Scattering from Metal Surfaces," Surface Sciences, Vol. 172, June 1986, pp. 121-150.

[22]Tenner, A.D., Gillen, K.T., Horn, T.C.M., Los, J., and Kleyn, A.W., "Energy and Angular Distributions for Scattering of K+ from W(110) at Normal Incidence," Surface Sciences, Vol. 172, June 1986, pp. 90-120.

[23]Boring, J.W. and Humphris, R.R., "Momentum Transfer to Solid Surfaces by N_2 Molecules in the Energy Range 7-200 eV," Proceedings of the 6th International Symposium on Rarefied Gas Dynamics, Vol. II, edited by L. Trilling and H. Wachman, Academic, New York, 1969, pp. 1303-1310.

[24]Doughty, R.O. and Schaetzle, W.J., Experimental Determination of Momentum Accommodation Coefficients at Velocities up to and Exceeding Earth Escape Velocities," Proceedings of the 6th International Symposium on Rarefied Gas Dynamics, Vol. II, edited by L. Trilling and H. Wachman, Academic, New York, 1969, pp. 1035-1054.

[25]Knechtel, E.D. and Pitts, W.C., "Experimental Momentum Accommodation on Metal Surfaces of Ions Near and Above Earth Satellite Speeds," Proceedings of the 6th International Symposium on Rarefied Gas Dynamics, Vol. II, edited by L. Trilling and H. Wachman, Academic, New York, 1969, pp. 1257-1266.

[26]Kolodney, E., Amirav, A., Elber, R., and Gerber, R.B., "Large Energy Transfer in Hyperthermal Heavy-Atom-Surface Scattering: A Study of Hg/MgO (100)," Chemical Physics Letters, Vol. 113, January 1985, pp. 303-306.

[27]Kolodney, E. and Amirav, A., "Collision Induced Dissociation of Molecular Iodine on Sapphire," Journal of Chemical Physics, Vol. 79, November 1983, pp. 4648-4650.

[28]Hays, W.J., Rogers, W.E., and Knuth, E.L., "Scattering of Argon Beams with Incident Energies up to 20 eV from a (111) Silver Surface," Journal of Chemical Physics, Vol. 56, 1972, pp. 1652-1657.

[29]Subbarao, R.B. and Miller, D.R., "Velocity Distribution Measurements of 0.06-1.4 eV Argon and Neon Atoms Scattered from the (111) Plane of a Silver Crystal," Journal of Chemical Physics, Vol. 58, 1973, pp. 5247-5257.

[30]Miller, D.R. and Subbarao, R.B., "Scattering of 0.06-2.5 eV Neon Argon Atoms from a Silver Crystal," Journal of Chemical Physics, Vol. 52, 1970, pp. 425-431.

[31]Alcalay, J.A. and Knuth, E.L., "Experimental Study of Scattering in Particle Surface Collisions with Particle Energies of the Order of 1. eV," Proceedings of the 5th International Symposium on Rarefied Gas Dynamics, Vol. I, edited by C.L. Brundin, Academic, New York, 1967, pp. 253-268.

[32]Jih, C.T.R. and Hurlbut, F.C., "Time of Flight Studies of Argon Beams Scattered from a Silver (111) Crystal Surface," Proceedings of the 10th International Symposium on Rarefied Gas Dynamics, Vol. 51, edited by J.L. Potter, AIAA, New York, 1976, pp. 539-554.

[33]Fisher, S.S. and Pjura, G.A., "Intensity and Velocity Distribution of 300°K He and Ar Aerodynamic Beams Scattered from a Vapor Deposited Ag (111) Surface," Proceedings of the 7th International Symposium on Rarefied Gas Dynamics, Vol. 1, edited by Dino Dini, Editrice Technico Scientifca, Piso, 1971, pp. 291-310.

[34]Bishara, M.N. and Fisher, S.S., "Observed Intensity and Speed Distribution of Thermal Energy Argon Atoms Scattered from the (111) Face of Silver," Journal of Chemical Physics, Vol. 52, 1970, pp. 5661-5675.

[35]Ronney, M.J. and Anderson, J.B., "Scattering of 0.05-5 eV Argon from the (111) Plane of Silver," Journal of Chemical Physics, Vol. 51, 1969, pp. 2490-2496.

[36]Sau, R. and Merrill, R.P., "The Scattering of Hydrogen, Denterium and the Rare Gases from Silver (111) Single Crystals," Surface Sciences, Vol. 34, January 1973, pp. 268-288.

[37]Smith, J.N. Jr.. O'Keefe, D.R., and Palmer, R.L., "Rare Gas Scattering from LiF: Correlation with Lattice Properties II," Journal of Chemical Physics, Vol. 52, 1970, pp. 315-320.

[38]Jakus, K. and Hurlbut, F.C., "Gas Surface Scattering Studies Using Nozzle Beams and Time of Flight Techniques," Proceedings of the 6th International Symposium on Rarefied Gas Dynamics, Vol. II, edited by L. Trilling and H. Wachman, Academic, New York, 1969, pp. 1171-1186.

[39]Jackson, D.P. and French, J.B., "High Energy Scattering of Inert Gases from Well Characterized Surface 1, Experimental," Proceedings of the 6th International Symposium on Rarefied Gas Dynamics, Vol. II, edited by L. Trilling and H. Wachman, Academic, New York, 1969, pp. 1119-1134.

[40]Janda, K.C., Hurst, J.E., Becker, C.A., Cowin, J.P., Wharton, L., and Auerbach, D.J., "Direct Inelastic and Trapping-Desorption Scattering of N_2 from Polycrystalline W: Elementary Steps in the Chemisorption of Nitrogen," Surface Sciences, Vol. 93, 1980, pp. 270-286.

[41]Janda, K.C., Hurst, J.E., Becker, C.A., Cowin, J.P., Auerbach, D.J., and Wharton, L., "Direct Measurement of Velocity Distributions in Argon Beam-Tungsten Surface Scattering," Journal of Chemical Physics, Vol. 72, February 1980, pp. 2403-2410.

[42]Hurst, J.E., Becker, C.A., Cowin, J.P., Janda, K.C., Wharton, L., and Auerbach, D.J., "Observations of Direct Inelastic Scattering in the Presence of Trapping-Desorption Scattering: Xe on Pt (111)," Physical Review Letters, Vol. 43, October 1979, pp. 1175-1177.

[43]Hurst, J.E., Wharton, L., Janda, K.C., and Auerbach, D.J., "Direct Inelastic Scattering Ar from Pt (111)," Journal of Chemical Physics, Vol. 78, February 1983, pp. 1559-1581.

[44]Hurlbut, F.C., "Current Experiments and Open Questions in Gas Surface Scattering," Proceedings of the 9th International Symposium

on Rarefied Gas Dynamics, edited by M. Becker and M. Fiebig, DFVLR, Porz-Wahn, FRG, 1974, pp. AX-E.3-1-E.3-23.

[45]Tully, J.G., "Dynamics of Gas-Surface Interactions: 3-D Generalized Langevin Model Applied to fcc and bcc Surfaces," Journal of Chemical Physics, Vol. 73, August 1980, pp. 1975-1985.

[46]Tully, J.G., "Theories of the Dynamics of Inelastic and Reactive Processes at Surfaces," Annual Review of Physics and Chemistry, Vol. 31, 1980, pp. 319-343.

[47]Liu, S.M., "Satellite Drag Coefficients Calculated from Measured Spatial and Energy Distributions of Reflected Helium Atoms," AIAA Journal, Vol. 17, December 1979, pp. 1314-1319.

[48]Knuth, E.L., "Free-Molecule Normal-Momentum Transfer at Satellite Surfaces," AIAA Journal, Vol. 18, May 1980, pp. 602-605.

[49]Schamberg, R.E., "A New Analytic Representation of Surface Interaction for Hyper-Thermal Free Molecule Flow with Application to Neutral-Particle Drag Estimates of Satellites," USAF Project Rand, RM-2313, 1959.

[50]Nocilla, S., "On the Interaction Between Stream and Body in Free-Molecule Flow," Proceedings of the Second International Symposium on Rarefied Gas Dynamics, edited by L. Talbot, Academic, New York, 1961, pp. 169-208.

[51]Hurlbut, F.C., "Sensitivity of Hypersonic Flow Over a Flat Plate to Wall/Gas Interaction Models Using DSMC," AIAA Paper 87-1545, June 1987.

[52]Cercignani, C. and Lampis, M., "Kinetic Models for Gas-Surface Interactions," Transport Theory and Statistical Physics, Vol. 1, 1971, pp. 101-114.

[53]Cercignani, C. and Frizzotti, A., "Numerical Simulation of Supersonic Rarefied Gas Flows Past a Flat Plate: Effects of the Gas-Surface Interaction Model on the Flowfield," this volume.

Sensitivity of Energy Accommodation Modeling of Rarefied Flow Over Re-Entry Vehicle Geometries Using DSMC

T. J. Bartel*
Sandia National Laboratories, Albuquerque, New Mexico

Abstract

This paper deals with the influence of energy accommodation in the modeling of the gas/wall interaction for rarefied flow over a typical re-entry vehicle geometry using the Direct Simulation Monte Carlo (DSMC) algorithm. The reflected molecule's energy accommodation was varied while a uniform angular distribution was assumed. The freestream conditions were M = 27.5 and altitudes of 76.2 and 91.4 km. The ratio of the freestream stagnation temperature to the wall temperature was approximately 100. Surface pressure, shear stress, and heat transfer distributions, as well as drag coefficients, are presented for varying degrees of accommodation. These will be compared to the results from classical specular and diffuse models. Unfortunately, no experimental data at the re-entry conditions were available to compare with these simulations.

Nomenclature

A	= base area
c_p	= specific heat
C_d	= drag coefficient, axial force / $1/2\,\rho_\infty U_\infty^2 A$
C_f	= skin friction coefficient, $\tau_w / 1/2\,\rho_\infty U_\infty^2$
Kn	= Knudsen number, λ / l
l	= characteristic length
L	= body length
M	= Mach number
n	= number density

*Member of Technical Staff, Computational Aerodynamics Division.

$\dot{q}$ = heat flux
R_n = Nose radius
St = Stanton number, $\dot{q}_w / [c_p \rho_\infty U_\infty (T_o - T_w)]$
T = temperature
U = velocity
x = axial location along centerline from nose
α_T = thermal accommodation coefficient
λ = mean free path
ρ = mass density
τ = shear stress

Subscripts

∞ = refers to freestream conditions
o = refers to stagnation conditions
w = evaluated at the wall

Introduction

In the past few years, several new types of vehicles and missions involving flight in the upper reaches of the Earth's atmosphere have been proposed. These include the Aero-assisted Orbital Transfer Vehicle (AOTV), the National Aerospace Plane (NASP), and various proposed maneuvering re-entry vehicles. The design of such vehicles necessitates a more detailed knowledge of their aerothermodynamic environment than has previously been available. Their flight profile typically includes the *transitional* rarefied flow regime, where densities are too low for the flow to be considered a continuum but not low enough to qualify as truly free-molecular flow. Unfortunately, no experimental facilities exist that can simultaneously simulate the very high Mach numbers, stagnation enthalpies and chemical species, along with the low Reynolds numbers: all of which characterize such flows. Hence, this class of vehicles is likely to rely more heavily on computational fluid dynamic (CFD) predictions than has been typical in the past.

This envelope of hypersonic speeds and very low densities is generally defined as: Mach numbers 15 to 30 and altitudes from 60 to 130 km. Currently, the Direct Simulation Monte Carlo (DSMC) approach developed by Bird[1-4] is the only viable method for realistically modeling these problems. In this method, tens of thousands of *simulation* molecules, each representing from 10^{11} to 10^{14} *actual* molecules, are tracked in the flow field of interest. The simulation molecules undergo representative collisions with each other as well as solid boundaries or surfaces. This method is widely known and, because it is used in the present investigation without substantive changes except for the molecule/surface interaction model, no further explanation will be given.

A major uncertainty with this method is in the modeling of the gas/wall interaction; typically, a simple combination of Maxwell's specular and diffuse models has been used. The classical specular model assumes a "billiard ball" collision with the surface; that is, no change in the tangential momentum, and no thermal accommodation with the surface. The classical diffuse model assumes a uniform probability for the reflected molecule's angular distribution and complete thermal accommodation of the incident molecule to the surface temperature. Usually, a probabilistic combination is used to obtain intermediate value; that is, a fraction of the collisions are modeled as diffuse and the remainder as specular. This model will be referred to as the *standard DSMC model* in the present work. Several investigators have evaluated the effects of these two classical models and the standard DSMC model (see for example Ref. 5). Hermina[6] found that a coefficient of 0.8 for the standard DSMC model, an 80% diffuse and 20% specular mix, produced results that agreed with experimental data for a flat plate. Unfortunately, the ratio of the freestream stagnation to surface temperature was only 1.02 for this data set. For typical re-entry conditions, this ratio is approximately 100. Thus, one would expect the role of the thermal accommodation to be greater for these conditions.

Recently, Hurlbut[7] explored the sensitivity of different surface interaction models using the same flat plate data as Hermina; he decoupled the angular reflection model from the thermal accommodation. Temperature, velocity, and density distributions normal to the surface were presented. However, the temperature ratio was not representative of re-entry conditions and the effect of varying the surface thermal accommodation for these conditions is not known.

The present investigation is concerned with the influence of the gas/surface interaction model for typical re-entry vehicles and flight conditions. The surfaces are engineering materials and are not polished metal surfaces; a diffuse model for the angular distribution seems to be a reasonable approximation. However, the degree of thermal accommodation is not known. In this paper, the reflected simulation molecule's energy accommodation is varied, while a uniform distribution is assumed for the reflected angle. Surface pressure, shear stress, and heat flux distributions for a representative re-entry vehicle geometry are presented for typical hypervelocity flow conditions; that is, conditions in which the surface temperature is much lower than the freestream stagnation temperature. Unfortunately, experimental data were not available to compare with the results; thus, only the relative sensitivity will be discussed.

Surface Reflection Model

The present model assumes that the angular distribution of the reflected molecule is uniform. This is the same assumption made by the classical diffuse model; however, the thermal accommodation of the incident simulation molecule during the surface interaction is modeled differently. In the present model, the degree of accommodation is varied from 0.0, for no accommodation, to 1.0 for full equilibrium with the surface temperature. A thermal accommodation coefficient, α_T, of 1.0 thus represents the classical diffuse surface interaction model. The translational, rotational, and vibrational energies of the reflected molecule are determined by the same coefficient; no attempt is made to model whether one energy mode accommodates by a different amount than another.

Two methods were used to obtain intermediate values of the mode energies: a linear interpolation and a probabilistic method. The linear method determines the reflected mode energy as a simple linear interpolation between the incident molecule's energy and the value it would have if it completely thermally accommodated with the surface (a classical diffuse reflection). This method calculates a mean value for the reflected molecule; that is, for an α_T of 0.5, each reflected molecule would have the average of the incident energy and the equivalent surface energy for each of the three modes. The second method determines the reflected molecule's energy in a probabilistic manner; that is, the accommodation coefficient determines the probability of the incident molecule accommodating to the surface temperature. For the above example, the energy of the reflected molecule would be the incident value for 50% of the collisions and the equivalent surface temperature for the remainder. The reflected energy distribution using the linear model would be a single value, whereas the probabilistic model would yield a bimodal distribution. The expected value from the probabilistic distribution should be equal to that from the linear model. Unless specified, the present results were generated using the linear model.

Test Cases

The present sensitivity study was done for the spherically blunted biconic geometry shown in Fig. 1. The nose radius and body length are 15.24 cm and 457.2 cm, respectively. Although no experimental data are available for this configuration, it is considered representative of many re-entry type profiles of current interest. Solutions were obtained for $M_\infty = 27.5$ and conditions corresponding to two different altitudes, 76.2 km and 91.4 km. Air was the freestream fluid. In the first case, $T_\infty = 206$ K, $n_\infty = 6.92 \times 10^{20}$ mol/m^3, and in the sec-

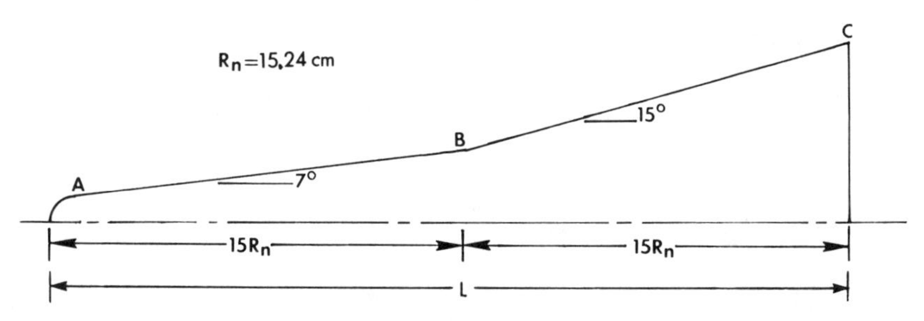

Fig. 1 Body geometry.

ond, T_∞ = 187 K, and $n_\infty = 5.5 \times 10^{19}$ mol/m^3. In both calculations, a constant wall temperature of 277 K was assumed. The freestream perfect gas stagnation temperature of the 76.2 km case was 31,300 K, giving a ratio of the stagnation to wall temperature of 113; the 91.4 km case had a perfect gas stagnation temperature of 28,500 K and temperature ratio of 103. The Knudsen number for these calculations, based on freestream conditions and the nose radius, are $Kn \sim 0.016$ and 0.2, respectively.

Five species, O_2, N_2, O, N, and NO, and 23 chemical reactions were modeled. The DSMC calculations for this geometry used approximately 3000 cells and tracked 70,000 simulation molecules. As used in prior DSMC simulations,[8] the cell dimension in the normal direction near the body was approximately $\lambda/2$, where λ is the mean free path of molecules reflected from the surface. This was modeled from the relation[9]

$$\frac{\lambda_w}{\lambda_\infty} = \frac{4}{\sqrt{\pi\gamma}}\left(\frac{T_w}{T_\infty}\right)^{1/2}\frac{1}{M_\infty} \tag{1}$$

where γ is the usual specific heat ratio. Equation (1) produces mean free paths considerably less than the freestream value. For example, the present case at 91.4 km had a freestream λ of 0.025 m and an emitted molecule λ of 0.002 m.

Bird's two-dimensional DSMC code[10] was executed on a Cray XMP until the surface pressure, heat flux, and skin friction had achieved steady-state values; approximately 6000 base time steps were used for the steady state average. Each calculation for a given set of model parameters required 2 - 3 CPU hours.

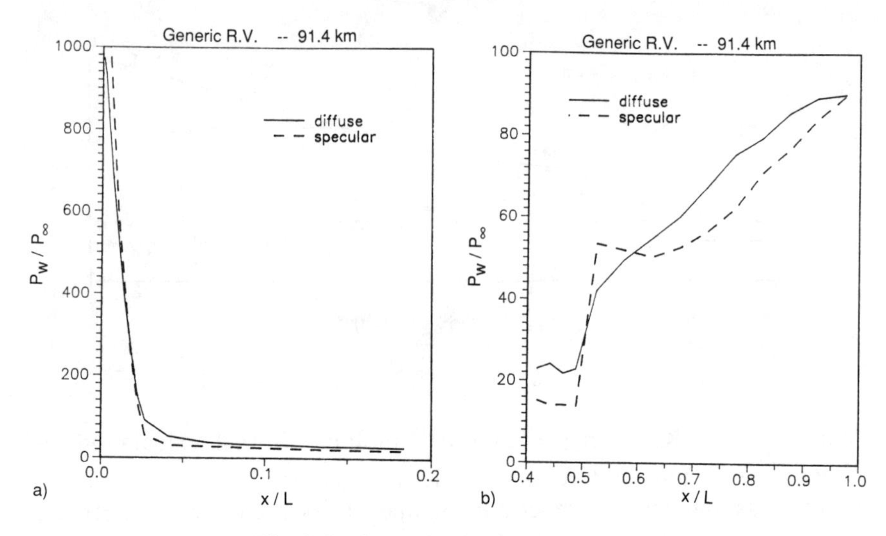

Fig. 2 Surface pressure distribution.

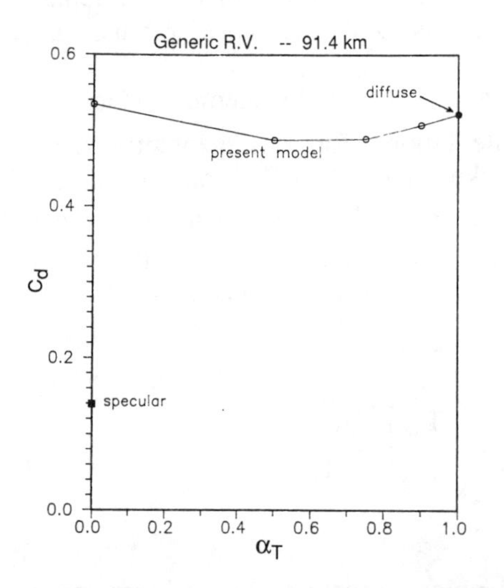

Fig. 3 Drag coefficient for different gas/surface models.

Results

Figures 2a and 2b compare the surface pressure distributions for the 91.4 km conditions from the classical diffuse and specular surface interaction models. These are presented only to show the typical DSMC model results. We note that the specular model predicts a higher surface pressure at both the nose and the conic juncture point. Skin friction and surface heat transfer comparisons cannot be made because, by definition, the classical specular model predicts that the incident molecule will not change its tangential momentum nor have any thermal accommodation with the surface during the collision. This, of course, is nonphysical. The differences in the body drag coefficient for these models can be seen in the next figure.

Figure 3 shows the drag coefficients for the classical specular case as well as 5 calculations with the present model. Recall that for an accommodation of 1.0, the present model is identical to the classical diffuse model. The present model is not identical to the classical specular model at an accommodation of 0.0 because the specular model assumes an elastic collision whereas the present model assumes a uniform angular distribution for the reflected molecule. The drag coefficients for the classical specular and diffuse cases are 0.14 and 0.52, respectively. This large difference is due to the higher surface pressure from the diffuse model as shown in Fig. 2 integrated over the body area and the lack of any skin friction predicted by the specular model. The standard DSMC model for determining a surface interaction between the classical diffuse and specular models can thus produce a very large variation in drag. The drag coefficient was rather insensitive to different degrees of thermal accommodation with the present model. The increase in drag when α_T is zero will be discussed later.

Figures 4a-c present the surface pressure, skin friction, and heat flux distributions for the forward portion of the vehicle for varying thermal accommodation. This forward portion shows the stagnation region of the nose and the first conic section; only this portion is shown to allow an expanded abscissa axis for clarity. The same trends continued down the body. Again, the complete accommodation case, $\alpha_T = 1$, is equivalent to the standard diffuse model. The pressure distribution predicted using the classical specular model is included for reference on Fig. 4a. Note that the surface pressure distributions for the partial accommodation calculations are slightly higher than the complete accommodation. One possible reason for this higher surface pressure might be that the molecules that are reflected upstream have a higher kinetic energy with the partial accommodation model and, thus, when they recollide with the surface, they will impart a greater normal momentum transfer. Koppenwallner and Legge[11] have also postulated this enhancing effect for drag from first collisions in near-free molecular flow ($Kn \approx 1$). This coupling of the reflected and freestream molecule might also explain the decrease in skin friction at the spherical nose as shown in Fig. 4b. The differences in surface pressure and skin friction obtained from varying the accommodation in the present model almost cancel each other out, as indicated by the small difference in total drag as was shown in Fig. 3.

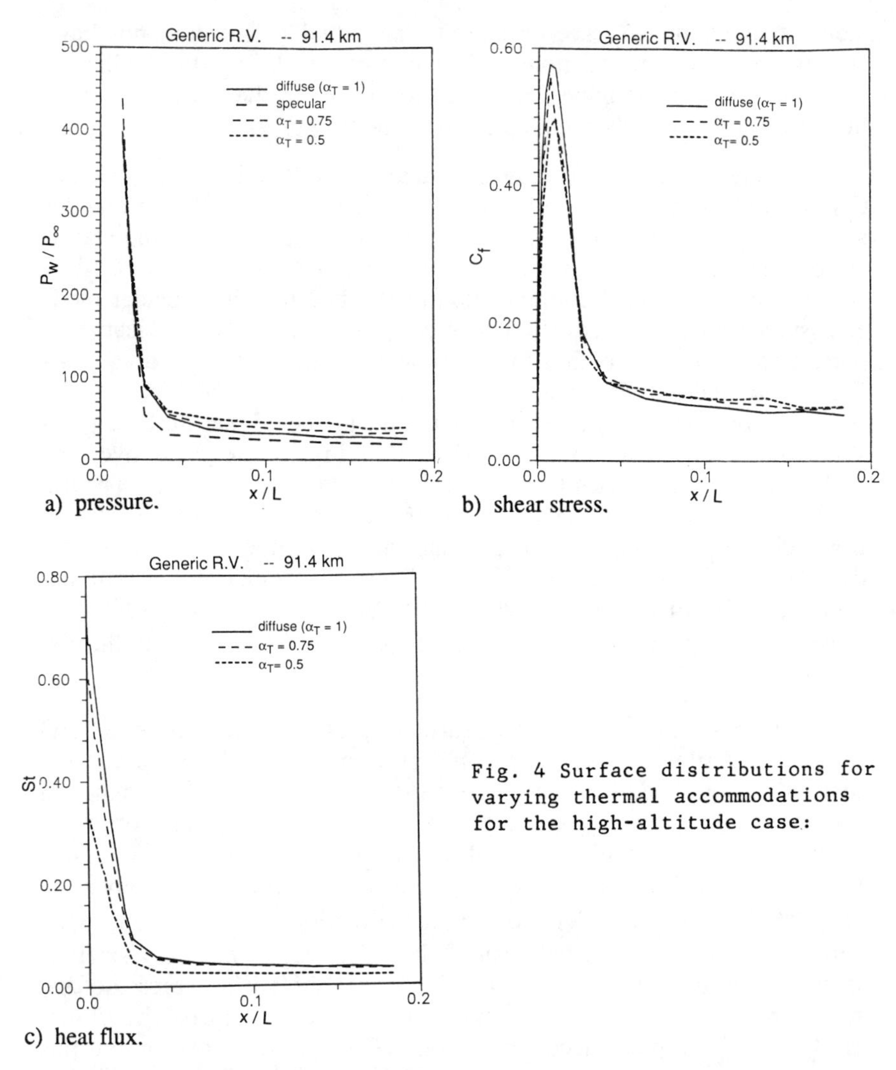

Fig. 4 Surface distributions for varying thermal accommodations for the high-altitude case:

The trends shown in Fig. 4c are expected because the lower the thermal accommodation, the less the energy transfer to the surface. Hurlbut[7] notes that the DSMC method with a classical diffuse model overpredicts the heat transfer along a windward stagnation line for the Shuttle at high altitudes by a factor of about 2.7. If the standard DSMC surface interaction model (with a coefficient of approximately 0.5 - 0.7) is used to match this heat transfer data, the computed drag would be much lower than predicted with the classical diffuse model as shown in Fig 3. Note that the abscissa of Fig. 3, α_T, is not equivalent to the coefficient used in the standard DSMC model. The present model can predict this observed behavior by using a thermal accom-

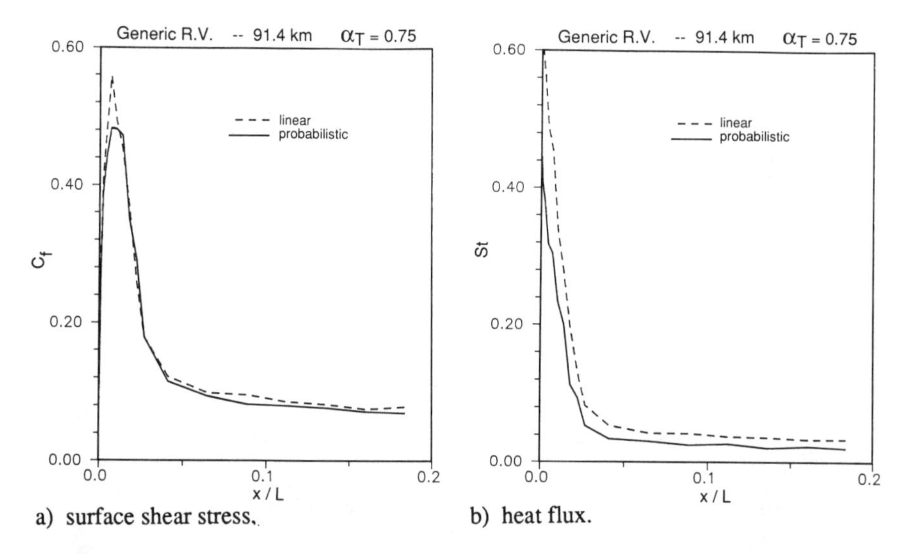

a) surface shear stress. b) heat flux.

Fig. 5 Comparisons between both thermal accommodation methods in the present model.

modation of approximately 0.5 without predicting the drastically reduced drag coefficient. This is simply because the momentum change and thermal accommodation modeling have been decoupled in the present model.

Figures 5a and 5b show the differences in skin friction and heat transfer between the current linear and probabilistic methods for obtaining partial energy accommodation. The pressure distributions of the two methods were almost equal and the drag coefficients were identical. Note that the major difference in skin friction occurs at the nose region; the linear model consistently predicts a higher heat flux than the probabilistic model. The linear model is based more on a physical intuition of the surface collision process, whereas the probabilistic model identically satisfies the principle of detailed balancing since both the specular and diffuse models do. Unfortunately, no detailed experimental data were available to verify either method; however, except near the nose region, both methods produce similar results.

Figures 6a and 6b show the surface distributions of pressure and heat flux for the lower altitude (Kn = 0.016) case. The results are for the linear accommodation model. We note that the surface pressure results from all the models tend to converge as Kn is decreased; the heat fluxes are almost identical for the two levels of acccomodation. This is the same trend noted by Hurlbut,[7] except that it occurs at a lower altitude for the present case. This is probably due to the smaller body geometry being investigated here: a re-entry vehicle vs

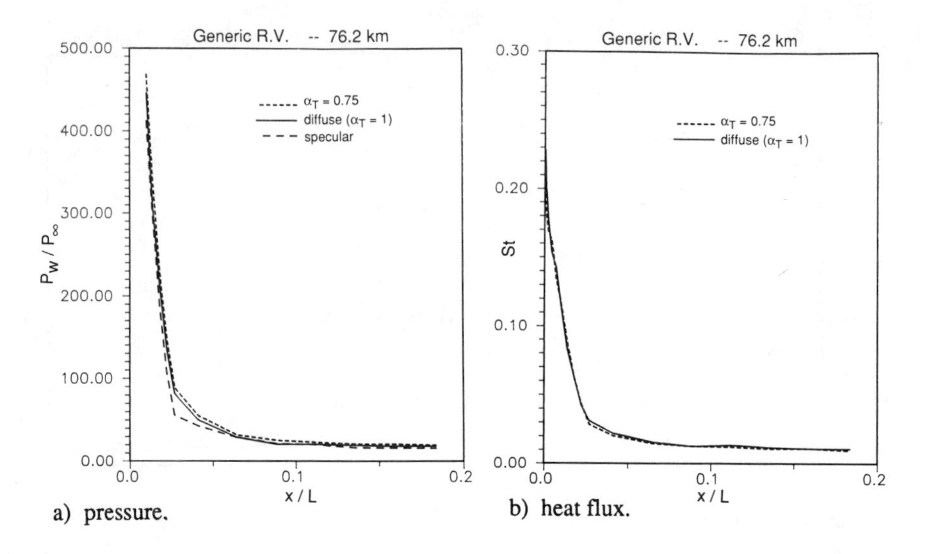

a) pressure. b) heat flux.

Fig. 6 Surface distributions for varying thermal accommodations for the low altitude case.

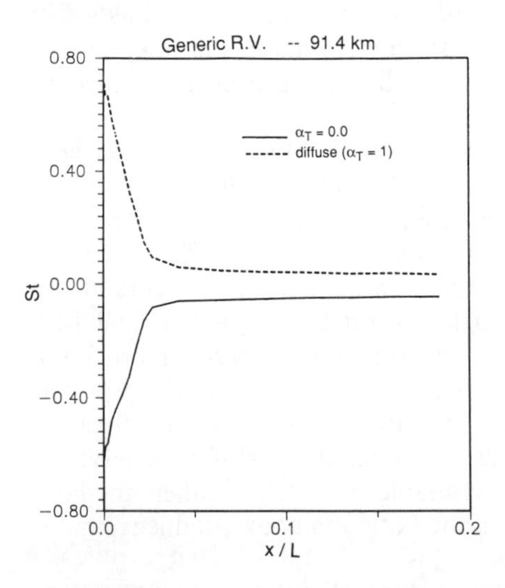

Fig. 7 Surface heat flux distributions for two thermal accommodations using the linear model.

the Shuttle. The present re-entry vehicle would have the same *Kn* as the Shuttle at a lower altitude.

Figure 7 shows the surface heat flux distribution for a solution using complete accommodation and one without any accommodation. It illustrates the problem with the present accommodation model that was alluded to earlier; that is, the increase in the predicted drag with zero accommodation. This problem exists for both the linear and probabilistic method for determining the intermediate accommodation values. We note the nonphysical behavior of a surface cooling effect in this figure when the thermal accommodation is zero, that is, no accommodation. With no accommodation, there should be no surface heat flux. But, because an inelastic scattering behavior is specified, the model predicts an energy exchange and thus the erroneous surface cooling. A surface interaction model without any thermal accommodation implies an elastic scattering and, therefore, a specular surface interaction model should be used. This errant behavior of the present model occurs only when the accommodation is 0.0; it is not a practical problem because zero thermal accommodation is not physically meaningful for the engineering surfaces and conditions being simulated.

Summary

This paper has presented results from different gas/wall interaction models in the DSMC method for a generic re-entry vehicle at typical freestream conditions. Because an engineering surface was investigated, a uniform angular distribution for the reflected simulation molecule was assumed; the degree of thermal accommodation was varied. The following conclusions can be drawn from the work presented herein:

1) The surface pressure results predicted using the present model were not bounded by the classical diffuse or specular models; the present model is not simply a dial to obtain values between the two models.

2) The observed trend that the surface heat flux at larger Kn is lower than that predicted by the classical diffuse model can easily be accounted for with the present model without the accompanying large decrease in drag. The momentum exchange and thermal accommodation have been decoupled in the current model

3) As the Kn is decreased, the results from the classical diffuse and the varying thermal accommodation converge. This is not true for the classical specular model which does not predict any surface shear stress or heat transfer.

4) The drag predictions using the present model are much less sensitive to the degree of thermal accommodation than the variation from the classical diffuse to specular modeling. Force and moment coefficients are very important quantities for designing a re-entry vehicle.

5) As a statement of the obvious, experimental data for engineering materials at these freestream conditions are needed to resolve discrepancies between gas/wall interaction models. Any dependence on geometry also needs to be investigated.

Acknowledgment

This work performed at Sandia National Laboratories, supported by the U. S. Department of Energy under Contract No. DE-AC04-76DP00789.

References

[1]Bird, G. A., *Molecular Gas Dynamics*, Oxford University Press, London, 1976.

[2]Bird, G. A., "Monte Carlo Simulation of Gas Flows," in *Annual Review of Fluid Mechanics*, Vol. 10, 1978, pp. 11-31.

[3]Bird, G. A., "Monte Carlo Simulation in an Engineering Context," in *Proceedings of the 12th International Symposium on Rarefied Gas Dynamics*, AIAA, New York, 1980, pp. 239-255.

[4]Bird, G. A., "Low Density Aerothermodynamics," AIAA Paper 85--0994, 1985.

[5]Moss, J. N. and G. A. Bird, "Direct Simulation of Transitional Flow for Hypersonic Reentry Conditions," AIAA Paper 84-0223, 1984.

[6]Hermina, W. L., "Monte Carlo Simulation of Rarefied Flow Along A Flat Plate", AIAA Paper 87-1547, 1987.

[7]Hurlbut, F. C., "Sensitivity of Hypersonic Flow Over a Flat Plate to Wall/Gas Interaction Models Using DSMC," AIAA Paper 87-1545, 1987.

[8]Bartel, T. J., Homicz, G. F. and Walker, M. A., "Comparisons of Monte-Carlo and PNS Calculations for Rarefied Flow over Reentry Vehicle Configurations," AIAA Paper 88-0465, 1988.

[9]Cox, R. N. and Crabtree, L. F., *Elements of Hypersonic Aerodynamics*, English Universities Press, London, 1965.

[10]Bird, G. A., "General Programs for Numerical Simulation of Rarefied Gas Flows," G.A.B. Consulting Pty, 1988.

[11]Koppenwallner, G. and Legge, H., "Drag of Bodies in Rarefied Hypersonic Flow," in *Thermophysical Aspects of Re-Entry Flows*, Moss and Scott editors, AIAA, New York, 1986.

Determination of Momentum Accommodation from Satellite Orbits: An Alternative Set of Coefficients

R. Crowther* and J. Stark†
University of Southampton, Southampton, England, United Kingdom

Abstract

The objective of this work is to show that the widely used normal and tangential momentum accommodation coefficients are unsuitable for certain applications. An alternative set of coefficients for adoption in the determination of momentum accommodation from the analysis of satellite orbits is introduced. Previous authors have represented the variation of momentum coefficients with incidence by empirical relations; the Schaaf and Chambre coefficients are found to be unstable at low incidence and, therefore, inappropriate for analysis of satellites with large planar appendages. Resolving the momentum relative to the incident flow vector rather than the surface allows the adoption of simpler relationships, leading to a more robust behavior. In addition, the new set of coefficients is found to provide a better indication of the nature of the gas-surface interaction. This is the result of what is effectively a decoupling of the bulk gas velocity component from the thermal velocity component for the coefficients. The example of the ANS-1 satellite is taken to demonstrate the advantage of these new coefficients.

Introduction

In an effort to complement molecular beam research investigating the nature of the interaction between high-velocity neutral atmospheric species and spacecraft surfaces, several areas of research, which attempt to

---

*Research Assistant, Department of Aeronautics and Astronautics.
†Senior Lecturer, Department of Aeronautics and Astronautics.

determine indirectly the nature of these gas-surface interactions, have evolved. In particular, observation of a satellite's motion under the influence of aerodynamic forces when in orbit about the Earth has been studied for many years.[1] The first example of this passive approach was the work of Reiter and Moe[2], who sought to derive the nature of the interaction from the observed change in the spin rate of the paddlewheel satellite Explorer 6. This technique was developed further by Imbro and Moe[3] and later by Karr.[4,5]

An alternative approach, which embraces the disciplines of both rarefied gasdynamics and astrodynamics, also was developed at this time, this was orbital analysis.

The orbit of a satellite in low Earth orbit is predominantly perturbed from a Keplerian trajectory by forces of gravitational origin (Earth, lunar, and solar), free molecular aerodynamic forces caused by the interaction between the spacecraft surface and neutral species in the atmosphere, and solar radiation pressure. Given a set of accurately determined orbits, and through the application of the Lagrange planetary equations,[1] it is possible to relate the changes in the orbit to the perturbing forces acting on the satellite. Once the effects of the other perturbing sources have been removed, one can relate the resulting changes in the orbit to the aerodynamic lift and drag forces acting on the spacecraft. Representing these aerodynamic forces with a form of re-emission model allows one to derive further the macroscopic nature of the gas-surface interaction from the analysis.

Although kinetic theory has proved successful in defining incident free molecular flows, no theory has yet become available that can account exactly for the nature of the gas-surface interaction. Therefore, it is necessary to adopt an empirical model of the observed behavior. One such approach is the use of momentum accommodation coefficients to provide the link between the incident and re-emitted fluxes.

The widely used normal and tangential momentum accommodation coefficients will be considered. An alternative set of coefficients then will be introduced, and the relative merits of each set compared for the orbital analysis application.

Normal and Tangential Momentum Accommodation Coefficients

These coefficients were introduced first by Schaaf and Chambre[6] and measure the degree of momentum accommodation in directions normal and tangential to a surface. They are defined by

$$\sigma_n = \frac{P_i - P_r}{P_i - P_w} \qquad \sigma_\tau = \frac{\tau_i - \tau_r}{\tau_i}$$

where the terms P and τ refer to the momentum flux acting normal and tangential to the surface per unit area per unit time, the subscripts i and r refer to the incident and re-emitted components, and w refers to the component that would be produced by a diffuse re-emission at the temperature of the surface (see Fig. 1).

These coefficients are used widely as a means of translating results from molecular beam experiments to aerodynamic coefficients of convex bodies. It is found in practice that σ_n is very dependent on the angle of incidence of the surface to the flow, especially at low incidence.[7] This does not represent a problem when applying gas-surface interaction results to the prediction of aerodynamic coefficients (lift or drag) of bodies if the variation of σ_n is known from experiments. However, since the nature of gas-surface interaction of appropriate orbital velocities is poorly known from laboratory experiments, the exact prediction of coefficients for lift and drag poses problems.

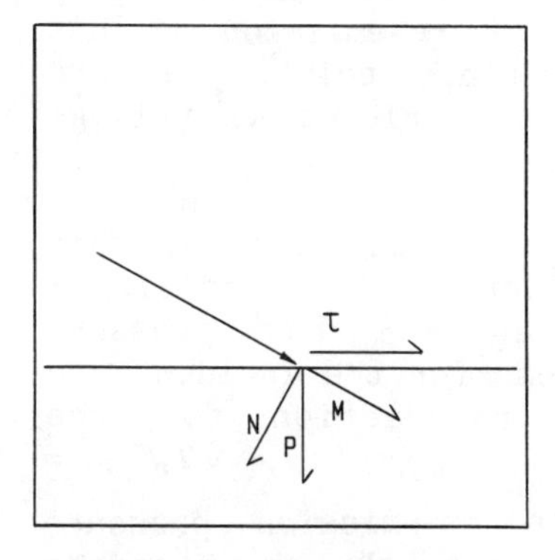

Fig. 1 Resolution of momentum at surface.

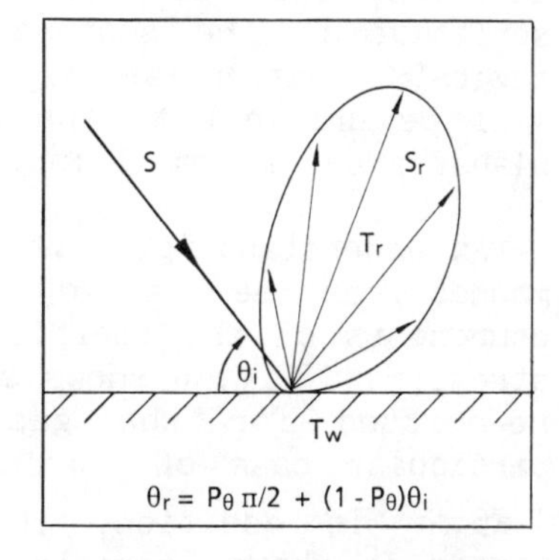

Fig. 2 Nocilla model of reemitted flux

A further problem arises when one attempts to reverse the process, i.e., to determine the nature of the gas-surface interaction from the aerodynamic coefficients, as is the case with orbital analysis. In order to derive the nature of the interaction from the aerodynamic forces, first, it is necessary to define the variation of the momentum accommodation coefficients with angle of incidence.

A parametric study serves to illustrate the behavior of the Schaaf and Chambre coefficients with incidence and nature of re-emission. As an example, the Nocilla model[9] is used herein to represent the re-emitted flux with appropriate values for the re-emission parameters (see Fig. 2) suggested by experiment for a predominantly atomic oxygen incident flux onto a polished carbon surface.[10,11] Although these results are for a fixed-incidence case, they are the first direct measurements of an atomic oxygen flow onto a surface at the correct energy and, as such, provide a valuable guide for characterization of the flow.

The Nocilla parameters are varied for a range about these values for incidence angles between 0 and 90 deg for flat-plate elemental areas. The re-emission speed ratio S_r varies between 0 and 2, the directional factor P_θ varies between 0 and 1, and the temperature ratio $\sqrt{T_r/T_w}$ varies between 1 and 2. The results are shown for $\sqrt{T_r/T_w}$; these are characteristic of those for S_r and P_θ and, thus, only the one set is shown. Figure 3 demonstrates that, at low incidence, the component normal to the surface is a very sensitive function of the angle of incidence. However, referring to Fig. 4, it can be seen that the tangential coefficient is insensitive to the angle of incidence. In addition, this tangential component is not particularly sensitive to the supposed nature of re-emission. This suggests that relatively small surface pointing errors will be critical to the analysis of vehicles with large planar areas at small incidence angles.

An understanding of why the coefficients behave in this manner can be gained by considering the individual components of the coefficients. Figure 5 shows pressure stress, and Fig. 6 shows shear stress for the incident and re-emitted for the gas-surface interaction for the particular case of $S = 9$, $S_r = 0.5$, $P_\theta = 0.5$, $\sqrt{T_r/T_w} = 1.25$. In addition, the diffuse re-emission pressure stress is shown (clearly, in this case, the shear stress is zero).

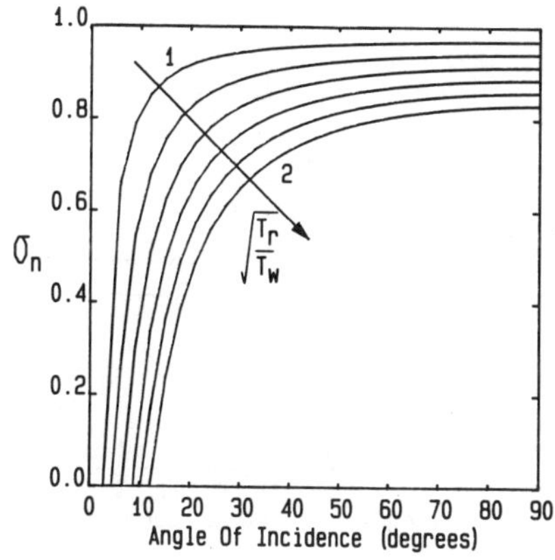

Fig. 3 σ_n variation with θ.

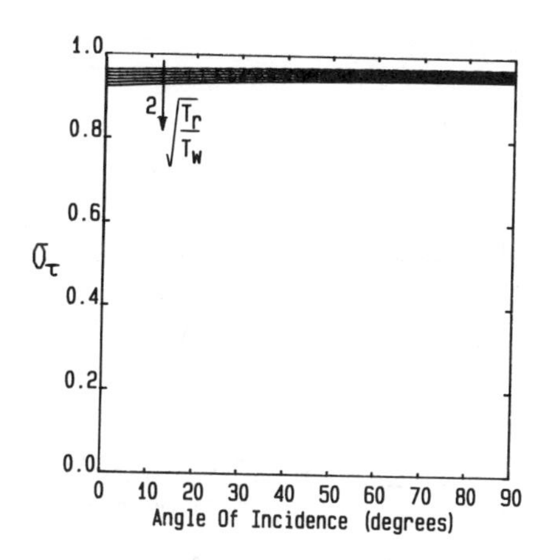

Fig. 4 σ_τ variation with θ.

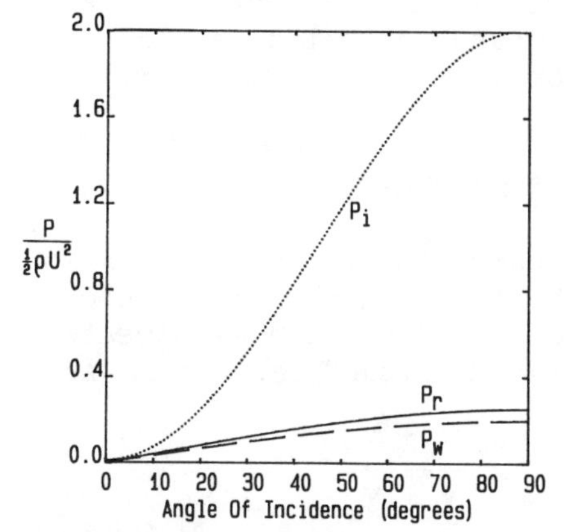

Fig. 5 Components of momentum normal to surface.

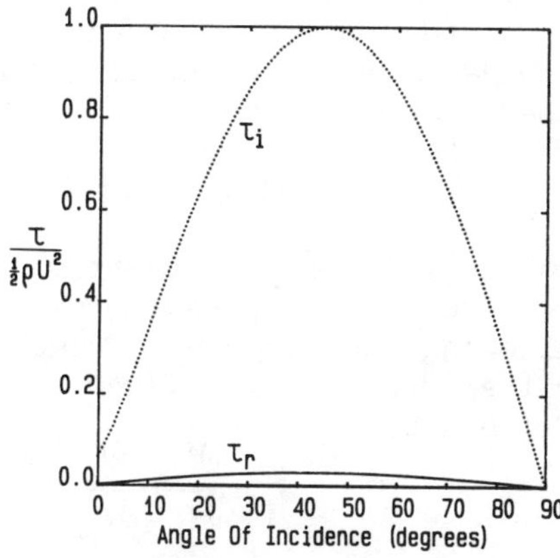

Fig. 6 Components of momentum tangenital to surface.

It is apparent that the magnitude of the incident components is the major influence on the values of the coefficients. The extreme variability of P_i with incidence is reflected in the variation of σ_n with incidence. The observed relative insensitivity of σ_τ to the nature of re-emission also can be seen as a result of the magnitude of τ_i as compared with τ_r. Thus, the Schaaf and Chambre set of coefficients is inappropriate for the orbital analysis application, since it is more sensitive to vehicle attitude errors than to the nature of the gas-surface interaction process for vehicles having large appendages flying at low incidence to the flow.

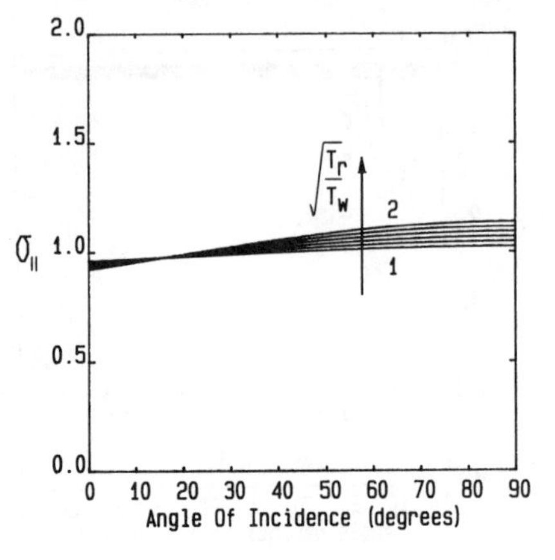

Fig. 7 $\theta_{\parallel}$ variation with θ.

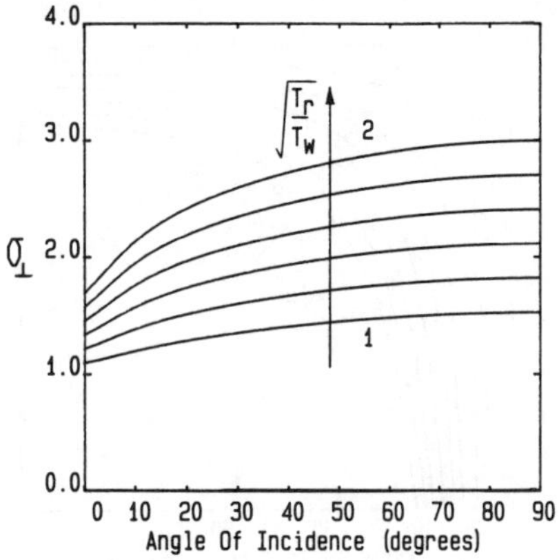

Fig. 8 $\sigma_{\perp}$ variation with θ.

In an attempt to reduce the variability with incidence, it has been found that a new set of coefficients, defined with the momentum transfer measured relative to the incident velocity vector rather than the surface, may be used in these cases.

Alternative Set of Coefficients

The new set of momentum accommodation coefficients is resolved parallel and normal to the incident flow vector, allowing the reference frame to move with the flow vector rather than remain fixed within the surface. They are defined by

$$\sigma_{\parallel} = \frac{M_i + M_r}{M_i + M_w} \qquad \sigma_{\perp} = \frac{N_i + N_r}{N_i + N_w}$$

where the terms N and M refer to the momentum flux per unit area per unit time acting normal and tangential to the flow vector, and the subscripts i, r, and w are consistent with the preceding (see Fig. 1).

The same parametric investigation that was applied to the Schaaf and Chambre coefficients was carried out using the new set of coefficients. Again, the results for $\sqrt{T_r/T_w}$ are characteristic of those for S_r and P_θ and are shown in Fig. 7 for $\sigma_{\parallel}$ and in Fig. 8 for $\sigma_{\perp}$. It is evident from Fig. 7 that the dependence on the angle of incidence noted in $\sigma_{\parallel}$ is greatly reduced. Furthermore, from Fig. 8, the coefficient that measures the momentum normal to the flow vector is now significantly more sensitive to the

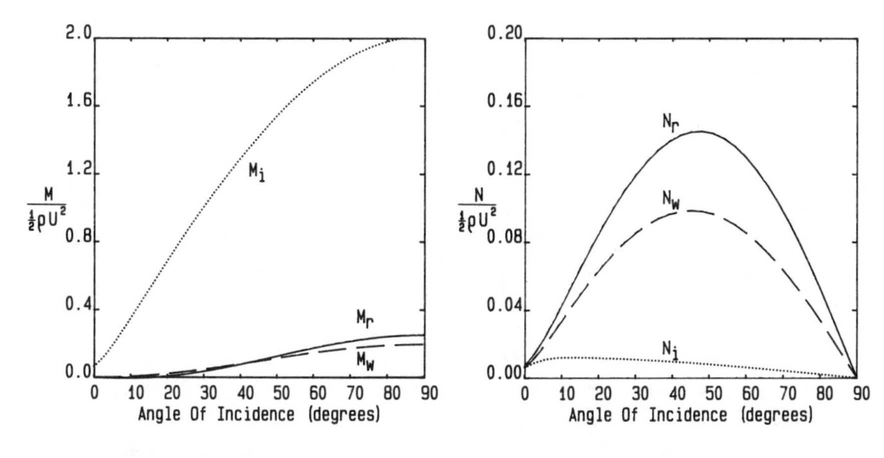

Fig. 9 Components of momentum tangenital to flow.

Fig. 10 Components of momentum normal to flow.

nature of re-emission than is the case for σ_τ while remaining relatively insensitive to the flow angle.

Consider the individual components of the coefficients shown in Figs. 9 and 10. It is seen that the coefficient that measures the momentum parallel to the flow vector is again dominated by the magnitude of the incident component, which would account for its relative insensitivity to the nature of the gas-surface interaction. For the second coefficient, the opposite is found to be true; namely, the re-emitted component is much larger than the incident component. This phenomenon arises from an effective decoupling of the bulk gas velocity component from the thermal velocity component of the incident momentum. The re-emitted momentum flux is then compared to the much smaller thermal velocity component, resulting in the observed sensitivity to the nature of re-emission.

This would also account for the greater sensitivity of lift coefficients as compared to drag coefficients, which was observed by Stark[12] in his investigation of spacecraft aerodynamics. The incident fluxes provide only a small contribution to the overall lift coefficient as opposed to drag coefficients where incident fluxes tend to dominate.

Application of Coefficients to ANS-1

The ANS-1 satellite was launched on August 30, 1974, at 14h:7min Universal Time. The satellite had a total mass of 129.6 kg, measuring 123 cm in height, 74 cm in depth, and 61 cm in width. The deployment of the two solar arrays increased the width to 144 cm.[13] The nominal orbit of ANS-1 was intended to be near circular, sun-synchronous at an altitude of 500 km. However, malfunction on the launch vehicle led to the spacecraft being injected into an orbit defined by the following: semimajor axis, 7098 km; eccentricity, 0.0640; inclination, 98.04 deg; right ascension, 245.20 deg; argument of perigee, 210.89 deg; and mean anomaly, 320.87 deg.

This near-polar sun-synchronous orbit about the Earth led to the solar array presenting a low angle of incidence relative to the flow vector varying between approximately 0 and 30 deg. The inclination of the orbit decreased by approximately 5×10^{-6} deg per orbit and the semimajor axis by approximately 20 m per orbit; these effects being predominantly caused by aerodynamic forces.

In order to compare the relative merits of the two sets of coefficients for the ANS-1 application, a theoretical data set representing the aerodynamic behavior of the vehicle was generated assuming that the nature of re-emission could be represented by the Nocilla model. The specific parameters chosen for this were $S = 9$, $S_r = 1$, $\sqrt{T_r/T_w} = 2$, and $P_\theta = 0.5$, being characteristic of a near-diffuse re-emission and, thus, in general agreement with most investigations. A model orbit was generated using these parameters so that a predicted "Nocilla perturbed orbit" could be evaluated. This model set then was analyzed to ascertain what values of σ_n, σ_τ, $\sigma_\parallel$, and $\sigma_\perp$ would best fit the data.

This data set represented the full incidence range from 0 to 90 deg. Using the empirical relations suggested by Moore and Sowter,[8] a least-squares determination procedure was applied to calculate the values of the constants σ_{n_0}, σ_{n_1}, and σ_{τ_0}, which provide a best fit to the generated data.

The relations chosen by Moore and Sowter for their investigation of the ANS-1 satellite using the Schaaf and Chambre set of coefficients were as follows:

$$\sigma_n = \sigma_{n_0} - \sigma_{n_1} \operatorname{cosec}\theta \qquad \sigma_\tau = \sigma_{\tau_0}$$

Table 1 Influence of incidence range on derived variations of σ_n.

σ_n variation derived using surface-fixed coefficients,	rms error	σ_n variation derived using flow-fixed coefficients	rms error	Incidence range of data, deg
0.8914 - 0.0341 cosec θ	0.0033	0.9467 - 0.0583 cosec θ	0.0071	3-80
0.8914 - 0.0398 cosec θ	0.0021	0.9418 - 0.0544 cosec θ	0.0053	3-60
0.8395 - 0.0339 cosec θ	0.0066	0.9343 - 0.0490 cosec θ	0.0030	3-40
0.6766 - 0.0198 cosec θ	0.0100	0.9238 - 0.0429 cosec θ	0.0007	3-20

In our analysis of ANS-1, the influence of incidence was effected by the progressive reduction of the incidence range and the introduction of a pointing error into the data by sampling an error angle $\Delta\theta$ from a normal distribution, distributed between ± 1 deg for the solar array. Such an error may occur easily in practice for the array on such a vehicle.

The process was repeated using the new set of coefficients represented by relationships of the form

$$\sigma_{\parallel} = \sigma_{\parallel_0} + \sigma_{\parallel_1} \sin\theta \qquad \sigma_{\perp} = \sigma_{\perp_0} + \sigma_{\perp_1} \sin\theta$$

and the results transformed back into the surface fixed reference frame for the purposes of comparison. The results are summarized in Table 1 and plotted in Figs. 11 and 12.

On the basis of the assumed Nocilla model, the value of σ_n should be as follows:

$$\sigma_n = 0.9152 - 0.0420 \ \ \text{cosec}\theta$$

This is the value to which (ideally) the least-squares analysis should converge. It is immediately apparent that, as the incidence range appropriate to the ANS-1 mission is approached, the new set of coefficients more closely represents the model data. The implication of

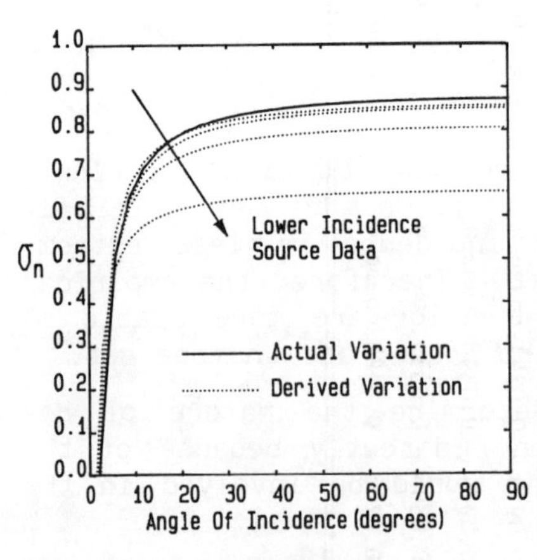

Fig. 11 σ_n variation derived using (σ_n, θ_τ).

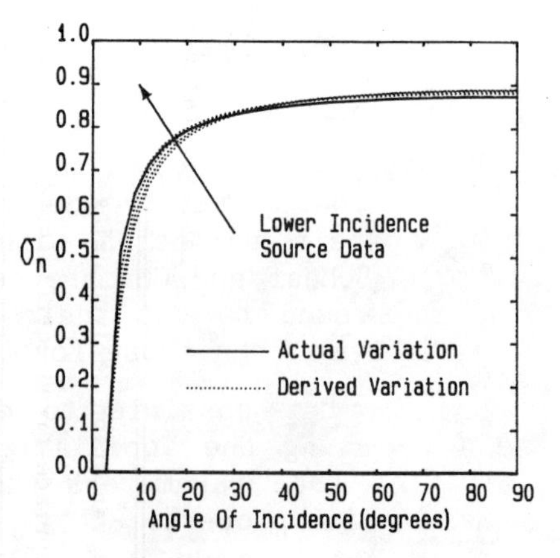

Fig. 12 σ_n variation derived using $(\sigma_\parallel, \theta_\perp)$.

using the surface resolved set in this particular case is that the degree of momentum accommodation is seriously underestimated.

However, in the case of the Moore analysis, the degree of accommodation appears to have been overestimated. The results obtained from their analyses were the following:

$$\sigma_n = 1.37 - 0.155 \operatorname{cosec}\theta \quad \sigma_\tau = 0.73$$

This suggests that the molecules are being re-emitted at a temperature below that of the surface, a somewhat improbable situation.

A possible explanation for these anomalous results is the extreme variation of σ_n at low incidence and a biased pointing error. A test case was again set up with a positive 1 deg pointing error for the array introduced into the determination procedure. The Nocilla parameters chosen were $S = 9$, $S_r = 1$, $P_\theta = 0.5$, and $\sqrt{T_r/T_w} = 1.5$.

The results derived using the two sets of coefficient samplings from an incidence range of 3-30 deg are as follows: σ_n variation derived using surface-fixed coefficients: $1.144 - 0.166 \operatorname{cosec}\theta$; σ_n variation derived using flow-fixed coefficients: $1.019 - 0.116 \operatorname{cosec}\theta$; and σ_n variation derived without pointing error (3-89 deg): $0.998 - 0.115 \operatorname{cosec}\theta$.

This behavior has serious implications for the application of the Schaaf and Chambre coefficients to the ANS-1 case.

Discussion

The results suggest that the new set of coefficients is more robust than the Schaaf and Chambre set. This is because their variation with incidence is less extreme than the Schaaf and Chambre set. Therefore, the empirical relations used to fit their behavior are more stable at low incidence than those for the Schaaf and Chambre set.

It is not possible to determine the nature of re-emission using the Nocilla model directly because of the number of free parameters that would be involved in the determination process.

The overestimation of momentum accommodation suggested by Moore and Sowter's results can be accounted for by a bias in pointing error. In practice, one would expect the nature of the pointing error to be biased rather than evenly distributed, as was assumed in the first test case.

Conclusions

The surface-resolved set of momentum accommodation coefficients is found to be more sensitive to the angle of incidence than to the nature of re-emission at low angles of incidence. This presents a problem in application to orbital analysis since the technique requires a priori definition of their parametric variation with incidence.

Application of the new set of coefficients, which is resolved relative to the incident flow vector rather than the surface, is found to reduce the errors encountered when pointing errors are involved.

As a result of decoupling of the incident bulk gas velocity and thermal velocity components, the new set of coefficients also proves to be more indicative of the re-emission process, suggesting their adoption in future applications. In a future paper, the authors will apply the new set of coefficients to the orbital data obtained for ANS-1.[14]

Acknowledgment

The first author would like to acknowledge the support of a Science and Engineering Research Council award.

References

[1]King-Hele, D. G., Satellite Orbits in an Atmosphere, Blackie & Son, Glasgow, 1987.

[2]Reiter, G. S. and Moe, K., "Surface Particle Interaction Measurements Using Paddlewheel Satellites," Proceedings of 6th Rarefied Gas Dynamics Symposium, Held in Cambridge, Mass., published by Academic Press Inc. NY. 1969, pp. 1543-1555.

[3]Imbro, D. R. and Moe, M. M., "On Fundamental Problems in the Deduction of Atmospheric Densities from Satellite Drag," Journal of Geophysical Research, Vol. 80, No. 22, 1975, pp. 3077-3086.

[4]Karr, G. R., "Analysis of Effects of Gas-Surface Interaction on Spinning Convex Bodies with Application to Satellite Experiments," Ph.D. Thesis, Univ. of Illinois, Urbana, 1969.

[5]Karr, G. R., "Analysis of Satellite Drag and Spin Decay Data," NASA CR-178609, 1985.

[6]Schaaf, S. A. and Chambre, P. L., Flow of Rarefied Gases, Princeton Aeronautical Paperbacks, Princeton University Press, Princeton, NJ, 1961.

[7]Knechtel, E. D. and Pitts, W. C., "Normal and Tangential Momentum Accommodation for Earth Satellite Conditions," Astronautica Acta, Vol. 18,1973, pp. 171-184.

[8]Moore, P. and Sowter, A., "The Use of Momentum Accommodation for Gas Surface Interaction at Satellite Altitudes with Application to ANS-1 (1975-70A)," paper to be presented at AIAA/AAS Astrodynamics Conference, 1988, Paper No. AIAA 88-4292.

[9]Nocilla, S. "The Surface Re-emission Law in Free Molecular Flow," Proceedings of 3rd Symposium on Rarefied Gas Dynamics, held in Paris, France, published by Academic Press Inc., NY.,1963, pp. 327-346.

[10]Karr, G. R., Gregory, J. C. and Peters, P. N., "A Measurement of the Angular Distribution of 5eV Atomic Oxygen Scattered Off a Solid Surface in Earth Orbit," Proceedings of 15th Symposium on Rarefied Gas Dynamics, held in Grado, Italy, published by B.G. Teubner, Stuttgart, 1986, pp. 609-617.

[11]Gregory, J. C. and Peters, P. N., "Free Molecule Drag and Lift Deduced from Shuttle Flight Experiment," Proceedings of 15th Symposium on Rarefied Gas Dynamics, held in Grado, Italy, published by B.G. Teubner, Stuttgart, 1986, pp. 646-656.

[12]Stark, J., "Aerodynamic Modelling of Spacecraft for Precise Orbit Determination," Proceedings of the 2nd International Symposium on Spacecraft Flight Dynamics, ESA-SP-255, 1986, pp. 239-246.

[13]Wakker, K. F., "Orbit Prediction for the Astronomical Netherlands Satellite," Journal of the British Interplanetary Society, Vol. 31, 1978, pp. 387-397.

[14]Crowther, R. and Stark, J., "Determination of Momentum Accommodation at Satellite Altitudes Using a New Set of Coefficients," Paper submitted to the AIAA 27th Aerospace Sciences Meeting, Reno, NV, Jan. 1989, Paper No. AIAA 89-0456.

Upper Atmosphere Aerodynamics: Gas-Surface Interaction and Comparison with Wind-Tunnel Experiments

M. Pandolfi* and M. G. Zavattaro*
Politecnico di Torino, Torino, Italy

Abstract

A gas-surface interaction model derived by the Nocilla wall re-emission model is considered. The distribution function of the particles leaving the surface is "approximated" by a Maxwellian distribution with mass velocity related to the incoming distribution by suitable identification parameters. The proposed mathematical model, on the basis of classical aerodynamic methods in kinetic theory, allows evaluation of the aerodynamic coefficients. Comparisons with experimental results in the hypersonic rarefied wind tunnel of DFVLR laboratories have been performed.

Introduction

It is well known that the crucial problem of upper atmosphere aerodynamics[1,2] consists in correlating, in the free-molecule regime, the distribution function of the gas particles scattered from the outer surface of a space vehicle to the distribution function of the particles hitting the surface. In general, this correlation is realized by using mathematical models that contain suitable identification parameters identified by comparing theory and experiments. It is also quite well understood that by adjusting the identification parameters (usually two or more), one can

---

* Associate Professor, Dipartimento di Matematica.

recover experimentally measured physical macroscopic quantity (e.g. an aerodynamic coefficient).

On the other hand, the experiments realized in the DFVLR laboratories of Goettingen and presented in Ref. 3 have obtained the simultaneous measurements of two macroscopic quantities, the heat and drag coefficients. This experiment has shown consistent difficulty[4] for the mathematical models available in the literature for gas-surface interaction to reproduce such experimental information.

In this paper, we consider a gas-surface interaction model derived by the Nocilla wall re-emission model[5].

Using the present model, the distribution function of the particles leaving the surface is "approximated" by Maxwellian distribution with mass velocity such that the characteristics of the re-emitted distribution function are related to those of the incoming distribution through suitable identification parameters.

This model is certainly very flexible and easily computable for aerodynamic calculations. In addition, this paper shows how "relatively" accurate results (as far as comparisons with the experiments of Ref. 3 are concerned) can be obtained by such a "relatively" simple model.

The general problem cannot, however, be regarded as already solved. In fact, accuracy in reproducing experimental results certainly can be improved, and one has to keep in mind that the proposed model still suffers some inconsistency with the so-called "reciprocity law"[2] proved by Cercignani and Kuscer independently.

Althouh this aspect does not appear relevant for engineering calculations, and in spite of the several advantages involved because of the simplicity of the model, it is hoped that a more rigorous approach will be attempted.

The following section describes the gas-surface interaction model in detail. The next section deals with aerothermodynamic calculations and performs the aforementioned comparisons between theory and experiments.

Description of the Model

Consider a flat plate in a molecular stream at temperature T_∞ with number density n_∞ and mass velocity $\vec{V}_\infty$, as shown in Fig. 1.

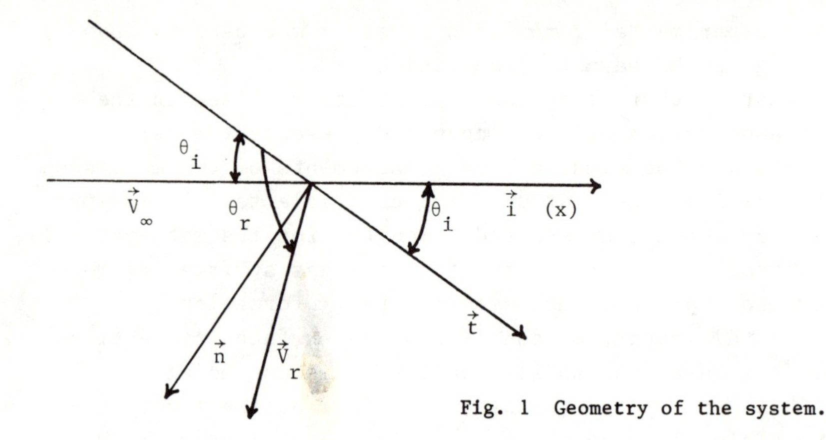

Fig. 1 Geometry of the system.

The mathematical model proposed herein is described by the following set of axioms:

1) The gas particles' mean free path is large with respect to the dimensions of the body, i.e., free molecular flow. The observation time of the system is large with respect to the mean time between two collisions. The flow conditions are time invariant and the adsorption-desorption kinetics on the surface are negligible with respect to the almost instantaneous scattering kinetics.

2) The gas particle velocity distribution function, preceding the surface interaction, is a Maxwellian distribution with mass velocity $\vec{V}_\infty$ and angle of attack θ_i

$$f_i(\vec{v})=f_i(n_\infty,T_\infty,V_\infty,\theta_i)=n_\infty\left(\frac{m}{2\pi kT_\infty}\right)^{3/2}\exp\left(-\frac{m}{2kT_\infty}(\vec{v}-\vec{V}_\infty)^2\right) \quad (1)$$

3) The gas particle velocity distribution function, following surface interaction, is "approximated" by a Maxwellian distribution with mass velocity $\vec{V}_r$ and temperature T_r

$$f_r(\vec{v})=f_r(n_r,T_r,V_r,\theta_r)=n_r\left(\frac{m}{2\ kT_r}\right)^{3/2}\exp\left(-\frac{m}{2kT_r}(\vec{v}-\vec{V}_r)^2\right) \quad (2)$$

where $\vec{V}_r$ lays in the plane of $\vec{V}_\infty$ with $\vec{n}$ outward normal to the surface, and $T_r \neq T_\infty$.

4) The physical quantity n_r is identified by the continuity equation corresponding to the condition of impermeability

$$N_i = N_r = N \tag{3}$$

where $N_{i,r}$ is a number flux (i = incoming, r = reflected).

The first step toward the main object of this paper consists of the calculation of the number and energy flux of the incoming and re-emitted molecules.

If we consider the surface element of Fig. 1, characterized by the normal and tangential unit vectors $\vec{n}$ and $\vec{t}$, respectively, with $\vec{t}$ oriented so that $\vec{V}_\infty \cdot \vec{t} > 0$, classical calculations in the kinetic theory of gases[5] give the following equations:

$$N_{i,r} = \frac{n_{\infty,r}}{2\sqrt{\pi}}\sqrt{\frac{2kT_{\infty,r}}{m}}\,\mathcal{N}_{i,r} \tag{4}$$

$$\vec{Q}_i = P_i\vec{n} + \tau_i\vec{t} \tag{5a}$$

$$\vec{Q}_r = P_r(-\vec{n}) + \tau_r\vec{t} \tag{5b}$$

where

$$P_{i,r} = -\frac{\rho_{\infty,r}}{2\sqrt{\pi}S^2_{\infty,r}}V^2_{\infty,r}\,\mathcal{P}_{i,r} \tag{6}$$

$$\tau_{i,r} = \frac{\rho_{\infty,r}V_{\infty,r}}{2\sqrt{\pi}S_{\infty,r}}\cos\theta_{i,r} \tag{7}$$

$$E_{i,r} = \frac{n_{\infty,r}}{2\sqrt{\pi}}\;\frac{2kT_{\infty,r}}{m}\;kT_{\infty,r}\,\mathcal{E}_{i,r} \tag{8}$$

with

$$\mathcal{N}_{i,r} = \exp(-S^2_{\infty,r}\sin^2\theta_{i,r}) + \sqrt{\pi}S_{\infty,r}\sin\theta_{i,r}$$

$$\times(1+\mathrm{erf}(S_{\infty,r}\sin\theta_{i,r})) \tag{9}$$

$$P_{i,r} = S_{\infty,r} \sin\theta_{i,r} \exp(-S^2_{\infty,r} \sin^2\theta_{i,r}) + \\ +\sqrt{\pi}(1/2+S^2_{\infty,r} \sin^2\theta_{i,r})(1+\mathrm{erf}(S_{\infty,r} \sin\theta_{i,r})) \quad (10)$$

$$E_{i,r} = (S^2_{\infty,r}+2)\exp(-S^2_{\infty,r} \sin^2\theta_{i,r})+\sqrt{\pi}(S^2_{\infty,r}+5/2) \\ \times(S_{\infty,r} \sin\theta_{i,r}(1+\mathrm{erf}(S_{\infty,r} \sin\theta_{i,r}))) \quad (11)$$

Note that Eqs. (5-8) have been extended to the re-emitted flow, since the outgoing molecules can be considered as incoming to the same element with outward normal $-\vec{n}$ instead of $\vec{n}$. For large values of $S_{\infty,r}$ (hypersonic flow) and for $\theta_{i,r}$ larger than zero, the following approximations hold for Eqs. (9-11):

$$N_{i,r} \cong 2\sqrt{\pi} S_{\infty,r} \sin\theta_{i,r} \quad (12)$$

$$P_{i,r} \cong 2\sqrt{\pi} S^2_{\infty,r} \sin^2\theta_{i,r} \quad (13)$$

$$E_{i,r} \cong 2\sqrt{\pi} S^3_{\infty,r} \sin\theta_{i,r} \quad (14)$$

Before using the above-described model, the crucial point to be solved is identification of re-emission parameters V_r and θ_r, as depending on the properties of the incoming stream, so that the model can be regarded as a transformation of the Nocilla wall reflection model into a gas-surface interaction model. It is well known that this model does not satisfy the reciprocity law[2]. On the other hand, one has to look at it as an "approximation" of physical reality for technical applications.

It will be shown that by the proposed model one can obtain satisfactory results (compared with those obtained by other models[1]), with the further advantage of easily computable calculations.

Aerodynamical Calculations and Comparisons with Experiments

The mathematical model described previously allows, on the basis of classical aerodynamical methods in kinetic theory[6], one to face the aerodynamical problem: calculation of aerothermodynamic coefficients.

The method developed for a flat plate can be easily extended to both convex bidimensional axisymmetric bodies and more complex geometries.

Consider now the flat plate shown in Fig. 1. The drag and heat coefficients, C_D and C_H, are obtained by normalizing the component in the $\vec{i}$ direction of $\vec{Q}$, as well as the total energy transfer $E = E_i - E_r$

$$C_D = \frac{-(P_i + P_r)\sin\theta_i + (\tau_i - \tau_r)\cos\theta_i}{1/2\,\rho_\infty V_\infty^2 \sin\theta_i} \tag{15}$$

$$C_H = \frac{E_i - E_r}{1/2\,\rho_\infty V_\infty^3 \sin\theta_i} \tag{16}$$

Analogous calculations can be realized for the lift coefficients. The drag and heat coefficients assume the following expressions in the hypersonic approximation:

$$C_D = 2(1+\sqrt{T_r/T_\infty}\; S_r/S_\infty(\sin\theta_r \sin\theta_i - \cos\theta_r\cos\theta_i)) \tag{17}$$

$$C_H = 1 - T_r/T_\infty\; S_r^2/S_\infty^2 \tag{18}$$

Let us now characterize the model by the following identification parameters:

$$\alpha = \sin\theta_r/\sin\pi/2 \tag{19}$$

$$\beta = S_r/S_\infty \sqrt{T_r/T_\infty} \tag{20}$$

It will be assumed that $\pi/2 \leq \theta_r \leq \pi$ for physical plausibility, so that

$$C_D = C_D(\alpha,\beta;\theta_i) = 2(1+\beta\,(\alpha \sin\theta_i + \cos\theta_i\sqrt{1-\alpha^2})) \tag{21}$$

$$C_H = C_H(\beta) = 1 - \beta^2 \tag{22}$$

Equations (21) and (22) show that the mathematical model is well defined by comparison with experimental results. In the more general case, Eqs. (9-11), referred to as the re-emitted flow, assume the following form:

$$N_r(\alpha,\beta,T_r/T_\infty) = \exp(-S_\infty^2\alpha^2\beta^2 T_\infty/T_r) + \sqrt{\pi}\, S_\infty\, \alpha\beta\sqrt{T_\infty/T_r}\,(1+\mathrm{erf}(S_\infty\alpha\,\beta\sqrt{T_\infty/T_r})) \tag{23}$$

$$P_r(\alpha,\beta,T_r/T_\infty) = S_\infty\alpha\,\beta\sqrt{T_\infty/T_r}\exp(-S_\infty^2\alpha^2\beta^2 T_\infty/T_r) + \sqrt{\pi}\,(1/2+S_\infty^2\alpha^2\beta^2 T_\infty/T_r\,(1+\mathrm{erf}(S_\infty\,\alpha\beta\sqrt{T_\infty/T_r})) \tag{24}$$

$$E_r(\alpha,\beta,T_r/T_\infty) = (S_\infty^2\beta^2 T_\infty/T_r+2)\exp(-S_\infty^2\alpha^2\beta^2 T_\infty/T_r) + \sqrt{\pi}(S_\infty^2\beta^2 T_\infty/T_r+5/2)S_\infty\alpha\beta\sqrt{T_\infty/T_r}\,(1+\mathrm{erf}(S_\infty a\beta\sqrt{T_\infty/T_r})) \tag{25}$$

Consequently,

$$C_D(\alpha,\beta,T_r/T_\infty) = \frac{1}{\sqrt{\pi}\,S_\infty^2}\,(P_i+N_i\sqrt{T_r/T_\infty}\,P_r/N_r + S_\infty N_i\,\frac{\cos\theta_i}{\sin\theta_i}\,(\cos\theta_i + \beta\sqrt{1-\alpha^2})) \tag{26}$$

$$C_H(\alpha,\beta,T_r/T_\infty) = \frac{1}{2\sqrt{\pi}\,S_\infty^3\sin\theta_i}(E_i - N_i T_r/T_\infty\, E_r/N_r) \tag{27}$$

These equations can be used when the flow velocity is moderate. A crucial problem that arises when using Eqs. (23-25) is how to characterize the re-emitted flow temperature T_r. Since it does not seem convenient to introduce a third parameter, the first possibility of relating T_r to the incoming flow temperature T_∞ may be to assume that wall collisions do not modify the energy level of each, both incoming and re-emitted streamline.

As is well known, in steady adiabatic flow conditions, the relation, valid for each streamline, that expresses the total temperature conservation may be written as

$$\frac{T_r}{T_\infty} = \frac{1 + ((\gamma-1)/\gamma) S_\infty^2}{1 + ((\gamma-1)/\gamma) S_r^2} \tag{28}$$

Equation (28) is a consequence of the wall behavior, which is assumed to preserve the total energy, distributed into enthalpic and kinetic energy.

Taking into account Eq. (20), Eq. (28) becomes

$$\frac{T_r}{T_\infty} = 1 + \frac{\gamma-1}{\gamma} S_\infty^2 (1-\beta^2) \tag{29}$$

However, this problem does not seem so relevant for practical calculations. In fact, the aerodynamic speeds are sufficiently large to allow Eqs. (21-22) to be sufficiently accurate so that the model can be regarded as identified by two parameters only (α, β).

Let us now perform some comparisons between this theory and some experiments for the simultaneous measurement

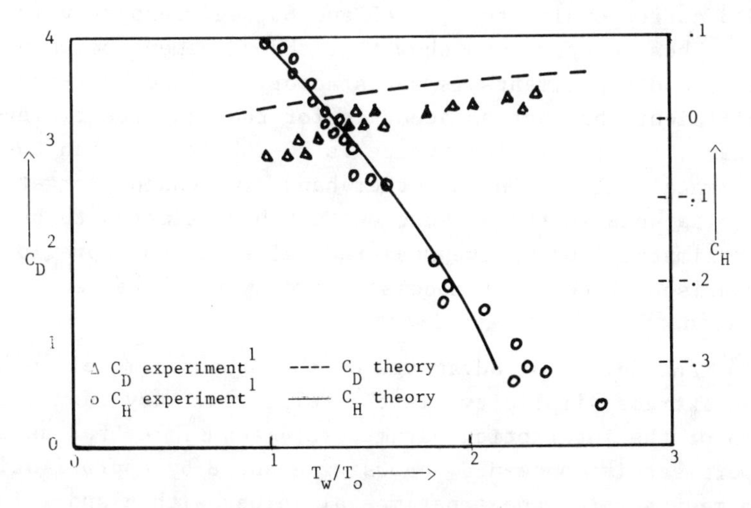

Fig. 2 C_D and C_H vs wall and stagnation temperature ratio, theory and experiment[1]; θ_i = 45 deg.

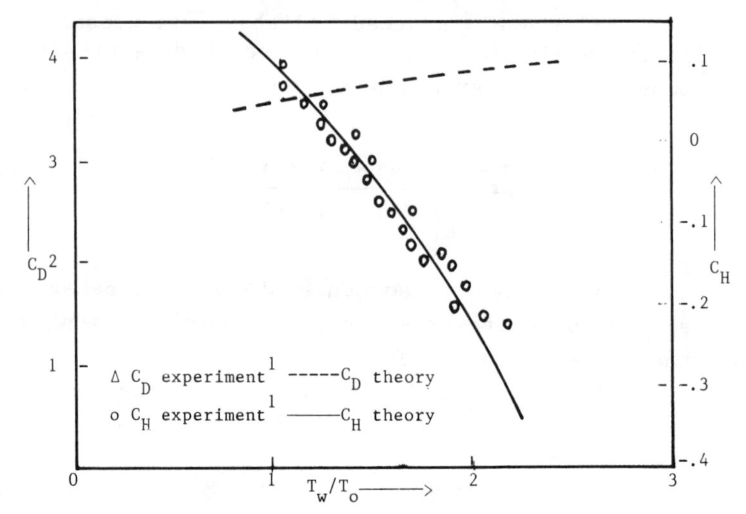

Fig. 3 θ_i = 60 deg. C_D and C_H vs wall and stagnation temperature ratio, theory and experiment[1].

of heat and drag coefficients realized in the DFVLR laboratories of Goettingen, some time ago[1].

Comparisons have been realized in the framework of the hypersonic approximation for experimental values of C_D and C_H corresponding to different temperature ratios T_w/T_o for two angles of attack. Some comparisons are shown in Figs. 2 and 3 corresponding to θ_i = 45 and 60 deg, respectively.

These comparisons show that the agreement between theory and experiments is satisfactory for the heat-transfer coefficient, but not as accurate for the drag coefficient. In fact, the theoretical value is 10-20% larger than the experimental value. On the other hand, one cannot forget that the clearance in the measurement can be estimated to be approximately 10% of the measured value and that previous comparisons with other models[1] showed worse agreement between theory and experiments.

An additional advantage of the proposed model, other than extreme simplicity, is the relatively easy identification of the interaction parameters α and β. As far as this experiment is concerned, choosing α and β by approximating (in mean square) the experimental values with α and β independent of the attack angle, i.e., depending only on T_w/T_o,

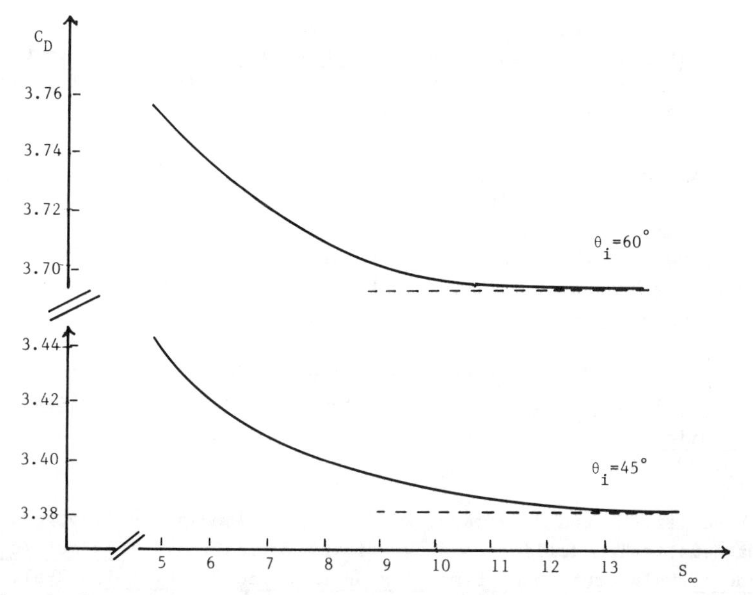

Fig. 4 Drag-theoretic coefficient vs S_∞ for two different angles of attack.

one obtains

$$\alpha = 1, \quad \beta = \beta_o + bT_w/T_o$$

with $\beta_o = 0.896$ and $b = 0.054$.

The preceding results are rather satisfactory since they show good agreement with the asymptotic behavior of the drag coefficients evaluated by applying the general equations [(26) and (29)], as shown in Fig. 4. Therefore one can conclude that the proposed model is accurate enough for aerodynamical calculations.

Acknowledgment

This work has been partially supported by the National Council for the Research, project Applied Mathematics in Industry and Technology of Gruppo Nazionale della Fisica Matematica, and by the Ministery of Education.

References

[1] Bellomo, N., Dankert, C., Legge, H., and Monaco, R., "Drag, Heat Flux, and Recovery Factor Measurements in Free Molecular Hypersonic

Flow and Gas-Surface Interaction Analysis," Proceedings of the 14th International Symposium on Rarefied Gas Dynamics, Vol. 1, Edited by O.M. Belotserkovskii, M.N. Kogan, S.S. Kutateladze and A.K. Rebrov, Plenum Publishing Corp., N.Y., 1985, pp. 421-430.

[2]Cercignani, C., Theory and Application of the Boltzmann Equation, Scottish Academic, Edinburgh and London 1975.

[3]Koppenwalner, G., "Freimolekulare Aerodynamik fur Satelliten-Anwendung," DFVLR, Institut fur Experimentelle Stromungsmechanik, Goettingen, Federal Republic of Germany 1982.

[4]Nocilla, S., "The Surface Re-emission Law in Free Molecular Flow," Proceedings of the Third International Symposium on Rarefied Gas Dynamics, Vol. 1, edited by J. Laurmann, Academic Press, N.Y., 1963 pp. 327-346.

[5]Pandolfi, M. and Zavattaro, M. G., "Upper Atmosphere Aerodynamics Mathematical Modelling and Experiment Validation," Proceedings of the 15th International Symposium on Rarefied Gas Dynamics, Vol. 1, edited by V. Boffi and C. Cercignani, B.G. Teubner Stuttgart 1985, pp. 618-626.

[6]Shidlovskij, V.P., Introduction to Dynamics of Rarefied Gases, American Elsevier, N.Y. 1967.

Nonreciprocity in Noble-Gas Metal-Surface Scattering

K. Bärwinkel* and S. Schippers†
University of Osnabrück, Osnabrück, Federal Republic of Germany

Abstract

Generalities of the scattering-kernel formalism are sketched and a reciprocity criterion is proposed. A reinterpretation of experimental results on energy accommodation for the gas-surface systems Ar-W, Kr-W, and Xe-W is then provided. Being a hint to nonreciprocity, our result implies severe consequences concerning the interaction mechanism. A threshold velocity for trapping can be drawn from the analysis. This threshold is explained on the basis of charge transfer and electronic relaxation during scattering.

Phenomenology of Gas-Surface Interaction

The phenomenological theory of gas-surface scattering of monatomic gases is concerned with the boundary condition for the one-particle distribution function f at the gas-substrate interface. If $\vec{n}$ is the unit vector in the direction of the surface normal pointing into the gas phase, let

$$\phi^{-}(\vec{c}) := -(\vec{c}\cdot\vec{n})f \tag{1}$$

for $\vec{c}\cdot\vec{n} < 0$ denote the flux distribution in the velocity space of the incident particles; whereas, for $\vec{c}\cdot\vec{n} > 0$,

$$\phi^{+}(\vec{c}) := (\vec{c}\cdot\vec{n})f \tag{2}$$

describes the particles coming from the wall. Excluding more complicated situations[1,2] we restrict ourselves to the

*Professor of Theoretical Physics.
† Student.

usual linear boundary condition

$$\phi^{+}(\vec{c}) = \int_{-} \phi^{-}(\vec{c}')P(\vec{c}' \to \vec{c})\mathrm{d}\vec{c}' \tag{3}$$

where the subscript under the integral sign hints to integration over the appropriate velocity half-space. As a probability density, the scattering kernel $P(\vec{c}' \to \vec{c})$ is nonnegative and we assume conservation of the number of particles:

$$\int_{+} P(\vec{c}' \to \vec{c})\mathrm{d}\vec{c} = 1 \tag{4}$$

A measure in velocity space

$$\mu(\mathrm{d}\vec{c}) := M(\vec{c})\mathrm{d}\vec{c} \tag{5}$$

may be defined, where

$$M(\vec{c}) = (2/\pi)c_T^{-3}|\vec{c}\cdot\vec{n}/c_T| \exp -(c/c_T)^2 \tag{6}$$

stands for the Maxwellian distribution normalized to unity. Here

$$c_T = (2k_\mathrm{b}T/m)^{1/2} \tag{7}$$

is the thermal velocity with T the solid-body temperature and m the mass of a gas particle. Only one half-space, the "+" half-space, say, needs to be considered after setting

$$\phi^{-}(\vec{c}) = M(\vec{c})\varphi(\vec{c}_\mathrm{R}) \tag{8}$$

with the reflected velocity

$$\vec{c}_\mathrm{R} = \vec{c} - 2\vec{n}(\vec{c}\cdot\vec{n}) \tag{9}$$

Doubly Stochastic Scattering Operator

A convenient operator notation is introduced by the definition

$$(P\varphi)(\vec{c}) := (M(\vec{c}))^{-1} \int_{+} \mu(\mathrm{d}\vec{c}')\varphi(\vec{c}')P(\vec{c}'_\mathrm{R} \to \vec{c}) \tag{10}$$

for $\varphi \in L^p(\mu)$, where any $p > 1$ may be chosen. This choice of the domain of P will be discussed later. Evidently

$$P\varphi \geq 0 \quad \text{for} \quad \varphi \geq 0 \tag{11}$$

and with the notation

$$<\ldots>:=\int_{+}\mu(\mathrm{d}\vec{c})\ldots \tag{12}$$

for taking an expectation value, Eq. (4) can be restated as

$$<P\varphi>=<\varphi> \tag{13}$$

The preceeding specification of the domain of the scattering operator as L^p guarantees finiteness of the entropy change functional $\delta S(\varphi)$, which we are going to explain next.

Let δS be the entropy change in the total gas-solid system, which is caused by gas-surface interaction, per incident gas particle. If the entropy flux within the gas is defined in accordance with Boltzmann's H theorem and if the solid substrate is considered to act as a heat bath that remains locally in thermal equilibrium, then the functional dependence of δS on the incident flux distribution via φ is obtained readily by simple thermodynamic reasoning (see, for example, Ref. 3, where the more complicated situation of a liquid substrate has been considered). Our most concise representation of the result is in terms of the nonnegative convex function

$$C(x)=x\ln x-(x-1) \tag{14}$$

which vanishes only for $x=1$:

$$\delta S(\varphi)=k_{\mathrm{b}}\,(<C(\varphi/<\varphi>)>-<C(P\varphi/<\varphi>)>) \tag{15}$$

Making use of Eq. (14) and invoking Jensen's inequality, it becomes trivial to prove equivalence of the entropy principle

$$\delta S(\varphi)>0 \quad \text{for any } \varphi \tag{16}$$

with

$$P1=1 \tag{17}$$

which may be called the equilibrium property, because it means that an incoming Maxwellian distribution is transformed into an outgoing Maxwellian. This last property is therefore, in addition to Eqs. (11) and (13) adopted as a third condition on the scattering operator, which accordingly is (an extension to L^p of) a doubly stochastic operator.

Matrix elements $<fPg>$ may be considered with $f\in L^q, g\in L^P$ where $p^{-1}+q^{-1}\leq 1$. The adjoint operator is then determined by the scattering kernel according to

$$(P^*f)(\vec{c}') := \int_+ P(\vec{c}'_R \to \vec{c})f(\vec{c})\mathrm{d}\vec{c} \tag{18}$$

Equation (4) or (13) can now be replaced by

$$P^*1 = 1 \tag{19}$$

and double stochasticity may be defined by the conditions $P1 = P^*1 = 1$ and $P\varphi \geq 0$ if $\varphi \geq 0$.

A convenient way to express accommodation coefficients is in terms of matrix elements.[4] In the following we shall use "AC" as an abbreviation for "accommodation coefficient". The most important quantity for the present article is the isothermal energy AC which is written α_{44} in Kuščer's notation.[4] With $Q = Q(\vec{c})$ denoting the kinetic energy of a gas particle we have

$$\alpha_{44} = 1 - < Q^{\perp}PQ^{\perp} > / < Q^{\perp 2} > \tag{20}$$

where

$$Q^{\perp} := Q - < Q > \tag{21}$$

Reciprocity vs Nonreciprocity

Another physical property that is often realized is reciprocity. For P as an operator on L^2, reciprocity may be stated as

$$P^* = RPR \tag{22}$$

where R is the representation of a rotation through π around the surface normal, or much as well, a reflection at the normal:

$$(Rf)(\vec{c}) := f(-\vec{c}_{\mathrm{R}}) \tag{23}$$

$P^* = P$ in case of reflection invariance of the solid surface.

Reciprocity can be inferred from the time reversal invariance of the microscopic dynamics of the scattering for a single (!) atom, provided that the solid body is in thermal equilibrium prior to the scattering event.[1] The seeming general validity of these assumptions has led Kuščer to dare the statement: "All gas-surface kernels obey the reciprocity relation."

Yet another presupposition should be made explicit: Microscopic time reversal invariance holds only for closed systems. In the usual experiments, however, a myriad of gas-surface collisions occur more or less simultaneously. Therefore, derivation of the reciprocity relation requires several conditions to be fulfilled:

1) For every incident gas atom, it must be possible to separate a dynamical subsystem in the solid which, together with the incident atom, remains isolated during the time of scattering . Such a subsystem can be envisaged to consist of sufficiently localized quasiparticles; e. g., wavepackets formed of phonon states or of Bloch states of electrons. It should be a macroscopic system; i. e., large enough for the notion of thermal equilibrium to make sense. On the other hand, it should be comparatively small and centered around the atom's point of incidence. Let us assign a width l or a surface element of area

$$\mathrm{d}A = l^2 \tag{24}$$

to the collisional subsystem thus chosen. For the quasiparticle system, some type of kinetic theory (Boltzmann equation, Boltzmann-Peierls equation, or related theories) will be applicable that allows the definition of a mean free path λ_0 and a relaxation time τ for establishment of local equilibrium. Disturbances will propagate with some velocity v, the order of magnitude of which is

$$v \approx \lambda_0/\tau \tag{25}$$

2) The collisional subsystem initially must be in thermal equilibrium. Consequently, at the corresponding point of incidence, the disturbing influence of neighboring collisions must have been eliminated by dissipation. The mean distance l between neighboring collisions, which defines the extent of our collisional subsystem, is therefore required to exceed the mean free path λ_0:

$$l > \lambda_0 \tag{26}$$

3) The condition of initial equilibrium must hold for every new scattering event; i. e., equilibrium must be reestablished quickly enough so that

$$\tau < t_{\mathrm{inc}} \tag{27}$$

where

$$t_{\mathrm{inc}} = (< \varphi > l^2)^{-1} \tag{28}$$

with $< \varphi >$ as the total flux of impinging particles, is the mean time between collision events on an area $\mathrm{d}A = l^2$. In view of Eqs. (25-27) we then arrive at a genuine condition:

$$\lambda_0/v < (<\varphi> \lambda_0^2)^{-1} \tag{29}$$

which may be fulfilled or violated. For phonons, λ_0 takes very large values with the temperature tending to zero so that reciprocity will be violated at a low enough temperature. Possibly this effect accounts for the disagreement[5] between measured and quantum-mechanically calculated energy ACs at low temperatures.

4) The collisional subsystem must remain dynamically separated from the rest of the solid body during the time t_{int} of its interaction with the gas atom. A reasonable ansatz for t_{int} is 10^{-12}s. Considering τ also as the time of dynamical separation, we have

$$10^{-12}\text{s} \approx t_{\text{int}} < \tau \approx \lambda_0/v \tag{30}$$

which means that irreversible effects in the collisional subsystem cannot be detected if the time of probing is too short.

Thus we have arrived at the two necessary conditions [Ineqs. (29) and (30)] for reciprocity. The latter inequality will be violated if electronic excitation and relaxation in metals are involved. The corrosponding relaxation time according to undergraduate textbook knowledge is

$$\tau \approx \sigma m_e / n e_0^2 \tag{31}$$

where σ, m_e, n and e_0 denote, respectively, the specific conductivity, the electronic mass, the density of conduction electrons, and the elementary charge. Values such as 10^{-13}s and less for τ are then nothing extraordinary. We exspect electronic effects of this type to become important if the scattering process involves charge transfer between the atom and the metallic substrate. The possibility of charge transfer between inert gases and metals is not a new idea. Charge rearrangement maps for such systems have been calculated.[6] The much discussed harpooning process is also a good candidate for nonreciprocal gas-surface scattering.

To conclude this section, let us cast our necessary conditions in a handier form. Using

$$<\varphi> = p/\sqrt{2\pi m k_b T} \tag{32}$$

with p and T as pressure and temperature of the gas and m the mass of the gas atoms, we find as the necessary and sufficient condition for reciprocity

$$10\text{Å} \cdot \frac{v}{\text{km/s}} < \lambda_0$$
$$< 5.7 \times 10^2\text{Å} \left(\frac{v}{\text{km/s}} / \frac{p}{10^{-2}\text{Torr}}\right)^{1/3} \left(\frac{T}{100\text{K}} \cdot \frac{m}{m_{\text{Ar}}}\right)^{1/6} \qquad (33)$$

where m_{Ar} denotes the mass of an argon atom.

If only phonons have to be considered, one may use the elementary formula for the phonon heat conductivity

$$\kappa \approx n c_V v \lambda_0 / 3 \qquad (34)$$

Here n is the number density of lattice atoms and c_V the specific heat (at constant volume) per lattice atom. The formula, which can also be justified as a crude approximation by kinetic theory, yields

$$\lambda_0/\text{Å} \approx 72 \frac{\kappa}{\text{W/(m·K)}} \Big/ \left(\frac{v}{\text{km/s}} \frac{n}{10^{22}/\text{cm}^3} \frac{c_V}{3k_\text{b}}\right) \qquad (35)$$

from which it is easy to understand that Ineq. (33) and, thus, reciprocity hold in many of the usual experimental situations.

Kuščer[4] has pointed out that the reciprocity relation is not necessary for the preservation of the Maxwellian. Indeed, for the derivation of the entropy change functional [Eq. (15)] in a stationary situation, it is sufficient to consider only the bulk of the solid as a heat bath and to allow for a thin surface layer where nonequilibrium may occur. Then the reciprocity relation fails but $P1 = 1$ is still valid. We hope to supply some evidence for such a situation in the sections that follow.

Menzel Plot and Implications

The experimental data that have stimulated our interest are those of Menzel and Kouptsidis[7], who have measured the isothermal energy AC over a considerable range of temperatures: $\alpha_{44} = \alpha(T)$. What we call the Menzel plot is the graphical representation of $\ln[1 - \alpha_{44}(T)]$ as a function of $1/T$. For each of the systems mentioned, Menzel and Kouptsidis found some dimensionless constant $\widetilde{c}, (0 < \widetilde{c} < 1)$ and a characteristic temperature $\widetilde{T}$ such that

$$\ln[1 - \alpha_{44}(T)] = -(\widetilde{T}/T) - \widetilde{c}$$

for approximately $0.3 \leq \widetilde{T}/T \leq 1$.

They also gave an interpretation of $\widetilde{c}$ in terms of an effective mass ratio and of $\widetilde{T}$ as a trapping threshold without, however, offering a description within the frame of the scattering-kernel formalism. This is what we are going to do next.

Our aim is to develop a simple but, hopefully, reasonable model theory for the scattering kernel in two versions, reciprocal and nonreciprocal, and to decide between the two on the basis of the Menzel plot. We are aware of the fact that knowledge of the temperature dependence of a single accommodation coefficient can never result in a rigorous decision against reciprocity because Maxwell's model

$$P(\vec{c}\,' \to \vec{c}) = (1 - \alpha(T))\delta(\vec{c} - \vec{c}\,'_{\mathrm{R}}) + \alpha(T)M(\vec{c}) \tag{36}$$

is clearly reciprocal with any prescribed $\alpha(T)$.

Model Description of Scattering and Trapping

We consider gas-surface scattering in a stationary situation with the possibility of temporary trapping in the long-range potential well near the surface and assume that a trapped fraction desorbs in a completely accommodated way; i. e., with a distribution ϕ^+_{acc} which up to normalization does not depend on the incident distribution. Let

$$\phi^+_{\mathrm{acc}}(\vec{c}) = M(\vec{c})\varphi^+_{\mathrm{acc}}(\vec{c}) = <\varphi^+_{\mathrm{acc}}> M(\vec{c})r(\vec{c}) \tag{37}$$

with

$$<r> = 1 \tag{38}$$

The remainder of the re-emitted flux is called directly scattered and will be described in the usual way with the aid of a scattering operator P_{sc} so that

$$\phi^+(\vec{c})/M(\vec{c}) = (P_{\mathrm{sc}}\varphi)(\vec{c}) + <\varphi^+_{\mathrm{acc}}> r(\vec{c}) \tag{39}$$

where, however,

$$w^*(\vec{c}) := (P^*_{\mathrm{sc}}1)(\vec{c}) \tag{40}$$

which is the scattering probability for a particle with velocity $\vec{c}_{\mathrm{R}}$, takes values between 0 and 1. The analogous definition

$$w(\vec{c}) := (P_{sc}1)(\vec{c}) \tag{41}$$

is also introduced. Evidently,

$$< w > = < w^* > \tag{42}$$

We now want to rewrite the right-hand side of Eq. (39) as $(P\varphi)(\vec{c})$ with a doubly stochastic P, called the total scattering operator. This is easily achieved making use of $P^*1 = P1 = 1$ and of the independence of $r(\vec{c})$ from φ. We get

$$P = P_{sc} + P_{acc} \tag{43}$$

where P_{acc} can be defined in terms of its kernel analogously to Eq. (10):

$$P_{acc}(\vec{c}_R' \to \vec{c}) = M(\vec{c}) \frac{(1 - w^*(\vec{c}'))(1 - w(\vec{c}))}{1 - < w >} \tag{44}$$

The last step in the specification of our model will be the choice of P_{sc}. All particles undergoing one or more internal reflections are considered to be accounted for by P_{acc} so that P_{sc} describes only the first scattering generation. Our reasoning for direct scattering then is as follows: The total potential energy of a gas particle approaching a solid surface is approximately described as a sum of a rigid potential well (long-range interaction with the bulk), which is independent of the coordinates parallel to the surface, and the potential of interaction with individual lattice atoms in the vicinity of the point of incidence. Certainly, the transfer of tangential momentum to the surface will be small in the first scattering generation. We neglect it completely, thus holding only P_{acc} responsible for tangential momentum transfer, by making the ansatz

$$P_{sc}(\vec{c}_R' \to \vec{c}) = \delta(c_2 - c_2')\delta(c_3 - c_3')P_1(u' \to u) \tag{45}$$

where u' and u are the normal velocity coordinates and the indices 2,3 refer to the tangential ones. The one-dimensional scattering operator P_1 now remains to be chosen.

Failure of a Reciprocal Model

An obvious idea is to describe the approaching gas particles as falling down a potential step, to assign to them the corresponding change of velocity, to describe the scattering process within the attractive well by a doubly stochastic and reciprocal scattering kernel $U(u' \to u)$, and to apply the inverse velocity change for the first scattering generation surmounting the potential step. Because of Liouville's theorem,

$$P_1(u' \to u) = U(-(u'^2 + v^2)^{1/2} \to (u^2 + v^2)^{1/2})u/(u^2 + v^2)^{1/2} \quad (46)$$

with $v^2/2m$ as the depth of the potential well, must be inserted into Eq. (45). Clearly, P_1 obeys the reciprocity relation if U does so.

A reasonable ansatz for in-well scattering is the hard-cube model, which introduces the effective mass m_s of a hypothetical surface particle or, equivalently, the mass ratio $q = m/m_s$ as a new parameter. The corresponding scattering kernel is known from Ref. 2.

It is worth while to work out the formulas for the hard-cube ACs of interest. They are temperature independent functions of the mass ratio. The hitherto unpublished formula for the isothermal energy AC reads

$$\alpha_{44}^{hc} = \frac{2q}{(1+q)^2}\left\{\frac{15q}{(1+q)^2} + \frac{(3-q)(1-q)}{(1+q)^{3/2}} - 2\right\} \quad (47)$$

According to our equations so far, our reciprocal model is now completely specified. Following the ideas of Ref. 7, we had expected to find a sufficiently extended region of approximate linearity in the theoretical Menzel plot. Dissapointingly enough, the numerically calculated Menzel plot is not similar to the experimental one: There is always to much curvature. Also the slope vanishes for $q \to 0$. This last fact is easily inferred from the observation that, in the limit of $m_s \to \infty$, the hard-cube model simply means specular, i. e., elastic, scattering.

Nonreciprocal Model

Our aim is to improve the forgoing model by introducing irreversibility. As before, we consider the classical trajectory $x = x(t)$ of a gas particle, now in a rigid potential well (x^{-3}-attraction, x^{-12}-or x^{-9}-repulsion, or similar). The hard-cube model will be built in later. At a distance a from the position x_0 of the potential's minimum, we assume that the incident particle takes up and the reflected particle loses a point charge δe, which may be a fraction of the electronic charge e_0:

$$\delta e = \epsilon e_0 \quad (48)$$

Admittedly, this is only a crude picture of charge transfer. In view of more elaborate results,[6] however, $a/\text{Å}$ might range from -0.5 to 0 and $|\epsilon| = 0.1$ might be a reasonable value.

The dynamical effect of charge transfer is now accounted for by the usual image force. Accordingly, the equation of motion for the gas particle is

$$m\ddot{x} = -V'(x) \qquad \text{for} \quad x > x_0 + a \tag{49}$$

$$m\ddot{x} = -V'(x) - \epsilon^2 k_c/(x - x_i)^2 \quad \text{for} \quad x < x_0 + a \tag{50}$$

with $k_c = 14.4\text{eVÅ}$. The quantity

$$x_i = -x(t) + 2(X + x_0) + d \tag{51}$$

is the position of the ficticious image point charge being retarded by the misfit d and $X + x_0$ denotes the position of the mirror plane.

Dissipative effects in the solid-body electron gas are then modeled by our relaxation ansatz

$$\dot{d} = (\dot{d})_{x_i=\text{const}} + (\dot{d})_{\text{relax}} = \dot{x} - d/\tau \tag{52}$$

Throughout our calculations so far, we have put $d(0) = 0$. The relaxation time τ is an additional parameter that should turn out to be similar to the electronic relaxation time.

The numerical solution of the system of Eqs. (49-52) with the initial condition $x(t) \to \infty$ and $\dot{x} \to u'$ for $t \to -\infty$ shows that, for a given reasonable potential function (of the type indicated), there exists a nonnegative threshold velocity

$$u_* = u_*(a, X, \epsilon, \tau) \tag{53}$$

such that the scattered particle belongs to the first scattering generation if, and only if, $|u'| > u_*$. In other words, the scattering probability is

$$w^*(|u'|) = \Theta((|u'|/u_*) - 1) \tag{54}$$

Such a threshold velocity does not exist for the reciprocal model. Furthermore, we attain the asymptotic velocity u for $t \to +\infty$ as a function of $u' : u = f(|u'|)$. A typical result for what we call the threshold factor

$$s(|u'|) := f(|u'|)/|u'| \tag{55}$$

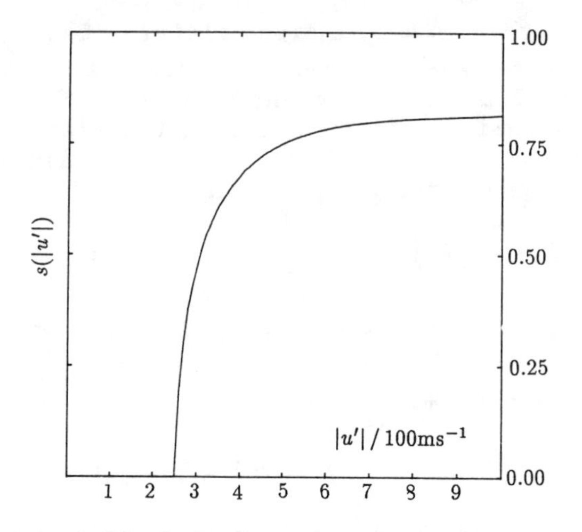

Fig. 1 Threshold factor vs incident velocity for argon. Parameters: $\epsilon = 0.13, a = -0.2\text{Å}, \tau = 0.06\text{ps}, X = -1.7\text{Å}, u_* = 254\text{m/s}$.

is represented in Fig. 1, and the underlying potential function has been adapted to Ref. 8. The shape of the threshold factor is, however, widely unaffected by mild changes in the potential.

Evidently, $s < 1$, which means inelastic scattering in the first generation. Neglect of inelasticity means

$$s(|u'|) = w^*(|u'|) = \Theta((|u'|/u_*) - 1) \tag{56}$$

An ambitious microscopic theory of gas-surface scattering would now have to deal with the simultaneous influence of relaxation in the electron gas and of lattice dynamics. We content ourselves with defining our conclusive model by

$$P_1(u' \to u) = \Theta((|u'|/u_*) - 1)P_1^{\mathrm{hc}}(u' \to u) \tag{57}$$

which is manifestly nonreciprocal. In applying our model, we consider the mass ratio q and the threshold velocity u_* as parameters to be fitted. From the interpretation of u_*, according to Eq. (53), we can then draw information about relaxational effects in the disturbed electron gas during the very short time of the most intimate contact between the gas atom and the substrate. Inelasticity in the first scattering generation is thus ascribed only to lattice dynamical effects accounted for by the hard-cube model. Interference between relaxational and lattice-dynamical effects is neglected.

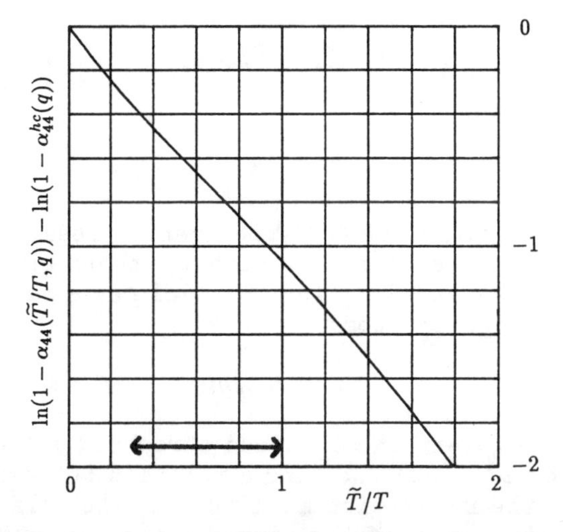

Fig. 2 Normalized Menzel plot. The arrow indicates the region of linearity covered by the experimental data of Ref. 7.

Application of the Nonreciprocal Model

Numerical calculations now yield Menzel plots exhibiting a region of approximate linearity, just as required by the measured data of Ref. 7. The analysis of experimental results will be facilitated by what we call the normalized Menzel Plot. This is the representation of

$$N(x) := \ln(1-\alpha_{44})_{\rm model} - \ln(1-\alpha_{44}^{\rm hc}) \tag{58}$$

as a function of

$$x := \psi(q) \cdot T_*/T \tag{59}$$

where the slope factor ψ is chosen so as to adjust the derivative $N'(x)$ to -1 in the region of linearity. Figure 2 shows the normalized Menzel plot for $q = 0.1$. For $x < 1$, however, $N(x)$ remains nearly the same for all values of $q, 0 \leq q \leq 1$.

Fitting the model parameters to the experimental data is now achieved by comparing the linear parts of the respective Menzel plots:

$$\ln(1-\alpha_{44})_{\rm exp} = -\widetilde{c} - \widetilde{T}/T \tag{60}$$

and

$$\ln(1-\alpha_{44})_{\rm model} - \ln(1-\alpha_{44}^{\rm hc}) = -c(q) - \psi(q) T_*/T \tag{61}$$

whence

$$T_* = \widetilde{T}/\psi(q) \tag{62}$$

with q determined from

$$\widetilde{c} = c(q) - \ln(1 - \alpha_{44}^{\mathrm{hc}}) \tag{63}$$

Our reinterpretation of the experimental results of Ref. 7 is summarized in Table 1. The number z there gives the mass of the hard-cube model surface particle as a multiple of the mass of one tungsten atom:

$$z := m_{\mathrm{s}}/m_{\mathrm{W}} \tag{64}$$

If we simply explain the obtained threshold velocities as indicated by Eq. (53), we find a relation between the charge fraction ϵ, the relaxation amplitude a, the relaxation time τ, and the position X of the mirror plane. It turns out that the threshold velocities in Table 1 can be explained with reasonable values for these parameters. The fact that our model theory does not apply to the lighter noble gases seems to be understandable in view of the potential curves given by Ossicini.[8] The turning point for the lighter gases is considerably farther from the substrate than for the heavy gases. Therefore, no charge transfer of importance is expected, which excludes nonreciprocity.

Table 1 Evaluation of Menzel plots

	Experimental data		Resulting model parameters			
System	$\widetilde{c}$	$\widetilde{T}$	q	z	T^*/K	u^*/ms^{-1}
Ar-W	0.0856	63	0.0275	7.9	155	254
Kr-W	0.1744	106	0.0652	7.0	250	224
Xe-W	0.3011	237	0.1361	5.2	498	251

Conclusion

We have discussed the possibility of nonreciprocity. A criterion for the validity of the reciprocity relation is proposed. The reciprocity relation is likely to become invalid for low enough temperatures and whenever charge transfer between gas and metal substrate becomes important.

The Menzel plot obtained from experiment is reproduced by a model theory within the frame of the scattering-kernel formalism. The scattering-kernel model is nonreciprocal and yields, via fitting, an effective mass ratio and a threshold velocity. Despite the inherent neglections of the model, a simple relaxation ansatz can explain the threshold velocity, thereby relating various properties of the metal surface and the mechanism of its interaction with the heavy noble gases.

A more basic study of the phenomena is desirable. We deplore the paucity of accessible data, which would allow Menzel plots to be attempted for metals other than tungsten.

References

[1] Kuščer, I. , "Reciprocity in Scattering of Gas Molecules by Surfaces," Surface Science, Vol. 25, 1971, pp. 225-237.

[2] Bärwinkel, K. , and Schmidt, H. J. , "Fixed Point Properties of Gas-Surface Scattering Operators," Progress in Astronautics and Aeronautics: Rarefied Gas Dynamics, Vol. 74, Part I, edited by Sam S. Fisher, AIAA, New York, 1981, pp. 150-166.

[3] Bärwinkel, K. , and Monaco, R. , "Equivalence of the Entropy Principle with the Equilibrium Condition for Scattering of Simple Gases on Solid and Liquid Substrates," Workshop on Mathematical Aspects of Fluid and Plasma Dynamics, edited by C. Cercignani, S. Rionero, and M. Tessarotto, Università degli Studi di Trieste, Italy, 1984, pp. 33-44.

[4] Kuščer, I. , "Phenomenology of Gas-Surface Accommodation," Rarefied Gas Dynamics, Proceedings of the 9th International Symposium, edited by M. Becker, M. Fiebig, DFVLR, Porz-Wahn, Federal Republic of Germany, Vol. II, 1974, pp. E1.1-E1.21.

[5] Goodman, F. O. , "Theoretical Aspects of Atom Surface Diffraction, Inelastic Scattering and Accommodation Coefficients," Progress in Astronautics and Aeronautics: Rarefied Gas Dynamics, Vol. 74, Part I, edited by Sam S. Fisher, AIAA, New York, 1981, pp. 3-49.

[6] Lang, N. D. , "Interaction between Closed-Shell Systems and Metal Surfaces," Physical Review Letters, Vol. 46, 1981, pp. 842-845.

[7] Menzel, D. , and Kouptsidis, J. , "Energy Transfer by Single Collisions and by Trapping in Thermal Accommodation," Rarefied Gas Dynamics, Proceedings of the 9th International Symposium 1974, edited by M. Becker, M. Fiebig, DFVLR-Press, Porz-Wahn, Federal Republic of Germany, Vol. II, pp. E14.1-E14.11.

[8] Ossicini, S. , "Interaction potential between rare-gas atoms and metal surfaces," Physics Review B, Vol. 33, 1986, pp. 873-878.

Studies of Thermal Accommodation and Conduction in the Transition Regime

Lloyd B. Thomas,* C. L. Krueger,† and S. K. Loyalka‡
University of Missouri, Columbia, Missouri

Abstract

Thermal conductivity measurements are reported for krypton and argon, taken in cells of cylindrical symmetry. The data were taken under extreme clean filament conditions so that the thermal accommodation coefficient remains constant over the full range of pressures, ∿0.005-20 torr. Both the accommodation coefficient and thermal conductivity are determined. The data are used to test a number of theoretical treatments of the complex problem of heat transfer in the transition regime of kinetic theory.

Introduction

The thermal accommodation coefficients (AC or α) found in the literature have been obtained primarily from thermal conduction measurements on gases in one of two pressure regions: the low-pressure free-molecule regime, or the higher-pressure transition regime. If the Knudsen number with respect to the central cylinder of the conductivity cell is 20 or above for the measurement, we treat the data as in the free-molecule range and will designate the AC values as FM or α_{FM}. The theory used to extract α_{FM} is well established, simple and direct, and does not appear to be seriously questioned. The problem with determining α_{FM} is principally in precise

---

* Professor Emeritus, Physical Chemical Laboratory.

† Research Associate, Physical Chemical Laboratory.

‡ Professor, Nuclear Engineering Program, and Particulate Systems Research Center, College of Engineering.

determination of the low gas pressures involved. The pressures commonly involved in AC determination in the transition regime range up to 100 torr or more, some 10,000 times those used in the FM range. Extracting AC values from transition regime measurements, which we will designate as TR values or α_{TR}, is much more complicated than in the FM case. Numerous theoretical treatments of this problem have been offered, varying from phenomenological treatments of the temperature jump to various applications of the Boltzmann equation to the intermingling of the streams of molecules about the central cylinder. As far as we are aware, there has been no previous experimental work done in such a way or with sufficient accuracy to provide an adequate test of these theories and perhaps to assign relative merit as to their abilities to produce the results as observed. This paper describes experiments and results of part of our efforts to provide such a test.

Four decades ago we noted that many α_{TR} values in the literature for given solid-gas systems were on the extreme high side of those reported, and we questioned whether the AC values calculated by the TR theories then applied would agree with the FM results. The experiments designed to make this test[1] made determinations of the AC of He, Ne, and CO_2 on Pt by both methods in the same conductivity cell, being careful not to unnecessarily alter the adsorbed condition of the platinum surface between runs. Temperature jump theory generally as described in Kennard[2] yielded the α_{TR} values in remarkable agreement with the chronologically interspersed α_{FM} values. The conclusion was made that weakness in the TR theory could not account for the high α_{TR} values found in the literature. However, we could not be sure that the adsorbed condition of the gas covered filament did not change with the wide range of pressures used when working in the two regimes, and this was a fundamental weakness. A second attempt[3] on this problem was made after we had discovered the "getter" method of maintaining clean vacuum conditions for AC determination in the FM regime. It was found that a bare tungsten surface could be maintained through a series of TR measurements by showing that the α_{FM} value for He before and after the TR series, without an intervening flash of the filament, remained the same at $\sim$0.0168, indicating that no adsorbtion had occurred. Thus the doubt mentioned earlier, with regard to interpretation of the Ref. 1 results, could be eliminated. Reference 3 reports comparison of results of application of three TR theories for extraction of the AC values from TR measurements, now made with the surface of the filament maintained bare, with the AC values determined on bare tungsten in the FM regime in the same cell. The result was extraordinary agreement of the α_{TR} values by all three theories with the α_{FM} values. The failure of these experiments to discriminate between theories was attributed to the low AC values of He and Ne on the bare tungsten, 0.0168 and 0.0444 respectively. It was shown that the ability to discriminate between the results from the theories applied rises rapidly with the value of the AC of the gas under study. Bare surfaces and hence constant AC values through both FM and TR ranges can be maintained on tungsten for Ar, Kr, and Xe. Results with Ar (α=0.23) and Kr (α = 0.38) on clean W, and with Kr (α = 0.73) on Ta contaminated W, but with α constant through numerous flashes, are reported and analyzed in this paper.

Experimental Methods and Results

The experiments were performed in thermal conductivity cells, or tubes, of a type shown in Fig. 1. The antechamber shown to the right and partly above the cell proper connects the cell to the vacuum system through a double U-tube trap which is immersed in liquid N_2 at all times to exclude mercury and other condensable materials from the tube. The antechamber walls can be coated with the titanium getter from the Ti-Ta alloy wire shown. The antechamber serves to intercept adsorbable impurities diffusing toward the tube from the vacuum system, to provide a large-diameter passage through the temperature gradient from the ambient temperature to the somewhat higher bath temperature (thus minimizing the small thermal transpiration pressure), and to intercept the interchange of radiation between the filament and ambient, which has been found to be "piped" within the glass tubing when the cell is directly connected to the vacuum system. The cell proper is immersed in a thermostat bath, held, for the data reported, at 308.15°K within ±0.01°K. The central tungsten filaments used are approximately 0.002 and 0.004 cm in radius and 25-30 cm in length and have been electrolytically polished. The filament is held taut between slightly flexed tungsten current leads which are backed as shown by potential leads. The envelopes are of Pyrex glass of internal radius 2.35 cm. The two getter wires shown in the cell are about halfway out on the radius from the central filament and are maintained taut during evaporation by flexing the 0.035 inch leads a proper amount when the wires are spot welded to the leads, which are already in place. The vacuum systems used,

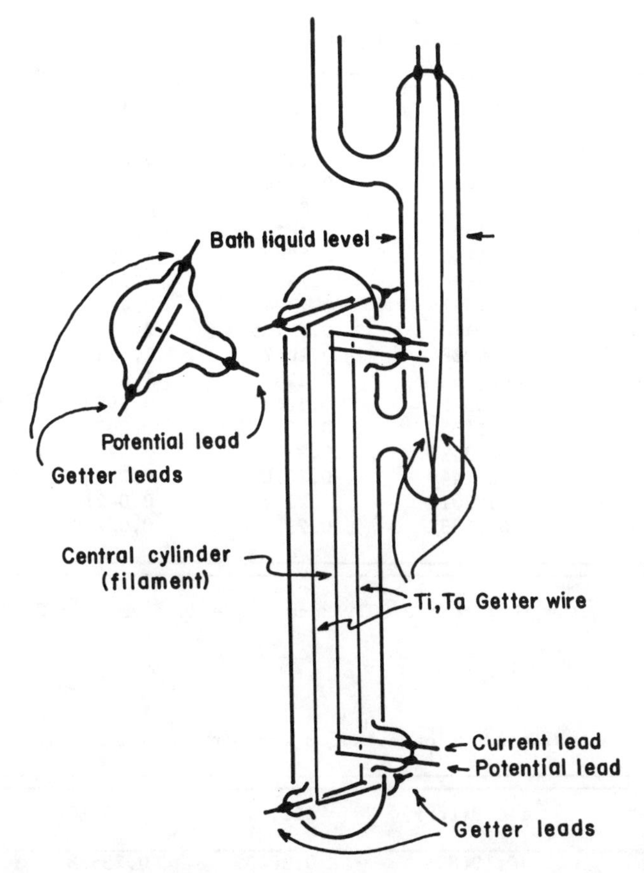

Fig. 1 Thermal conductivity cell with getter evaporating filaments and antechamber. See text for dimensions.

vacuum preparation for the clean surface work, getter evaporation, and electrical measurements required are described elsewhere[4-6] and will not be repeated here. Pressure measurements are made with an extended range McLeod gauge and are corrected for the mercury pumping effect by the method and data supplied by this Laboratory.[7] Filament end-loss corrections are made by a method due to Brown and Thomas.[8] The equations used for extraction of the α_{EM} values are well known and may be found in Refs. 4 and 6. Data reported are all for 10°C temperature difference between filament and wall (bath), i.e. $\Delta T=T_f-T_w=10°C$. These are given in Tables 1, 2, and 3 in four columns: Column I gives the run number in chronological order; Column II gives the pressure of the gas in torr; ColumnIII gives α_{app}, the "apparent" accommodation coefficient; and Column IV gives the cell

Table 1 Thermal conduction data for argon*

I Run	II P(torr)	III α_{app}	IV $\Lambda x10^3$
1	0.01645	0.2409	0.03464
2	19.542	0.05845	9.986
3	12.686	0.07956	8.824
4	8.200	0.10365	7.430
5	5.4116	0.1282	6.065
6	3.5650	0.1511	4.709
7	2.2997	0.1724	3.466
8	1.4412	0.19165	2.415
9	0.90224	0.2047	1.615
10	0.5651	0.2183	1.0785
11	0.3523	0.2263	0.6970
12	0.1435	0.2331	0.2924
13	0.03695	0.2360	0.0761
14	0.02440	0.2370	0.05055
15	0.02012	0.2396	0.04214
16	0.01331	0.2329	0.02710

*T_w = 308.15; ΔT = 10°; r_1 = 0.001963 cm; r_2 = 2.35 cm; α = 0.2357.

Table 2 Thermal conduction data for krypton on clean tungsten*

I. Run	II. P(torr)	III. α_{app}	IV. $\Lambda x10^3$
1	0.10682	0.36489	0.23528
2	0.067512	0.36539	0.14890
3	0.014209	0.37030	0.31759
4	0.002865	0.37930	0.006559
5	17.343	0.058226	6.0952
6	15.332	0.064550	5.9739
7	13.180	0.073188	5.8223
8	8.8903	0.099360	5.3319
9	6.0885	0.12929	4.7515
10	4.1159	0.16512	4.1022
11	2.9235	0.19723	3.4803
12	1.9381	0.23485	2.7484
13	1.2000	0.27542	1.9949
14	0.71990	0.30617	1.3304
15	0.44694	0.33512	0.90406
16	0.30234	0.34774	0.63459
17	0.076188	0.36772	0.16910
18	0.019981	0.37783	0.045568
19	0.014335	0.37804	0.032710
20	0.01036	0.37689	0.23060

*T_w = 308.15; ΔT = 10°; r_1 = .001963 cm; r_2 = 2.35 cm.; α = .3787.

Table 3 Thermal conduction data for Kr on contaminated tungsten*

I. Run	II. P(torr)	III. α_{app}	IV. $\Lambda x10^3$
1	0.089872	0.69673	0.37795
2	0.059703	0.69433	0.25021
3	0.014152	0.71172	0.060796
4	0.009894	0.71790	0.04287
5	1.2344	0.42512	3.1674
6	0.77516	0.50223	2.3499
7	0.48415	0.56833	1.6609
8	0.30050	0.62283	1.1297
9	0.17734	0.65520	0.70134
10	11.428	0.092210	6.3606
11	10.006	0.10279	6.2084
12	6.6566	0.14638	5.8814
13	4.2914	0.20757	5.3767
14	2.7428	0.28214	4.6710
15	1.7597	0.36428	3.8693
16	0.15135	0.67988	0.62112
17	0.032496	0.71520	0.14028

*T_w = 308.15; ΔT = 10°; r_1 = 0.001963 cm; r_2 = 2.35 cm, α = 0.7284.

conductance, per unit area of filament and per degree of ΔT at the pressure specified. Table 1 is for argon and is taken from Ref. 6. Tables 2 and 3 are for krypton on clean W and on W contaminated with the nonvolatile Ta, showing the AC of Kr about double the value on clean W, but holding constant through the runs. The experimental quantities, α_{app}, are determined by dividing the power conducted from the filament by the gas by that calculated as though the process were a free molecular one with the molecules impinging at the wall temperature upon the filament where they have unit AC. In the limit as the gas pressure approaches zero, α_{app} ----> α_{FM}, and becomes the true AC. Of course, as seen in the tables, α_{app} falls off rapidly as P increases, bringing the true impingement temperature of the molecules on the filament nearer to the filament temperature.

Discussion and Observation on the Clean Surface Data

Two ways of plotting the data have been found useful for extracting respectively, the free molecule, clean surface AC, α or α_{FM}, and the limiting conductance of the cell in the absence of the temperature jump, and hence the thermal conductivity K of the gas. To obtain α we plot

$1/\alpha_{app}$ vs P, and find that the points fall typically very well on a straight line. The Table 2 data for Kr plot out linearly particularly well with all 20 points, up to 17 torr, falling very closely on the line (on a large scale 20x20 in. graph). The Y intercepts of these plots are the reciprocals of the true α. We believe these intercepts (located by a least squares method) are capable of giving better values of α, when applied to clean surface data, than can be obtained from low pressure measurement because of the difficulty in measuring precisely the low pressures required. We have not seen this technique used elsewhere, and it would only be feasible when applied to data for which the true α remains constant, or possibly if it varies in a proportional manner with P.

The second way mentioned earlier is the conventional one, plotting $1/\Lambda$ vs $1/P$ and using the Y intercept as the limiting high pressure value for $1/\Lambda$. The temperature discontinuity should vanish as $(1/P) \longrightarrow 0$, and the thermal conductivity of the gas is derived from the value of the intercept and the cell radii, r_1 and r_2. The data of Tables 1-3 plot out in this manner quite well as straight lines, again as long as α remains constant for all pressures (see Ref. 1, Fig. 3), and there is no appreciable heat transfer from the filament by convection. Incidence of convection is indicated by the points breaking below the straight line on this type of plot at the low 1/P end (high P). None of this is observed in the data presented here, but it is observed on the larger filament (0.004125 cm radius) tube for both Ar and Kr, beginning at ∿15 torr, and becoming more evident as P increases. The equations for the lines, six in all, for the ($1/\alpha$ app) vs P and for the $(1/\Lambda)$ vs $(1/P)$ plots of data in Tables 1-3 are given below. These are in the form $y=m\chi+b$, and the values of α and K derived from the y-intercepts are given:

Table 1: $(1/\alpha_{app}) = 0.65956P+4.24165$; $\alpha = 0.2357$

Table 2: $(1/\alpha_{app}) = 0.83713P+2.64061$; $\alpha = 0.3787$

Table 3: $(1/\alpha_{app}) = 0.79714P+1.37291$; $\alpha = 0.7284$

Table 1: $(1/\Lambda) = 485.184(1/P)+75.433$; $K = 1844.4\text{x}10^{-7}$

Table 2: $(1/\Lambda) = 434.419(1/P)+139.368$; $K = 998.3\text{x}10^{-7}$

Table 3: $(1/\Lambda) = 225.682(1/P)+134.544$; $K = 1034.1\text{x}10^{-7}$

where the units of K are W/cm x deg.

The α values are quite insensitive to temperature over the small range applicable here. The preceding α values should be characteristic of the impingement temperature. The K values increase with temperature, about 4.8×10^{-7}/deg for Ar and 3×10^{-7} for Kr. The above K values are characteristic of the midtemperature of T_w and T_e, where T_e is the gas temperature just off the filament surface. For the K values at T_w, 308.15°K, from consideration of several sets of our measurements on each gas, we have selected the values 1819×10^{-7} for Ar and 994×10^{-7} for Kr. The former is about 0.3% above the value selected for a National Bureau of Standards survey[9] and the latter is 3% higher than an approximate value calculated by a method and constants given in Hirschfelder et al.[10]

Two distinct processes occur in experiments such as these: the conduction of heat by the gas and the acquisition of heat from the surface of the filament by the impinging molecules. It is axiomatic that the heat crossing each coaxial cylindrical surface of unit length per second is the same in the steady-state of heat flow. Thus the power taken by the molecules in the accommodation process from the filament surface is identical to that conducted through the gas. It may be of interest to some to examine each of these processes separately in our data, applying the Knudsen concept of accommodation to the first above, and the Fourier equation of heat conduction respectively to these processes. The heat conducted by the molecules per second from unit area of the filament or by the gas per length of cylinder, L, required for unit area of filament is 10Λ since the data is all for 10° ΔT. To put this power radially through the gas from r_1 to r_2 requires a temperature difference ΔTe, with the inner temperature Te and the outer Tw. The effective thermal conductivity of our nonhomogeneous gas is taken as that of the mid-temperature (Te+Tw)/2, with little error. The integrated Fourier equation gives $\Delta Te = 10\Lambda(\ln r_2/r_1)/2\pi LK$. The ΔTe values for runs 2-10 of Table 1, calculated using values of K (based on 1819×10^{-7} at 308.15°K) at the midtemperature as mentioned, are plotted vs P in Fig. 2. Then the impingement rate at Te multiplied by our α (0.2357) and the heat capacity (2k) is divided into the same 10Λ for each point to find the temperature difference, ΔT_i, offered to the impinging molecules, i.e. $T_f - T_i$. Except for the slight dependence of the impingement rate on the temperature near the filament surface (a 1.6% decrease with 10° rise from 308°K), this calculation of ΔT_i seems independent of the ΔTe calculation. Another

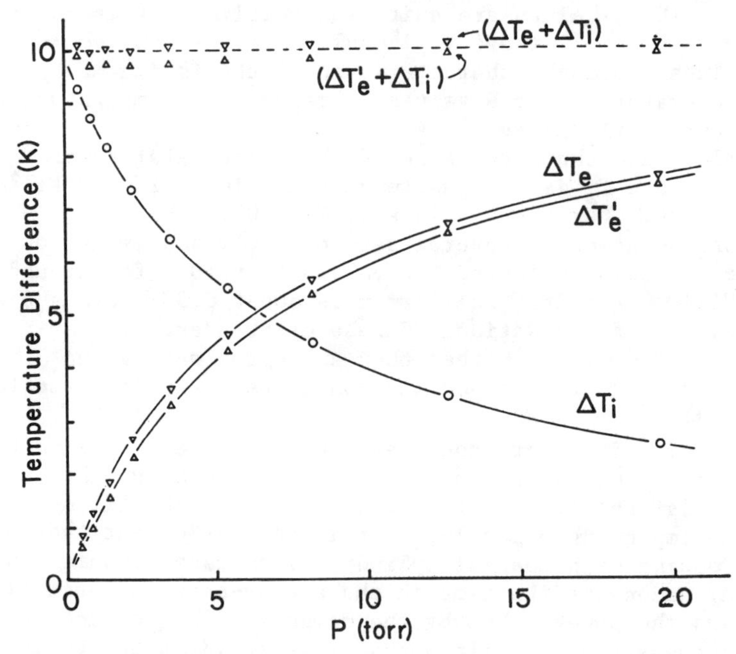

Fig. 2 Plots of the temperature differences, ΔT_e and ΔT_i, required to cause the observed heat flows through argon (Table 1), and across the interface at the filament surface, respectively. See text for details.

approach gives the same values for ΔTi as that described, i.e. $\Delta T'_i = (\alpha_{app}/\alpha)\sqrt{T_i/T_w}$. This simply recognizes the fact that α apparent would become the true α if one used the temperature of the impinging molecules instead of Tw to subtract from T_f in its calculation. (Also there is the factor $\sqrt{T_i/T_w}$ to adjust for the increase in the temperature controlling the impingement rate from T_w to T_i.) Of course, ΔT_i is the temperature difference offered to the impinging molecules, but they are able to accept only the fraction α of this, thus, the stream of molecules just emerging from collisions with the inner cylinder surface should be at an average energy described by Te+αΔTi, or by $T_f-(1-\alpha)\Delta Ti$. A remarkable feature of the ΔTe and ΔTi values calculated is that their sum, for all the points, is very close to 10°, the experimental (T_f-Tw). The largest deviation is 0.1° for Run 10 and the mean deviation is 0.05°. These sums for each run are plotted along the top, 10° ΔT ordinate of Fig. 2. We also show on Fig. 2 the $\Delta T'_e$ calculated by the same Fourier equation for the temperature in the gas one mean free path off the filament surface, i.e.,at $r_1+\ell$ where $\ell = 1/\sqrt{2}\, n\pi\sigma^2$

and σ for Ar is taken as 3.42 A. These $\Delta T_e'$ values are taken as the impingement temperatures in the mean free path method of extracting α from transition regime thermal conduction measurements.[3] Using our α value (0.2357) as before we would get the same ΔT_i values as before, and the sum $\Delta T_i + \Delta T_e'$ would be less than the experimental 10°ΔT as shown on Fig. 2. If we take $T_f - T_e' = \Delta T_i$ for the mean free path method, somewhat smaller values of α are found than 0.2357. This is simply an observation and is not meant to be a test of the mean free path method.

The calculations and observations displayed in Fig. 2 do seem to substantiate a constancy of the accommodation coefficient throughout the range of pressures. They also show in a simple manner how the phenomenon of temperature jump arises, and how this is affected by the AC, and how the magnitude of the temperature jump brings the two mechanisms of heat transfer to balance. When one encounters the ingoing and outgoing streams of molecules at the filament surface, one realizes the complexity of the problem and why the various attempts at its analysis, as tested in the following section, show variations in success in describing the phenomonon.

Test of Various Theories Against Experiments

For cells, with large ratio of outer radius (r_2) to the inner radius (r_1), as discussed in Ref. 6, we can write

$$\alpha = \left[\frac{1}{\alpha_{app}} - \frac{1-\hat{Q}(R,1)}{\hat{Q}(R,1)}\right]^{-1} \qquad (1)$$

where

$$\alpha_{app} = \dot{Q}(R,\alpha)/\dot{Q}_{FM}(R,1)$$

and

$$\hat{Q}(R,1) = \dot{Q}(R,1)/\dot{Q}_{FM}(R,1)$$

where $\dot{Q}(R,\alpha)$ is the experimentally determined heat loss rate from the filament due to gas conduction; $\dot{Q}_{FM}(R,1)$ is the theoretical heat loss rate, calculated for all pressures as if the gas were at the free molecular (fm) limit with impingement temperature, Tw, and with α = 1; $\dot{Q}(R,1)$ is the calculated heat loss rate if α = 1, obtained by solving the Boltzmann equation appropriate to the problem. Thus α_{app} is essentially determined from the experiments, and $\hat{Q}$ is obtained through theoretical calculations as a function of the inverse Knudsen number R, where,

$$R = r_1/\ell_t$$

and ℓ_t is the mean free path defined as

$$\ell_t = (4KT/5P)(m/2kT)^{\frac{1}{2}}$$

in which K is the thermal conductivity, P is the pressure, T is the temperature, m is the molecular mass of the gas, and k is the Boltzmann constant.

In Ref. 6, we have summarized results for $\dot{Q}(R,1)$ as provided by several of the available methods (variational, moments, boundary conditions, etc.). We have also presented $\hat{Q}(R,1)$ in forms such that with experimental values of α available, α_{app} can be calculated quite conveniently. Knowledge of K of the gas is required for calculation of the ℓ_t. The methods that we have considered include

1) the moments method of Lees and Liu[11];
2) the variational (integral) method of Bassanini, Cerignani, and Pagani[12];
3) the variational (integro-differential) method of Lang and Loyalka[13];
4) the boundary conditions method of Miklavcic and Kuscer applied to BGK and rigid spheres models[14];
5) the variational expressions for Maxwell molecules (hybrid)[15]; and
6) the variational expression for rigid sphere molecules (hybrid).[15]

The experimental data of Tables 2 and 3 together with the theoretical treatments indicated have been used to calculate values of the accommodation coefficients from each run, using Eq. (1). The resulting values from Table 2 are given in Table 4 and are shown plotted in Fig. 3. The values for the Table 3 data are given in Table 5 and are shown plotted in Fig. 4. The curves in Figs. 3 and 4 are numbered according to the method of computation as listed above.

The first few rows of Tables 4 and 5 show results for data in the free molecule regime and the theoretical methods give essentially identical values of α for each pressure. Differentiation in the α values becomes distinct and increases in the lower rows of the tables (as P increases). Figs. 3 and 4 show the results of the higher pressure runs. It must be noted that the vertical positions of the four curves are quite sensitive to the value of thermal conductivity used for the calculations; higher values than those indicated in the tables move the curves downward and vice versa, but their relative positions remain intact. Actually using K(308.15) = 964 shows the variational results to be the best, whereas

Table 4 Accommodation coefficient of krypton on bare tungsten (data of Table 2), as calculated by use of several theoretical treatments

		Accommodation Coefficient					
Pressure torr.	Inverse Knudsen No.	1 MM	2 BC, BGK	3 BC,RS	4 V,BGK	5 H,Maxmol	6 H,RS
0.01013	0.002677	0.3781	0.3780	0.3778	0.3780	0.3784	0.3781
0.01433	0.003786	0.3798	0.3796	0.3793	0.3796	0.3801	0.3797
0.01998	0.005277	0.3803	0.3800	0.3797	0.3800	0.3807	0.3801
0.07618	0.02011	0.3766	0.3758	0.3749	0.3756	0.3782	0.3762
0.3023	0.07975	0.3816	0.3791	0.3764	0.3777	0.3849	0.3792
0.4469	0.1178	0.3836	0.3802	0.3767	0.3780	0.3869	0.3797
0.7199	0.1895	0.3760	0.3713	0.3669	0.3678	0.3783	0.3696
1.2000	0.3156	0.3815	0.3926	0.3771	0.3689	0.3812	0.3701
1.9381	0.5089	0.3802	0.3573	0.3444	0.3626	0.3755	0.3622
2.9235	0.7665	0.3820	0.3338	0.3225	0.3591	0.3719	0.3565
4.1159	1.0779	0.3835			0.3559	0.3681	0.3507
6.0885	1.5923	0.3787			0.3468	0.3576	0.3383
8.8903	2.3222	0.3795			0.3431	0.3525	0.3307
13.1800	3.4393	0.3758			0.3362	0.3438	0.3202
15.3320	3.9993	0.3707			0.3309	0.3377	0.3137
17.3430	4.5230	0.3697			0.3291	0.3354	0.3109

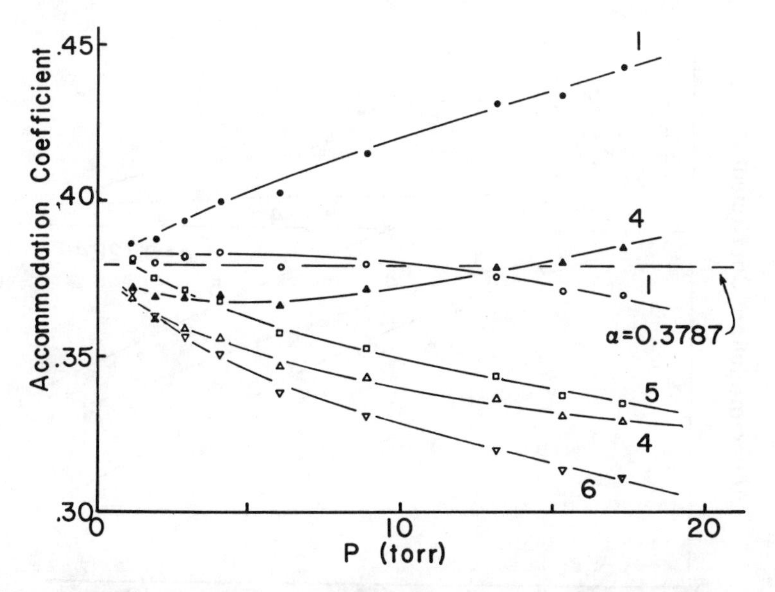

Fig. 3 Accommodation coefficients extracted from the data of Table 2. Curves are numbered the same as the methods listed in the text. The thermal conductivity at 308.15°K is taken as 994 for the open symbols and 964 for the solid symbols.

Table 5 Accommodation coefficient of kryton on tantalum coated tungsten (data of Table 3), as calculated by use of several theoretical treatments.

		Accommodation Coefficient					
Pressure torr.	Inverse Knudsen No.	1 MM	2 BG,BGK	3 BC,RS	4 V,BGK	5 H,Maxmol	6 H,RS
0.00574	0.001515	0.7331	0.7328	0.7323	0.7328	0.7336	0.7330
0.00989	0.002612	0.7222	0.7217	0.7210	0.7217	0.7231	0.7220
0.01415	0.003736	0.7178	0.7171	0.7162	0.7171	0.7191	0.7175
0.03250	0.008582	0.7295	0.7280	0.7262	0.7279	0.7323	0.7288
0.05970	0.01576	0.7195	0.7171	0.7143	0.7166	0.7241	0.7182
0.08987	0.02372	0.7356	0.7321	0.7280	0.7311	0.7422	0.7334
0.1513	0.03993	0.7445	0.7390	0.7329	0.7369	0.7541	0.7405
0.1773	0.04676	0.7264	0.7205	0.7139	0.7179	0.7363	0.7218
0.3005	0.07920	0.7395	0.7302	0.7205	0.7251	0.7520	0.7306
0.4841	0.1274	0.7398	0.7264	0.7133	0.7177	0.7520	0.7241
0.7751	0.2037	0.7467	0.7277	0.7101	0.7133	0.7548	0.7197
1.2344	0.3239	0.7600	0.7968	0.7353	0.7105	0.7579	0.7147
1.7597	0.4612	0.7877			0.7199	0.7719	0.7202
2.7428	0.7178	0.8013			0.7100	0.7619	0.7016
4.2914	1.1214	0.8129			0.6962	0.7445	0.6753
6.6566	1.7370	0.7846			0.6553	0.6941	0.6223
10.066	2.6241	0.7498			0.6156	0.6451	0.5728
11.428	2.9792	0.7610			0.6189	0.6471	0.5713

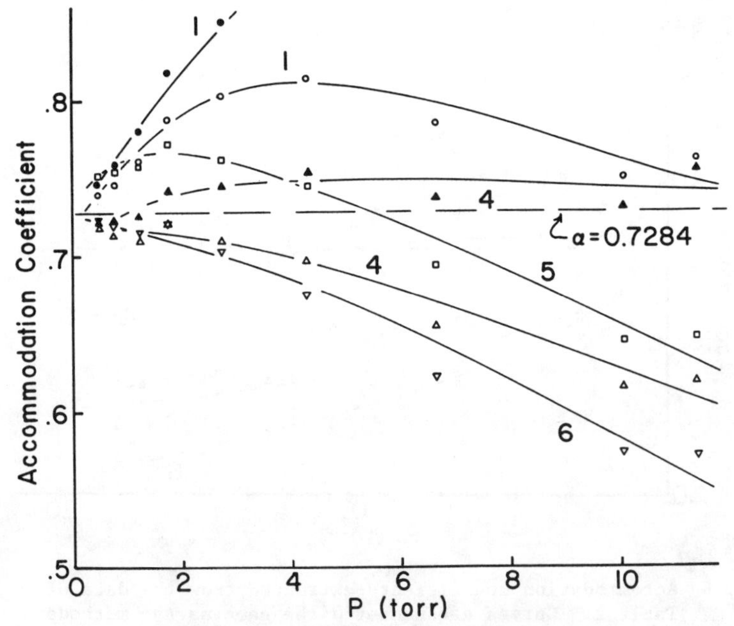

Fig. 4 Accommodation coefficients extracted from the data of Table 3. Curves are numbered the same as the methods listed in the text. Symbols are as in Fig. 3.

using K(308.15) = 994 shows the moments method expression yielding the more nearly constant value of α.

We can extract both α and K, regarding these as adjustable parameters, from the data of Tables 2 and 3, using the results of the various theoretical treatments for Q in Eq. (1) and optimizing the fits. This is particularly simple with the moments method since, like the jump regime expressions, it leads to linear regression. We have carried out such analyses for both the moments and the variational methods, and we will be reporting results of these in the near future.

Considering the importance of K in the analysis we note that it could be determined precisely if the present data could be corrected for convection in the high pressure region. Investigations in this area (modeling heat and mass transfer in the tube with jump and slip) would be helpful in resolving remaining uncertainties. Also, efforts will continue towards acquisition of expanded data on kyrpton and on xenon, and on improvement of basic theoretical description of the combination of processes involved in this investigation.

Acknowledgment

We wish to express appreciation for support of this work by the National Science Foundation Heat Transfer Program, the Senior Professor Research Fund of the Chemistry Department, and the Graduate School of the University of Missouri through its Weldon Springs and Research Council Grants.

References

[1] Thomas, L. B. and Golike, R. C., "A Comparative Study of Accommodation Coefficients by the Temperature Jump and Low Pressure Methods and Thermal Conductivities of He, Ne, and CO_2," Journal of Chemical Physics Vol. 22, Feb. 1954, pp. 300-305.

[2] Kennard, E. H., Kinetic Theory of Gases, McGraw-Hill, New York, 1938.

[3] Roach, D. V. and Thomas, L. B., "Comparative Study of Accommodation coefficients of Helium and Neon on Clean Tungsten under Transition, Temperature Jump, and Free Molecule Conditions," Journal of Chemical Physics, Vol. 59, Sept. 1973, pp. 3395-3402.

[4] Thomas, L. B., "Thermal Accommodation of Gases on Solids," Fundamentals of Gas-Surface Interactions, editors H. Saltsburg, J. N. Smith, and M. Rogers, Academic Press, New York, 1967, pp 346-369.

[5]Thomas, L. B., "Thermal Accommodation of Gases on Solid Surfaces--Description of Experimental Methods and of Some Recently Completed Investigations," International Symposium on Rarefied Gas Dynamics, Vol. 1, Plenum, New York, 1985.

[6]Thomas, L. B., Krueger, C. L., and Loyalka, S. K., "Heat Transfer in Rarefied Gases: Critical Assessments of Thermal Conductance and Accommodation of Argon in the Transition Regime," Physics of Fluids, Vol. 31, Oct. 1988, pp. 2854-2864.

[7]Thomas, L. B., Harris, R. E., and Krueger, C. L., "The Mercury Vapor Pumping Effect for ^3He, ^4He, H_2, D_2, Ne, Ar, Kr, Xe, N_2, O_2, CO_2, CH_4, C_2H_4, and C_2H_6: Its Precise Measurement and Conveient Application in McLeod Gauge Metrology," Proceedings of the Royal Society of London, Vol. A397, 1985, pp. 311-339.

[8]Brown, R. E., and Thomas, L., "The Assignment of the Heat Flow from Electrically Powered Filaments to Gas Conduction for Accurate Measurement of Thermal Accommodation," Rarefied Gas Dynamics, edited by C. Cercignani, D. Dini, and S. Nocilla, Editrice Technico, Scientifica, Pisa, Italy, p. 347.

[9]Powell, R. W., Ho, C. Y., and Liley, P. E., Thermal Conductivities of Selected Materials, National Bureau of Standards, Superintendent of Documents, U.S. Gov't. Printing Office, Wash. D.C., NSRDS-NBS-8,1966.

[10]Hirschfelder, J. O., Curtiss, C. F., and Bird, R. B., The Molecular Theory of Gases and Liquids, Wiley, New York, 1954, pp. 534 and 575.

[11]L. Lees and C. Y. Liu, "Kinetic Theory Description of Heat Transfer from a Fine Wire," Physics of Fluids, Vol. 5, 1962, p. 1137; see also F.C. Hurlbut, "Note on Conductive Heat Transfer from a Fine Wire," Physics of Fluids Vol. 7, 1964, p. 904.

[12]Cercignani, C. and Pagani, C. D. "Rarefied Flows in Presence of Fractionally Accommodating Walls," in Rarefied Gas Dynamics, edited by L. Trilling and H.Y. Wachman, Academic, New York, 1969, p. 269; see also Bassanini, P., Cercignani, C., and Pagani, C.D. "Influence of the Accommodation Coefficient on the Heat Transfer in a Rarefied Gas," International Journal of Heat and Mass Transfer, Vol. 11, 1967, p. 1359.

[13]H. Lang and S. K. Loyalka, "Application of the Integro-Differential Variational Principle to Poiseuille Flow and Heat Transfer in Cylindrical Geometries," Max Planck Institute fur Stromungsforschüng, Gottingen, FRG, 1970, Vol. 5/1972.

[14]Miklavcic, M. and Kuscer, I. "Pressure Corrections in Measurements of Accommodation Coefficients," International Journal of Heat and Mass Transfer, Vol. 23, 1980, p. 1297.

[15]Cipolla, J. W. and Cercignani, C. "Effect of Molecular Model and Boundary Conditions on Linearized Heat Transfer," Rarefied Gas Dynamics, edited by C. Cercignani, D. Dini, and S. Nocilla, Editrice Technico Scientifica, Pisa, Italy, Vol. II, p. 767.

Large Rotational Polarization Observed in a Knudsen Flow of H_2-Isotopes Between LiF Surfaces

L. J. F. Hermans* and R. Horne†
Leiden University, Leiden, The Netherlands

Abstract

Molecular angular momentum polarization produced in gas-surface collisions is investigated by the magnetic field effect on a Knudsen flow between parallel surfaces in the temperature range of 78-695 K. The effect for H_2 isotopes scattered by lithium fluoride LiF(001) crystal surfaces are found to be two to three orders of magnitude larger than previously observed for metal surfaces. This is attributed to the large corrugation of the LiF crystal face, since the polarization type involved is produced by in-plane forces. Both first and second rank polarizations ("orientation" and "alignment") are observed, the production matrix elements being of order 10^{-1} to 10^{-2}. The off-center rotation of HD is found to play only a minor role in the non-spherical interaction responsible for these polarizations. This phenomenon gives contributions on the order of 1% to the drag in ordinary Knudsen flow of polyatomic molecules through highly corrugated ducts.

I. Introduction

In molecule-surface scattering, changes in the molecular angular momentum $\underline{J}$ can be produced by nonspherical molecule-surface interaction. One aspect of this change, viz., the change in the magnitude of $\underline{J}$, has been given a great deal of attention in the form of studies into internal energy accommodation. Information has been obtained from transport properties and, more recently, from molecular beams. The other aspect, viz., changes in the direction of

---

* Graduate Student.

† Associate Professor.

$\underline{J}$, has become subject of a much newer field of research. These studies involve measurement of nonisotropic angular momentum distributions or rotational polarizations, which are produced in nonequilibrium situations. Such measurements can be a fruitful source of information on the molecule-surface interaction. Indeed, since angular momentum is a vector quantity, such data may yield more detailed information than do data on the scalar rotational energy distribution. Also here, one has two very different sources of information. First, one can probe the role of these polarizations in Knudsen transport phenomena by using a magnetic field, as was first demonstrated by Borman et al. (for a review see Ref. 2). Second, one can study polarizations in the scattered distribution of a molecular beam experiment, using some optical detection scheme like laser-induced fluorescence or multiphoton ionization.[3-5]

In this paper, polarizations are investigated by measuring the magnetic field effect on a Knudsen particle flux between parallel plates. Here the polarization is produced by the coupling through nonspherical interaction between tangential velocity and angular momentum. In an elementary classical picture, the role of the polarization in the particle flux is easily visualized by considering two successive collisions. At the first collision (the "polarizing" collision), part of the tangential motion is converted into angular momentum (cf. a tennis ball obliquely hitting a rough surface). At the next collision (the "analyzing" collision) the reverse process will take place: The outgoing velocity will be influenced by the incoming rotational state of the molecule. This will affect the flow of the gas as a whole. This contribution to the drag may be subtle and could be easily swamped by the drag associated with the spherical part. However, when a magnetic field is applied, precession of the angular momenta in between the two collisions will change or even completely randomize the polarization, and the resulting change in the drag will be exclusively caused by the polarization. Note that molecules that have been temporarily trapped, even if they would be polarized on desorption, will not contribute, since the outgoing trajectories can be assumed to carry no memory for the flow direction.

This technique is applicable to any molecular species, since each rotating molecule carries a magnetic moment $\underline{\mu} = g\mu_N \underline{J}/\hbar$, where μ_N is the nuclear magneton and g the molecular Landé factor (g = 0.88 for H_2, 0.66 for HD and 0.44 for D_2). The kinetic theory of this effect is well established for the case of classical rigid rotors.[6,7] So far, experiments were performed[2,8-11] for various gases on surfaces of Au, Pt, and mica, which were presumably dominated by adsorbates. The effects were found to be quite small, the relative change in the Knudsen flow being only on the order of 10^{-5} to 10^{-4}.

However, the production of the relevant polarizations depends on in-plane forces acting during the molecule-surface collisions. In fact, the responsible matrix elements[7] of the surface scattering operator can be shown to vanish if the parallel velocity is conserved. Such is the case for collisions with a smooth surface. In contrast, for particles colliding with a sinusoidal surface, the average change in the parallel momentum is found to be proportional to the square of the corrugation amplitude. One should therefore expect larger effects for surfaces having a large corrugation amplitude.

The (001) plane of LiF provides such a surface. From molecular beam studies[12] it is known to have a corrugation amplitude for thermal particles of about 0.3 Å in the [100] direction. In addition, the trapping probability is small, the attractive potential well being only about 40 meV for hydrogen isotopes.[13] Therefore, we have performed such experiments[14] for LiF(001) surfaces, the flow being in the [100] direction, which has the largest corrugation. Data were taken for the hydrogen isotopes H_2, D_2, and HD. These molecules represent an especially interesting case; they are chemically identical, and the spherical parts of the interaction are very much alike. Yet, the differences in rotational level splitting and the rotation around an off-center axis in the case of HD make them rather different in the context of the present experiment.

II. Theory

Following Ref. 7, we consider a Knudsen flow Φ of classical rigid rotors between two parallel plates separated by a gap b. The flow will be in the x direction, and the surface normal will be in the z direction.

The dependence of the Knudsen flow on field strength B is governed by the average precession angle $\omega\tau$, where $\tau = b[m/(2kT)]^{1/2}$ is an average time of flight between the plates and $\omega = (g\mu_N B)/\hbar$ is the precession frequency for linear molecules (with g the rotational Landé factor and μ_N the nuclear magneton). The precession frequency is independent of J and is thus uniform throughout the gas. Consequently, a magnetic field will rotate a polarization as a whole. However, the flight time is nonuniform due to the Maxwellian distribution of velocities, changing the polarization by a factor $1-\cos[(k\omega\tau)/v_z)]$ weighted with a Maxwellian flux, where v_z represents the dimensionless z velocity of a molecule. The k denotes single (k = 1) or double (k = 2) frequency behavior, depending on whether the polarization involved has a C_1 or a C_2 symmetry in the plane of rotation. The resulting field induced change in the flow shows a damped oscillatory behavior, saturating for high fields:

$$\frac{\delta\Phi(B)}{\Phi} \propto F(k\omega\tau) = 2 \int_0^\infty v_z e^{-v_z^2} \left(1 - \cos \frac{k\omega\tau}{v_z}\right) dv_z \qquad (1)$$

with k = 1,2 (see Fig. 1).

The magnitude of the field effect in saturation is determined by squares of matrix elements. These matrix elements, denoted by c_{vj} in the language of Ref. 7, represent the production of polarization from tangential velocity v_x and vice versa. The matrix elements associated with the four simplest polarizations produced in a Knudsen flow are listed in Table 1 for linear molecules. Note that the polarizations are actually x components of true vectors in the combined velocity-angular momentum space. This is required by the Curie principle, since the driving thermodynamic force is a true vector in the x direction, whereas the surface scattering operator $\hat{P}$ is a scalar. Note also that, for a matrix element to be nonzero, the tangential velocity v_x must change in the scattering process. Consequently, the action of in-plane forces during the collision (e.g., due to surface corrugation) is essential.

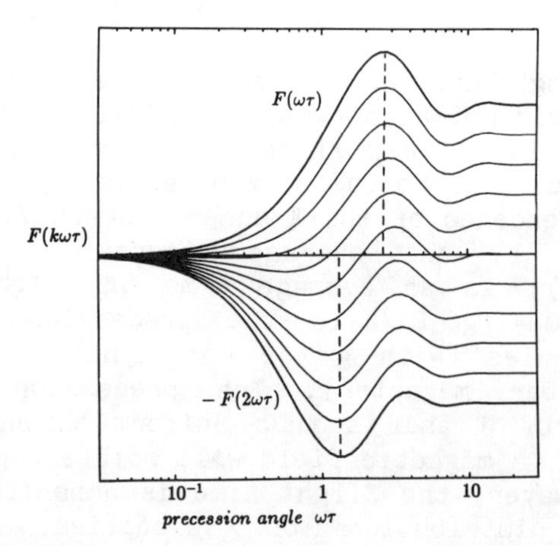

Fig. 1 Theoretical curves of $F(\omega\tau)$ and $-F(2\omega\tau)$ [see Eq. (1)], and linear combinations of the two at 10% intervals.

Table 1 Four lowest rank polarizations produced in a Knudsen flow (see Ref. 7)

Type of polarization	Matrix elements	Contribution to field effect (πb_0 omitted)	
		B_x	B_y
$(\underline{v} \times \underline{n})_x(\underline{J} \cdot \underline{n}) = v_y J_x$	$c_{v1} = -2\sqrt{3}\langle v_y J_x \vert \hat{P} v_x \rangle$	$+c_{v1}^2 F(\omega\tau)$	$+c_{v1}^2 F(\omega\tau)$
$(\underline{J} \times \underline{n})_x = J_y$	$c_{v2} = \sqrt{6}\langle J_y \vert \hat{P} v_x \rangle$	$+c_{v2}^2 F(\omega\tau)$	0
$v_x\{3(\underline{J}.\underline{n})^2 - J^2)\} = v_x(3J_x^2 - J^2)$	$c_{v3} = -\sqrt{\frac{5}{2}}\langle v_x(3J_x^2 - J^2) \vert \hat{P} v_x \rangle$	$-\frac{3}{4}c_{v3}^2 F(2\omega\tau)$	$-\frac{3}{4}c_{v3}^2 F(2\omega\tau)$
$J_x(\underline{J} \cdot \underline{n}) = J_x J_x$	$c_{v4} = \sqrt{15}\langle J_x J_x \vert \hat{P} v_x \rangle$	$-c_{cv4}^2 F(\omega\tau)$	$-c_{v4}^2 F(2\omega\tau)$

See Fig. 2 for the coordinate system. Note that the polarizations are actually x components of true vectors[7], e.g., $J_y = (\underline{J} \times \underline{n})_x$, with $\underline{n}$ the surface normal. Compare table with Eq. (2).

The sign of the effect in saturation is determined by the tensorial rank of the polarization.[7] For a system invariant under space inversion, as is the case in the present experiment, it follows that polarizations odd in J (i.e., orientations) give an increase in the particle flow (+ sign) when the field is switched on, whereas polarizations even in J (i.e., alignments) make the flow decrease (- sign).

The dependence of the field effect in saturation on field orientation is determined by the spatial structure of the polarizations.[15]

These features allow one to disentangle the four separate contributions from the experimental data. In the present experiments, this was achieved by taking data as a function of field strength ($\propto \omega\tau$) for the field along x axis and along y axis. As can be derived from Ref. 7, the change in the flow by a field in the x-y plane can be written as follows:

$$\frac{\delta\Phi(\underline{B})}{\Phi} = \pi b_0 \Big[c_{v1}^2 F(\omega\tau) + c_{v2}^2 \cos^2\phi F(\omega\tau) - \frac{3}{4} c_{v3}^2 F(2\omega\tau) - c_{v4}^2 \{\cos^2\phi F(\omega\tau) + \sin^2\phi F(2\omega\tau)\} \Big] \quad (2)$$

with $b_0 = (\ln Kn + 0.45)^{-1}$ and $Kn \equiv \bar{\ell}/b$ with $\bar{\ell}$ the mean free path. The result for $\phi = 0$ ($\underline{B}$ // x axis) and $\phi = \pi/2$ ($\underline{B}$ // y axis) is given in Table 1. By making use of the single-frequency contribution $\propto F(\omega\tau)$ and the double-frequency contribution $\propto F(2\omega\tau)$ for either field orientation, four independent combinations of the c^2_{Vj} are obtained from which the separate values can then be determined.

III. Experimental Setup

The experiments were performed in a setup similar to that described in Ref. 8. They consist of observing the field-induced change in the flow resistance of a flat channel which is part of a Wheatstone bridge flow circuit. The channel (length 20 mm, width 10 mm, gap 0.17 mm) was constructed from two LiF(001) faces with two 0.17-mm Pt wires serving as spacers. It was assembled in a dry N_2 atmosphere at about 900 K with mixtures of LiF and PbF_2 serving as sealing agent. Details of the construction will be given elsewhere.[16] The LiF channel thus constructed proved vacuum tight at temperatures between 78 and 695 K.

In the experiment, a gas flow at pressures around 1 Pa or 10^{-2} torr was established through the channel, yielding Knudsen numbers (ratio of mean free path to gap thickness) of around 50. Magnetic fields up to B = 2.2 T were applied. This produces precession angles $\omega\tau$ on the order of 10. The field could be oriented parallel to the flow, $\underline{B} = (B_x, 0, 0)$, and perpendicular to both flow and surface normal, $\underline{B} = (0, B_y, 0)$ (cf. the coordinate system in Fig. 2).

In the temperature-dependent studies, care was taken to avoid spurious effects arising from the field effect on thermal creep. This was achieved by shifting the part of the flow circuit containing the temperature gradient to a position well outside the field. Pressure values needed for calibration purposes were corrected for thermal creep by making use of a numerical relation derived by Loyalka and Storvick[17] for intermediate Knudsen numbers.

IV. Results and Discussion

The resulting changes in the flow at constant pressure drop, $\delta\Phi/\Phi$, are given as a function of precession angle $\omega\tau \propto B$ in Figs. 2-4 for H_2, D_2, and HD at temperatures of T = 293 and 695 K. These data are compared with the theoretical behavior given in Eq. (1) by fitting to the data linear combinations of $F(\omega\tau)$ and $-F(2\omega\tau)$ as shown in Fig. 1. The weight of the two contributions was chosen such as to yield the best fit. Note that the position of the curves along the $\omega\tau$ axis is fixed by the known values of B, g, and b, and does not contain any adaptable parameter. The data are found to be very well described by the theoretical curves, especially the B_y data, which show an almost pure

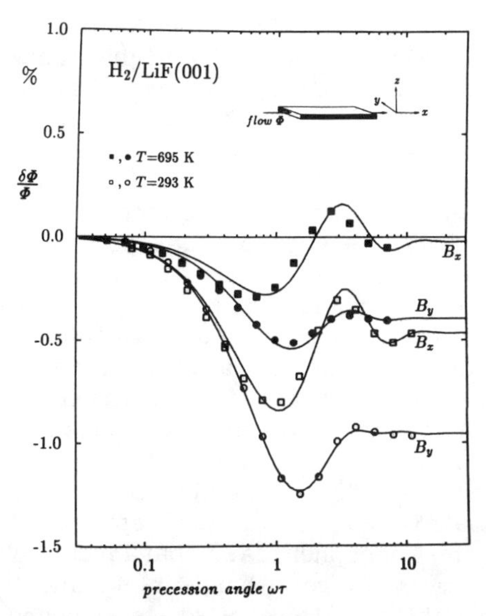

Fig. 2 Observed field-induced change in the particle flux $\delta\Phi/\Phi$ as a function of precession angle $\omega\tau$ for H_2 at 293 K (solid symbols) and 695 K (filled-in symbols) with the field along x axis (squares) or y axis (circles). The solid lines represent superpositions of the field functions $F(\omega\tau)$ and $-F(2\omega\tau)$ as given in Fig. 1, with weight adjusted as to yield the best fit. Note that $\omega\tau$ is fixed by the known values of field strength (up to 2.2 T), rotational Landé factor (g = 0.883), and gap width (0.17 mm). The data were taken at Kn = 47 and 112 at 293 and 695 K, respectively.

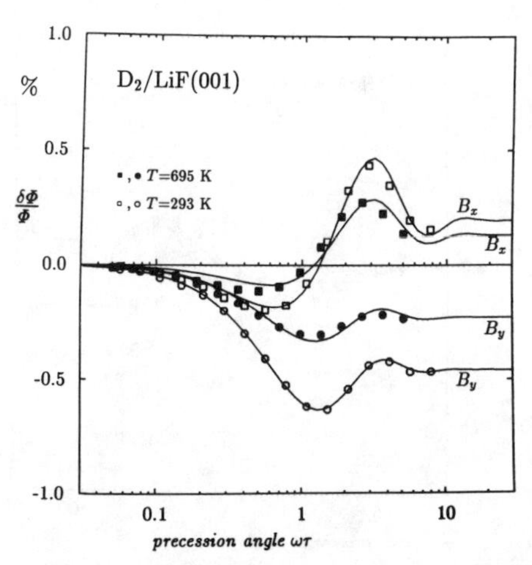

Fig. 3 As in Fig. 1, for D_2 with Kn = 51 and 112 at T = 293 and 695 K, respectively, and with g = 0.443.

double frequency behavior. For some B_x data we note a slight discrepancy that may be caused by higher rank polarizations, which can produce a triple-frequency behavior $\propto F(3\omega\tau)$.

The magnitude of the effects is found to be two to three orders of magnitude larger than that observed earlier for mica and Au at room temperature.[8-11] The relative change in the flow, $\delta\Phi/\Phi$, is now on the order of 1%, which means that the matrix elements describing production of polarization are on the order 10^{-1}. These large effects provide a rather dramatic illustration of the crucial role played by surface corrugation in the production of rotational polarization from tangential velocity.

A second striking feature seen in Figs. 2-4 is that HD shows effects that are smaller than those for H_2 and D_2 at 293 K. This is somewhat unexpected. In HD, the center of mass is shifted away from the H-D bond center, making it a strongly asymmetric rotor ("loaded sphere" character). As a consequence, the nonspherical character of HD should be expected to be much larger than for H_2 and D_2. In addition, it has the selection rule $\Delta J = \pm 1$ as opposed to $\Delta J = \pm 2$ for the homonuclears H_2 and D_2, giving it a smaller effective rotational level splitting. Indeed, rotational excitation for HD both in the gas phase[18] and in collisions with metal surfaces[19] as well as insulating surfaces[20] is known to be much stronger than for the homonuclears. Therefore, the small effects for HD compared to H_2 and D_2 indicate that collisions with $\Delta J = 0$ and $\Delta m_J \neq 0$ give the major contribution to the production of polarization; here m_J is the z projection of the rotational quantum number.

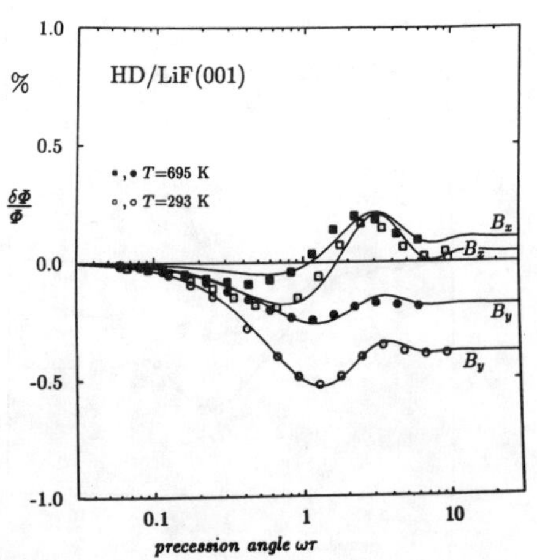

Fig. 4 As in Fig. 1, for HD with Kn = 59 and 143 at T = 293 and 695 K, respectively, and with g = 0.663.

Table 2 Values in units 10^{-2} for squares of the matrix elements (cf. Table 1) for H_2, D_2, and HD at 293 and 695 K, as derived from the present experiments

	Orientation type polarization				Alignment type polarization			
	c_{v1}^2		c_{v2}^2		c_{v3}^2		c_{v4}^2	
	293 K	695 K	293 K	695 K	293 K	695 K	293 K	695 K
H_2	≈ 0	≈ 0	0.67	0.61	1.75	1.10	≈ 0	0.13
D_2	≈ 0	0.10	0.91	0.59	1.00	0.64	≈ 0	≈0
HD	≈ 0	0.10	0.61	0.49	0.74	0.45	≈ 0	≈0

Error margins are up to ± 0.20 (see text)

For a more meaningful comparison between the three H_2 isotopes and their temperature dependence, the data have been decomposed into contributions from the four polarizations given in Table 1. The results are given in Table 2. This analysis shows that, for the gases considered here, only two polarizations contribute significantly, viz., the orientation type J_y and the alignment type $v_x(3J_z^2 - J^2)$. However, also polarizations other than the four considered here make small contributions, making the error margins in Table 2 relatively large in some cases. Even so, the data indicate clearly that the contribution from the alignment decreases more rapidly with increasing temperature than the contribution from the orientation. This suggests that an increasing fraction of rotationally inelastic collisions, having $\Delta J \neq 0$, affects the production of alignment more than that of orientation. This picture seems to be corroborated by the dominance of the alignment for H_2 at 293 K, where the rotationally inelastic events are suppressed by the large effective level splitting. Another factor playing a role in the behavior of H_2 is that, at room temperature, H_2 has still an appreciable fraction of elastic collisions, as can be seen from the relatively large value of the Debye-Waller factor. This is much less the case for the heavier isotopes, since the exponent in the Debye-Waller factor is proportional to the mass of the incident molecule. It is recalled that collisions in which the velocity is randomized do not contribute to the effect studied here. Such behavior is favored by inelastic collisions, in which phonon exchange with the surface takes place. The fact that the alignment type $v_x(3J_z^2 - J^2)$ seems to suffer more from the velocity randomization than the orientation J_y may be related to the extra velocity involved in that type of polarization.

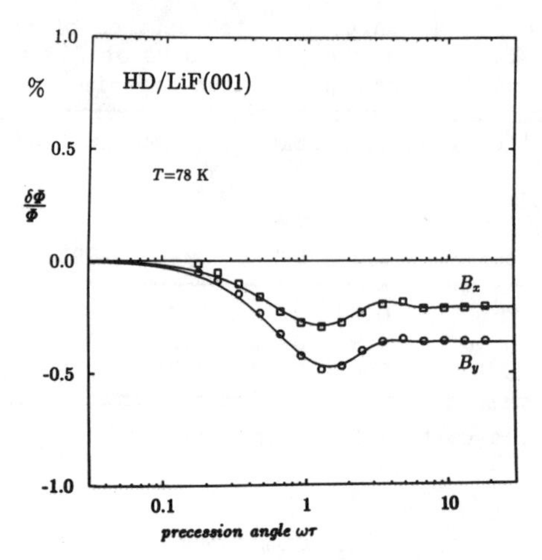

Fig. 5 As in Fig. 1, for HD at 78 K with Kn = 17 and g = 0.663.

To examine the possible role of adsorbates in our system, data were taken for HD at temperatures in the range of 78-695 K. The results at the lowest temperature of 78 K are shown in Fig. 5. Whereas the results at 293 K and 695 K do not differ substantially (see Fig. 4), the results at 78 K are clearly different, being dominated by the production of alignment. Detailed analysis shows that the alignment is predominantly of type $v_z(3J_z^2 - J^2)$ with $c_{v3}^2 = 0.32 \times 10^{-2}$, and for a smaller part of type J_xJ_z with $c_{v4}^2 = 0.07 \times 10^{-2}$. A significant contribution is still given by the orientation of type J_y with $c_{y2}^2 = 0.16 \times 10^{-2}$. The temperature dependence of the field effect in saturation is shown in Fig. 6. The gradual change in effect between 250 and 700 K for both field orientations confirms that in this temperature region no major changes in surface conditions occur. This is in sharp contrast to the behavior of the mica surface, where a steep change in effect was observed around 450 K.[21]

Given the large polarizations found in these experiments, the question arises whether this technique may provide a means to produce molecular beams having a reasonable degree of polarization. The actual degree of alignment and orientation in a given flow can be deduced from the corresponding matrix element responsible for its production by multiplication of the matrix elements by flow velocity $\bar{v}_x$ divided by average thermal velocity $\bar{v}$, e.g., $\langle J_y \rangle_{n.e.} \approx 4/\sqrt{\pi} \langle J_y \mid \hat{P}\, v_x \rangle \bar{v}_x/\bar{v}$. Here, the nonequilibrium average $\langle J_y \rangle_{n.e.}$ is taken over the half space of molecules leaving a surface. Note that for the

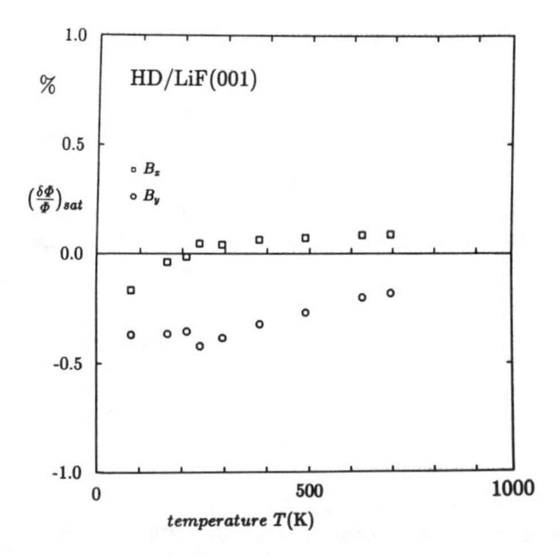

Fig. 6 Saturation value for the observed field induced change in the flow, $(\delta\Phi/\Phi)_{sat}$, for HD as a function of temperature for the field along x axis (squares) and y axis (circles).

equilibrium average one has $\langle J_y \rangle = 0$; in the limit of perfect orientation of all angular momenta in the +y direction one finds, using the normalization of Ref. 7, $\langle J_y \rangle = \sqrt{\pi}/2$ if the distribution over rotational energy is assumed to remain Maxwellian. Since in a nozzle expansion, $\bar{v}_x/\bar{v}$ may be of order unity, one must conclude that polarizations on the order of 10% may be produced.

Several conclusions can be drawn from these results. First, they demonstrate that relatively large polarizations, which depend entirely on the existence of in-plane forces, can be produced for diatomic molecules at highly corrugated surfaces like LiF(001). This includes the production of both alignment and orientation. Second, they show that the off-center rotation of HD plays a much smaller role in the nonspherical interaction responsible for these polarizations than one might expect. Third, they indicate that the major contribution to the production of polarization is given by rotationally elastic, reorienting collisions having $\Delta J = 0$, $\Delta m_J \neq 0$. Next, they show the need for detailed quantum mechanical molecular dynamics calculations of light diatoms colliding with a corrugated surface. In addition, they show that in the kinetic description of ordinary Knudsen flow for polyatomic molecules, drag contributions arising from polarization may not be negligible if the surface of the duct is highly corrugated. Finally, these experiments suggest that it may be feasible to produce molecular beams having an appreciable angular momentum polarization by using appropriate nozzles.

Acknowledgments

The authors are indebted to L. van As, E. de Kuyper, and R.M. van Leeuwen for technical assistance; and to B.J. Mulder, A.C. Levi, I. Kuščer, J.J.M. Beenakker, B.E. Nieuwenhuys, J.F. van der Veen, and H. Batelaan for most helpful discussions. This work is part of the research program of the Stichting voor Fundamenteel Onderzoek der Materie (FOM), which is financially supported by the Nederlandse Organisatie voor Wetenschappelijk Onderzoek (NWO).

References

[1]Borman, V.D., Laz'ko, V.S., and Nikolaev, B.I., "Effect of a Magnetic Field on Heat Transport in Tenuous Molecular gases," Soviet Physics-JETP, Vol. 39, 1974, pp. 657-660.

[2]Hermans, L.J.F., "Non-spherical Molecule-surface Interaction and Transport Phenomena in the Knudsen Regime," Proceedings of the 14th International Symposium on Rarefied Gas Dynamics, Vol. 1, Univ. of Tokyo, Japan, 1984, pp. 333-348.

[3]Kleyn, A.W., Luntz, A.C., and Auerbach, D.J., "Rotational Polarization in NO Scattered from Ag(111)," Surface Science, Vol. 152, 1985, pp. 99-105.

[4]Sitz, G.O., Kummel, A.C., and Zare, R.N., "Population and Alignment of N_2 Scattered from Ag(111)," Journal of Vacuum Science and Technology A, Vol. 5, 1987, pp. 513-517.

[5]Sitz, G.O., Kummel, A.C., and Zare, R.N., "Alignment and Orientation of N_2 Scattered from Ag(111)," Journal of Chemical Physics, Vol. 87, 1987, pp. 3247-3250.

[6]Borman, V.D., Krylov, S.Y., Nikolayev, B.I., Ryabov, V.A., Troyan, V.I., and Frolov, B.A.," The Effect of a Magnetic Field on the Knudsen Gas Flow," Physics Letters, Vol. 79A, 1980, pp. 315-317.

[7]Knaap, H.F.P. and Kuščer, I., "Transport Phenomena in Molecular Knudsen Gas," Physica, Vol. 104A, 1980, pp. 95-114.

[8]Van der Tol, J.J.G.M., Krylov, S.Y., Hermans, L.J.F., and Beenakker, J.J.M., "Experimental Determination of Angular Momentum Polarization Produced in a Knudsen Gas Flow," Physics Letters, Vol. 99A, 1983, pp. 51-53; Van der Tol, J.J.G.M., Hermans, L.J.F., Krylov, S.Y., and Beenakker, J.J.M., "Molecular Angular Momentum Polarization Produced in a Knudsen Flow," Physica, Vol. 131A, 1985, pp. 545-569.

[9]Van der Tol, J.J.G.M., Hermans, L.J.F., and Beenakker, J.J.M., "Angular Momentum Polarization in a Knudsen Flow; Results for Various Gases on a Au-Surface," Physica, Vol. 134A, 1985, pp. 216-230.

[10]Van der Tol, J.J.G.M., Hermans, L.J.F., and Beenakker, J.J.M., "Angular Momentum Polarization in a Knudsen Flow; Results for Different Surfaces," Physica, Vol. 139A, 1986, pp. 28-40.

[11]Hermans, L.J.F., Van der Tol, J.J.G.M., and Beenakker, J.J.M., "Angular Momentum Polarization Produced by Molecule-surface Collisions in a Knudsen Flow," Journal of Chemical Physics, Vol. 84, 1986, pp. 1029-1032.

[12]Boato, G., Cantini, P., and Mattera, L., "A Study of the (001)LiF Surface at 80 K by means of Diffractive Scattering of He and Ne Atoms at Thermal Energies," Surface Science, Vol. 55, 1976, pp. 141-178.

[13]Boato, G., Cantini, P., and Mattera, L., "Elastic and Rotationally Inelastic Diffraction of Hydrogen Molecular Beams from the (001) Face of LiF at 80 K," Journal of Chemical Physics, Vol. 65, 1976, pp. 544-549.

[14]Horne, R. and Hermans, L.J.F., "Large Rotational Polarization Observed for H_2, D_2 and HD Scattered from LiF(001)," Physical Review Letters, Vol. 60, 1988, pp. 2777-2780.

[15]Horne, R., Hermans, L.J.F., Krylov, S.Y., and Kuščer, I., "Orientational Dependence of Magnetic Field Effects in a Molecular Knudsen Gas", Physica, Vol. 151A, 1988, pp. 341-348.

[16]Horne, R., Van As, L., De Kuyper, E., Hermans, L.J.F., and Mulder, B.J., "Temperature Resistant and Vacuum Tight LiF Connections," Journal of Physics E, to be published.

[17]Loyalka, S.K. and Storvick, T.S., "Kinetic Theory of Thermal Transpiration and Mechanocaloric Effect. III Flow of a Polyatomic Gas Between Parallel Plates," Journal of Chemical Physics, Vol. 71, 1979, pp. 339-350.

[18]Hermans, P.W., Hermans, L.J.F., and Beenakker, J.J.M., "A Survey of Experimental Data Related to the Nonspherical Interaction of the Hydrogen Isotopes and Their Mixtures with Noble Gases", Physica, Vol. 122A, 1983, pp. 173-211.

[19]Cowin, J.P., Yu, C.F., Sibener, S.J., and Wharton, L., "HD Scattering from Pt(111): Rotational Excitation Probabilities," Journal of Chemical Physics, Vol. 79, 1983, pp. 3537-3549.

[20]Rowe, R.G. and Ehrlich, G., "Rotationally Inelastic Diffraction of Molecular Beams: H_2, D_2, HD from (001) MgO," Journal of Chemical Physics, Vol. 63, 1975, pp. 4648-4665.

[21]Horne, R., "On Rotational Polarization in Nonequilibrium Knudsen Gases," Ph.D. Dissertation Leiden University, Leiden, The Netherlands, 1988; Horne, R. and Hermans, "Rotational Polarization in a Knudsen Flow; Various Gases on Mica Between 293 and 666 K, Physica, accepted for publication.

Internal State-Dependent Molecule-Surface Interaction Investigated by Surface Light-Induced Drift

R. W. M. Hoogeveen,* R. J. C. Spreeuw,† G. J. van der Mee,†
and L. J. F. Hermans‡
Leiden University, Leiden, The Netherlands

Abstract

Experimental results are reported on surface light-induced drift in a one-component gas. This effect originates from velocity-selective excitation followed by state-dependent molecule-surface interaction. The experiments are performed on gaseous CH_3F, rovibrationally excited by a narrow-band CO_2 laser tuned within the Doppler-broadened absorption profile. The accommodation coefficient for parallel momentum α is found to depend much more strongly on the rotational than on the ν_3 vibrational state, the change in α on excitation being in the range of 10^{-3} to 10^{-2}. The experimental results suggest that α increases stronger than linearly with the rotational quantum number. The orientation of the angular momentum with respect to the figure axis is found to be of minor importance in this process. At higher pressures, where molecule-molecule collisions dominate, light-induced drift of a different origin is observed. This effect is caused by a state-dependent collision rate in combination with nonuniform light intensity. The results are in good agreement with the theoretical description. The change in collision rate is found to be primarily caused by the vibrational excitation, depending only weakly on the rotational sublevels involved.

I. Surface Light-Induced Drift

Velocity-selective excitation in a gas can be achieved by tuning a narrow-band laser within the Doppler-broadened

---

* Graduate Student.

† Student.

‡ Associate Professor.

absorption profile of some optical transition. This can give rise to a deformation of the velocity distribution and thus to a new class of kinetic effects, if there is a difference in interaction between excited and ground state particles with respect to their collision partner. In the case that the collision partner is a buffer-gas particle, a state-dependent collision rate gives rise to a drift of the optically absorbing particles through the buffer gas. This phenomenon, predicted in 1979 by Gel'mukhanov and Shalagin,[1] has been demonstrated in both atomic and molecular systems (see Ref. 2 and the references therein). It can take spectacular forms, as shown by Werij et al. (optical piston,[3] optical machine gun[4]). However, in the case that the collision partner is a solid surface, state-dependent molecule-surface interaction can also produce a drift of the molecules along the wall. This surface light-induced drift was predicted by Ghiner et al.[5] We have recently observed this effect by detecting the resulting pressure difference in a closed tube.[6]

The principle of the experiment is shown in Fig. 1. A Knudsen gas in a capillary is velocity selectively excited by a laser beam propagating along the capillary axis (x axis). Molecules are excited in a narrow velocity class around $v_{xL} = \Delta\omega/k$, where $\Delta\omega = 2\pi(\nu_L - \nu_0)$ is the detuning of the laser with respect to the absorption line center and k the magnitude of the wave vector. If the accommodation coefficient for transfer of parallel momentum is state-dependent, a net momentum transfer to the wall with result. In an open tube, this will give rise to a drift velocity

$$v_d = -\frac{n_e}{n}\left(\frac{\alpha_e - \alpha_g}{\alpha}\right) v_{xL} \quad (1)$$

with n_e the number density of excited molecules in the selected velocity class around v_{xL}, n the total density, and $\alpha_{e,g}$ the accommodation coefficients for parallel momentum transfer of excited and ground state particles,

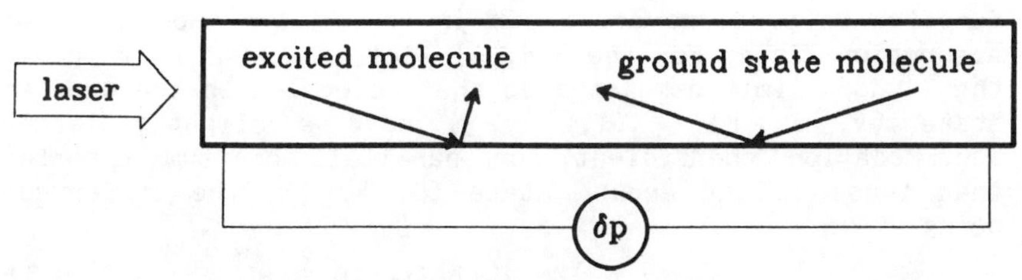

Fig. 1 Principle of surface light-induced drift experiments.

respectively. In a closed tube, this drift will produce a pressure difference $\delta p = L\nabla_x p = p_{out} - p_{in}$. In the stationary state, one finds for δp in the Knudsen limit

$$\frac{\delta p}{p} \approx \frac{3}{2}\frac{L}{R}\frac{v_d}{\bar{v}}\alpha \approx -\frac{3}{2}\frac{L}{R}\frac{n_e}{n}\frac{v_{xL}}{\bar{v}}(\alpha_e - \alpha_g) \qquad (2)$$

where $\bar{v} = (8kT/\pi m)^{1/2}$ is the mean thermal speed and L/R is the ratio of length to radius of the capillary. Note that the sensitivity to detect small values of v_d (and thus small differences in α) is greatly enhanced by the apparatus geometry L/R, which has the value 400 in this experiment.

The observations were made for CH_3F, rovibrationally excited by a tunable CO_2 laser having a grating for line selection and a piezocrystal for fine tuning.[7] By using different coincidences between the available laser lines and molecular transitions in $^{12}CH_3F$ or $^{13}CH_3F$, it is possible to select different rotational transitions with $\Delta J = -1, 0, +1$ (P, Q, R lines, respectively) within the vibrational $v = 0 \rightarrow 1$ absorption band (ν_3 mode or C-F stretch). Results have been obtained for various lines and different surfaces. An example is shown in Fig. 2,[8,9] where the pressure difference δp between the ends of the tube is recorded while the laser is scanned through the absorption profile. Note that δp is odd in the detuning $\Delta\nu$, in accordance with Eq. (2), with a slight asymmetry caused by the neighboring absorption line R(4, 2) and the (small) effect of light pressure.

For a quantitative analysis of the observed pressure difference, the maximum of the effect at positive detuning, $\delta p_m/p$ (cf. Fig. 2b), is normalized by the excited fraction n_e/n. To this end, n_e is determined from the number of photons absorbed in the gas per unit volume, which equals $\Delta P/h\nu V$, with ΔP the absorbed power, $h\nu$ the photon energy, and V the cell volume. The excited state density n_e is then found by multiplying this number by the kinetic collision time τ_c for which the Knudsen limit value $\tau = 2R/\bar{v}$ is used. (Note that the radiative life time is long, $\approx$ 1s.)

Values of $(\delta p_m/p)(n_e/n)^{-1}$ for a quartz capillary as a function of pressure are shown in Fig. 3 for two different absorption lines. For the R(4, 3) line, the nonzero value in the Knudsen limit demonstrates that molecules in the excited state (v, J, K) = (1, 5, 3) have a slightly larger accommodation coefficient for parallel momentum transfer than those in the ground state (0, 4, 3), the difference being

$$(\alpha_e - \alpha_g)/\alpha = +1.3 \times 10^{-3} \qquad (3)$$

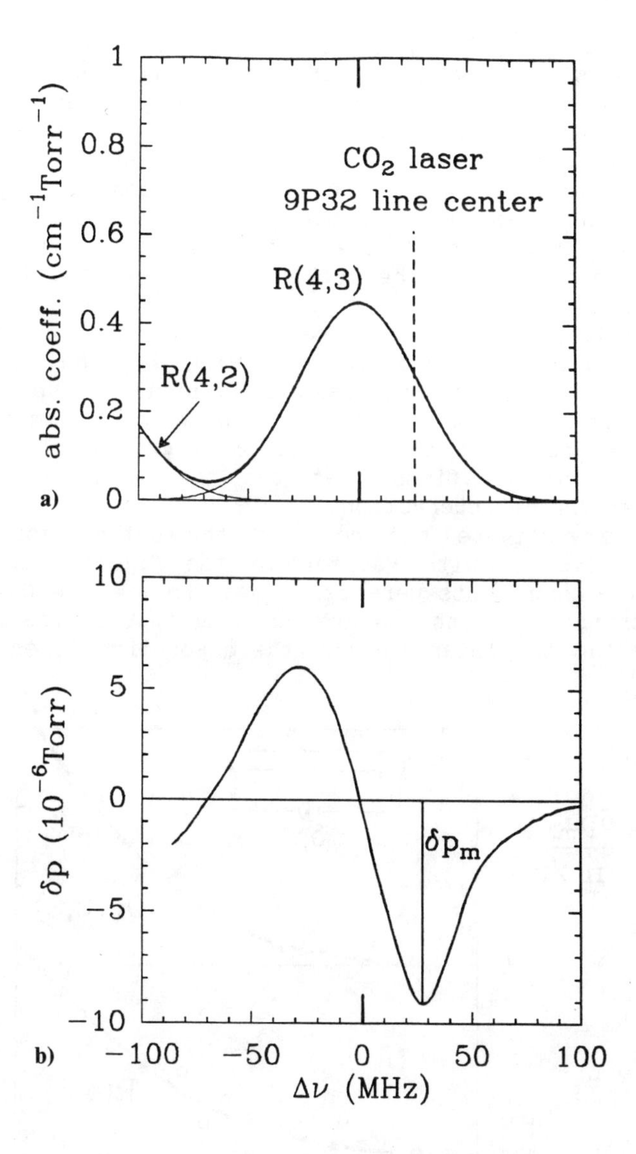

Fig. 2 a) Doppler-broadened absorption profile of $^{13}CH_3F$ near the 9P32 laser line. This spectrum was computed from the data given in Ref. 8; the position of the 9P32 line was taken from Ref. 9. At the left, one notices the tail of the R(4, 2) line, centered at -131 MHz.

b) Typical recorder trace of the observed pressure difference vs laser detuning $\Delta\nu \propto v_{xL}$, relative to the center of the absorption line R(4,3). Pressure p = 0.030 Torr, laser intensity I = 5 W/cm^2. The data are for a quartz capillary with radius 0.75 mm.

for a quartz capillary (K is the projection of J on the figure axis). In contrast, this difference is found to be very much smaller for the Q(12, 2) transition, for which only the vibrational quantum number is changed while the rotational state J = 12, K = 2 remains unaffected. These results indicate that the accommodation for parallel momentum transfer is much more strongly dependent on the rotational than on the vibrational state. This is confirmed by additional data on the Q(12, 3) and P(24, 13) lines (cf. Fig. 7).

Similar experiments were performed for a stainless steel and for a Teflon-coated cell with the same dimensions. The results were found to essentially similar, the magnitude in the Knudsen limit being a factor of 2 smaller for stainless steel and a factor of 2 larger for Teflon (cf. Fig. 4.) This confirms that the phenomenon arises from molecule-surface interaction.

To investigate the role of the orientation of the angular momentum with respect to the figure axis of the molecule, experiments were performed for the v = 0 → 1, J = 4 → 5 transition with K = 0-4 unchanged. This was achieved by scanning the laser through the absorption lines R(4, K)

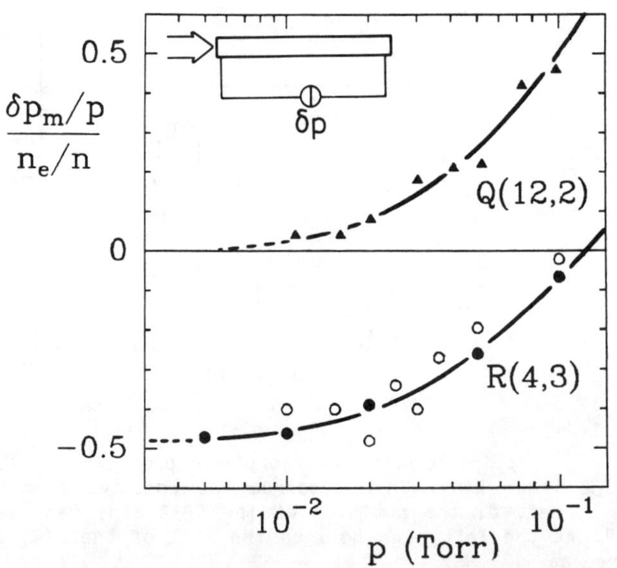

Fig. 3 Values of the maximum observed pressure difference for positive detuning [cf. Fig. 2b], normalized by the excited fraction, as a function of pressure, for the transitions R(4,3) and Q(12,2). The data are for a quartz capillary (L = 300 mm, R = 0.75 mm). Laser intensity I = 110 W/cm^2 (triangles), I = 60 W/cm^2 (filled circles), and I = 5 W/cm^2 (open circles).

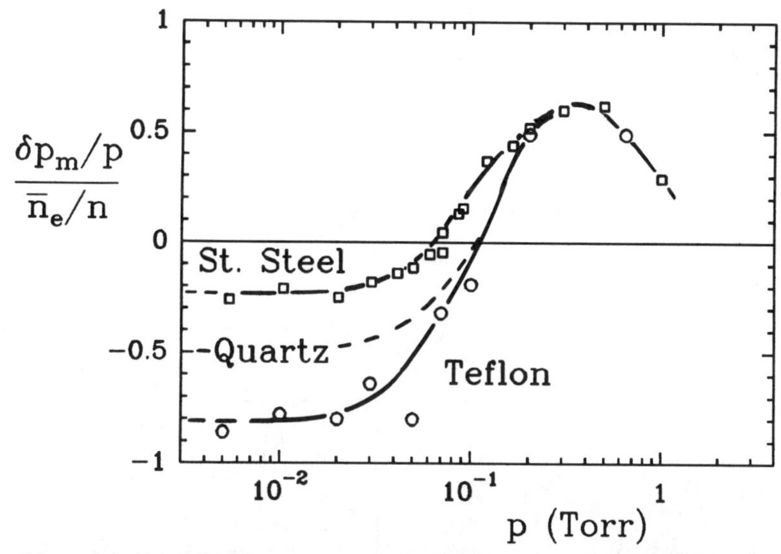

Fig. 4 Maximum of the observed pressure difference, normalized by the excited state fraction, as a function of pressure. The data are for the R(4, 3) transition in a stainless steel (squares) and a Teflon capillary (circles). The results for quartz (Fig. 3) are given for comparison. For the high-pressure part of the data, see Sec. II.

(see Fig. 5a). To cover the entire range spanned by these absorption lines, the center of the laser line was up- or down-shifted by 180 MHz using an Acousto-Optic Modulator, before tuning the laser over its 260-MHz scanning range by a piezocrystal. Although the single R(4, K) lines are overlapping for K = 0, 1, 2, these data do give information on the role of K/J in the accommodation coefficient α. If α is independent of K/J, the observed pressure difference should follow the derivative of the envelope of the absorption profile.[10] (This is a consequence of the so-called Bakarev-Folin theorem, which would also apply in this case; see Ref. 10.) This theoretical behavior is shown in Fig. 5b for a ratio of homogeneous linewidth to Doppler width of 0.2. The experimental observations for absorption and pressure difference are shown in Figs. 5c and 5d. It is seen that the experimental results agree very well with the theoretical behavior displayed in Figs. 5a and 5b. Obviously, the magnitude of the effect is not proportional to K^2. This seems to exclude the possible explanation of the surface light-induced drift effect put forward by Chapovsky.[11] In this picture, the change in accommodation coefficient upon excitation by linearly polarized laser

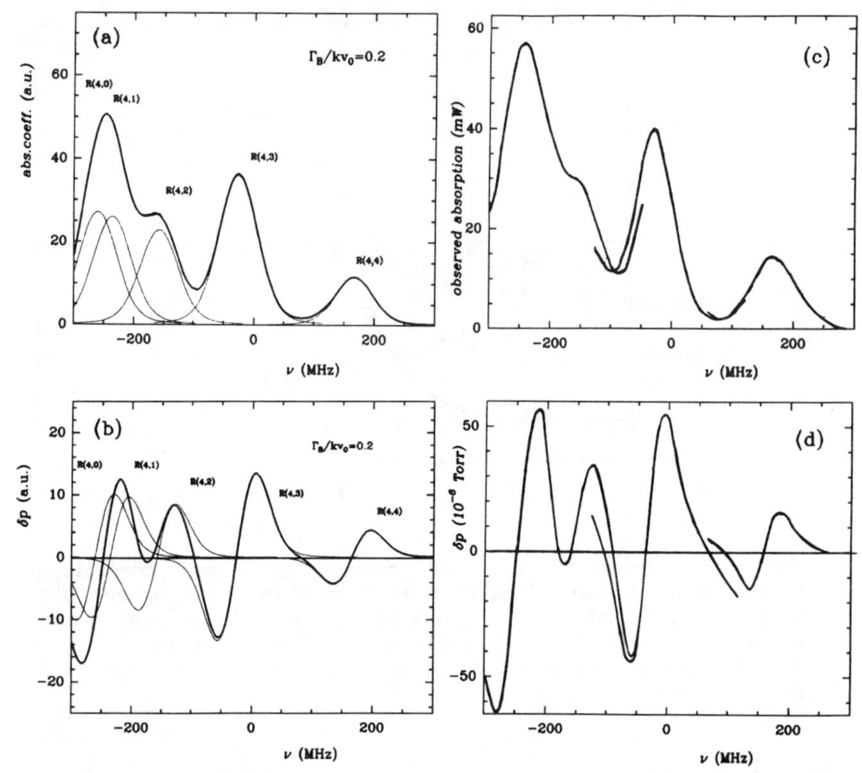

Fig. 5 a) Calculated absorption profile corresponding to the R(4, K) transitions (i.e., $v = 0 \rightarrow 1$, $J = 4 \rightarrow 5$, K unchanged) in $^{13}CH_3F$, relative to the 9P32 laser line.
b) Expected SLID signal following from the absorption profile with $\Delta\alpha/\alpha$ assumed constant.
c) Observed absorption profile in terms of absorbed power in mW.
d) Observed SLID signal in terms of δp. The experiments were performed at p = 0.05 Torr and laser power 0.5 W (intensity ≈ 30 W/cm^2), with the laser line center positioned at three different frequencies, 180 MHz apart (see text).

light results from a change in the dipole-induced dipole part of the molecule-surface interaction, which is proportional to $K^2M^2/J^2(J+1)^2$, with M the projection of J on the surface normal. This interaction should vanish for $K/J \rightarrow 0$, which seems contradictory to the observations. It must therefore be concluded that the dipole moment does not play a major role in the accommodation of tangential momentum in collisions of CH_3F with a quartz surface. This conclusion is supported by the observation that vibrational excitation alone (Q(12, 2) transition), which changes the dipole by 5%,[12] does not produce a measurable effect (cf. Fig. 3).

In summary, it is found from the Knudsen limit results that the accommodation coefficient for tangential momentum in CH_3F colliding with a quartz surface is changed by 10^{-3} to 10^{-2} upon vibrational excitation accompanied by a rotational transition (P and R branches). Vibrational excitation alone (Q branch) does not produce a detectable change and the K/J value also does not seem to be of much influence. The results suggest that the accommodation coefficient increases stronger than linearly with the rotational quantum number J. Further experiments on this dependence are underway.

II. Light-Induced Viscous Flow in the Near-hydrodynamic Regime

The behavior of the data at higher pressure cannot be accounted for by state-dependent molecule surface interaction alone. This is most obvious for the Q(12, 2) and Q(12, 3) data, where the results from the low-pressure limit indicate a negligible change in accommodation upon vibrational excitation; therefore, the pressure differences observed at higher pressure must originate in the bulk of the gas. In view of momentum conservation, such an effect may seem somewhat unexpected in a pure gas since photon momentum can be neglected. However, as will be shown below, the combination of a state-dependent collision cross section and nonuniform illumination produces stresses in the gas that give rise to a drift. Although phenomenologically this effect bears some resemblance to light-induced pulling observed by Atutov et al.,[13] it is an essentially different

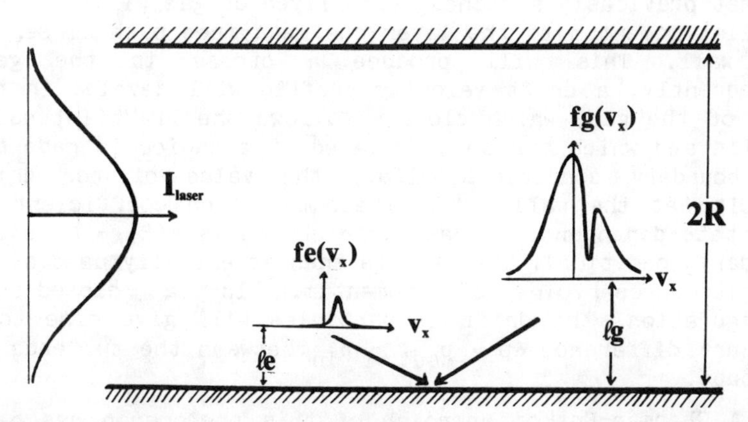

Fig. 6 Principle of the mechanism of light-induced viscous flow in a pure gas due to a state-dependent mean free path combined with nonhomogeneous light intensity.

phenomenon, since, in our experiment, there is no buffer gas involved.

The principle of the effect is shown in Fig. 6. Let a one-component gas be contained in a capillary cell with a diameter of 2R, at pressures such that $\bar{\ell} \ll R$, with $\bar{\ell}$ the mean free path of the molecules. The laser beam profile is assumed to be cylindrically symmetric but nonuniform in the transverse direction, such that it has the highest intensity at the axis of the tube. Because of velocity-selective excitation, a dip in the ground state distribution will appear around the resonant velocity, and a complementary peak will appear in the excited state distribution. The magnitude of the peak and dip depends on the axial coordinate r, since the illumination is nonuniform. For an elementary mean free path picture of this effect, let us first consider the boundary layer near the wall. Because of their larger kinetic cross section,[14] excited particles colliding with the wall have suffered their last collision, on the average, in a darker region of the cell than ground state particles. Consequently, the peak in the excited state distribution is smaller than the dip in the ground state distribution, and the total velocity distribution (ground state plus excited state) will show a dip around v_{xL}. This means that there is a net parallel momentum transfer to the wall, even if the accommodation coefficients of the levels involved are equal. As a result, a net drift will appear, the direction being that of the excited particles, i.e., the particles having the largest cross section. However, this process is not confined to the boundary layer within one mean free path from the wall. In the same elementary picture as that previously sketched, each layer of gas will transfer parallel momentum to the layer one mean free path closer to the wall. This will produce a stress in the gas. Consequently, a drift velocity profile will develop in the bulk of the gas, which closely follows the light intensity profile and which can be calculated if a choice is made for the boundary conditions, i.e., the value of the drift velocity at the wall. If the accommodation coefficient is not state-dependent, a reasonable choice is v(R) = 0 (stick boundary condition). The wall is then essentially used as an infinite reservoir of momentum. In a closed-tube configuration, the drift of particles will give rise to a pressure difference $\delta p = p_{out} - p_{in}$ between the two ends of the tube.

A Chapman-Enskog approach of this phenomenon has been worked out.[15] The results of this treatment is a drift velocity profile for an open tube given by

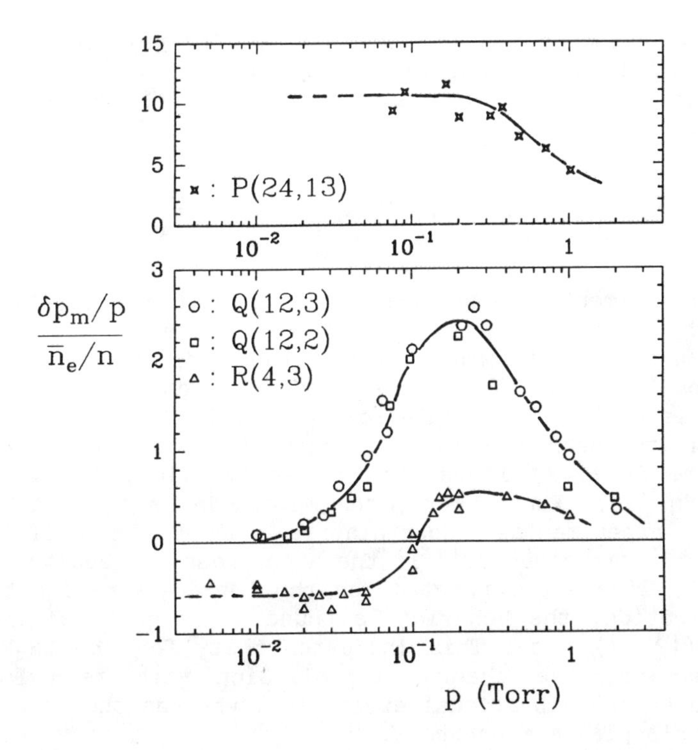

Fig. 7 The maximum of the observed pressure differences for positive detuning, normalized by the excited state fraction, as a function of pressure for a quartz capillary of 0.75 mm radius. The rovibrational excitation involves $\Delta v = +1$ with $\Delta J = -1, 0, +1$ (P, Q, R lines, respectively).

$$v(r) = \pi v_{xL} \frac{\Delta \nu}{\nu} \left\{ \frac{n_e}{n}(r) - \frac{n_e}{n}(R) \right\} \tag{4}$$

To derive an expression for the total flow in an open tube, or the corresponding pressure difference in a closed tube, Eq. (4) has to be integrated over the tube cross section. This requires knowledge of the radial dependence of n_e/n. Experimental determination of this dependence proved difficult, since the radial light intensity profile was found to vary appreciably along the tube axis. However, knowledge of $n_e/n(r)$ is unnecessary if it is assumed that the light intensity (or more precisely, the excited state fraction) approaches zero at the wall. In that case, the experimental results in terms of $(\delta p_m/p)(\bar{n}_e/n)^{-1}$ become independent of the profile, since both δp and the cross section averaged excited fraction $\bar{n}_e/n$ (as derived from the measured absorbed power) contain an integration over the same radial dependence. In this approximation. i.e., $n_e/n(R) = 0$, one finds for the observed pressure difference δp in a closed tube

$$\frac{\delta p/p}{\bar{n}_e/n} = +16\,\frac{L}{R}\cdot\frac{\bar{\ell}}{R}\cdot\frac{v_{xL}}{\bar{v}}\cdot\frac{\Delta\nu}{\nu} \qquad (5)$$

Experimental results for all transitions investigated are shown in Fig. 7 as a function of pressure. For these data, $\bar{n}_e$ was determined from the absorbed power as previously outlined, but now with τ_c determined by both molecule-wall and molecule-molecule collisions. It is found that in the near-hydrodynamic regime ($p > 0.3$ torr) the data for the Q(12, K) lines show a 1/p behavior, in agreement with Eq. (5). Furthermore, the magnitude is found to agree within experimental uncertainties with Eq. (5) if $\delta\nu/\nu = 1.1 \times 10^{-2}$ is assumed.[14] If the high-pressure results of the R(4, 3) line are corrected for the low-pressure limit SLID contribution, the behavior is found to be quite similar to the Q(12, K) case. This indicates that, for the CH_3F-CH_3F interaction, the change in collision rate is primarily caused by the vibrational excitation, whereas the rotational sublevels play a minor role.

Acknowledgments

The authors wish to acknowledge most fruitful discussions with A.V. Ghiner, I. Kuščer, G. Nienhuis and J.P. Woerdman. This work is part of the research program of the Stichting voor Fundamenteel Onderzoek der Materie (FOM), which is financially supported by the Nederlandse Organisatie voor Wetenschappelijk Onderzoek (NWO).

References

[1] Gel'mukhanov, F.Kh. and Shalagin, A.M., "Light-Induced Diffusion of Gases," Pis'ma Zhurnal Eksperimental'noi i Teoreticheskoi Fiziki, Vol. 29, 1979, pp. 773-776; JETP Letters., Vol. 29, 1979, pp. 711-713.

[2] Gel'mukhanov, F.Kh., Il'ichov, L.V., and Shalagin, A.M., "Kinetic Theory of Light-Induced Drift of Gas Particles," Physica, Vol. 137A, 1986, pp. 502-530; Werij, H.G.C., Haverkort, J.E.M., and Woerdman, J.P., "A Study of the Optical Piston," Physical Review Section A, Vol. 33, 1986, pp. 3270-3281.

[3] Werij, H.G.C., Woerdman, J.P., Beenakker, J.J.M., and Kuščer, I., "Demonstration of a Semipermeable Optical Piston," Physical Review Letters, Vol. 52, 1984, pp. 2237-2240.

[4]Werij, H.G.C., Haverkort, J.E.M., Planken, P.C.M., Eliel, E.R., Woerdman, J.P., Atutov, S.N., Chapovskiĭ, P.L., and Gel'mukhanov, F.Kh., "Light-Induced Drift Velocities in Na-noble Gas Mixtures," Physical Review Letters, Vol. 58, 1987, pp. 2660-2663.

[5]Ghiner, A.V., Stockmann, M.I., and Vaksman, M.A., "Surface Light-Induced Drift of a Rarefied Gas," Physics Letters, Vol. 96A, 1983, pp. 79-82.

[6]Hoogeveen, R.W.M., Spreeuw, R.J.C., and Hermans, L.J.F., "Observation of Surface Light-Induced Drift," Physical Review Letters, Vol. 59, 1987, pp. 447-449.

[7]Hoogeveen, R.W.M., Van den Oord, R.J., and Hermans, L.J.F., "Light-Induced Kinetic Effects in Molecular Gases," Proceedings of the 15th International Symposium on Rarefied Gas Dynamics, Teubner, Stuttgart, FRG, 1986, Vol. I, pp. 321-329.

[8]Lee, S.K., Schwendeman, R.H., and Magerl, G., "Infrared Microwave Sideband laser of the ν_3 and $2\nu_3 \leftarrow \nu_3$ Bands of $^{13}CH_3F$," Journal of Molecular Spectroscopy, Vol. 117, 1986, pp. 416-434.

[9]Bradley, L.C., Soohoo, K.L., and Freed, C., "Absolute Frequencies of Lasing Transitions in Nine CO_2 Isotopic Species," IEEE Journal of Quantum Electronics", Vol. QE-22, 1986, pp. 234-267.

[10]Bakarev, A.E. and Folin, A.K., "New Phenomenological Relations in the Theory of Light-Induced Drift," Optika i Spektroskopiya Vol. 62, 1987, 475-477; Opt. Spectrosc. (USSR), Vol. 62, 1987, pp. 284-285.

[11]Chapovsky, P.L., "Surface Light-Induced Drift of Molecules," preprint, USSR Academy of Science, Novosibirsk, 1988.

[12]Freund, S.M., Duxbury, G., Römheld, M., Tiedje, J.T., and Oka, T., "Laser Stark Spectroscopy in the 10 µm Region: the ν_3 Bands of CH_3F," Journal of Molecular Spectroscopy, Vol. 52, 1974, pp. 38-57.

[13]Atutov, S.N., Podjachev, S.P., and Shalagin, A.M., "Diffusion Pulling of Na-vapor into the Light Beam," Optics Communications, Vol. 57, 1986, pp. 236-238.

[14]Panfilov, V.N., Strunin, V.P., and Chapovsky, P.L., "Light-Induced Drift of CH_3F Molecules," Soviet Physics-JETP, Vol. 58, 1983, pp. 510-516, Zhurnal Eksperimental'noi i Teoreticheskoi Fiziki,. Vol. 85, 1983, pp. 881-892.

[15]Hoogeveen, R.W.M., Van der Meer, G.J., Hermans, L.J.F., Ghiner, A.V., and Kuščer, I., "Light-Induced Viscous Flow of a One Component Molecular Gas," Physical Review Section A, to be published.

Models for Temperature Jumps in Vibrationally Relaxing Gases

Raymond Brun,* Simon Elkeslassy,† and Ilan Chemouni†
Université de Provence-Centre Saint Jérôme, Marseille, France

Abstract

A semimacroscopic model for the description of the interaction between a nonequilibrium gas and a wall is proposed. The possibility for the gas of interchanging kinetic and internal energy is taken into account during its interaction with the wall. Thus, exchange coefficients are defined, as well as accommodation coefficients for each specific mode which differ from the usual definition. Expressions for the translational and internal temperature jumps are then obtained depending explicitly on these coefficients. An application to the case of an end-wall boundary layer developing behind a reflected shock in vibrational nonequilibrium conditions is also presented; wall heat transfer and wall temperature evolutions are calculated, and the importance of the intermode exchange is analyzed.

Nomenclature

c = specific heat
E = kinetic energy flux
e = molecule internal energy
$\bar{e}$ = equilibrium internal energy
f = distribution function
K = efficient exchange coefficient
k = Boltzmann constant
L = exchanged energy flux
m = molecular mass
n = molecular density
q = wall heat transfer
Q = partition function
T = temperature

---

*Head, Department of SETT, Milieux Hors d'Equilibre.
†Graduate Student, Department of SETT, Milieux Hors d'Equilibre.

t = time
$\underset{\sim}{V}$ = molecular velocity
x = coordinate normal to the wall
β = accommodation coefficient
ε = internal energy flux
λ = conductivity
ρ = density

Subscripts

a = internal level
e = internal mode
f = frozen conditions
i = incident molecules entering the interaction zone
r = reflected molecules
s = wall material
t = translational mode
v = vibrational mode
w = wall conditions

Introduction

As is well known, the influence of the catalytic nature of a wall is of preponderant importance on the heat transfer between a reactive gas and this wall. Furthermore, in the transitional regime between continuum and free molecular flows, the accommodation effects are also significant; thus, the action of catalyticity and accommodation must be simultaneously examined. The modeling of these effects consists in proposing explicit expressions for temperature and concentration jumps at the wall in order to use them as boundary conditions for the Navier-Stokes equations governing the reactive flow.

The solving of Boltzmann equation in the Knudsen layer and the detailed knowledge of the interaction between the impinging gas molecules and the solid wall are, still now, complicated problems in reactive flows; hence, it is necessary to build semimacroscopic models in order to give a global but physically correct picture of the accommodation and catalyticity phenomena. In fact, there are only a few kinetic models of this kind[1,2] that attempt to describe the interaction of a hot nonequilibrium gas and a cold wall where recombination or de-excitation may occur. These models have common points such as the characterization of gas-wall and gas-gas transfers by coefficients called accommodation and exchange coefficients. However, they have been constructed for different purposes: model 1 describes the interaction of a wall with a rotationally excited gas in the free molecular regime,[1] and model 2 has been built for a mixture of reacting gases in the slip flow regime.[2]

We propose here a different model, although also using macroscopic coefficients, the basis of which has been presented in Ref. 3 and which is adapted to the case of an undissociated gas of polyatomic molecules in vibrational nonequilibrium able to interchange their translational (+ rotational) energy with a vibrational one during the interaction process with the wall. The expressions of jumps for translational and vibrational temperatures finally obtained are then applied to the concrete case of a gas close to the end wall of a tube and successively submitted to an incident and a reflected shock wave. As far as it is possible, comparisons with results given by model 2 are presented.

Hypotheses and Modelization

The gas is assumed to have no macroscopic velocity, and we consider for gas molecules a translational energy mode and one internal mode only, the extension to several internal modes being straightforward.

The incident molecules entering the interaction zone possess a known nonequilibrium first-order Chapman-Enskog distribution[3,4] defined with two temperatures T_t and T_e, i.e :

$$f_a = f_a^0 \left[1 + \frac{2}{5} \frac{\lambda_t}{nkT_t} \frac{m}{kT_t} \left(\frac{5}{2} - \frac{mV^2}{2kT_t} \right) V_x \frac{dT_t}{dx} + \frac{\lambda_e}{nkT_t} \frac{m}{c_e T_e} \frac{\bar{e} - e_a}{kT_e} V_x \frac{dT_e}{dx} \right] \quad (1)$$

where f_a^0 is the zeroth-order nonequilibrium function equal to

$$n \left(\frac{m}{2\pi kT_t} \right)^{3/2} \exp \left(\frac{mV^2}{2kT_t} \right) \frac{\exp\left(e_a / kT_e \right)}{Q\left(T_e\right)} \quad (2)$$

Now, specific accommodation coefficients are defined for each energy mode β_t and β_e, and intermode exchange terms are also defined; i.e., $\mathbf{L}_t$ is the translational energy flux transfered to the internal one and $\mathbf{L}_e$ the reverse in the interaction zone. It is then assumed that one part of these terms $\mu_t\mathbf{L}_t$ and $\mu_e\mathbf{L}_e$ remains available for a further exchange with the wall ($0 < \mu_t, \mu_e < 1$). It is also assumed[5] that $\mathbf{L}_t$ and $\mathbf{L}_e$ are proportional to the maximum available fluxes $E_i - E_w$ and $\varepsilon_i - \varepsilon_w$, respectively:

$$\mathbf{L}_t = \alpha_t \ (E_i - E_w) \tag{3a}$$

$$\mathbf{L}_e = \alpha_e \ (\varepsilon_i - \varepsilon_w) \tag{3b}$$

and

$$E_i = \sum_a \int_{V_x > 0} \frac{mV^2}{2} V_x f_a \, d_3V,$$

$$E_w = \sum_a \int_{V_x > 0} \frac{mV^2}{2} V_x f^0_{aw} d_3V \tag{4a}$$

$$\varepsilon_i = \sum_a \int_{V_x > 0} \varepsilon_a V_x f_a \, d_3V,$$

$$\varepsilon_w = \sum_a \int_{V_x > 0} \varepsilon_a V_x f^0_{aw} \, d_3V \tag{4b}$$

We can now express the energy balances in the interaction zone. Thus, for the wall, the real incident energy fluxes E_{i_p} and ε_{i_p} are respectively

$$E_{i_p} = E_i - \mu_t \mathbf{L}_t + \mu_e \mathbf{L}_e \tag{5a}$$

$$\varepsilon_{i_p} = \varepsilon_i - \mu_e \mathbf{L}_e + \mu_t \mathbf{L}_t \tag{5b}$$

In the same way, the reflected fluxes (Fig. 1) are

$$E_r = E_i - q_{t_w} - \mathbf{L}_t + \mathbf{L}_e \tag{6a}$$

$$\varepsilon_r = \varepsilon_i - q_{e_w} - \mathbf{L}_e + \mathbf{L}_t \tag{6b}$$

so that, for the wall, the real reflected energy fluxes are respectively,

$$E_{r_p} = E_r + (1 - \mu_t)\mathbf{L}_t - (1 - \mu_e)\mathbf{L}_e \tag{7a}$$

$$\varepsilon_{r_p} = \varepsilon_r + (1 - \mu_e)\mathbf{L}_e - (1 - \mu_t)\mathbf{L}_t \tag{7b}$$

As a consequence, the accommodation coefficients β_t and β_e may be written, respectively,

$$\beta_t = \frac{E_{i_p} - E_{r_p}}{E_{i_p} - E_w} = \frac{q_{t_w}}{E_{i_p} - E_w} \tag{8a}$$

$$\beta_e = \frac{\varepsilon_{i_p} - \varepsilon_{r_p}}{\varepsilon_{i_p} - \varepsilon_w} = \frac{q_{e_w}}{\varepsilon_{i_p} - \varepsilon_w} \tag{8b}$$

Comparing the expressions (10) with the ones[3,5] corresponding to the case without exchange, coupling terms

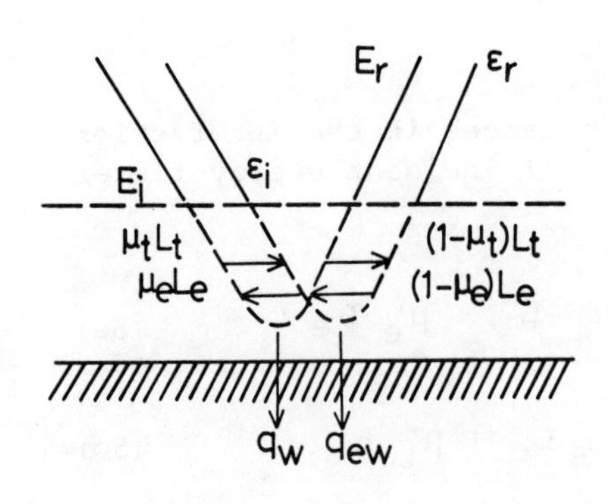

Fig. 1 Scheme of interaction

appear in expressions (10) where each temperature jump depends on both temperature gradients.

Application : Reflection of a Shock Wave at an End Wall of a Shock Tube.

The expression (10) may be used as boundary conditions for the Navier-Stokes equations describing situations where accommodation and catalytic phenomena are nonnegligible, for example, the case for the boundary layer developing at an end wall of a tube just after the reflection of a shock wave on this end wall. The shock wave is assumed strong enough to provoke vibrational non equilibrium; of course, we have in this particular case $Le \sim 0$, and we can apply the previous results, if we consider that T_t is the translational-rotational temperature, the rotational mode being assumed in equilibrium with the translational one and with $T_e = T_v$.

Now, as a first approximation and in order to retain only the present specific effects, we consider the very first instants after the reflection. Thus, the gas may be considered as frozen outside and inside the boundary layer and energy balance equations may be easily solved after a Von Mises transformation and with the assumption of constant Prandtl and Lewis numbers[6,7] (A complete computation taking into account the relaxation before and behind the reflected shock as well as in the boundary layer is in course and will be reported further.)

Thus, the evolution of translational-rotational and vibrational temperature profiles are obtained in the boundary layer, and the total heat flux $q_w = q_{tw} + q_{ew}$ may be calculated. Examples of $q_w(t)$ are represented in Figs. 2a and 2b for different values of β_t, β_e, and K_t.

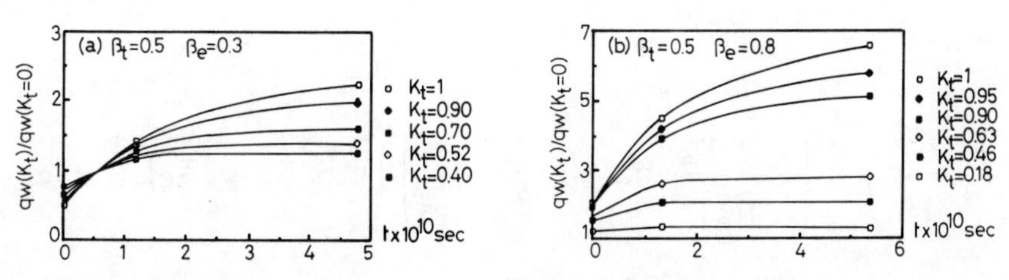

Fig. 2 Evolution of wall heat-transfer behind a reflected shock-wave (CO_2, M_s = 3,6).

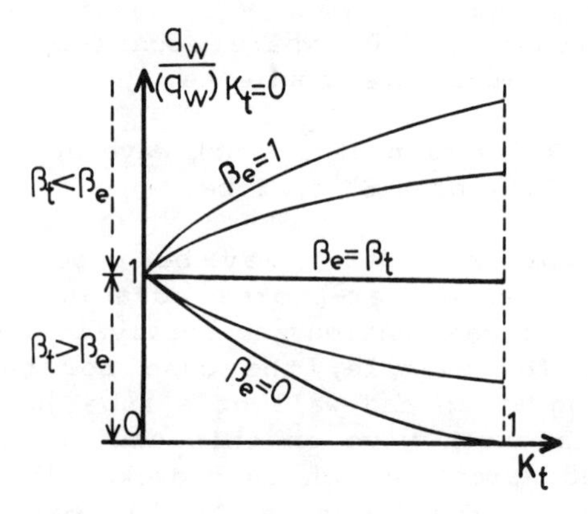

Fig. 3 Influence of the exchange parameter K_t on the wall heat flux behind a reflected shock wave : Universal curves ($t \rightarrow 0$).

Thus, the influence of these parameters can be easily deduced from these figures which correspond to a CO_2 gas submitted to an incident shock wave with a Mach number of 3,6.

However, a more significant result may be drawn from Fig. 3 in which the ratio $q_w/(q_w)_{K_t} = 0$ is represented for the first instants following the reflection ($t \rightarrow 0$). It may be seen that, when $\beta_t > \beta_e$, i.e, when the gas exchanges more easily translational-rotational energy than vibrational energy with the wall, then the intermode exchange decreases the wall heat flux, the result being inversed when $\beta_t > \beta_e$. Finally, if we call K_t and K_e the efficient exchange coefficients, we have

$$E_i - E_w = \left(\frac{1 - K_t}{\beta_t} q_{t_w} - \frac{K_e}{\beta_e} q_{e_w} \right) \left(1 - K_t - K_e\right)^{-1} \quad (9a)$$

$$\varepsilon_i - \varepsilon_w = \left(\frac{1 - K_e}{\beta_e} q_{e_w} - \frac{K_t}{\beta_t} q_{t_w} \right) \left(1 - K_t - K_e\right)^{-1} \quad (9b)$$

with $K_t = \mu_t \alpha_t$ and $K_e = \mu_e \alpha_e$.

Now, if E_i and E_w are expressed with Eqs. (4), (1), and (2), we obtain for the temperature jumps

$$T_t - T_w = r_t \left(\frac{dT_t}{dx}\right)_w + r_e' \left(\frac{dT_e}{dx}\right)_w \tag{10a}$$

$$T_e - T_w = r'_t \left(\frac{dT_t}{dx}\right)_w + r_e \left(\frac{dT_e}{dx}\right)_w \tag{10b}$$

where

$$r_t = -\frac{\lambda t}{A} f(K_t, K_e, \beta_t), \qquad r_e' = -\frac{\lambda e}{A} g(K_t, K_e, \beta_e)$$

$$r_t' = -\frac{\lambda t}{B} f(K_e, K_t, \beta_e), \quad r_e = -\frac{\lambda e}{B} g(K_e, K_t, \beta_t)$$

and with

$$A = kn_w \left(\frac{2kT_w}{\pi m}\right)^{1/2}, \qquad B = \frac{c_e n_w}{2} \left(\frac{2kT_w}{\pi m}\right)^{1/2}$$

$$f(K_t, K_e, \beta_t) = \frac{2(1 - K_e) - \beta_t(1 - K_t - K_e)}{2\beta_t(1 - K_t - K_e)}$$

$$g(K_t, K_e, \beta_e) = \frac{K_e}{\beta_e(1 - K_t - K_e)}$$

and similar forms for $f(K_e, K_t, \beta_e)$ and $g(K_e, K_t, \beta_t)$

From an experimental point of view, the measured quantity is the increase of wall temperature after the shock-wave reflection ΔT_w. This quantity may be obtained from the heat flux with the following expression[6]:

$$q_w = \left(\rho_s \lambda_s c_s / \pi\right)^{1/2} \int_0^t \frac{d}{du}\left(\Delta T_w\right) \frac{du}{(t-u)^{1/2}} \tag{11}$$

Thus, a wall temperature increase ΔT_w is represented in Fig. 4 for the same conditions as in Fig. 2 and for a glass wall. It should be noted that the

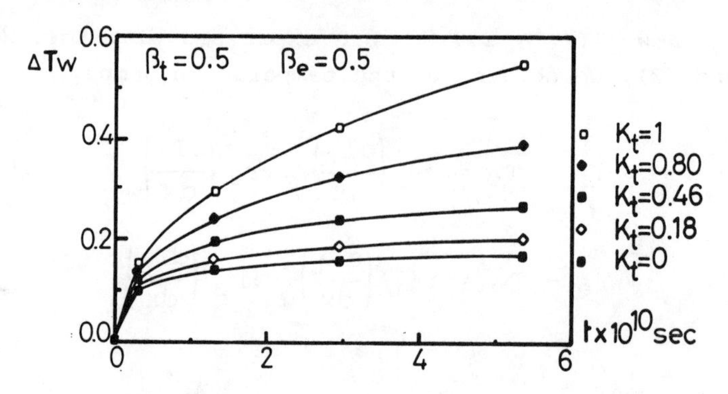

Fig. 4 - Evolution of wall temperature behind a reflected shock wave (CO_2, M_s = 3,6, glass wall).

present computation is only valid for $t \ll \tau_v$ (frozen case), which is the case represented here ($t \sim 10^{-9}$ s). Thus, the curves corresponding to an exchange ($K_t > 0$) are above the curve for $K_t = 0$; hence, there is a maximum for the ΔT_w curves, since, when $t \rightarrow \infty$, ΔT_w tends to the value corresponding to the continuum regime obtained for $T_w = C^{te}$ ($K_t = 0$). However, the inviscid flow does not remain frozen, and definitive conclusions concerning the behavior of ΔT_w will be drawn only when a complete computation is made.

Comparison with Scott's Model

If we extend Scott's model[2] to the case of vibrational nonequilibrium, we may write the following balance for the translational flux:

$$q_{t_w} = E_i - \left(1 - \theta_t\right)E_i - \left(\theta_t - \gamma_t\right)E_w \qquad (12)$$

where θ_t is an accommodation coefficient (different from β_t) and γ_t an exchange coefficient (different from K_t) and with $\gamma_t \leq \theta_t$. A similar expression is used for q_{ew}. As in the present model, the assumption $\gamma_e \ll \gamma_t$ is used. Then, the temperature jumps are easily obtained:

$$T_t - T_w = - \frac{\gamma_t}{\theta_t} T_w + \frac{\lambda_t}{A} \frac{2 - \theta_t}{\theta_t}\left(\frac{dT_t}{dx}\right)_w \qquad (13a)$$

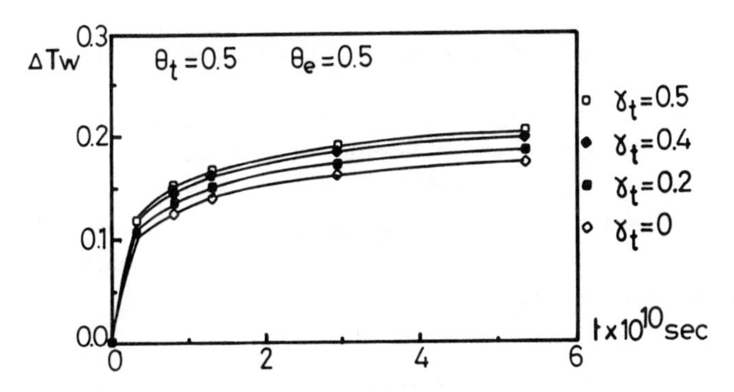

Fig. 5 - Evolution of wall temperature behind a reflected shock-wave : Scott's model (CO_2, M_s = 3,6, glass wall)

$$T_e - T_w = -\frac{\gamma_t}{\theta_e} T_w + \frac{\lambda_e}{B}\frac{2-\theta_e}{\theta_e}\left(\frac{dT_e}{dx}\right)_w \qquad (13b)$$

The coupling appears in a way different from this given by Eqs. (10).

A computation of the wall heat transfer behind a reflected shock has been also made, and an example of the wall temperature evolution is represented in Fig. 5.

The behavior of ΔT_w is similar to this resulting from the present model (Fig. 4).

Conclusions

The use of the present model avoids having to solve the Boltzmann equation in the interaction zone. Such a calculation has been made in Ref. 3 for the case of no interchange, and it has been shown that the form of the distribution function for the incident molecules given by Eq. (1) is a good approximation.

The present model is now being extended to the realistic case of the reflection of a shock wave taking into account the vibrational relaxation before and behind this shock, as well as in the boundary layer. A similar computation has been previously made for the case of a catalytic wall[7] ($T = T_v = T_w$). The results that will be obtained will remain valid for any time and will not be limited to the very first instants after the reflection.

At last, the present model depends on unknown parameters so that measurements are foreseen in a shock tube for different wall materials and gases. It seems that the very few number of parameters to determine (β_t, β_e, K_t) may contribute to obtain significant quantitative values for these parameters.

References

[1] Larina, I.N. and Rykov, V.A. "The Boundary Condition on the Body Surface for a Diatomic Gas," Proceedings of the 15th International Symposium on Rarefied Gas Dynamics Teubner, Stuttgart, FRG, 1986,Vol. I, pp. 635-643.

[2] Scott, C.D., "Wall Boundary Equations with Slip and Catalysis for Multicomponent, Non-Equilibrium Gas Flows," NASA TM X-58111, 1973.

[3] Larini, M. and Brun, R., "Discontinuités de Températures Pariétales dans un Gaz Polyatomique Hors d'Equilibre," International Journal of Heat and Mass Transfer, Vol. 16, 1973, pp. 2189-2203.

[4] Brun, R. and Zappoli, B. "Model Equations for a Vibrationally Relaxing Gas," Physics of Fluids, Vol. 20, N° 9, Sept. 1977, pp. 1441-1448.

[5] Kogan, M.N. Rarefied Gas Dynamics, Plenum, New-York, 1969.

[6] Brun, R. and Meolans, G.J. "Introduction au Problème de l'Echange d'Energie entre un Gaz Polyatomique et une Paroi", Entropie, Vol. 39, May-June 1971, pp. 15-20.

[7] Brun, R., Duran, G., Philippi, P.C., Dourieu, M.F. and Tosello, R. "Linearized Kinetic Models for Polyatomic Gases and Mixtures of Gases: Application to Vibrationally Relaxing Flows," Flames, Lasers, and Reactive Systems, Vol. 88 edited by J.R. Bowen, N. Manson, A.K. Oppenheim, and R.I. Soloukin, Progress in Astronautics and Aeronautics : AIAA, New York, 1983, pp. 293-304.

Variational Calculation of the Slip Coefficient and the Temperature Jump for Arbitrary Gas-Surface Interactions

C. Cercignani*
Politecnico di Milano, Milano, Italy
and
M. Lampis†
Università di Udine, Udine, Italy

Abstract

The aim of this paper is to compute the slip and temperature jump coefficients for a rarefied gas having an arbitrary interaction with a solid surface by means of a variational technique. This problem has become fashionable again in space research, but was considered in 1972 by Klinc and Kuščer . They used a variational principle for the integral version of the Boltzmann equation. Here we use a variational method for the integro-differential version of the Boltzmann equation, proposed by Cercignani in 1969. With the simplest trial functions, one obtains general formulas that look simpler than those proposed by Klinc and Kuščer but reduce to the latter when all of the accommodation coefficients are equal. Numerical values compare favorably with existing numerical solutions.

Introduction

Interest in the study of upper atmosphere aerodynamics has returned in the last few years following the U.S. Shuttle missions and the projected launch of the European shuttle Hermes. On one hand, this has led to the need for more reliable atmospheric data, especially in the range between 70 and 120 km of altitude and, on the other hand, to better numerical analyses of rarefied gas flows past a space vehicle. Part of these analyses refers to the so-called slip regime of rarefied gasdynamics, where the compressible Navier-Stokes equations yield an adequate

---

*Professor, Department of Mathematics.

†Professor, Department of Theoretical and Applied Mechanics.

description of the situation, provided they are supplemented by suitable boundary conditions. Thus, the old problems of computing the slip and temperature jump coefficients[1-4] have been taken up again.[5] Currently, we are more aware of the necessity of describing more accurately the gas-surface interaction, in order to simulate the flow in the neighborhood of the surface and compute the related values of surface stresses and heat transfer. Accordingly, it appears of interest to obtain reasonably accurate estimates of the slip and temperature jump coefficients for arbitrary models of gas-surface interaction. An efficient method of computing these coefficients was recognized many years earlier[6] based on a variational technique. Although the first variational method in rarefied gasdynamics was based on the use of the Bhatnagar, Gross and Krook (BGK) model[3,4] in the integral form,[6] it was later recognized[7] that another variational technique based on the integro-differential form of the Boltzmann equation has a wider range of application and leads to simpler calculations with simple-looking results. Despite that, the only existing calculations of the slip and temperature jump coefficients for arbitrary gas-surface interaction existing at present, to the best of the authors' knowledge, are based on the use of a variational technique for the integral form of the Boltzmann equation.[8]

Herein we use the variational method proposed by Cercignani[7] in 1969 for the integro-differential form of the Boltzmann equation and arrive at new expressions for the slip and temperature jump coefficients for an arbitrary gas-surface interaction.

Basic Equations for Evaluating the Slip Coefficient

In order to consider the problem of computing the slip coefficient, the distribution function $f(x,\underline{c})$ is written in the following form:

$$f = f_o[1 + 2kxc_z + 2kL^{-1}(c_xc_z) + h] \qquad (1)$$

where k is the velocity gradient and f_o the wall Maxwellian. $\underline{c}$ is the velocity vector with components c_x, c_y and c_z, and L^{-1} is the inverse of the linearized collision operator defined on the space orthogonal to the collision invariants.[3,4] All of the velocities are made nondimensional through $(2RT_o)^{1/2}$, where R is the gas constant and T_o the wall temperature. The perturbation h

due to the presence of the wall satisfies (if we are only interested in terms linear in k) the linearized Boltzmann equation[3,4]:

$$c_x \frac{\partial h}{\partial x} = Lh \tag{2}$$

with the boundary condition

$$h^+(0,\underline{c}) = h_o + Ah^- \tag{3}$$

$$h_o = -2kL^{-1}(c_x c_z) + A(2kL^{-1}(c_x c_z)) \tag{4}$$

where h^+ and h^- are the restrictions of h to positive and negative values, respectively, of c_x. A is an operator describing the gas surface interaction.[3,4]

Using the technique described in Ref. 7, we define the following functional of a trial function h:

$$J(\hat{h})=((\hat{h},R(D-L)\hat{h})) + (\hat{h}^+-A\hat{h}^- -2h_o,R\hat{h}^-)_B \tag{5}$$

where R is the operator changing $h(\underline{c})$ into $h(-\underline{c})$, $D=c_x(\partial/\partial x)$, and

$$((g,h)) = \iint ghf_o d\underline{c}dx \tag{6}$$

$$(g,h)_B = \int c_x g^+ h^+ f_o d\underline{c} \tag{7}$$

If we compute the first variation δJ of J, we find that it is zero if, and only if, h=h, where h is the solution of Eq. (2) with boundary condition (3). In addition, one can show that the slip coefficient ζ is related to the value of J(h) by

$$J(h)= 2k^2\zeta\ell\pi^{-1/2} + 2k(h_o,L^{-1}(c_x c_z))_B \tag{8}$$

where ℓ is the mean free path related to the viscosity coefficient by the following relation:

$$\ell =(\mu/\rho)(\pi/2RT)^{1/2} \tag{9}$$

Since the second term on the right-hand side of Eq. (8) is known in principle, the value of the functional J(h) for h=h gives the value of ζ. This is the basis for a variational (and, hence, accurate) evaluation of the slip coefficient.

Calculation of the last integral in Eq. (8) is easy for Maxwellian molecules but a difficulty is met for other models when computing an integral of the type shown in Eq. (7) with

$$g = -2kL^{-1}(c_x c_z) + 2kA(L^{-1}(c_x c_z)) \tag{10a}$$

$$h = L^{-1}(c_x c_z) \tag{10b}$$

In order to obtain simple results for the general case, we resort to the trick of replacing $L^{-1}(c_x c_z)$ with its first Chapman-Enskog approximation, i.e., letting

$$L^{-1}(c_x c_z) = -\beta c_x c_z \tag{11}$$

where $\beta > 0$ is independent of $\underline{c}$ and is related to the first approximation of the viscosity coefficient.

This approximation, which already has been used by Klinc and Kuščer,[8] leads to a simple expression for the last integral in Eq. (8) and, hence, to the following expression for the functional J(h):

$$J(h) = 2k^2\zeta\ell\pi^{-1/2} + \pi^{-1/2}k^2\beta^2(\alpha_{55}-2) \tag{12}$$

Here the accommodation coefficients α_{ik} are defined in the same manner as in Ref. 8, i.e.,

$$\alpha_{ik} = \frac{(Q_i(\underline{c}), Q_k(\underline{c}) - \bar{Q}_k(\underline{c}))_B}{(Q_i(\underline{c}), Q_k(\underline{c}) - N_k)_B} \tag{13}$$

where $Q_k(\underline{c})$ are monomials in the components of $\underline{c}$, and N_k and $\bar{Q}_k$ are defined as follows:

$$N_k = \frac{(Q_k, Q_0)_B}{(Q_0, Q_0)_B} \tag{14}$$

$$\bar{Q}_k = A(Q_k(-c_x, c_y, c_z)) \tag{15}$$

The list of the first few monomials Q_k reads as follows:

$$Q_0 = 1, \quad Q_1 = c_x, \quad Q_2 = c_y, \quad Q_3 = c_z,$$
$$Q_4 = c^2, \quad Q_5 = c_x c_y, \quad Q_6 = c_x c_z, \quad Q_7 = c_x c^2 \tag{16}$$

Evaluation of the Slip Coefficient

In order to proceed to the actual computation of the slip coefficient, we have adopted a simple trial function h given by

$$\hat{h} = a c_z \tag{17}$$

where a is a constant to be determined.

The result of a simple calculation gives the following expression for $J(\hat{h})$:

$$J(\hat{h}) = -Ca^2 + Ba \tag{18}$$

where

$$C = \alpha_{22}/4\sqrt{\pi} \tag{19a}$$

$$B = k\beta - k\beta\alpha_{25}/2 \tag{19b}$$

The derivative of $J(\hat{h})$ with respect to a vanishes for a = B/2C. This is the optimal value of a in the simple trial function we have chosen, Eq. (17). Plugging this value into Eq. (18), we find the optimal value of the functional J(h) and arrive at the following result for the slip coefficient:

$$\zeta = \ell(\mu_1/\mu)^2[(2-\alpha_{55})(2/\pi) + (2-\alpha_{25})^2/(2\alpha_{22})] \tag{20}$$

where μ is the viscosity coefficient and μ_1 its first approximation in the Chapman-Enskog method.[9] The preceding expression is based on a variational estimate[8] of the constant β in terms of μ and μ_1. The result in Eq. (20) has a simpler shape than the one obtained by Klinc and Kuščer. However, the two results are coincident when all of the accommodation coefficients are equal (Maxwell's boundary conditions).

Furthermore, if all of the accommodation coefficients are unity (complete accommodation), the exact result for the BGK model[10] differs from the preceding result by less than 1%.

We remark that explicit expressions for the accommodation coefficients α_{ik} in correspondence with the most used models of gas-surface interaction (such as the Maxwell model, the Cercignani-Lampis model, and the diffuse-elastic scattering model) have been found by Klinc and Kuščer.[8]

Basic Equations for Evaluating the Temperature Jump Coefficient.

In order to consider the problem of computing the temperature jump coefficient, the distribution function $f(x,\underline{c})$ is written in the following form:

$$f = f_o(1 + kx(c^2-5/2) + kL^{-1}(c_x(c^2-5/2)) + h) \quad (21)$$

where k is now the asymptotic temperature gradient (divided by the wall temperature). The perturbation h satisfies (if we are interested only in terms linear in k) the linearized Boltzmann equation, Eq. (2), with the boundary condition (3), where now h_o is given by

$$h_o = -kL^{-1}(c_x(c^2-5/2)) + A(kL^{-1}(c_x(c^2-5/2))) \quad (22)$$

Using the same technique as used earlier, we introduce the functional defined by Eq. (5), where now, of course, h_o is defined by Eq. (22).

Again, the first variation δJ of J is zero if, and only if, h=h, where h is the solution of Eq. (2) with boundary condition (3). In addition, the temperature jump coefficient τ is related to the value of J(h) by

$$J(h) = -5k^2\tau\ell/(2\pi^{1/2}Pr) - k(h_o, L^{-1}(c_x(c^2-5/2)))_B \quad (23)$$

Since the second term on the right hand side of Eq. (23) is known in principle, the value of the functional J(h) for h=h gives the value of τ. This is the basis for a variational (and, hence, accurate) evaluation of the temperature jump coefficient.

As before, the calculation of the last integral in Eq. (23) is easy for Maxwellian molecules but difficulty is encountered for other models when computing an integral of the type shown in Eq. (7) with

$$g = -kL^{-1}(c_x(c^2-5/2)) + kA[L^{-1}(c_x(c^2-5/2))] \quad (24a)$$

$$h = L^{-1}(c_x(c^2-5/2)) \quad (24b)$$

In order to obtain simple results for the general case, we resort to the same trick as before; i. e., we replace $L^{-1}(c_x(c^2-5/2))$ by its first Chapman-Enskog approximation, by letting

$$L^{-1}(c_x(c^2-5/2)) = -\gamma c_x(c^2-5/2) \tag{25}$$

where γ is independent of $\underline{c}$ and is related to the first approximation to the heat conductivity. This approximation was used by Klinc and Kuščer[8] as well and leads to a simple expression for the last integral in Eq. (23).

Evaluation of the Temperature Jump Coefficient

In order to proceed to the actual computation of the temperature jump coefficient, we have adopted a simple trial function h given by

$$\hat{h} = a(c^2 - 5/2) \tag{26}$$

where a is a constant to be determined.

The result of a simple calculation gives the following expression for $J(\hat{h})$:

$$J(\hat{h}) = Ca^2 - Ba \tag{27}$$

where

$$C = \alpha_{44}/\sqrt{\pi} \tag{28a}$$

$$B = (5/2)k\gamma[1-(3/4)\alpha_{47} + (1/4)\alpha_{41}] \tag{28b}$$

Here the accommodation coefficients α_{ik} are defined as in the earlier evaluation of the slip coefficient. The derivative of J(h) with respect to a vanishes for a = B/2C. This is the optimal value of a in the simple trial function chosen, Eq. (26). Plugging this value into Eq. (27), we find the optimal value of the functional J(h) and arrive at the following result for the temperature jump coefficient:

$$\tau=(2\ell/5\mathrm{Pr})(\kappa_1/\kappa)^2[13/(2\pi)+24(1-\alpha_{77})/\pi+(25/8)(\alpha_{77}+\alpha_{11}-2\alpha_{17})- -(30/\pi)(1-\alpha_{17})+25(1-\alpha_{11})/2\pi+(25/\alpha_{44})(1/2-3\alpha_{47}/8+\alpha_{41}/8)^2] \tag{29}$$

where κ_1 is the first approximation[9] to the heat conductivity κ and Pr is the Prandtl number. Again, the result is in agreement with Ref. 8 when all of the α's coincide. Note that, in the case of Maxwellian molecules, $\mu_1/\mu = \kappa_1/\kappa = 1$ and Pr = 2/3.

Concluding Remarks

We have computed the slip and temperature jump coefficients for a monatomic gas satisfying boundary conditions of a general nature. The same approach could be used to deal with polyatomic gases and mixtures. The authors remark that the traditional way of computing the coefficients in the boundary conditions, which goes back to Maxwell,[1] produces coefficients with errors of 15% or more and its use can lead to incorrect estimates of the accommodation coefficients. In spite of this, the old approach was used again quite recently[11] to compute the slip-boundary equations for multicomponent nonequilibrium airflow.

Acknowledgment

This work was supported by Ministero della Pubblica Istruzione and performed in the frame of the activity of Gruppo Nazionale di Fisica Matematica del Consiglio Nazionale delle Ricerche.

References

[1]Maxwell, J. C., "On Stresses in Rarified Gases Arising from Inequalities of Temperature", Philosophical Transactions of the Royal Society of London, Vol. 170, 1879, pp. 231-256.

[2]Cercignani, C., "Elementary Solutions of the Linearized Gas Dynamics Boltzmann Equation and Their Application to the Slip Flow Problem," Annals of Physics (New York), Vol. 20, 1962, pp. 219-233.

[3]Cercignani, C., Mathematical Methods in Kinetic Theory, Plenum, New York, 1969.

[4]Cercignani, C., The Boltzmann Equation and its Applications, Springer, New York, 1987.

[5]Golse, F., "Applications of the Boltzmann Equation within the Context of Upper Atmosphere Aerodynamics," preprint, 1987.

[6]Cercignani,C. and Pagani, C. D., "Variational Approach to Boundary Value Problems in Kinetic Theory," Physics of Fluids, Vol. 9, 1966, pp. 1167-1173.

[7] Cercignani, C., "A Variational Principle for Boundary Value Problems in Kinetic Theory," Journal of Statistical Physics, Vol. 1, 1969, pp. 297-311.

[8] Klinc, T. and Kuščer, I., "Slip Coefficients for General Gas-Surface Interaction," Physics of Fluids, Vol. 15, 1972, pp. 1018-1022.

[9] Chapman, S. and Cowling, T. G., The Mathematical Theory of Non-Uniform Gases, Cambridge Univ. Press, London, 1940.

[10] Albertoni, S., Cercignani, C., and Gotusso, L., "Numerical Evaluation of the Slip Coefficient," Physics of Fluids, Vol. 6, 1963, pp. 993-996.

[11] Gupta, R.N., Scott, C. D., and Moss, J. N., "Slip-Boundary Equations for Multicomponent Nonequilibrium Airflow," NASA TP 2452, Nov. 1985.

Author Index

PROGRESS IN ASTRONAUTICS AND AERONAUTICS SERIES VOLUMES

VOLUME TITLE/EDITORS

*1. **Solid Propellant Rocket Research** (1960)
Martin Summerfield
Princeton University

*2. **Liquid Rockets and Propellants** (1960)
Loren E. Bollinger
The Ohio State University
Martin Goldsmith
The Rand Corporation
Alexis W. Lemmon Jr.
Battelle Memorial Institute

*3. **Energy Conversion for Space Power** (1961)
Nathan W. Snyder
Institute for Defense Analyses

*4. **Space Power Systems** (1961)
Nathan W. Snyder
Institute for Defense Analyses

*5. **Electrostatic Propulsion** (1961)
David B. Langmuir
Space Technology Laboratories, Inc.
Ernst Stuhlinger
NASA George C. Marshall Space Flight Center
J.M. Sellen Jr.
Space Technology Laboratories, Inc.

*6. **Detonation and Two-Phase Flow** (1962)
S.S. Penner
California Institute of Technology
F.A. Williams
Harvard University

*7. **Hypersonic Flow Research** (1962)
Frederick R. Riddell
AVCO Corporation

*8. **Guidance and Control** (1962)
Robert E. Roberson
Consultant
James S. Farrior
Lockheed Missiles and Space Company

*9. **Electric Propulsion Development** (1963)
Ernst Stuhlinger
NASA George C. Marshall Space Flight Center

*10. **Technology of Lunar Exploration** (1963)
Clifford I. Cummings
Harold R. Lawrence
Jet Propulsion Laboratory

*11. **Power Systems for Space Flight** (1963)
Morris A. Zipkin
Russell N. Edwards
General Electric Company

12. **Ionization in High-Temperature Gases** (1963)
Kurt E. Shuler, Editor
National Bureau of Standards
John B. Fenn, Associate Editor
Princeton University

*13. **Guidance and Control—II** (1964)
Robert C. Langford
General Precision Inc.
Charles J. Mundo
Institute of Naval Studies

*14. **Celestial Mechanics and Astrodynamics** (1964)
Victor G. Szebehely
Yale University Observatory

*15. **Heterogeneous Combustion** (1964)
Hans G. Wolfhard
Institute for Defense Analyses
Irvin Glassman
Princeton University
Leon Green Jr.
Air Force Systems Command

16. **Space Power Systems Engineering** (1966)
George C. Szego
Institute for Defense Analyses
J. Edward Taylor
TRW Inc.

17. **Methods in Astrodynamics and Celestial Mechanics** (1966)
Raynor L. Duncombe
U.S. Naval Observatory
Victor G. Szebehely
Yale University Observatory

18. **Thermophysics and Temperature Control of Spacecraft and Entry Vehicles** (1966)
Gerhard B. Heller
NASA George C. Marshall Space Flight Center

*19. **Communication Satellite Systems Technology** (1966)
Richard B. Marsten
Radio Corporation of America

*Out of print.

20. Thermophysics of Spacecraft and Planetary Bodies: Radiation Properties of Solids and the Electromagnetic Radiation Environment in Space (1967)
Gerhard B. Heller
NASA George C. Marshall Space Flight Center

21. Thermal Design Principles of Spacecraft and Entry Bodies (1969)
Jerry T. Bevans
TRW Systems

22. Stratospheric Circulation (1969)
Willis L. Webb
Atmospheric Sciences Laboratory, White Sands, and University of Texas at El Paso

23. Thermophysics: Applications to Thermal Design of Spacecraft (1970)
Jerry T. Bevans
TRW Systems

24. Heat Transfer and Spacecraft Thermal Control (1971)
John W. Lucas
Jet Propulsion Laboratory

25. Communication Satellites for the 70's: Technology (1971)
Nathaniel E. Feldman
The Rand Corporation
Charles M. Kelly
The Aerospace Corporation

26. Communication Satellites for the 70's: Systems (1971)
Nathaniel E. Feldman
The Rand Corporation
Charles M. Kelly
The Aerospace Corporation

27. Thermospheric Circulation (1972)
Willis L. Webb
Atmospheric Sciences Laboratory, White Sands, and University of Texas at El Paso

28. Thermal Characteristics of the Moon (1972)
John W. Lucas
Jet Propulsion Laboratory

29. Fundamentals of Spacecraft Thermal Design (1972)
John W. Lucas
Jet Propulsion Laboratory

30. Solar Activity Observations and Predictions (1972)
Patrick S. McIntosh
Murray Dryer
Environmental Research Laboratories, National Oceanic and Atmospheric Administration

31. Thermal Control and Radiation (1973)
Chang-Lin Tien
University of California at Berkeley

32. Communications Satellite Systems (1974)
P.L. Bargellini
COMSAT Laboratories

33. Communications Satellite Technology (1974)
P.L. Bargellini
COMSAT Laboratories

34. Instrumentation for Airbreathing Propulsion (1974)
Allen E. Fuhs
Naval Postgraduate School
Marshall Kingery
Arnold Engineering Development Center

35. Thermophysics and Spacecraft Thermal Control (1974)
Robert G. Hering
University of Iowa

36. Thermal Pollution Analysis (1975)
Joseph A. Schetz
Virginia Polytechnic Institute

37. Aeroacoustics: Jet and Combustion Noise; Duct Acoustics (1975)
Henry T. Nagamatsu, Editor
General Electric Research and Development Center
Jack V. O'Keefe, Associate Editor
The Boeing Company
Ira R. Schwartz, Associate Editor
NASA Ames Research Center

38. Aeroacoustics: Fan, STOL, and Boundary Layer Noise; Sonic Boom; Aeroacoustic Instrumentation (1975)
Henry T. Nagamatsu, Editor
General Electric Research and Development Center
Jack V. O'Keefe, Associate Editor
The Boeing Company
Ira R. Schwartz, Associate Editor
NASA Ames Research Center

39. Heat Transfer with Thermal Control Applications (1975)
M. Michael Yovanovich
University of Waterloo

40. Aerodynamics of Base Combustion (1976)
S.N.B. Murthy, Editor
Purdue University
J.R. Osborn, Associate Editor
Purdue University
A.W. Barrows
J.R. Ward, Associate Editors
Ballistics Research Laboratories

41. Communications Satellite Developments: Systems (1976)
Gilbert E. LaVean
Defense Communications Agency
William G. Schmidt
CML Satellite Corporation

42. Communications Satellite Developments: Technology (1976)
William G. Schmidt
CML Satellite Corporation
Gilbert E. LaVean
Defense Communications Agency

43. Aeroacoustics: Jet Noise, Combustion and Core Engine Noise (1976)
Ira R. Schwartz, Editor
NASA Ames Research Center
Henry T. Nagamatsu, Associate Editor
General Electric Research and Development Center
Warren C. Strahle, Associate Editor
Georgia Institute of Technology

44. Aeroacoustics: Fan Noise and Control; Duct Acoustics; Rotor Noise (1976)
Ira R. Schwartz, Editor
NASA Ames Research Center
Henry T. Nagamatsu, Associate Editor
General Electric Research and Development Center
Warren C. Strahle, Associate Editor
Georgia Institute of Technology

45. Aeroacoustics: STOL Noise; Airframe and Airfoil Noise (1976)
Ira R. Schwartz, Editor
NASA Ames Research Center
Henry T. Nagamatsu, Associate Editor
General Electric Research and Development Center
Warren C. Strahle, Associate Editor
Georgia Institute of Technology

46. Aeroacoustics: Acoustic Wave Propagation; Aircraft Noise Prediction; Aeroacoustic Instrumentation (1976)
Ira R. Schwartz, Editor
NASA Ames Research Center
Henry T. Nagamatsu, Associate Editor
General Electric Research and Development Center
Warren C. Strahle, Associate Editor
Georgia Institute of Technology

47. Spacecraft Charging by Magnetospheric Plasmas (1976)
Alan Rosen
TRW Inc.

48. Scientific Investigations on the Skylab Satellite (1976)
Marion I. Kent
Ernst Stuhlinger
NASA George C. Marshall Space Flight Center
Shi-Tsan Wu
The University of Alabama

49. Radiative Transfer and Thermal Control (1976)
Allie M. Smith
ARO Inc.

50. Exploration of the Outer Solar System (1976)
Eugene W. Greenstadt
TRW Inc.
Murray Dryer
National Oceanic and Atmospheric Administration
Devrie S. Intriligator
University of Southern California

51. Rarefied Gas Dynamics, Parts I and II (two volumes) (1977)
J. Leith Potter
ARO Inc.

52. Materials Sciences in Space with Application to Space Processing (1977)
Leo Steg
General Electric Company

53. Experimental Diagnostics in Gas Phase Combustion Systems (1977)
Ben T. Zinn, Editor
Georgia Institute of Technology
Craig T. Bowman, Associate Editor
Stanford University
Daniel L. Hartley, Associate Editor
Sandia Laboratories
Edward W. Price, Associate Editor
Georgia Institute of Technology
James G. Skifstad, Associate Editor
Purdue University

54. Satellite Communications: Future Systems (1977)
David Jarett
TRW Inc.

55. Satellite Communications: Advanced Technologies (1977)
David Jarett
TRW Inc.

56. Thermophysics of Spacecraft and Outer Planet Entry Probes (1977)
Allie M. Smith
ARO Inc.

57. Space-Based Manufacturing from Nonterrestrial Materials (1977)
Gerard K. O'Neill, Editor
Princeton University
Brian O'Leary, Assistant Editor
Princeton University

58. Turbulent Combustion (1978)
Lawrence A. Kennedy
State University of New York at Buffalo

59. Aerodynamic Heating and Thermal Protection Systems (1978)
Leroy S. Fletcher
University of Virginia

60. Heat Transfer and Thermal Control Systems (1978)
Leroy S. Fletcher
University of Virginia

61. Radiation Energy Conversion in Space (1978)
Kenneth W. Billman
NASA Ames Research Center

62. Alternative Hydrocarbon Fuels: Combustion and Chemical Kinetics (1978)
Craig T. Bowman
Stanford University
Jorgen Birkeland
Department of Energy

63. Experimental Diagnostics in Combustion of Solids (1978)
Thomas L. Boggs
Naval Weapons Center
Ben T. Zinn
Georgia Institute of Technology

64. Outer Planet Entry Heating and Thermal Protection (1979)
Raymond Viskanta
Purdue University

65. Thermophysics and Thermal Control (1979)
Raymond Viskanta
Purdue University

66. Interior Ballistics of Guns (1979)
Herman Krier
University of Illinois at Urbana-Champaign
Martin Summerfield
New York University

***67. Remote Sensing of Earth from Space: Role of "Smart Sensors"** (1979)
Roger A. Breckenridge
NASA Langley Research Center

68. Injection and Mixing in Turbulent Flow (1980)
Joseph A. Schetz
Virginia Polytechnic Institute and State University

69. Entry Heating and Thermal Protection (1980)
Walter B. Olstad
NASA Headquarters

70. Heat Transfer, Thermal Control, and Heat Pipes (1980)
Walter B. Olstad
NASA Headquarters

71. Space Systems and Their Interactions with Earth's Space Environment (1980)
Henry B. Garrett
Charles P. Pike
Hanscom Air Force Base

72. Viscous Flow Drag Reduction (1980)
Gary R. Hough
Vought Advanced Technology Center

73. Combustion Experiments in a Zero-Gravity Laboratory (1981)
Thomas H. Cochran
NASA Lewis Research Center

74. Rarefied Gas Dynamics, Parts I and II (two volumes) (1981)
Sam S. Fisher
University of Virginia at Charlottesville

75. Gasdynamics of Detonations and Explosions (1981)
J.R. Bowen
University of Wisconsin at Madison
N. Manson
Université de Poitiers
A.K. Oppenheim
University of California at Berkeley
R.I. Soloukhin
Institute of Heat and Mass Transfer, BSSR Academy of Sciences

76. Combustion in Reactive Systems (1981)
J.R. Bowen
University of Wisconsin at Madison
N. Manson
Université de Poitiers
A.K. Oppenheim
University of California at Berkeley
R.I. Soloukhin
Institute of Heat and Mass Transfer, BSSR Academy of Sciences

77. Aerothermodynamics and Planetary Entry (1981)
A.L. Crosbie
University of Missouri-Rolla

78. Heat Transfer and Thermal Control (1981)
A.L. Crosbie
University of Missouri-Rolla

79. Electric Propulsion and Its Applications to Space Missions (1981)
Robert C. Finke
NASA Lewis Research Center

80. Aero-Optical Phenomena (1982)
Keith G. Gilbert
Leonard J. Otten
Air Force Weapons Laboratory

81. Transonic Aerodynamics (1982)
David Nixon
Nielsen Engineering & Research, Inc.

82. Thermophysics of Atmospheric Entry (1982)
T.E. Horton
The University of Mississippi

83. Spacecraft Radiative Transfer and Temperature Control (1982)
T.E. Horton
The University of Mississippi

84. Liquid-Metal Flows and Magnetohydrodynamics (1983)
H. Branover
Ben-Gurion University of the Negev
P.S. Lykoudis
Purdue University
A. Yakhot
Ben-Gurion University of the Negev

85. Entry Vehicle Heating and Thermal Protection Systems: Space Shuttle, Solar Starprobe, Jupiter Galileo Probe (1983)
Paul E. Bauer
McDonnell Douglas Astronautics Company
Howard E. Collicott
The Boeing Company

86. Spacecraft Thermal Control, Design, and Operation (1983)
Howard E. Collicott
The Boeing Company
Paul E. Bauer
McDonnell Douglas Astronautics Company

87. Shock Waves, Explosions, and Detonations (1983)
J.R. Bowen
University of Washington
N. Manson
Université de Poitiers
A.K. Oppenheim
University of California at Berkeley
R.I. Soloukhin
Institute of Heat and Mass Transfer, BSSR Academy of Sciences

88. Flames, Lasers, and Reactive Systems (1983)
J.R. Bowen
University of Washington
N. Manson
Université de Poitiers
A.K. Oppenheim
University of California at Berkeley
R.I. Soloukhin
Institute of Heat and Mass Transfer, BSSR Academy of Sciences

89. Orbit-Raising and Maneuvering Propulsion: Research Status and Needs (1984)
Leonard H. Caveny
Air Force Office of Scientific Research

90. Fundamentals of Solid-Propellant Combustion (1984)
Kenneth K. Kuo
The Pennsylvania State University
Martin Summerfield
Princeton Combustion Research Laboratories, Inc.

91. Spacecraft Contamination: Sources and Prevention (1984)
J.A. Roux
The University of Mississippi
T.D. McCay
NASA Marshall Space Flight Center

92. Combustion Diagnostics by Nonintrusive Methods (1984)
T.D. McCay
NASA Marshall Space Flight Center
J.A. Roux
The University of Mississippi

93. The INTELSAT Global Satellite System (1984)
Joel Alper
COMSAT Corporation
Joseph Pelton
INTELSAT

94. Dynamics of Shock Waves, Explosions, and Detonations (1984)
J.R. Bowen
University of Washington
N. Manson
Université de Poitiers
A.K. Oppenheim
University of California
R.I. Soloukhin
Institute of Heat and Mass Transfer, BSSR Academy of Sciences

95. Dynamics of Flames and Reactive Systems (1984)
J.R. Bowen
University of Washington
N. Manson
Université de Poitiers
A.K. Oppenheim
University of California
R.I. Soloukhin
Institute of Heat and Mass Transfer, BSSR Academy of Sciences

96. Thermal Design of Aeroassisted Orbital Transfer Vehicles (1985)
H.F. Nelson
University of Missouri-Rolla

97. Monitoring Earth's Ocean, Land, and Atmosphere from Space – Sensors, Systems, and Applications (1985)
Abraham Schnapf
Aerospace Systems Engineering

98. Thrust and Drag: Its Prediction and Verification (1985)
Eugene E. Covert
Massachusetts Institute of Technology
C.R. James
Vought Corporation
William F. Kimzey
Sverdrup Technology AEDC Group
George K. Richey
U.S. Air Force
Eugene C. Rooney
U.S. Navy Department of Defense

99. Space Stations and Space Platforms — Concepts, Design, Infrastructure, and Uses (1985)
Ivan Bekey
Daniel Herman
NASA Headquarters

100. Single- and Multi-Phase Flows in an Electromagnetic Field Energy, Metallurgical, and Solar Applications (1985)
Herman Branover
Ben-Gurion University of the Negev
Paul S. Lykoudis
Purdue University
Michael Mond
Ben-Gurion University of the Negev

101. MHD Energy Conversion: Physiotechnical Problems (1986)
V.A. Kirillin
A.E. Sheyndlin
Soviet Academy of Sciences

102. Numerical Methods for Engine-Airframe Integration (1986)
S.N.B. Murthy
Purdue University
Gerald C. Paynter
Boeing Airplane Company

103. Thermophysical Aspects of Re-Entry Flows (1986)
James N. Moss
NASA Langley Research Center
Carl D. Scott
NASA Johnson Space Center

104. Tactical Missile Aerodynamics (1986)
M.J. Hemsch
PRC Kentron, Inc.
J.N. Nielsen
NASA Ames Research Center

105. Dynamics of Reactive Systems Part I: Flames and Configurations; Part II: Modeling and Heterogeneous Combustion (1988)
J.R. Bowen
University of Washington
J.-C. Leyer
Université de Poitiers
R.I. Soloukhin
Institute of Heat and Mass Transfer, BSSR Academy of Sciences

106. Dynamics of Explosions (1986)
J.R. Bowen
University of Washington
J.-C. Leyer
Université de Poitiers
R.I. Soloukhin
Institute of Heat and Mass Transfer, BSSR Academy of Sciences

107. Spacecraft Dielectric Material Properties and Spacecraft Charging (1986)
A.R. Frederickson
U.S. Air Force Rome Air Development Center
D.B. Cotts
SRI International
J.A. Wall
U.S. Air Force Rome Air Development Center
F.L. Bouquet
Jet Propulsion Laboratory, California Institute of Technology

108. Opportunities for Academic Research in a Low-Gravity Environment (1986)
George A. Hazelrigg
National Science Foundation
Joseph M. Reynolds
Louisiana State University

109. Gun Propulsion Technology (1988)
Ludwig Stiefel
U.S. Army Armament Research, Development and Engineering Center

110. Commercial Opportunities in Space (1988)
F. Shahrokhi
K.E. Harwell
University of Tennessee Space Institute
C.C. Chao
National Cheng Kung University

111. Liquid-Metal Flows: Magnetohydrodynamics and Applications (1988)
Herman Branover, Michael Mond, and Yeshajahu Unger
Ben-Gurion University of the Negev

112. Current Trends in Turbulence Research (1988)
Herman Branover, Michael Mond, and Yeshajahu Unger
Ben-Gurion University of the Negev

113. Dynamics of Reactive Systems Part I: Flames; Part II: Heterogeneous Combustion and Applications (1988)
A.L. Kuhl
R & D Associates
J.R. Bowen
University of Washington
J.-C. Leyer
Université de Poitiers
A. Borisov
USSR Academy of Sciences

114. Dynamics of Explosions (1988)
A.L. Kuhl
R & D Associates
J.R. Bowen
University of Washington
J.-C. Leyer
Université de Poitiers
A. Borisov
USSR Academy of Sciences

115. Machine Intelligence and Autonomy for Aerospace (1988)
E. Heer
Heer Associates, Inc.
H. Lum
NASA Ames Research Center

116. Rarefied Gas Dynamics: Space-Related Studies (1989)
E.P. Muntz
University of Southern California
D.P. Weaver
U.S. Air Force Astronautics Laboratory (AFSC)
D.H. Campbell
The University of Dayton Research Institute

(Other Volumes are planned.)

Other Volumes in the Rarefied Gas Dynamics Series*

Rarefied Gas Dynamics, Proceedings of the 1st International Symposium, edited by F. M. Devienne, Pergamon Press, Paris, France, 1960.

Rarefied Gas Dynamics, Proceedings of the 2nd International Symposium, edited by L. Talbot, Academic Press, New York, 1961.

Rarefied Gas Dynamics, Proceedings of the 3rd International Symposium, Vols. I and II, edited by J. A. Laurmann, Academic Press, New York, 1963.

Rarefied Gas Dynamics, Proceedings of the 4th International Symposium, Vols. I and II, edited by J. H. deLeeuw, Academic Press, New York, 1965.

Rarefied Gas Dynamics, Proceedings of the 5th International Symposium, Vols. I and II, edited by C. L. Brundin, Academic Press, London, England, 1967.

Rarefied Gas Dynamics, Proceedings of the 6th International Symposium, Vols. I and II, edited by L. Trilling and H. Y. Wachman, Academic Press, New York, 1969.

Rarefied Gas Dynamics, Proceedings of the 7th International Symposium, Vols. I and II, edited by D. Dini, C. Cercignani, and S. Nocilla, Editrice Tecnico Scientifica, Pisa, Italy, 1971.

Rarefied Gas Dynamics, Proceedings of the 8th International Symposium, Edited by K. Karamcheti, Academic Press, New York, 1974.

Rarefied Gas Dynamics, Proceedings of the 9th International Symposium, Vols. I and II, edited by M. Becker and M. Fiebig, DFVLR-Press, Porz-Wahn, Germany, 1974.

Rarefied Gas Dynamics, Proceedings of the 10th International Symposium, Parts I and II of Vol. 51 of *Progress in Astronautics and Aeronautics,* edited by L. Potter, AIAA, New York, 1977.

Rarefied Gas Dynamics, Proceedings of the 11th International Symposium, Vols. I and II, edited by R. Campargue, Commissiariat a l'Energie Atomique, Paris, France, 1979.

Rarefied Gas Dynamics, Proceedings of the 12th International Symposium, Parts I and II of Vol. 74 of *Progress in Astronautics and Aeronautics,* edited by S. Fisher, AIAA, New York, 1981.

Rarefied Gas Dynamics, Proceedings of the 13th International Symposium, Vols. I and II, edited by O. M. Belotserkovskii, M. N. Kogan, C. S. Kutateladze, and A. K. Rebrov, Plenum Press, New York, 1985.

Rarefied Gas Dynamics, Proceedings of the 14th International Symposium, Vols. I and II, edited by H. Oguchi, University of Tokyo Press, Tokyo, 1984.

Rarefied Gas Dynamics, Proceedings of the 15th International Symposium, Vols. I and II, edited by V. Boffi and C. Cercignani, B. G. Teubner, Stuttgart, 1986.

Rarefied Gas Dynamics, Proceedings of the 16th International Symposium, Vols. 116, 117, and 118 in *Progress in Astronautics and Aeronautics,* edited by E. P. Muntz, D. Weaver, and D. Campbell, AIAA, Washington, DC, 1989.

---

*Copies may be purchased directly from the publisher in each case. The AIAA cannot fill orders for volumes published by other publishers. This list is provided for information only.